Geology: An Earth Science

Geology: An Earth Science

Editor: Cortez Ford

R CALLISTO
REFERENCE
www.callistoreference.com

Callisto Reference,
118-35 Queens Blvd., Suite 400,
Forest Hills, NY 11375, USA

Visit us on the World Wide Web at:
www.callistoreference.com

ISBN: 978-1-64116-079-7 (Hardback)

Cataloging-in-Publication Data

Geology : an earth science / edited by Cortez Ford.
  p. cm.
Includes bibliographical references and index.
ISBN 978-1-64116-079-7
1. Geology. 2. Earth sciences. I. Ford, Cortez.
QE26.3 .G46 2019
551--dc23

Table of Contents

Preface..IX

Chapter 1   **Extensional Seismotectonic Motion and its Dynamics in the Eastern Margin
of the Tibetan Plateau and its Surroundings**..1
Jiren Xu and Zhixin Zhao

Chapter 2   **A Geological Appraisal of Slope Instability in Upper Alaknanda Valley,
Uttarakhand Himalaya, India**..9
Sajwan KS and Sushil K

Chapter 3   **Evalauation of Groundwater Potential Zones using Electrical Resistivity
Response and Lineament Pattern in Uppodai Sub Basin, Tambaraparani River,
Tirunelveli District, Tamilnadu, India**..16
Jeyavel Raja Kumar T, Dushiyanthan C, Thiruneelakandan B, Suresh R,
Vasanth Raja S, Senthilkumar M and Karthikeyan K

Chapter 4   **Fold-thrust Style and Fluid Reservoir Potential of Eocene Sakesar
Limestone: Sothern Surghar Range, Trans-Indus Ranges, North Pakistan**.............21
Alam I

Chapter 5   **Interpreting Seismic Profiles in terms of Structure and Stratigraphy
with Implications for Hydrocarbons Accumulation, an Example from Lower
Indus Basin Pakistan**..27
Majid K, Shahid N, Munawar S and Muhammad H

Chapter 6   **Isolated Paleomagnetic Component and its Dating from the Malha Formation,
Southwestern Sinai, Egypt**..36
Khashaba A, Soliman SA, Takla EM, Farouk S and Mostafa R

Chapter 7   **Seismic Refraction Method to Study Subsoil Structure**...41
Fkirin MA, Badawy S and El deery MF

Chapter 8   **Roadside Air Pollutants along Elected Roads in Nairobi City, Kenya**.....................47
Shilenje ZW, Thiong'o K, Ongoma V, Philip SO Nguru P and Ondimu K

Chapter 9   **The Correlation of North Magnetic Dip Pole Motion and Seismic Activity**............57
Williams B

Chapter 10   **Volumetric Assessment through 3D Geostatic Model for Abu Roash "G"
Reservoir in Amana Field-East Abu Gharadig Basin-Western Desert-Egypt**...........61
Abu-Hashish MF and Ahmed Said

Chapter 11   **Longitudinal Dependence and Seasonal Effect on Equatorial Electrojet using
MAGDAS Data**...73
Ibrahim Khashaba A and Essam Ghamry

Chapter 12   **Continental Drift and Plate Tectonics vis-a-vis Earth's Expansion: Probing the Missing Links for Understanding the Total Earth System**...83
Sen S

Chapter 13   **Risk Assessment and Remedial Solutions of Coastal Flooding: Case Study of Hammam Lif Coastline, Northern Tunisia**...88
Abir B, Mohamed B, Samir M and Chokri Y

Chapter 14   **Analysis of Soil and Sub-Soil Properties around Veritas University, Obehie, Southeastern Nigeria**...99
Youdeowei PO and Nwankwoala HO

Chapter 15   **Application of Electrical Resistivity and Ground Penetrating Radar Techniques in Subsurface Imaging around Ajibode, Ibadan, Southwestern Nigeria**...103
Adelekan AO, Oladunjoye MA and Igbasan AO

Chapter 16   **Origin of the Arima-type and Associated Spring Waters in the Kinki District, Southwest Japan**...113
Hitomi Nakamura, Kotona Chiba, Qing Chang, Noritoshi Morikawa, Kohei Kazahaya and Hikaru Iwamori

Chapter 17   **Recognition of Lithostratigraphic Breaks in Undifferentiated Rock Units using Well Logs: A Flow Chart**...128
Shaaban FF and Al-Rashed AR

Chapter 18   **Geotechnical Evaluation of Foundation Conditions in Igbogene, Bayelsa State, Nigeria**...136
Nwankwoala HO and Adiela UP

Chapter 19   **An Overview of CBM Resources in Lower Indus Basin, Sindh, Pakistan**...141
Nazeer A, Habib Shah S, Abbasi SA, Solangi SH and Ahmad N

Chapter 20   **Characterization of Low Grade Natural Emerald Gemstone**...154
Reshma B, Sakthivel R and Mohanty JK

Chapter 21   **Land Magnetic Investigation on the West Qarun Oil Field, Western Desert-Egypt**...160
Khashaba A, Mekkawi M, Ghamry E and Abdel Aal E

Chapter 22   **Simulating Arsenic Mitigation Strategies in a Production Well**...166
Dara A. Goldrath, John A. Izbicki and Kathryn W. Thorbjarnarson

Chapter 23   **Lithofacies Superimposition in a Shallow Basin - Interplay of Tectonics and Sedimentation: Evidences from Kolhan Basin, Eastern India**...174
Smruti Rekha Sahoo and Subhasish Das

Chapter 24   **Oil and Gas Industrial and Ecosystem Mechanical Impacts of Environment**...182
Elosta F

Chapter 25   **Lithology Investigation of Shaly Sand Reservoir by using Wire Line Data, "Nubian Sandstone" SE Sirt Basin**...187
Ben Ghawar BM

Chapter 26   **Land Surface Temperature and Scaling Factors for Different Satellites Datasets**..................................192
Mukesh Singh Boori, Heiko Balzter, Komal Choudhary and Vit Vozenílek

Chapter 27   **Utility of Large Scale Photogrammetric Techniques for 3-D Mapping and**
Precision Iron Ore Mining in Open Pit Areas...198
Murali Krishna G and Nooka Ratnam K

Chapter 28   **Rare Earth Elements of the Arima Spring Waters, Southwest Japan:**
Implications for Fluid–Crust Interaction during Ascent of Deep Brine.................................202
Hitomi Nakamura, Kotona Chiba, Qing Chang, Shunichi Nakai,
Kohei Kazahaya and Hikaru Iwamori

Chapter 29   **Petrology and Geochemistry of Gabbros from the Andaman Ophiolite:**
Implications for their Petrogenesis and Tectonic Setting..210
Rasool QA, Ramanujam N and Biswas SK

Permissions

List of Contributors

Index

Preface

Geology is the scientific study of rocks that form the solid Earth and the changes which they undergo over time. It is a sub-discipline of Earth science. It is undergoing rapid development owing to the technological growth witnessed in the last decade. Geologists aim to study dynamics of earth's crust and understand phenomena like the evolution of earth, tectonic shifts, etc. The concepts of geology have applications in the fields of mining, engineering, hydrology, etc. This book is compiled to provide a general overview of the subject. It strives to provide a fair idea about this discipline and to help develop a better understanding of the latest advances within this field. It will serve as a valuable source of reference for graduate and post graduate students. Researchers, experts and all associated with the field of geology will benefit alike from this book.

This book is a comprehensive compilation of works of different researchers from varied parts of the world. It includes valuable experiences of the researchers with the sole objective of providing the readers (learners) with a proper knowledge of the concerned field. This book will be beneficial in evoking inspiration and enhancing the knowledge of the interested readers.

In the end, I would like to extend my heartiest thanks to the authors who worked with great determination on their chapters. I also appreciate the publisher's support in the course of the book. I would also like to deeply acknowledge my family who stood by me as a source of inspiration during the project.

Editor

Extensional Seismotectonic Motion and its Dynamics in the Eastern Margin of the Tibetan Plateau and its Surroundings

Jiren Xu* and Zhixin Zhao

Institute of Geology, Chinese Academy of Geological Sciences, Beijing 100037, China

Abstract

The seismotectonic motions and stress fields in the eastern margin of the Tibetan plateau and its surroundings were investigated by analyzing seismic data. The findings showed that volumes of normal faulting type events dislocating extensively along the N-S direction concentrated southwest of the Xianshuihe fault (SWXSH-NSNF). Moreover, volumes of normal faulting type events dislocating extensively along the E-W direction concentrated east of the lower reach of the Jinshajiang river (ELRJSJ-EWNF). The events are almost thrust faulting type ones along the Longmenshan fault. The P-axes aligned generally around the NE-SW direction in the western region and around the NW-SE direction in the eastern region in the eastern margin of the plateau, respectively. The seismogenic stress field in the eastern region in the eastern margin of the Tibetan plateau might be affected by the stress field on the South China block resulted by the collision between the Eurasian and the Philippine Sea plates in the Taiwan region and the subduction of the Philippine Sea plate along the Ryukyu trench. The longitudinal boundary of the P-axis orientations of stress field between the eastern and western regions in the eastern margin of the Tibetan plateau lay along about longitude 100°E-101°E, being within the eastern margin of the Tibetan plateau. The above stress boundary does not coincide with the tectonic boundary between the Tibetan plateau and the South China block.

Keywords: Normal faulting type event region; Normal fault event dislocating extensively along N-S direction; Normal fault event dislocating extensively along E-W direction; Eastern margin of the Tibetan plateau; principle compressive P-axis; South China block

Introduction

The tectonic motions in the Tibetan plateau are related to the Himalayan orogeny where exists mountain-perpendicular compressive stresses [1]. The northeastward motion of India is partitioned by strike-slip and trust faults in the plateau. Furthermore, a significant portion of the lateral slip on these faults is transferred to thrust faults in the margins of the plateau [2]. Large faults lie almost near the NW or NWW direction in the Tibetan plateau [3,4]. The large fault motion, whereas revealed various configurations on the eastern margin of the Tibet Plateau as shown in Figure 1 [5]. Many faults turn the strike direction to near N-S or NE-SW direction. The Jinshajiang fault, Nujiang fault and the Anning river fault, these strike-slip faults all extend along the N-S direction [6,7]. The strike-slip type Xianshuihe fault lies along the NW-SE direction [8]. The Jinpingshan-Yulongxueshan fault extends along in the NE-SW direction. The Longmenshan fault also extends along the NE-SW direction, bending somewhat along the northeastern edge of the Sichuan basin. Such a special strike direction of the fault may imply that the Sichuan basin obstructs the southeastward movement of the Tibetan plateau. Regional blocks cut by the above large faults also reveal tectonically complexity as illustrated in Figure 1. The eastern margin of the Tibetan plateau is on the western border of the South China block. The southern segment of the famous North-South Seismic Belt in China (NSSB) likely coincides with the boundary between the eastern margin of the Tibetan plateau and the South China block [9]. The events in the southern segment of NSSB occurred almost in the eastern margin of the Tibetan plateau and Yunnan region indeed. A few events occurred in the Sichuan basin in the South China block. The serious damages, 2008 M8 Wenchuan event and 2013 Lushan M7 event occurred in the Longmenshang fault in the study margin.

The tectonic deformation features of the India-Eurasia continental collision zone reveal with thrust compression, lateral extrusion and clockwise rotation. However, in the middle southern plateau, there is subregion with a series of NS striking normal faults [10,11]. Some interested results of seismotectonic motion have been reported [12-

15]. Many normal faulting type events concentrated on the high altitude region of the Tibetan plateau, i.e., in the region from the Gangdise mountains in the south to the Kunlun mountains in the north [16-18]. However, the thrust-faulting type event predominated the focus mechanism of earthquake occurrence in low altitude regions surrounding the Tibetan plateau [19]. The seismotectonic motions also show different regional characteristics in the eastern margin of the Tibetan plateau and its surroundings [20,21]. Most of events revealed the strike-slip faulting type there. Few thrust-faulting type events were reported in the eastern margin before the 2008 Wenchuan M8 event (31.1°N, 103.3°E). The serious disaster Wenchuan M8 event and the aftershocks formed a new series of events with thrust-faulting regime in the Longmenshan fault. The new events provided substantial amount of reliable data to investigate the tectonic stress field in the eastern margin of the Tibetan Plateau further [22].

In general the seismotectonic motions are closely related to the tectonic stress. The tectonic stress field in the eastern margin of the Tibetan plateau is related to the collision between the Eurasian Plateau and Indian plates [23,24]. The mountain-perpendicular compressive stresses exists along the Himalayas [1]. The eastward escape hypothesis for the lithosphere usually was reported as the tectonically dynamic causative mechanism of earthquake in the eastern margin of the plateau [25,26]. The seismogenic stress fields were quite complicated in the eastern margin of the Tibetan plateau and the surroundings indeed. Some P-axes aligned NW-SE direction in the eastern margin of the Tibetan plateau although the tectonic forces caused a wide distribution

---

*Corresponding author: Jiren Xu, Institute of Geology, Chinese Academy of Geological Sciences, Beijing 100037, China, E-mail: xujiren1125@aliyun.com

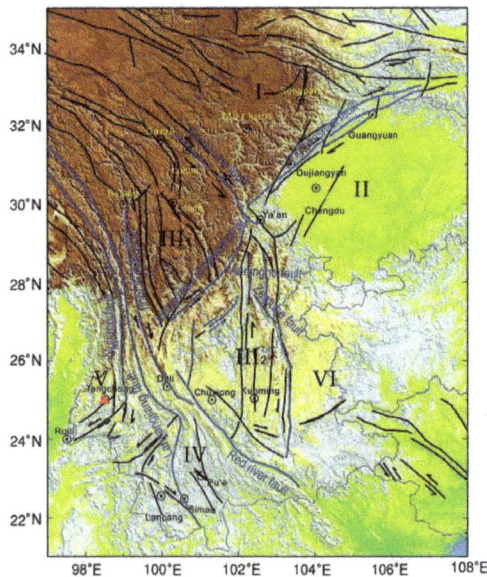

Figure 1: Tectonic outline in the eastern margin of the Tibetan planet and its surroundings. I) Longmenshan block. II) East of the Sichuan basin. III) Sichuan-Yunnan rhombic block. IV) Dianzhong block. V) Tenchong-Baoshan block. VI) East of Yunnan block. Red triangle: Tengchong volcano. Blue thick line: main fault.

of P-axes near NE-SW direction in most regions of the Tibetan plateau [16]. The temporal variations of seismicity in the eastern margin of the Tibetan plateau were not synchronous with that of either vicinity seismic zone, i.e., the Tibetan plateau or the Southern China block (Yangtze block). It was difficult to find the temporal relativity of seismicity between the southern segment of NSSB and either vicinity seismic zone based on statistic analyses of the seismic temporal series variation [27].

In the present analysis, the complicated seismotectonic motion and earthquake generating stress field in the eastern margin of the Tibetan plateau were further investigated on view of the tectonics on the boundary between the Tibetan plateau and the South China block employing seismicity and focal mechanism solution parameters. Specially, the extensional seismotectonic motions in the eastern margin and the dynamics are discussed in detail.

Seismic Data

Focal mechanism solutions of 239 earthquakes with magnitude equal to or greater than M5 on the eastern margin of the Tibetan plateau and its vicinities during the period between 1933 and 2008 were analyzed in this study. 85 fault plane solutions of events were determined form the distribution of the polarities of initial P-wave motion on the focal hemisphere by the authors referring to data from the Bulletin of the International Seismological Center (ISC). Another 77 solutions determined by P-wave first motions were selected from the other reports referring to data from the Chinese Seismic Net and the Bulletin of the International Seismological Center (ISC) using P-wave first motions [28-30]. All the solutions used in the analysis were well selected based on the contradictory rate of the plus to minus signs for the first polarities of P-wave motions. The rest of 77 CMT solutions were (determined by the Centroid Moment tensor solutions) selected from the Harvard University and USGS. The CMT solutions for events with magnitude equal to or greater than 5.0 from 1976 to 2003

were also used in this study [31,32]. Both of mechanisms published and derived from authors were employed to investigate the seismotectonic motion. The events used in this analysis occurred in the crust. The depths of most events are less than 70 km in the analysis. The earthquake catalogue for seismicity is drawn from the seismic networks of the Seismic Net Center of the China Earthquake Administration.

Earthquake Faulting Motion Regimes

The faulting motion regime of the focal mechanism solution shows the regional characteristics of the seismotectonic motion. In order to investigate the seismotectonic motions, Figure 2 shows the focal mechanism solutions of the equal area projection of events in the eastern margin region of the Tibetan plateau and its surroundings. The occurrence regimes of events revealed the distinguished regional characteristics although the mechanism diagrams of strike-slip faulting type events widely appeared in many areas in Figure 2. The events were almost strike-slip type ones in the northwestern segment of the Xianshuihe fault, whereas some events of strike-slip faulting with normal faulting component occurred in the southeastern segment of the Xianshuihe fault besides strike-slip ones as shown in Figure 2 e.g., the largest one along the Xianshuihe fault, the 1955 M7 Kangding event was the event of strike-slip faulting with normal faulting component occurring in the southeastern end of the Xianshuihe fault in Figure 2.

While many normal faulting type events occurred in the region south of the Xianshuihe fault, extending southward for hundreds of

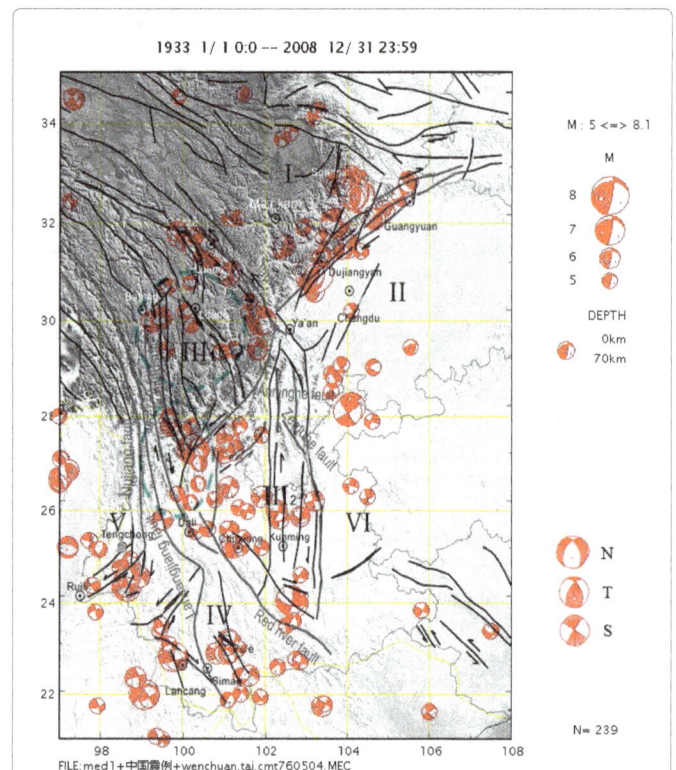

Figure 2: Distribution of the fault plane solution diagrams projected on the equal area in low hemisphere in the eastern margin of the Tibetan plateau and its surroundings. Green dash circle: N-S extension normal faulting type event region southwest of the Xianshuihe fault (SWXSH-NSNF), ranging about 98°-102°E, 29°-32°N; Green dash ellipse: E-W extension normal faulting type event region east of the lower reach of the Jinshajiang fault (ELRJSJ-EWNF), ranging about 98°-102°E, 26°-29°N. Diagrams in the legend in the lower right corner, N: normal faulting type event; T: thrust faulting type event; S: strike-slip faulting type event.

kilometers along the eastern side of the Jinshajiang river fault in Figure 2. These normal faulting type events might be taken as two groups based on the faulting dislocation regimes indeed. The first, the normal faulting type events concentrated on the region southwest of the Xianshuihe fault and east of the upper reach of the Jinshajiang river ranging about 99°E-102°E and 29°N-31°N, marked with the green dash circle in Figure 2, for hundreds of kilometers in the longitudinal and latitudinal directions. Both of the strike directions of the fault planes and the fault auxiliary planes of focal mechanism solutions of the normal faulting type events lay likely in the E-W and near E-W directions. It showed that the dislocations of these normal faulting type events extended along or near the N-S direction in the region southwest of the Xianshuihe fault. The north-south components of fault slips were all great. This region is taken as the N-S extensively normal faulting type event region southwest of the Xianshuihe fault (SWXSH- NSNF) marked with the green dash circle in Figure 2. The fifteen events occurring in the SWXSH-NSNF were all normal faulting type ones during the study period. Three events among these fifteen were equal to or greater than M6.0. The greatest one was the April 15, 1998 Litang, Sichuan Province M6.4 event (29.98°N, 99.24°E). The normal fault dislocating extensively along or near N-S direction in the normal faulting type event region southwest of the Xianshuihe fault as mentioned above, is a newly notable property on the seismotectonic motion. The property implies that the earthquake rupture regimes are predominated by the normal faulting type events with the N-S extensive dislocation in SWXSH -NSNF in the eastern margin of the Tibetan plateau. It is probably different from the traditionally eastward escape movement in the eastern Tibetan plateau.

Another normal faulting type event region marked by a green dash ellipse in Figure 2 was located in the region east of the lower reach of the Jinshajiang fault ranging about 99°E-101°E and 26°N-28°N, for hundreds of kilometers in the longitudinal and latitudinal direction. Both of the strike directions of the fault planes and their fault auxiliary planes of the normal faulting type events were almost along or near the N-S direction. It implied that the dislocations of the normal faulting type events extended along or near the E-W direction in the region east of the lower reach of the Jinshajiang fault. The E-W components of fault slips were all great. The earthquake rupture regime was predominated by the normal faulting type event that the dislocation extended along the E-W direction although the large tectonic faults with N-S strike direction lay on the western and eastern side of the normal faulting type event region. This region is taken as the E-W extensional normal faulting type event region east of the lower reach of the Jinshajiang fault (ELRJSJ-EWNF) marked by a green dash ellipse in Figure 2. The ELRJSJ-EWNF is particularly located near joint region between the Jinshajiang and Jinpingshan-Yulongxueshan faults in Figures 1 and 2. The seventeen normal faulting type events occurred in the region east of the lower reach of the Jinshajiang. Five of them were equal to or greater than M6.0 and the greatest one was the February 3, 1998 Lijiang, Yunnan Province M7.0 event (27.3°N, 100.21°E) in ELRJSJ-EWNF. The regimes of normal faulting dislocations along the E-W direction in the ELRJSJ- EWNF are quite different from those along the N-S direction in the SWXSH-NSNF although the ELRJSJ-EWNF is on the southern border of SWXSH- NSNF in Figure 2.

The Longmenshan fault with NE-SW strike direction crosses to the Xianshuihe fault with NW-SE strike direction in the eastern margin of the Tibetan plateau north of 28°N in Figure 1. The earthquake faulting regimes are obviously different each other in Figure 2. The majority of events occurring in the Longmenshan fault were the thrust faulting type ones although few thrust-faulting type events were reported in the

eastern margin before the 2008 Wenchuan M8 event. The Wenchuan M8 event in the Longmenshan fault and its strong aftershocks were thrust faulting events except only a strike-slip faulting one [33]. The 2013 Lushan M7 event was also a thrust one there. Moreover some thrust faulting type events occurred in the Longmenshan block in Figure 1, the northwestern neighborhood of the Longmenshan fault in Figure 2. The thrust faulting dislocations played an important role for the regimes of earthquake occurrence in the Longmenshan fault and its northwestern neighborhood.

As mentioned above, the thrust faulting type events occurred almost in the northeastern region in the eastern margin of the Tibetan plateau. The normal faulting type events occurred almost in the western region of the eastern margin in Figure 2. Except the two normal faulting zones and one thrust faulting zone listed above, the events almost revealed strike-slip faulting type in other zones in Figure 2, i.e., the wide Yunnan region and the southwestern edge of the Sichuan basin.

Analysis of Stress Field

The occurrence regimes of events showed evidently regional properties in the eastern margin of the Tibetan plateau as mentioned above. The regime of earthquake occurrence or seismotectonic motion property is likely related to the regional tectonic stress field. The eastern margin of the Tibetan plateau is located in the boundary between the Tibetan plateau and the South China block. The seismogenic stress field in and around the boundary was usually investigated in view of the plate tectonics. Figures 3 and 4 showed the distribution of horizontal projections for the principal compressive stress P-axes and extensional stress T-axes in the study region.

Regional stress field related to earthquake faulting type

With respect to the tectonic stresses for the normal faulting type event region southwest of the Xianshuihe fault ranging about 99°E-102°E and 29°N-31°N , the green circle region (SWXSH-NSNF), the P-axes of normal faulting events aligned almost in the ENE-WSW or near E-W direction in Figure 3. The major T-axes of normal faulting events aligned almost conformably in the NNW-SSE or near N-S direction with great horizontal component projection in Figure 4. In order to investigate completely the spatial orientation of stress axes of the normal faulting type events Figure 5a, 5b showed the distributions of the projections of P- and T- axes on the vertical profile AB which was the diameter of the green dash circle along the N-S orientation from the north (A) to south (B) in SWXSH- NSNF in Figure 2, respectively. It could be seen that most P-axes aligned almost along the vertical direction with great vertical components. Most T-axes lay horizontally on the profile along N-S direction with great horizontal components. These characteristics of stress field showed the seismogenic mechanism of the normal faulting type events dislocating extensively along the N-S direction in the region southwest of the Xianshuihe fault in Figure 2.

In the normal faulting type event region east of the lower reach of the Jinshajiang river (ELRJSJ-EWNF) ranging about 99°E-101°E and 26°N-28°N marked with the green dash ellipse in Figure 2, the horizontal component of P-axes of normal faulting events were very little in Figure 3. The horizontal projections of the T-axes of normal faulting type events were generally great as shown in Figure 4. Major T-axes orientated near NNE-SSW direction. Figure 6 showed the distributions of the projections of P- and T- axes on the vertical profile CD. CD is the axis of the green dash ellipse along the E-W orientation from the west (C) to east (D) in ELRJSJ-EWNF in Figure 2. The P-axes aligned almost near the vertical direction with great vertical components. Most T-axes lay horizontally on the profile along E-W direction with great

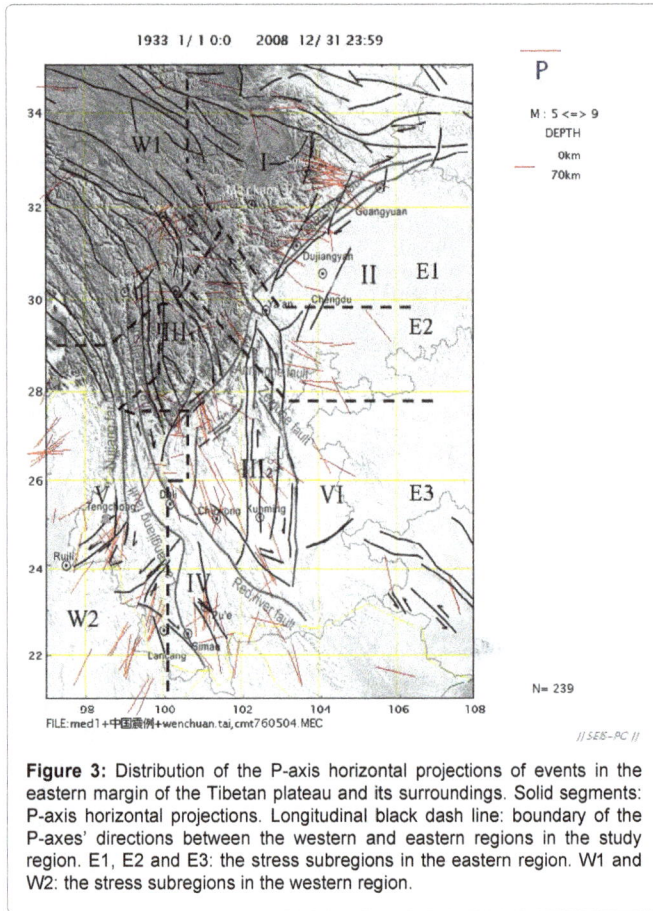

Figure 3: Distribution of the P-axis horizontal projections of events in the eastern margin of the Tibetan plateau and its surroundings. Solid segments: P-axis horizontal projections. Longitudinal black dash line: boundary of the P-axes' directions between the western and eastern regions in the study region. E1, E2 and E3: the stress subregions in the eastern region. W1 and W2: the stress subregions in the western region.

horizontal components. The characteristics of the stress field identified the dynamic mechanism of the extensional dislocation motion along the E-W direction of normal faulting type events in the region east of the lower reach of the Jinshajiang river.

The orientation of the P-axes in a general ENE-WSW direction in the northwestern segment of the Xianshuihe fault was different from those in a general E-W or WNW-ESE direction in the southeastern segment in Figure 3. The T-axes turn the general NW-SE or NNW-SSE direction in the northwestern segment into the near NNE-SSW direction in the southeastern segment of the Xianshuihe fault in Figure 4. The horizontal components of P- and T- axes, however were great. This was the typical mechanism of stress field of strike-slip faulting type event. In the thrust faulting type event region, the Longmenshan fault and its northwestern neighborhood, the horizontal components of P-axes of thrust faulting events were great and the P-axes aligned almost in the WNW-ESE or NW-SE direction in Figure 3. The T-axes of thrust faulting events aligned almost in the NNE-SSW or NE-SW direction with little horizontal component in Figure 4. Such stress characteristics showed the dynamic mechanism of the dislocation of thrust faulting type events. The azimuth of P-axis of the Wenchuan event ranged 290° to 302°, dip angle ranged about 6° to 8°, little [19,30]. The little dip angle of P-axis showed the stress field characteristic of thrust faulting type event occurrences in the Longmenshan fault and its northwestern neighborhood as shown in Figure 2 [34].

Stress field difference between the eastern and western regions

The P-axes' azimuths in the eastern region looked widely divergent from those in the western regions on the whole in Figure 3. Most of the

compressive stress P-axes aligned in a general NW-SE direction in the eastern region. The majority of P-axes in the western region, however, aligned along a general NE-SW direction in Figure 3. The boundary of the P-axis orientation between the eastern and western regions lay roughly along longitude 100°E in the region north of the about latitude 28°N and about longitude 101°E in the region south of the about

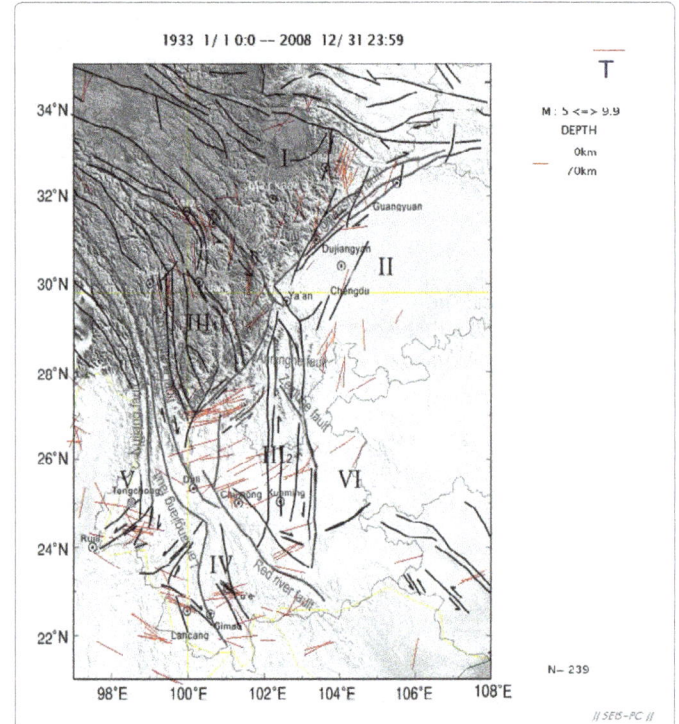

Figure 4: Spatial distribution of the T-axis horizontal projections of events in the eastern margin of the Tibetan plateau and its surroundings.

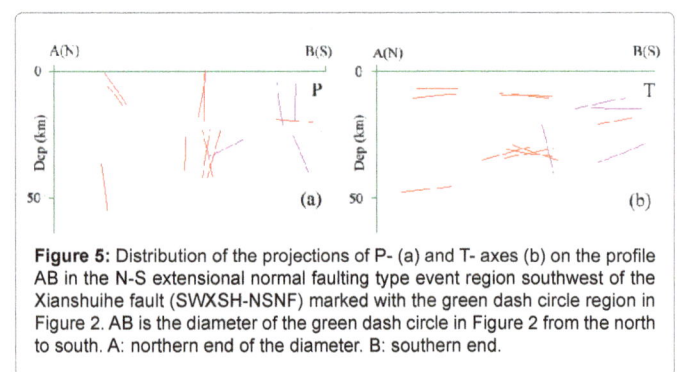

Figure 5: Distribution of the projections of P- (a) and T- axes (b) on the profile AB in the N-S extensional normal faulting type event region southwest of the Xianshuihe fault (SWXSH-NSNF) marked with the green dash circle region in Figure 2. AB is the diameter of the green dash circle in Figure 2 from the north to south. A: northern end of the diameter. B: southern end.

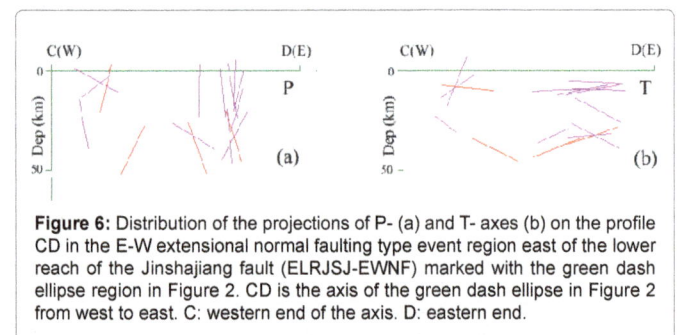

Figure 6: Distribution of the projections of P- (a) and T- axes (b) on the profile CD in the E-W extensional normal faulting type event region east of the lower reach of the Jinshajiang fault (ELRJSJ-EWNF) marked with the green dash ellipse region in Figure 2. CD is the axis of the green dash ellipse in Figure 2 from west to east. C: western end of the axis. D: eastern end.

latitude 28°N, displayed by a longitudinal black dashed line in Figure 6. In order to study the regional stress characteristics further, the study region was zoned into the five subregions in Figure 3. Figure 7 showed statistic rose diagrams of the azimuths of P- and T-axes of events in the eastern subregions Figure 6. The predominant direction of P-axes was about in the NNW-SSE direction in the southern subregion E3, and WNW-ESE direction in the northern subregion E1 in Figures 3 and 7, respectively. It looks as though the P-axes turn gradually the orientations from the NNW-SSE direction in the southern region E3 to the WNW-ESE direction in the northern region E1 through region E2. Many P-axes aligned near the E-W direction in subregion E2 in Figure 3 east of the normal faulting type event regions of SWXSH-NSNF in Figure 2. Figure 8 showed statistic diagrams of the azimuths of P- and T-axes of events in the western subregions of the study region. Similarly, P-axes turn gradually the orientations from the NNE-SSW in

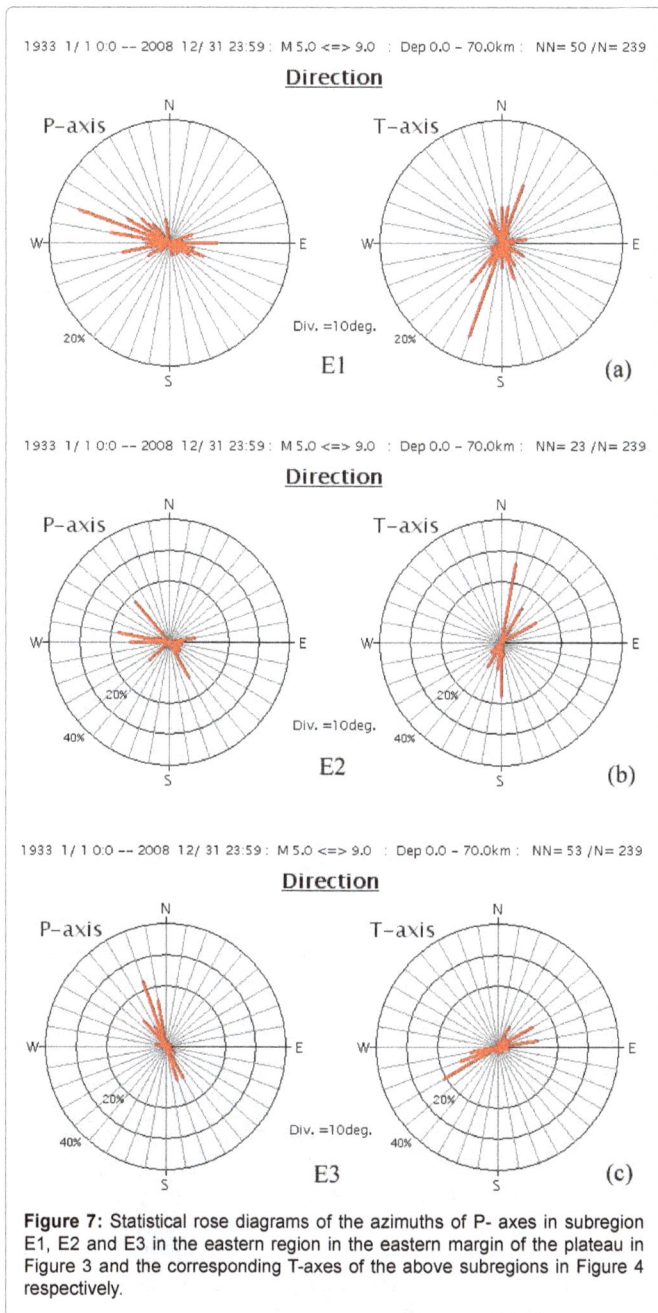

Figure 7: Statistical rose diagrams of the azimuths of P- axes in subregion E1, E2 and E3 in the eastern region in the eastern margin of the plateau in Figure 3 and the corresponding T-axes of the above subregions in Figure 4 respectively.

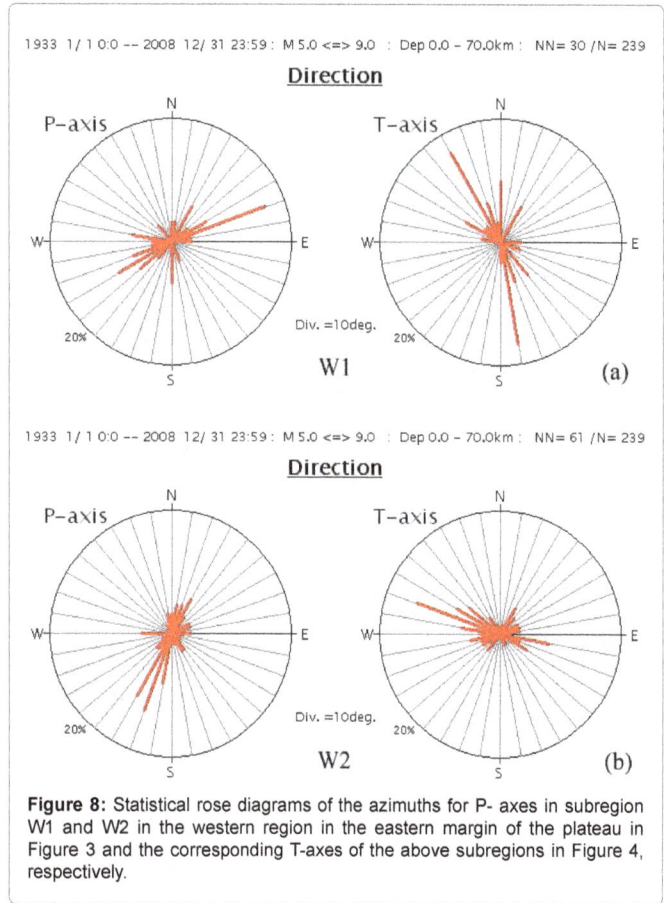

Figure 8: Statistical rose diagrams of the azimuths for P- axes in subregion W1 and W2 in the western region in the eastern margin of the plateau in Figure 3 and the corresponding T-axes of the above subregions in Figure 4, respectively.

the southern region W2 to the ENE-WSW in the northern region W1. i.e., the N-S components of P-axis horizontal projections decreased and the E-W components increased from the south to the north in Figure 3. The horizontal projections of P-axes got great E-W components in the region north of the 28°N and great N-S components in the Yungui plateau south of the 28°N. The predominant orientations of P-axes in the eastern region and western regions seem to cross into an obtuse angle in the region north of about 28°N and cross into a sharp angle south of about 28°N as shown in Figures 3 and 7. The regional characteristics of the T-axes' orientations in Figure 4 were investigated similarly referencing the stress longitudinal boundary between the eastern and western regions in Figure 3. In the western region of Figure 4, the T-axes' orientations aligned about in the WNW-ESE direction in the southern part (south of 28°N) and NNW-SSE direction in the northern part (north of 28°N). In the eastern region, the T-axes' orientations aligned about in the ENE-WSW direction in the southern part and NNE-SSW direction in the northern part in Figure 4.

In another view, the majority of T-axes likely had great horizontal component along near N-S direction in the region north of about 28°N and great horizontal component along near E-W direction in the region south of 28°N. Referencing results in Figure 3, the belt region along about latitude 28°N looks a latitudinal boundary for the orientation changes of P- and T-axes between the north and south parts of the study region, which is also near the southeastern boundary of the Tibetan plateau as shown in Figure 1.

Discussion

The findings of seismotectonic motions in the eastern margin of the

Tibetan plateau in the present analysis showed that the normal faulting type event region existed southwest of the Xianshuihe fault (SWXSH-NSNF) where the normal faulting type events dislocated extensively along the N-S direction. The above observed normal faulting type event region indicate that there are extensional motions along the N-S direction besides the eastward tectonic escape in the eastern margin of the Tibetan plateau [35]. The another normal faulting type event region, east of the lower reach of the Jinshajiang river (ELRJSJ-EWNF) where the dislocations of normal faulting type events extended along the E-W direction. The ELRJSJ-EWNF is on the southern border of the SWXSH- NSNF. The two normal faulting type event regions, are approximately located in the western study region. The events are almost thrust faulting type ones along the Longmenshan fault, in the eastern study region. The strike-slip faulting events occurred along the Xianshuihe fault and other wide regions. The faulting regimes of events show more complex seismo-tectonic motions in the eastern margin of the Tibetan plateau which is boundary between the south China block and the Tibetan plateau.

The extensional motion along the N-S direction due to the normal faulting type event took place southwest of the Xianshuihe fault. It is a new finding of the seismo-teconic motion in the present analysis. The dynamic cause of the N-S extensional motion probably is attributable to the regional tectonic motions and the stress fields of the subblocks around the N-S extensional normal faulting type event region southwest of the Xianshuihe fault (SWXSH-NSNF) in Figure 2. The Longmenshan block, region I in Figure 1 north of the Xianshuihe fault and northeast of the SWXSH-NSNF moved southeastward very slowly because of encountering strong obstructs from the Sichuan basin [36,37]. The channel flow of the deep crust may be inhibited by the rigid Sichuan Basin [38]. The compressive stress P-axes are also along WNW-SES direction here as shown in Figures 3 and 7. The horizontal projection of fault slip vector of the Wenchuan M8 earthquake in the Longmenshan fault is also southeastward. The strike direction, dip angle and rake angle of the M8 event fault were 230°, 30° and 128° respectively. The annual ratio of the surface displacement between Tibetan plateau and the south China block in near ESE-WNW direction was about 3 mm/y along the Longmenshan fault, very little based on GPS observations [39]. While the large faults lie along the N-S direction in the Sichuan-Yunnan rhombic block (region III, south of the Xianshuihe fault) as shown in Figure 1. The rhombic block moved southward with greater annual ratio than that in the Longmenshan block [39]. So the SWXSH-NSNF being located in the north end of the rhombic block might undergo extensional status southward. Furthermore the azimuths of many P-axes are in or near the E-W direction (greater than thirty percent between 270°and 280°) in the region E2 east of SWXSH- NSNF as shown in Figures 3 and 7. It implied that the SWXSH-NSNF was undergone strong compressive action in near E-W direction also. The T-axes aligning along N-S direction are greater than seventy percent in region E2 as shown in Figures 4 and 8. Such dynamic mechanism is evidently benefited to the extensional motion of normal fault in N-S direction.

The E-W extension normal faulting type event region east of the lower reach of the Jinshajiang fault (ELRJSJ-EWNF) is south of the SWXSH-NSNF in Figure 2. The northern end of the ELRJSJ-EWNF might be compressed by the southward extensional motion from the SWXSH-NSNF as motioned above. The southern end of ELRJSJ-EWNF is located northeast of the southern end of the Jinshajiang fault in Figures 1 and 2. The Jinshajiang fault along the N-S strike direction in the northern segment is turned into the NW-ES direction in the southern end of ELRJSJ-EWNF. This is an interesting tectonic

change for the normal faulting type events. It might imply that the ELRJSJ-EWNF in the southern end of the rhombic block (region III₁ in Figure 1) may encounter the tectonic obstruction from its southern neighborhoods. The stress field also shows that the ELRJSJ-EWNF is compressed by its southern vicinities. The P-axes aligned in the NNE or near the N-S direction in the stress subregion W3 southwest of the ELRJSJ-EWNF and in the NNW or near the N-S direction in the stress subregion E3 southeast of the ELRJSJ-EWNF as illustrated in Figures 3,7 and 8. It shows that the strong compressive stress distributed in the southwestern (W2 stress subregion) and southeastern neighborhood (E3 subregion) of the ELRJSJ-EWNF. Simultaneously, a strong extensional stress field existed in the eastern neighborhoods of ELRJSJ-EWNF as shown in Figure 4. Such surroundings for both of tectonic motions and stress fields imply the southern end of ELRJSJ-EWNF is undergoing the compression near the N-S direction and the extension near the E-W direction. The tectonic motion and the stress field benefits evidently the normal faulting extensional motion in E-W direction in ELRJSJ-EWNF.

Furthermore, the Indian plate subducts from the Myanmar region under the eastern end of the Himalayans in the Eurasian plate along near E-W direction. Based on seismic data this is an eastward subduction zone form the Myanmar region under the southeastern Tibetan plate and Yunnan region, China. The subduction slab extends for about 500 km from the southwest to northeast and the front reaches 150 km depth. The subduction zone ranges between about 20°N-27°N and 92°E-98°E [40]. The Tengchong volcano in Yunnan province, China is located east of the subduction zone in Figure 1. The normal faulting type event regions, SWXSH-NSNF and ELRJSJ-EWNF are about 200 km away from the subduction zone in the northeast and the east, respectively. The occurrences of the normal faulting type events in ELRJSJ-EWNF and SWXSH-NSNF are likely attributable to the extensional motion in the region back the subduction zone.

The directions of the P-axes in nearly NE-SW or NEE-SWW on the western region generally coincide with those on the Tibet Plateau. The seismogenic stress field on the Tibet Plateau and the plateau motion are caused by compression from the northward movement of the Indian plate [41-43]. The P-axes aligning in the NW-SE and WNW-ESE directions in the eastern regions E1, E2 and E3 are quite different from those in the NE-SW and ENE-WSW directions in the western regions W1 and W2. The orientations of the principle compressive stress P-axes reveal discontinuous change as crossing the longitudinal boundary of the P-axis as shown in Figures 3,7 and 8. It perhaps implies that another independent dynamic source being able to generate the compressive stress shown by the P-axes in the NW–SE and WNW-ESE directions might exist potentially in the eastern region in Figure 3. The principal compressive stress P-axes in a general NW-SE and WNW-ESE direction in the eastern region in the eastern margin of the Tibetan plateau approximate those in the South China and Taiwan region. The orientation of the P-axes in the eastern region in Figure 3 probably is affected by the stress field in the South China block and the Taiwan region [44]. This dynamic source of stress field in the eastern region in Figure 3 might be attributable to the transmission of the tectonic force resulting from subduction of the Philippine Sea plate along the Ryukyu trench and the collision between the Philippine Sea and Eurasian plates in the Taiwan region [45].

It is noticed that the longitudinal boundary of the P-axis between the eastern and western subregions lies within the Tibetan plateau and even passed through the large Xianshuihe fault. The longitudinal stress boundary in Figure 3 is also unrelated to the tectonic boundary between the Tibetan plateau and the South China block at the surface in

Figure 1. It might be relation with lateral differences in the lithospheric structure [46]. Based on the tomography results, a westward extension front of a high velocity zone beneath the South China block has crossed the Longmenshan fault by 20 km westward and arrived in the eastern margin of the plateau in the lower crust and the upper mantle at 50 km depth [13,47]. Aftershocks of the M8 Wenchuan earthquake were almost concentrated on the inner side of the high velocity zone front of the South China block [48]. This also implies that the tectonic stress in the South China block might have affected the earthquake generating stress field for the eastern region of the eastern margin of the plateau in Figure 3.

Conclusion

The results in the present analysis showed that the dislocation of normal faulting type events extended along the N-S direction in the normal faulting type event region southwest of the Xianshuihe fault (SWXSH-NSNF). Another, in normal faulting type event region was east of the lower reach of the Jinshajiang river (ELRJSJ-EWNF) the dislocations of normal faulting type events extended along the E-W direction. The ELRJSJ-EWNF is on the southern border of SWXSH-NSNF.

The P-axes aligned in the NE-SW or ENE-WSW directions in the western region and in the NW-SE or WNW-ESE directions in the eastern region. The seismogenic stress field in the eastern region in the eastern margin of the Tibetan plateau is related to the collision between the Eurasian and Philippine Sea plates along the Taiwan region through the South China block and the subduction from the Philippine Sea plate along the Ryukyu trench under the Eurasian plate. The longitudinal boundary of the P-axis of seismogenic stress field between the eastern and western regions lies within the eastern margin of the Tibetan plateau along about the longitude 100°E or 101°E. It does not coincide with the boundary between the Tibetan plateau and the South China block.

Acknowledgement

This study was supported partly by the Natural Science Foundation of China (Grant No. 41374052).

References

1. Meissner R, Mooney WD, Artemieva I (2002) Seismic anisotropy and mantle creep in young orogens. Geophys J Inter 149: 1-14.

2. Hetzel R (2013) Active faulting, mountain growth, and erosion at the margins of the Tibetan Plateau constrained by in situ-produced cosmogenic nuclides. Tectonophysics 582: 1-24.

3. Zhou LQ, Zhao CP, Xiu JG, Chen ZL (2008) Tomography of QLg in Sichuan-Yunnan Zone. Chin J Geophys 51: 1159-1167.

4. Blisniuk PM, Hacker BR, Glodny J (2001) Normal faulting in central tibet since at Least 13.5 Myr Ago. Nature 412: 628-632.

5. Hetzel R, Niedermann S, Tao M (2002) Low slip rates and long-term preservation of geomorphic features in Central Asia. Nature 417: 428-432.

6. Honglin He, Tsukuda E (2003) Recent progresses of active fault research in China. J Geograp 112: 489-520.

7. Lave J, Avouac JP, Lacassin R, Tapponnier P, Montgner JP (1990) Seismic anisotropy beneath Tibet –evidence for eastward extrusion of Tibetan lithosphere. Earth Planet Sci Lett 24: 1851-1854.

8. Xu G, Kamp P (2000) Tectonics and denudation adjacent to the Xianshuihe Fault, eastern Tibetan Plateau: Constraints from fission track thermochronology [Preview]. J Geophys Res 105: 19231-19251.

9. Zhang P (2013) A review on active tectonics and deep crustal processes of the Western Sichuan region, eastern margin of the Tibetan Plateau. Tectonophysics 584: 7-22.

10. Copley A (2008) Kinematics and dynamics of the southeastern margin of the Tibetan Plateau. Geophys J Inter 174: 1081-1100.

11. Elliott JR, Walters RJ, England PC, Jackson JA, Li Z, et al. (2010) Extension on the Tibetan plateau: recent normal faulting measured by InSAR and body wave seismology. Geophys J Inter 183: 503-535.

12. Clark MK, Royden LH, Whipple KX, Burchfiel BC, Zhang X, et al. (2006) Use of a regional, relict landscape to measure vertical deformation of the eastern Tibetan Plateau [Preview]. J Geophys Res 111: F03002.

13. Xu JR, Zhao ZX (2009) Geothermic activity and seismotectonics in the altitude of the Tibetan plateau. Earthq Scie 22: 651-658.

14. Chen Q, Papazachos C, Papadimitriou E (2002) Velocity field for crustal deformation in China Derived from seismic moment tensor summation of earthquakes. Tectonophysics 359: 29-46.

15. Liang S, Gan WJ, Shen CZ, Xiao GR, Liu J, et al. (2013) Three-dimensional velocity field of present-day crustal motion of the Tibetan Plateau derived from GPS measurements. J Geophys Res Solid Earth 118: 5722-5732.

16. Xu JR, Zhao ZX, Ishikawa Y, Oike K (1988) Properties of the stress field in and around west China derived from earthquake Mechanism solutions. Bull Disas Prev Res Inst Kyoto Univ 38: 49-78.

17. Levin V, Huang DG, Roecker S (2013) Crust–mantle coupling at the northern edge of the Tibetan plateau: Evidence from focal mechanisms and observations of seismic anisotropy. Tectonophysics 584: 221-229.

18. Xu X, Tan X, Yu G, Wu G, Fang W, et al. (2013) Normal- and oblique-slip of the 2008 Yutian earthquake: Evidence for eastward block motion, northern Tibetan Plateau. Tectonophysics 584: 152-165.

19. Zhang Y, Xu L, Chen Y (2009) Spatio-temporal variation of the source mechanism of the 2008 great Wenchuan earthquake. Chin J Geophys 52: 379-389.

20. Toda S, Lin J, Meghraoui M, Stein RS (2008) 12 May 2008 M=7.9 Wenchuan, China, earthquake calculated to increase failure stress and seismicity rate on three major fault systems. Geophys Res Lett 35: 6.

21. Xu JR, Zhao ZX, Ishikawa Y (2008) Regional Characteristics of Crustal Stress field and Tectonic Motions in and around Chinese Mainland. Chin J Geophy 51: 861-869.

22. Klinger Y, Ji C, Shen ZK, Bakun WH (2010) Introduction to the Special Issue on the 2008 Wenchuan, China, Earthquake. Bull Seism Soc Amer 100: 2353-2356.

23. Xu JR, Zhao ZX, Ishikawa Y (2005) Extensional stress field in the central and southern Tibetan plateau and dynamic mechanism of geothermic anomaly in the Yangbajin. Chin J Geophys 48: 861-869.

24. Hatzfeld D, Molnar P (2010) Comparisons of the kinematics and deep structures of the Zagros and Himalaya and of the Iranian and Tibetan plateaus and geodynamic implications. Revi Geophys 48: RG2005.

25. Arne D, Worley B, Wilson C, Chen She-Fa, Foster D, et al. (1997) Differential exhumation in response to episodic thrusting along the eastern margin of the Tibetan Plateau. Tectonophysics 280: 239-256.

26. Wang Z, Huang R, Pei S (2014) Crustal deformation along the Longmen-Shan fault zone and its implications for seismogenesis. Tectonophysics 610: 128-137.

27. Zhao ZX, Oike K, Matsumura K, Ishikawa Y (1990) Stress field in the continental part of China derived from temporal variations of seismic activity. Tectonophysics 178: 357-372.

28. Zhang ZC (2002) Chinese earthquake cases (1966-1975). Seismological Press, Beijing.

29. Xue QF (2002) Chinese earthquake cases (1997-1999). Seismological Press, Beijing.

30. Hu XP, Yu CQ, Tao K (2008) Focal mechanism solutions of Wenchuan earthquake and its strong aftershocks obtained from initial P-wave polarity analysis. Chin J Geophys 51: 1711-1718.

31. Dziewonski AM, Woodhouse JH (1983) An experiment in systematic study of geobal seismicity: centroid-moment tensor solution for 201 moderate and large earthquake of 1981. J Geophys Res 88: 3247-3271.

32. Dziewonski A, Friedman M, Giardini A, Woodhouse JH (1983) An experiment in

systematic study of Geobal seismicity of 1982: centroid-moment tensor solution for 308 earthquake. Phys Earth Planet Inter 33: 76-90.

33. Furuya M, Kobayashi T, TakadaY, Murakami M (2010) Fault source modeling of the 2008 wenchuan earthquake based on ALOS/PALSAR data. Bull Seism Soci Amer 100: 2750-2766.

34. Tom P, Ji Chen, Kirby E (2008) Stress changes from the 2008 Wenchuan earthquake and increased hazard in the Sichuan basin. Nature 454: 509-510.

35. Klemper SL (2006) Crustal flow in Tibet: geophysical evidence for the physical state of Tibetan lithosphere, and inferred patterns of active flow. In: Law RD, Searle MP, Godin L *"Channel flow, ductile extrusion and exhumation in continental collision zones"*. Geol Soc Lond Special Publication 268: 39-70.

36. Group of Crustal movement observation (2008) Coseismic displacement field for the 2008 Wenchuan Ms 8.0 earthquake determined by GPS. Scien Chin Series D: Earth 38: 1195-1206.

37. Chen Z, Burchfiel BC, Liu Y, King W, Royden LH, et al. (2000) Global positioning system measurements from eastern Tibet and their implications for India/Eurasia intracontinental collision. J Geophys Res 105: 16215-16227.

38. Clark MK, Bush JWM, Royden LH (2005) Dynamic topography produced by lower crustal flow against rheological strength heterogeneities bordering the Tibetan Plateau. Geophys J Inter 162: 575-590.

39. Wang Q, Zhang P, Freymueller JT, Bilham RL, Kristine M, et al. (2001) Present-day crustal deformation in china constrained by global positioning system measurements. Science 294: 574-577.

40. Xu Y, Huang RQ, Li ZW, Liu JH (2009) S wave velocity structure of the Longmenshan and Wenchuan earthquake area. Chin J Geophys 52: 329-338.

41. Keith P, Jackson J, Dan M (2008) Lithospheric structure and deep earthquakes beneath India, the Himalaya and southern Tibet. Geophys J Inter 172: 345-362.

42. Xu JR, Zhao ZX (2010) Normal faulting type earthquake activities in the tibetan plateau and Its tectonic implication. Acta Geologica Sinca 84: 135-144.

43. Xu ZH (2001) A present-day tectonic stress map for eastern Asia region. Acta Seismologica Sinica 14: 524-533.

44. Hsu YJ, Yu SB, Mark S, Kuo LC, Chen HY (2009) Interseismic crustal deformation in the Taiwan plate boundary zone revealed by GPS observations, seismicity, and earthquake focal mechanisms. Tectonophysics 479: 4-18.

45. Ching KE, Rau RJ, Johnson KM, Lee JC, Hu JC (2011) Present-day kinematics of active mountain building in Taiwan from GPS observations during 1995–2005. J Geophys Res 116: 9405.

46. Chen L, Berntsson F, Zhang Z, Wang P, Wu J, et al. (2014) Seismically constrained thermo-rheological structure of the eastern Tibetan margin: Implication for lithospheric delamination. Tectonophysics 627: 122-134.

47. Wu JP, Huang Y, Zhang TZ (2009) Aftershock distribution of the M_S 8 Wenchuan earthquake and three dimensional P-wave velocity structure in and around source region. Chin J Geophys 52: 320-328.

48. Zhao ZX, Xu JR (2009) Compressive tectonics around tibetan plateau edges. Journal of Earth Science 20: 477-483.

2

A Geological Appraisal of Slope Instability in Upper Alaknanda Valley, Uttarakhand Himalaya, India

Sajwan KS* and Sushil K

Disaster Mitigation and Management Centre (DMMC), Department of Disaster Management, Government of Uttarakhand, Uttarakhand Secretariat, Rajpur Road, Dehradun 248001, Uttarakhand, India

Abstract

Landslide is frequently occurring natural phenomenon causing land degradation in mountainous region of the world. The fragile Himalayan terrain of Uttarakhand often faces challenging and tough situation due to landslides, particularly in monsoon season. Change in rainfall pattern, anthropogenic activity, deforestation, construction on old landslide debris and Quaternary deposits and displacement of habitation from hill side to valley side are some significant increasing factors for the landslide susceptibility. Anthropogenic activities include very fast, unscientific and uncontrolled urbanization and these are considerably responsible to carved out the disaster viz., landslide, cloudburst and flash flood. Habitation and infrastructure development initiatives in close proximity of streams and rivers as also over Quaternary deposits and unplanned disposal of excavated rock and debris are observed to aggravate the fury of both, landslides and flash floods in the region. Mostly in monsoon season frequent disruption of road network by landslides cause hardship to tourists and pilgrims along with local people. During monsoon season torrential rainfall and cloudburst events are common which acts as basic factor for triggering the landslides in the area. It is therefore necessary to analyse the causes of landslides so as to suggest viable mitigation measures. The paper outlines slope instability and factors responsible for the landslide susceptibility in upper Alaknanda valley, Uttarakhand, India.

Keywords: Landslide; Upper Alaknanda valley; Uttarakhand Himalaya; Triggering factors; Causative factors

Introduction

Landslides are one of the most common and widespread natural hazards that affect at least 15% of the land area of our country, an area which exceeds 0.49 million km². Landslides of different types are frequent in geodynamically active domains in the Himalaya. Landslide is an important landform building process promoting soil formation and most habitations in the hills are located in the proximity of old stabilised landslides as these provide suitable land for agricultural operations. Direct losses due to landslides have been increasing and frequent disruption of transport network by landslides has been highlighted as a major hindrance in economic well-being of the region. Present study is an attempt to analyse the causes of landslides in the catchment of upper Alaknanda valley that meets Bhagirathi River at Devprayag to form Ganga River.

The study is initiated with the mapping of landslides in the upper Alaknanda valley and the same is followed by the correlation of these with different natural and man-made feature to assess the influence of these on the occurrence of landslides. This is intended to help in formulating of a viable strategy for minimising the menace of landslides. In the previous some years, mainly due to extreme precipitation events, landslides have however become major cause of concern. Losses due to landslides and flash floods in Uttarakhand in 2010, 2012 and 2013 testify this fact (Table 1).

Materials and Methods

The area under investigation lies in the upper Alaknanda valley that falls in lesser and Higher Himalayan terrain of Uttarakhand Himalaya (Figure 1). The study area lies within 78°47'45"E and 80°06'41"E longitudes and 31°05'28"N and 29°55'29"N latitudes. The area lies in seismic Zone - IV and V and receives heavy precipitation during monsoon season. The area is at the same time strategically important and well connected by road network of which National Highway 58 and 109 are an integral part. The area is visited by large number of people every year and the same contributes to the economy of the region. Rudraprayag town is located 140 kilometers from Rishikesh,

Sl. No.	Item	2010	2012	2013
1	Period of occurrence	August – September 2010	August – September 2012	Jun-13
2	Number of affected districts	13	2	13
3	Number of villages affected	9,162	129	1,603
4	Population affected (in million)	2.9	0.08	0.5
5	Permanent loss of land (in ha)	2,35,160	-	11,482
6	Cropped area affected (in ha)	5,02,741		10,899
7	Damaged houses	21,045	652	19,309
8	Human lives lost	214	81	169
9	Human beings missing	0	6	4,024
10	Persons injured	227	27	236
11	Animals lost	1,771	537	11,091
12	Damage to public properties (in million US $)	3526	134	2163

Table 1: Losses incurred in Uttarakhand due to the disasters of 2010, 2012 and 2013 [26-28].

on Rishikesh - Bardinath National Highway. Rishikesh is the nearest rail head while located in close proximity of Dehradun, the capital of Uttarakhand state, Jolly Grant airport also located in Dehradun.

Landslide distribution and classification

The terrain is characterized by predominance of high relative relief;

---

*Corresponding author: Sajwan KS, Disaster Mitigation and Management Centre (DMMC), Department of Disaster Management, Government of Uttarakhand, Uttarakhand Secretariat, Rajpur Road, Dehradun 248 001, Uttarakhand, India
E-mail: krishnasajwan05@gmail.com

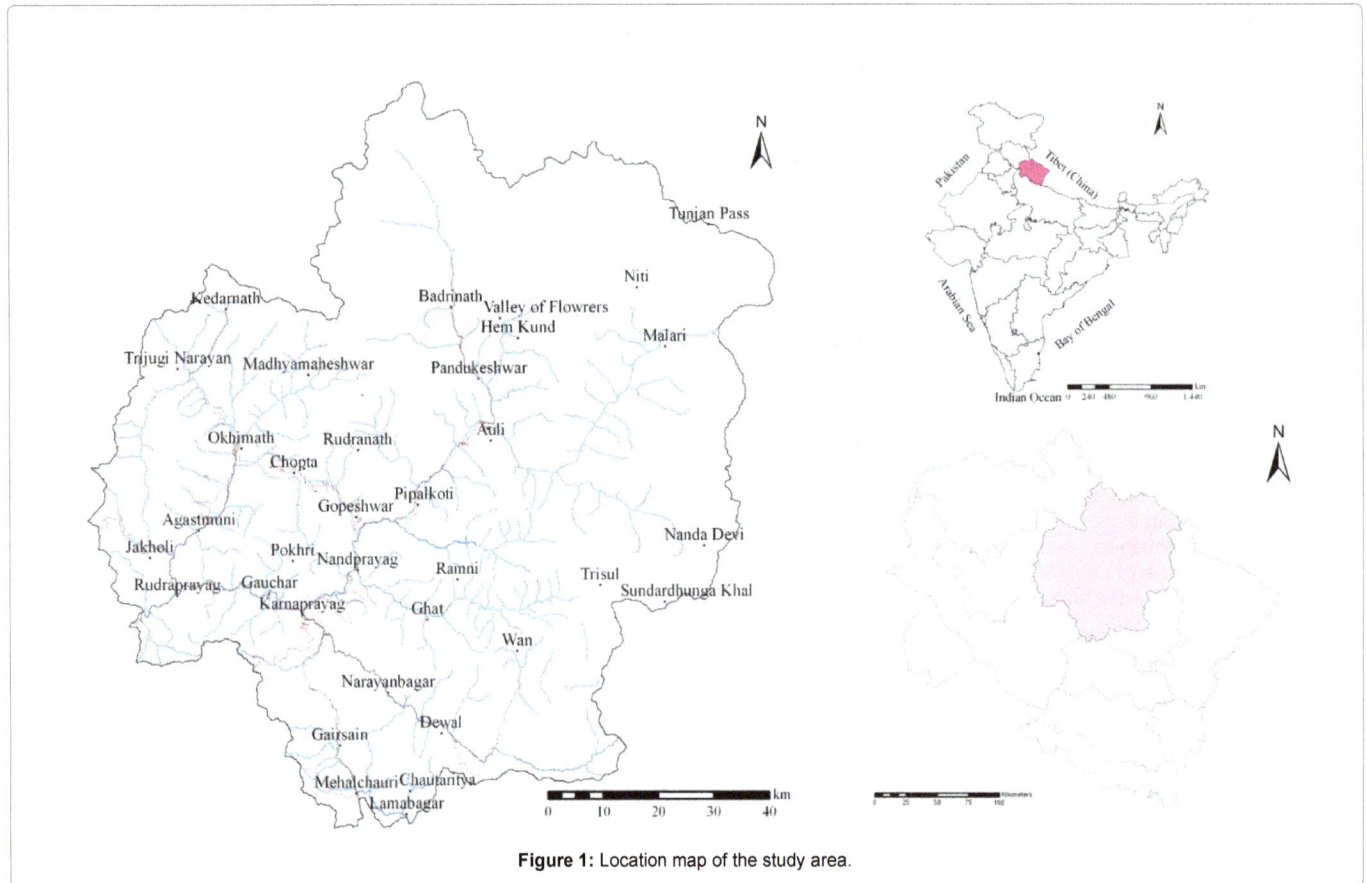

Figure 1: Location map of the study area.

monsoon rains further aggravate the problem of landslides. Various anthropogenic activities for infrastructure development further enhance the susceptibility of slope failure. These factors make the slope in the study area landslide prone. The macro-scale landslide inventory database of the area on 1:50000 scale has been prepared. A number of new landslides were observed to be initiated in the area due to flooding in Alaknanda and tributaries in June 2013. At many places old slides were observed to be reactivated.

A total of 510 landslides have been identified through field investigations in Upper Alaknanda valley (Figure 2). Observed landslides are classified on the basis of movement and rigidity of material comprising the slide mass; bed rock, debris and earth. The summary of the same is given in Table 2.

Landslide triggering factors

Heavy rainfall, bank erosion, road construction, lithological changes, presence of critical structural discontinuity and weathering are deducted to be the main causes of landslides observed in the area. For the assessment of landslide triggering factors proximity of the landslides to roads and streams was particularly analyzed together with other causative factors that include geological and structural set up of the area (Table 3).

Majority of landslides in this area (33.92%) are observed to be triggered by bank erosion by rivers and tributaries while 29.02% are caused by change in angle of repose, largely for road construction. Another 14.51% is triggered by both bank erosion and change in angle of repose. Large proportion of the landslides (22.55%) is caused by other factors that include heavy rainfall, cloudburst, seismicity, lithological

and structural condition of the bed rocks and sub-surface hydrological factors (springs and seepage).

Causative factors

In any area slope, topography and climate play important role for landslide susceptibility. Tectonic movements have resulted in intense shearing, fracturing and faulting of the rocks observed in the area. Some most dominant and considerable causative factors responsible for the landslides are described.

Regional tectonics: It is very important for better understanding of landslide mechanism to study and collect all the data about lithology and tectonics. Stability of a site is therefore often inferred from the geology that is considered a key parameter conditioning landslide occurrence as sensitivity to active geomorphological processes such as landslides is considered to vary with geology [1,2]. Based on the fieldwork done in the area geological set up of the area is reconstructed (Figure 3). While undertaking geological mapping the focus is kept on identifying lithological characteristics of the various rock slopes and landslide zones present in these.

In the study area Central Crystalline rocks are thrust over Garhwal Group of rocks along Main Central Thrust (MCT) which is a northerly dipping major tectonic discontinuity exposed across the Alaknanda River at Helang (Figure 3). The Garhwal Group of rocks of Lesser Himalaya are observed to comprise of low grade metasediments that are intruded by acidic and basic igneous rocks. These consist of thick succession of low grade metasediments made up of quartzite along with penecontemporaneous metabasics and carbonate rocks. The main

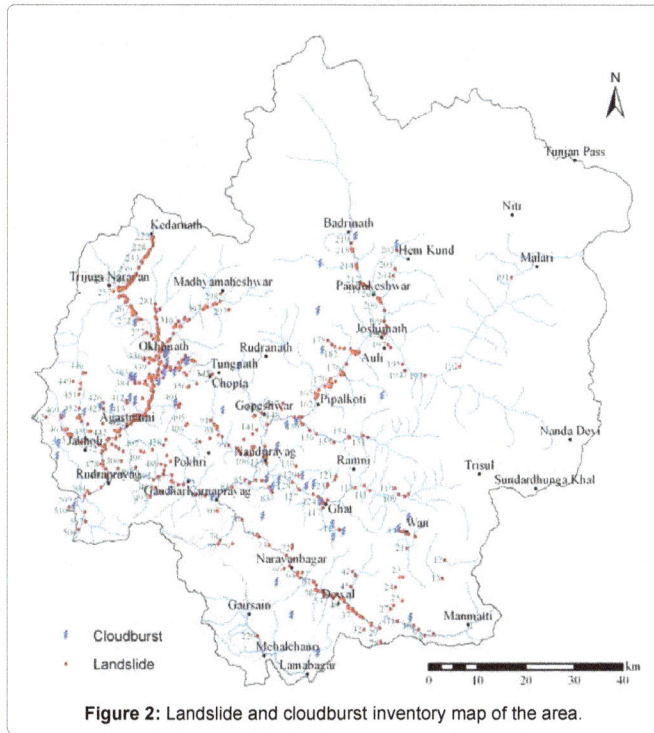

Figure 2: Landslide and cloudburst inventory map of the area.

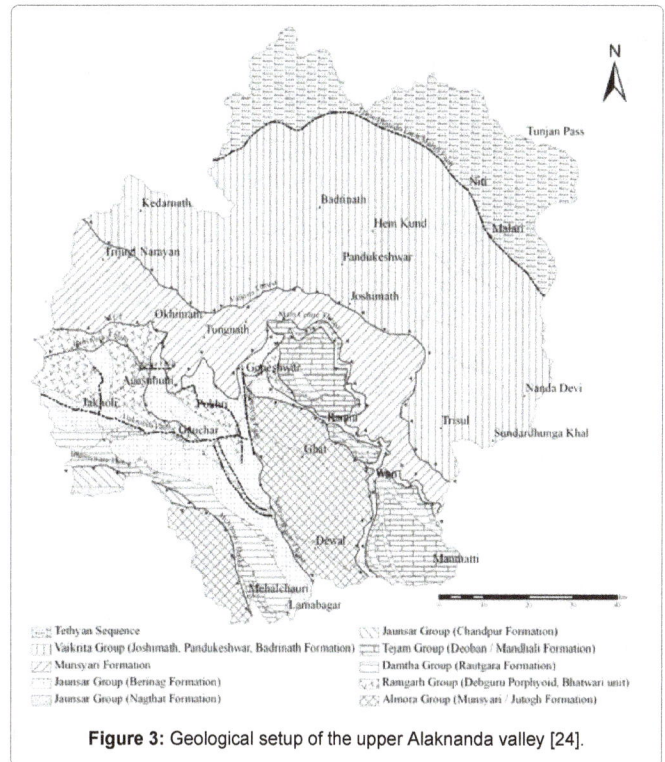

Figure 3: Geological setup of the upper Alaknanda valley [24].

Sl. No.	Landslide type	Total number of slides
1	Debris/bouldery debris slide	346
2	Rock cum debris slide	141
3	Rock slide/fall	23
	Total	**510**

Table 2: Summary of different type of landslides observed in the area.

Sl. No.	Triggering factors	Number of landslides
1	Bank erosion	173
2	Bank erosion and change in angle of repose	74
3	Change in angle of repose	148
4	Others	115
	Total	**510**

Table 3: Summary of landslide triggering factors in the area.

Tethyan Sequence			
Trans - Himadri Fault / Malari Fault			
Vaikrita Group	Central Crystallines	Badrinath Formation	Calc-silicate with sillimanite-kyanite-garnet-boitite gneiss and schist with granite (associated with migmatites).
		Pandukeshwer Formation	Biotite and muscovite rich quartzite intercalated with schist and subordinate gneiss.
		Joshimath Formation	Streaky and banded gneiss and kyanite-garnet rich mica schists.
		Vaikrita Thrust	
MCT zone		Helang/Munsiari Formation	Mylonitised porphyritic granites with subordinate amphibolite, chlorite-sericite schist and limestone.
Main Central Thrust / Munsiari Thrust			
Lesser Himalaya	Garhwal Group (Metasedimentary rocks)		

Table 4: Litho-tectonic succession of upper Alaknanda valley [24].

rocks observed in the area include phyllite, quartzite, limestone, slate, granite and metabasics.

The rocks of Central Crystallines are represented by variants of gneiss, schist, quartzite, crystalline limestone, marble and granite. The lithotectonic succession is given in Table 4.

Main central thrust is major structural discontinuity of Himalaya that traverses through central portion of the study area in northwest to southeast direction. Besides this, there also exist a number of local or regional weak planes. Rock mass in the vicinity of these is structurally weak and therefore these zones are identified as being highly vulnerable to mass wastage. Saturated debris together with fractured, jointed and sheared rocks on the hill slopes are easily displaced by moving water and gravity keeps them moving down.

Geomorphology: The area represents rugged topography characterized by high peaks, cliffs, moderate to steep slopes, elongated ridges, deep and narrow valleys of lesser and Higher Himalaya. The topography is highly precipitous, consisting of series of peaks like

Neelkanth (6,596 m), Nanda Devi (7,816 m), Kamet (7,756 m), Mana peak (7,272 m), Sumeru Parvat (6,350 m) and Kedarnath (6,940 m). The slopes of these peaks are covered with snow and glaciers.

The upper Alaknanda valley is highly prone to landslides; particularly area falls in Higher Himalayan crystallines, debris avalanche, debris fall and debris flow are common due to abundantly present overburden material by the glacial and glacio-fluvial action with steep gradient. The prominent glacial landforms present mostly up to Hanumanchatti and Rambara are moraines, cirque, glacial horns, glacial cones/fans, U-shaped glacial valleys, waterfalls and hanging valleys indicated a major phase of glacial activity in the area during Pleistocene period.

Downstream to Hanumanchatti and Rambara V-shaped valleys, river terraces, point bars, sand bars, narrow deep gorges, 'U', 'V' and 'S' shaped sinuosity or meandering have been observed in the area. The landforms present in the study area may be classified into glacial, glacio-fluvial, fluvial and denudational. The Quaternary deposits are present in the form of glacial cones/fans, moraines, alluvium, colluvium and fluvial terraces made up of river borne material (RBM) all along the valley.

The river and stream course of the area appears to be controlled by lithological and structural factors as well as climatic fluctuation. Landscape changes have been enhanced serious problem of slope instability in the region. Gentle and moderate slope have been transformed into steep/vertical slope due to several natural and manmade reason. Unconsolidated Quaternary deposits with less cohesion are highly susceptible to erosion. There exists high probability of slope instability being initiated in these deposits. Hitherto stable slopes were observed to have become unstable due to flooding in the streams.

Heavy rainfall/cloudburst: Change in rainfall pattern has been observed in recent years, rainfall intensity is increasing in place of rainfall amount. The study area experiences enough rainfall due to southwest monsoon during June to September. The precipitation mainly arises during these months. The scattered rains and snowfall arise during winter months. The heavy seasonal downpour on the geologically and structurally deformed rocks creates a variety of slope failures. The rainfall data of the study area was collected and summarized in Figure 4.

It is evident from Figure 4 that the rainfall in the area is highly variable due to its orographic and geographic disposition. The rainfall data reflect that the area has been recorded average maximum monthly rainfall 555.3 mm in July 2013 and minimum rainfall 1.5 mm in Nov. 2010 besides this 0% rainfall also recorded in some months. The maximum annual rainfall is about 1912.20 mm in the year of 2013 followed by 1628.80 mm in 2010 and minimum 562.70 mm in the year of 2009. Exceptionally high rainfall in June 2013 can also observe during five years (Figure 4).

During August and September 2010, Uttarakhand Himalaya witnessed large scale slope destabilization, particularly along the roads where widening work was in progress. The landslides killed about 220 people in the entire rainy season of 2010 [3]. According to National Remote Sensing Centre, ISRO, in the preliminary assessment, a total of 1356 landslides have been identified along the river valley of Alaknanda after Kedarnath disaster of June 2013. On the basis of landslide incidences, inundation and last five years rainfall data the triggering factor for the slope instability is exceptionally high rainfall in the area during the year of 2010 and 2013. The strength of the Indian monsoon determines the intensity of precipitation events in the Himalaya.

Changes in the intensity of the southwest monsoon, during years of strengthened atmospheric circulation correspond to an increase in the number of large landslides (>0.5 km^3) in the Late Pleistocene and Holocene periods [4,5].

Moreover, cloudburst is one of the most considerable factors responsible for the slope instability. Such events are related to extreme hydrometrological conditions, debris flow, mud flow and landslides along the tributaries/rivers are common due to this. Torrential rainfall/cloudburst are the natural phenomenon occurs during monsoon season over the regions dominated by orography like Himalaya. High intensity rainfall of more than 100 mm/hour within a limited geographical area of a few square kilometres is defined as cloudburst [6]. In this area total 52 cloudbursts location were identified and studied (Figure 2).

Based on field observation and information collected from local people mostly seasonal channel have been mostly inundated and debris flow also occurred along these, subsequently debris laden water accelerate the erosion capacity of major river (Figure 5). Slope failures and bank erosion are common during this phenomenon which result sedimentation and sometimes block the river course, turn them into big lake and create flood condition. For instance, the 1970, Alaknanda flood were caused on the destruction of a lake at Birahi, a tributary of Alaknanda [7]. Whereas, the 1978 Bhagirathi floods were the result of the breaking of the lake of Kandolia gad, a tributary of Bhagirathi [8]. Both these incidences have been associated with cloudburst induced landslide dammed lakes. Whereas, landslides in Ukhimath, dist. Rudraprayag and La-Jehkhla landslide dist. Pithoragad are recent example of cloudburst induced landslides in Uttarakhand. In Ukhimath hundreds of landslide scars riddled the ~5 km stretch between north of Chuni-Mangoli and up to Kimana village [9]. On 8th August, 2009 Lah- Jhekla landslide occurred due to cloudburst, the landslide wiped out two villages namely Jhakhla and Lah, claiming 43 lives [10,11].

If bed rocks present debris flow would be less, on the other hand unconsolidated material, old landslide debris, structurally weak area (traversed by fault/thrust and composed by highly weathered and jointed rocks) has been destroyed and damaged in a large scale (Figure 5). Downslope movement of debris was the main cause of devastation in all the previous cloudburst incidences in Mandakini valley that include Bheti-Paundar (August, 1998), Phata (July, 2001), Okhimath (September, 2012). Same was the case in Khirao Ganga (June, 2013) and Bhundar Ganga (June, 2013) in Alaknanda valley. It is noteworthy here that all these major cloudburst induced landslide incidences are concentrated in the Central crystalline domain or up to Main central thrust in upper Alaknanda valley.

Anthropogenic activities and increase in head load: Several anthropogenic factors are also responsible for making this area

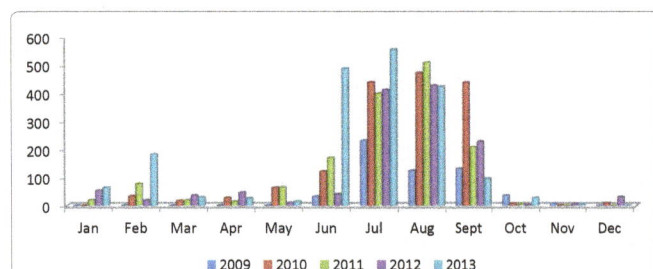

Figure 4: Average monthly rainfall (mm) in Upper Alaknanda Valley for last five years (Source data: IMD). Exceptionally high rainfall in July 2013 can be observed.

Figure 5: (A) Cloudburst induced landslides at Ukhimath area (2012); (B) Cloudburst incidence along Khirao Ganga valley (2013).

conduceive for mass movements. Encroachment into stream and river banks, steep cutting of slopes and left untreated, indiscriminate and unscientific construction of buildings, roads and dams are important anthropogenic activities. During the road construction, the removal of base support destablised the slope and promotes extensive landslides [12]. Newly constructed roads in Himalaya are ravaged by landslides and other processes of mass movements during the rains, thus producing collossal volume of debris [13]. Whereas, during construction phase of Hydro Electric Projects (HEPs) slope failure occur through excavation of slope mass in the construction of roads, tunnel and other project appurtenances. During field investigation a number of landslides were observed around Singroli - Bhatwari, Kali Ganga Stage – I and Son HEPs in Mandakini valley. Blasting with dynamite recklessly for the construction of dams and tunnels has triggered thousands of landslides. When the first rain comes, these landslides fill the riverbed with rubble [14]. Considering the extent of human interference, it is likely that the terrain sustainability is precariously balanced. A minor perturbation would have been enough to generate a cascading effect on the terrain instability. And the impetus was given by the unusual rain during 16 and 17 June 2013, which caused the Himalaya to respond violently against the unscientific human interference [15].

Semi village is situated just opposite to Okhimath town; it was faced worst condition of slope instability. National Highway (NH-109) going through lower, middle and upper part of the village, it was damaged and subsided at several portions (Figure 6). Large scale construction viz., buildings, hotels, shops, petrol pump and road were constructed on the Quaternary deposits. Singoli-Bhatwari hydro-electric project has also been constructed near the village and muck dumping sight was situated just below the village right bank of the Mandakini River. Moreover, absence of adequate drainage arrangement was also observed which exacerbates pore water pressure during heavy rainfall and resulting into failure of the slope. A number of buildings were observed to be damaged and large number of cracks was developed in many buildings. Overloading or increase in head load accompanied with bank erosion induced slope instability at Semi village and several locations in the area. Gaurikund, G.I.C. Tilaknagar, Kunjethi, etc. are some important locations encountered similar problem after June 2013 disaster.

Bank erosion: It is well-known that riverbank retreat occurs both by continuous fluvial erosion as well as abrupt bank failure [16,17]. River bank-toe erosion causes to increase the bank height and the slope of the bank to the extent that eventually riverbank mass failure occurs [18]. Bank erosion by streams and rivers is the major causative factor for the slope failure in this area. Old landslide debris, Quaternary deposits, colluvium, alluvium and overburden material at several locations makes the area vulnerable for the bank erosion. At a number of places in the field, bank erosion is observed to cause slope instability e.g., Pandukeswer, Gobindghat, Narayan bagar, Ganganagar, Vijaynagar, Chandrapuri, Tilwara etc. villages are under threat of landslides (Figures 7A-7F). Besides, lake formation was observed near Jaggi-Bedula slide, right bank of Madhyamaheswer river (Figure 7D) and the chronic kaliysaur landslide, left bank of Alaknanda river also triggered by bank/toe erosion.

Seismicity: The entire Himalayan region is geo-dynamically active and prone to earthquake along the existing thrust and faults. The epicentre of Chamoli earthquake (1999) is very close to Munsiari Thrust (MCT zone) across the Alaknanda River at Helong. The area falls under the high hazard and very high hazard categories (seismic zone IV and V) of the seismic zonation map of India. Moreover, habitations and infrastructure which are close vicinity to MCT zones are also susceptible to landslides (Figure 7E). Keefer [19] in a comprehensive study has identified 14 types of landslides associated with earthquake activity. The most common are rock falls, rock slides and disrupted soil slides. Earthquakes reduce stability by imparting both a shearing

Figure 6: Semi village was encountered worst condition of slope failure and multiple level of subsidence after June 2013 disaster. Overload on the crown part can easily observe, image within red dotted rectangle shows a collapsed house.

Figure 7: (A) Indiscriminate construction right bank of Nandakini river at Ghat town (B) NH-109 going on eroded terrace at Vijaynagar town, Mandakini river shifted left bank (10-15 m.) (C) Landslide along Karanprayag-Tharali road, generated by change in angle of repose (D) Lake formation observed near Jaggi-Bedula slide right bank of Madhyamaheswer river. (E) A major bouldery debris slide triggered by MCT at Thalla bend along Pokhri - Mohankhal road, approx. 200 m road stretch remarkably subsided. (F) View of severally damaged Barrage of Singoli - Bhatwari hydroelectric project at Kund after June 2013.

stress and a reduction in resistance to slope material. Uttarkashi (1991) and Chamoli earthquake (1999) and frequently occurring tremors of short duration have widened the joints of already highly fractured and jointed rocks of the area. Being tectonically active zone, the area is controlled by structural elements like thrust/faults and lineaments (Figure 3), subsequently responsible for slope failure during earthquake. Seismicity creates disturbance in these tectonic features and eventually causes landslide within their vicinity.

Deforestation: Deforestation has been a significant effect on soil erosion and run-off in the Himalaya due to moderate and steep slope. Large scale depletion of forests has drastically altered the hydrological conditions of the slopes. Infrastructure development and growing population in the study area is putting tremendous pressure on the agriculture land and forest land leading to deforestation. A large number of trees felling have been happened in the area for the road construction. Trees were uprooted either by dumping of excavated debris on the valley side or by landsliding occurring after the excavation. The vegetation covering the hill slopes also gets eroded in the process, which leads to further soil erosion as well as increase in run-off from the hill slopes [13].

Developmental activities like unscientific way of construction of houses, canals and unscientific blasts during road construction increase the probability of landslides [20]. The women led 'Chipko' movement (1973) started after the Alaknanda flood (1970), caused by logging in the Alaknanda valley, they connected the deforestation to landslide and flooding. Recent scientific studies suggest that landslide, flood and deforestation in the Alaknanda valley are closely related [21,22]. Apart from this forest fires have been increased in recent years, which lead to increased surface erosion and mass wasting.

Forest lands have been cleaned in a large scale for the construction and mining. The state governments own data shows that 1,608 hectares on the riverbeds were being mined in 2012. The state's forest department says that between 2000 and 2010, almost 4,000 hectares of land previously under its jurisdiction were diverted for mining [23]. As per data from the Union Ministry of Environment and Forests (MoEF), 44,868 ha of forest land have been diverted to non-forest use in Uttarakhand since 1980 [24]. The maximum number of development projects that required forest diversion has been approved in Chamoli district. A total of 1,767 ha forest land has been cleared in the district. Interestingly, the maximum forest area that has been cleared for hydel projects, roads and transmission lines is in Chamoli, Uttarkashi, Rudraprayag and Pithoragarh [25-28]. While Chamoli and Rudraprayag district are integral part of upper Alaknanda valley.

Discussion

Heavy and prolonged rainfall was the main triggering factor for landslides in Upper Alaknanda valley. The area is observed to be highly tectonised and the same has rendered the rocks of the area highly sheared, fractured and jointed. As many as three sets of joints are observed at most places. Field work carried out in the area suggests that landslides have been largely (in 77 percent cases) caused either by slope modification for infrastructure development or by toe/bank erosion by streams and rivers, mostly during spells of high discharge. Besides, zones of old landslides material, colluviums and alluvium deposits are identified as being most vulnerable to slope failure. Together with these the slopes occupied by unconsolidated material, loose soil and highly weathered and fractured rocks are observed to be vulnerable. Structural disposition of rocks is also observed to facilitate mass movement. The

decision to settle down over stable and firm ground at the same time minimized losses during an earthquake.

Indiscriminate and unscientific way of construction is most important anthropogenic activities accelerated deforestation and landslide incidences. Deforestation is the main contributing factor for the slope instability processes. The transportation network is the life line for the socio-economic development of hills and the hydropower potential provides the means to finance economic growth and social development. It is therefore necessary that any type of construction must be done according to the properly considered geo-environmental and ecological factors. Geo-environmental factors include geology/structure, slope, relative relief, landuse/landcover and hydrology; these are controlling factors for the landslides. It should be understood that infrastructure development if properly done it will be beneficial and auspicious for the human being but if not it will take toll in the form of disaster.

Conclusion

Landslide and flash flood are both induced by atmospheric precipitation and can well be predicted in case real time data on rainfall is available and thresholds are worked out for different catchments based upon slope, geology, distribution of Quaternary deposits, catchment area and landuse. Generation of timely warning, its timely dissemination in the area likely to be affected and prompt and effective response hold the key to saving lives. Metrological observation network in the area however needs to strengthen for the generation of site specific warnings.

In case of road construction, distance from the river and streams is also important. Mostly road disruption on large scale takes place at such location where toe/bank erosion occur frequently. We have to understand that farther the road from the river, safer it is. It may cost a little more in the beginning but it would prove out to be a better decision in long run. Road construction in the hills is necessarily accompanied by slope stabilisation and rainwater disposal measures. Durability of road network will be useful and strengthened response/relief, recovery, search and rescue operation after disaster. Besides, considering 68% debris slide in the study area, afforestation of the surroundings using site specific species, wire mesh and jute netting along with appropriate engineering measures can help in the stability of slope. Landslide hazard can be minimized by avoiding construction in landslide prone region. Moreover, in this area infrastructure development project should be allowed only after proper understanding of geological, geotechnical and environmental factors.

Landslide inventory map are useful for the planning of infrastructure development. Such maps also give ground level information about the sites and helpful in clear demarcation of safe and unsafe zones. Identification of such sites is very useful for the prohibition and regulation of future construction and developmental activities where cloud burst frequently occurred and active and old landslide zones situated. For better landslide risk reduction mass awareness on large scale should be conducted.

References

1. Brabb EE, Pampeyan EH, Bonilla MG (1972) Landslide susceptibility in San Mateo County, California. Miscellaneous Field Studies Map 360, US Geological survey publication.

2. Carrara A, Cardinali M, Detti F, Guzzetti F, Pasqui V, et al. (1991) GIS techniques and statistical models in evaluating landslide hazard. Earth Surf Process Landf 16: 427-445.

3. Sati SP, Sundriyal YP, Rana N, Dangwal S (2011) Recent landslides in Uttarakhand: nature's fury or human folly. Curr Sci 100: 1617-1620.

4. Bookhagen B, Thiede RC, Strecker MR (2005) Late Quaternary intensified monsoon phases control landscape evolution in the northwest Himalaya. Geology 33: 149-152.

5. Dortch JM (2009) Nature and timing of large landslides in the Himalaya and Transhimalaya of northern India. Quat Sci Rev pp: 1-18.

6. Das S, Ashrit R, Moncrieff MW (2006) Simulation of a Himalayan cloud burst event. J Earth Syst Sci 115: 299-313.

7. Nityanand, Prasad C (1972) Alaknanda tragedy: A Geomorphic appraisal. J Nat Geol 18: 206-212.

8. Prasad C, Rawat GS (1979) Bhagirathi flash floods. Himalayan Geology 9: 735-743.

9. Rana N, Sundriyal YP, Juyal N (2012) Recent cloudburst-induced landslides around Okhimath, Uttarakhand. Curr Sci 103.

10. Sarkar S, Kanungo DP (2010) Landslide disaster on Berinag–Munsiyari Road, Pithoragarh District, Uttarakhand. Curr Sci 98: 900-902.

11. Singh RA (2013) La-Jhekla landslides, Pithoragarh district, Uttarakhand, India. Landslides and Environmental Degradation. Gynodaya Prakashan, Naininital pp: 141-149.

12. Mithal RS (1982) The physical impact of Ramganga dam project on its drainage basin and catchment areas. Report-3 to CSIR, New Delhi p: 49.

13. Nainwal HC (1999) Impacts of road construction on environment in Garhwal Himalaya. Bull pure appl sci 18: 1-7.

14. Shiva V (2013) The Uttarakhand disaster: A wake call to stop the rape of our fragile Himalaya.

15. Rana N, Singh S, Sundriayal YP, Juyal N (2013) Recent and past floods in the Alaknanda valley: causes and consequences. Curr Sci 105: 171-174.

16. Thorne CR (1982) In: Hey RD, Bathurst JC, Thorne CR (Eds.) Processes and mechanisms of river bank erosion. Gravel-Bed Rivers, John Wiley and Sons, Chichester, England pp: 227-272.

17. Lawler DM, Couperthwaite J, Bull LJ, Harris NM (1997) Bank erosion events and processes in the Upper Seven Basin. Hydrol Earth syst Sci 1: 523- 534.

18. Carson MA, Kirkby MJ (1972) Hillslope form and process. Cambridge university press, New york 178: 1083-1084.

19. Keefer DK (1984) Landslides caused by earthquakes. Geo soc America Bull 95: 406-421.

20. Asthana AKL, Sah MP (2007) Landslides and cloudburst in the Mandakini basin of Garhwal Himalaya. Himalayan Geology 28: 59-67.

21. Kimothi MM, Juyal N (1996) Environmental impact assessment of a few selected watersheds of the Chamoli district (Central Himalaya) using remotely sensed data. Int J Remote Sensing 17: 1391-1405.

22. Wasson RJ, Juyal N, Jaiswal M, McCulloch M, Sarin MM, et al. (2008) The mountain-lowland debate: Deforestation and sediment transport in the upper Ganga catchment. J Environ Manage 88: 53-61.

23. Gupta J (2013) Climate change, poor policies multiply Himalayan flood effects.

24. Valdiya KS (1980) Geology of Kumaun Lesser Himalaya. Wadiya Institute of Himalayan Geology, Dehradun p: 291.

25. Srivastav KS (2013) Maximum devastation occurred in areas of maximum forestland diversion.

26. Government of Uttarakhand, Memorandum for Central Assistance (2010) Report of the Government of Uttarakhand Submitted to Ministry of Home Affairs, Government of India for Central Assistance.

27. Government of Uttarakhand, Memorandum for Central Assistance (2012) Report of the Government of Uttarakhand Submitted to Ministry of Home Affairs, Government of India for Central Assistance.

28. Government of Uttarakhand, Memorandum for Central Assistance (2013) Report of the Government of Uttarakhand Submitted to Ministry of Home Affairs, Government of India for Central Assistance.

3

Evalauation of Groundwater Potential Zones using Electrical Resistivity Response and Lineament Pattern in Uppodai Sub Basin, Tambaraparani River, Tirunelveli District, Tamilnadu, India

Jeyavel Raja Kumar T*, Dushiyanthan C, Thiruneelakandan B, Suresh R, Vasanth Raja S, Senthilkumar M and Karthikeyan K

Department of Earth Sciences, Annamalai University, Annamalai Nagar – 608002, India

Abstract

The present attempt has been made to identify potential groundwater zones using electrical resistivity and lineament pattern in the dry watershed of Uppodai. The area comprises Archean hard rocks of charnockite and biotite gneiss. The major types of soil are red sandy soil and black cotton soils. Depth to the water table varies from 1 m to 10 m below the ground level. About 23 Vertical Electrical Soundings were carried out at different locations with AB/2 spreading up to 100 m by Schlumberger configuration. The measured apparent resistivity values were interpreted by curve matching technique using RESIST87 software. Lineament map was prepared to understand its influence on the ground water system. A, K,AK and KH are the curve types obtained in the study area. The resistivity values for first , second and third layers are varied from 3.6 Ωm to 256 Ωm; 65.7 Ωm to 2022.3 Ωm and 161.5 Ωm to 2500 Ωm respectively. The layer thickness of first and second layers observed with the variation of 0.7 m to 41.5 m and 8.3 m to 65.6 m. As the resistivity value observed with a wide variation, the resistivity value below100 Ωm of second layer has been conveniently taken for potential groundwater zones. The low resistivity with high second layer thickness zones is very limited and promising zones were observed at some patches in the present study area. Two sets of lineament pattern have identified with NW-SE and NE-SW directions. The density and length of lineaments were comparatively high with NW-SE pattern in the study area and not seems a potential aquifer at shallow depths.

Keywords: Uppodai; Vertical electrical sounding; Groundwater potential; Tambaraparani river

Introduction

Our very own survival on earth depends on the basic resource of water, which is nature's valuable gift to mankind. This nature's gift is now becoming a crucial source in throughout the world. The total water resource of our world is estimated as 1.37 Million ha m, of these global water resources, about 97.2% are salt water, mainly in oceans and only 2.8% are available as fresh water at any time on the planet earth. However, the economically extracted groundwater with the present drilling technique is about 0.3% (41.1 ×10⁴ million ha m), remaining being unavailable as it is situated below depth of 800 m [1].

Groundwater condition in hard rock terrain is multivariate because of the heterogeneous nature of the aquifer due to varying composition, degree of weathering and density of fracturing. Now a day, the geophysical methods are widely used to determine the groundwater resources in any type of terrains. There are several geoelectrical sounding techniques for the groundwater investigations described in literature [2-5]. Geoelectric Resistivity method is one of the important methods used to investigate the nature of subsurface formations by studying the variations in their electrical properties. Generally, these methods have been used for the identification of geological contacts, tectonic and structural studies [6,7]. The conventional Schlumberger resistivity sounding is extensively used for routine groundwater investigations both in alluvial and hard rock terrain [8-10].

The objective of this method in the field of groundwater exploration is to locate groundwater bearing formations, to evaluate thickness and lateral extent of aquifers to find the depth to bedrock, etc. [11]. As groundwater is the largest available source of freshwater, it is necessity to identify the potential groundwater zone for continuous and sustainable development of both socio-economic and agricultural developments. Electrical resistivity survey provides much basic information to the hydrogeologist, like depth to water table, depth to the basement topography in hard rocks. Hence, the present attempt has been made to identify potential groundwater zones for development

activities in the dry watershed Uppodai where agriculture is the main occupation.

Geology and hydrogeology

The present study area Uppodai is one of the tributaries to Chittar river basin. It is located between the North latitude of 8°52' to 9°10' and the east Longitude of 77°35' to 77°55' is shown in the Figure 1. Red sandy soil and black cotton soil are found in the area. The area lies in a hot and dry climatic zone with a temperature varying from 25°C to 45°C. It falls under the semi-arid climate type [12]. The elevation is approximately of about 50 m above mean sea level. The average precipitation of 722.5 mm is recorded in Kayathar station during the year 1901 to 2003. The actual annual evapotranspiration is found to be 636 mm. The well depth level varied from 7 m to 9.5 m. It is observed that the top soil thickness varied from 1 m to 3 m. As this study area is a hard rock terrain, the secondary porosity could be the main source for ground water storage. In this basin, water occurs mainly under water table conditions in the weathered crystalline complex terrains. The pattern of precipitation is essentially of a tropical monsoon type where the effect of the winter monsoon is dominant.

The area comprises Archean hard rocks of charnockite and

*Corresponding author: Jeyavel Raja Kumar T, Department of Earth Sciences, Annamalai University, Annamalai Nagar – 608002, India
E-mail: tjeyavel@rediffmail.com

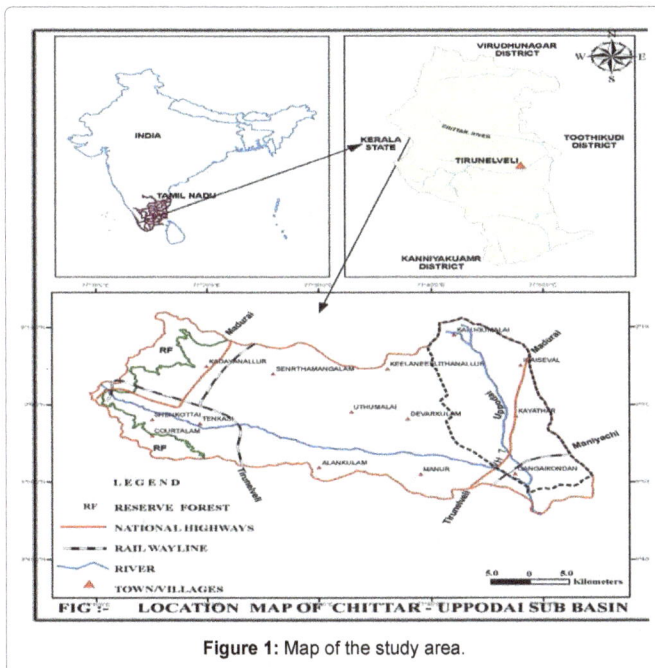

Figure 1: Map of the study area.

biotite gneissic rock types (Figure 2). The charnockites are acidic to intermediate in nature and are white to grey colour. Similarly, the biotite gneiss rock type varies in colour from light grey to grey. The gneissic area has been subjected to polyphase of tectonic activities resulting in high degree of metamorphism and multi-generation of folding and associated shearing and faulting. The lineaments are predominantly oriented along the north-west to south east directions (Figure 3). The joints are oriented along NW-SE, N-S and NE-SW directions. The secondary structures developed in the lithounits of the district are described in detail by Abdullah and Paranthaman and Abdullah [13,14].

The general geological succession of the region is presented given below (after Varadaraj) (Table 1) [15].

Methodology

Electrical resistivity method has gained considerable importance in the field of groundwater exploration because of its low cost, easy operation and efficiency to detect the water bearing formations. In the present study, about 23 Vertical Electrical Soundings (VES) have been carried out using a DDR-3 model resistivity meter in the study area (Figure 4). The Resistivity Meter is a specialized version of IGIS Resistivity meters designed for use in Resistivity surveys up to about 200 m depth. The equipment consists of two separate comportment-one for reading the current (G.Unit) and the other for directly reading the resistance/the potential (P-Unit), both housed in a single box. The equipment, powered by a 24V rechargeable battery can send highly stabilized currents up to 200 mA and read the resulting potential with a 100 micro volts resolution or ground resistance directly with 50 micro ohms resolution (www.igisindia.com) [16].

The AB/2 separations have measured up to 100 m by Schlumberger configuration. The measured apparent resistivity signals have been interpreted through curve matching technique using RESIST87 software [17]. The software interpreted resistivity and respective layer thicknesses are converted into one, two and three layers for anlayses. The software technique has been successfully used for prospecting ground water by many researchers [8,18-21]. The resistivities of

the rock types at different depth levels are used to understand the geoelectrical properties, because the behavior of conductivity mainly depends on the resistivity of lithology and age [3]. Apart from VES interpretation results, the lineament map also prepared to understand its influence on the ground water system. Because, lineaments are promising groundwater potential structures, and they act as the conduit of groundwater.

Result and Discussion

Electrical resistivity method is useful to investigate the nature of subsurface formations by studying the variations in their resistance to flow of electrical current and hence determine the occurrence of groundwater. As the present study area is a hard rock terrain, the response would have a high resistivity due to low flow of electrical current. The resistivity values obtained from the interpretation in the study area is given in the Table 2.

In general, curve type A will encounter in hard rock terrains.

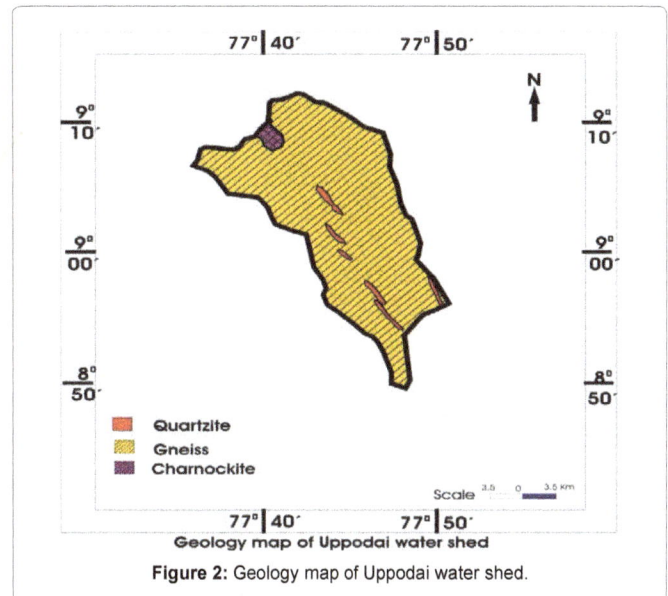

Geology map of Uppodai water shed

Figure 2: Geology map of Uppodai water shed.

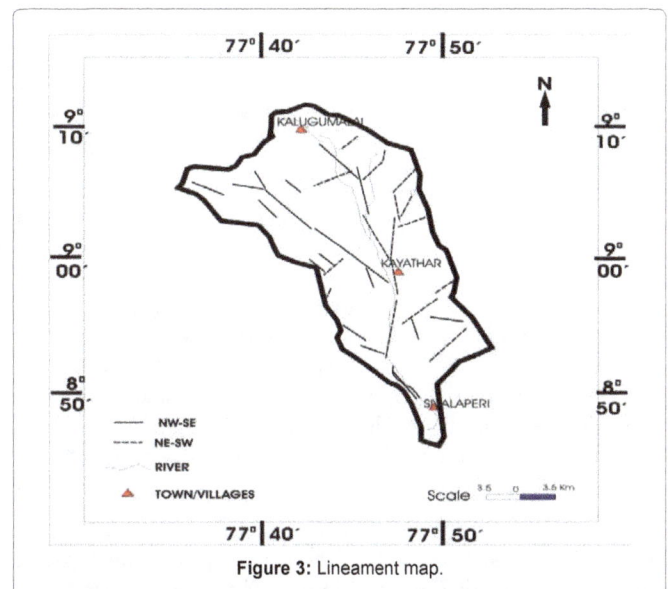

Figure 3: Lineament map.

Period	Lithogroup	Lithology
Quaternary	Recent	soils, alluvium, kankar, laterite
Western ghat super group	Acid intrusives	Pegmatites, Quartz vein, Pink granites, Grey granites, Leucogranites
A		
R		
C	Basic intrusives	Basic dykes, Basalt, Dolerites
H		
A	Migmatic complex	Garnetiferous quartzofeldspathic granulite gneisses
E		
A		
N		
	Charnockite group	Charnockites, Pyroxene granulites
	Khondalite group	Crystalline limestone, Calc granulites, Garnetiferous biotite-sillimanite quartzites

Table 1: The general geological succession of the region is presented given below.

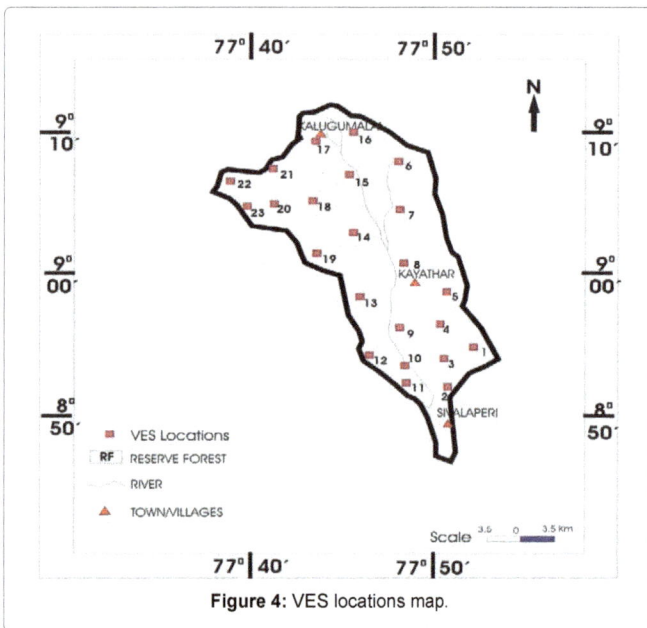

Figure 4: VES locations map.

Similarly, the present study interpretation results showed that about 15 VES curves were found with A curve type. The curve types of K, AK and KH were observed in 5,2 and 1 VES locations. Besides, out of of 23 VES interpretation, the three layer strata has been observed in 14 locatrons, 6 locations were shown 2 layer strata, and 3 locations were shown 4 layer strata. Some of the of the interpreted curve types of the study area are given in the Figure 5.

The resistivity values of first and second layers are varied from 3.6 Ωm to 256 Ωm and 65.7 Ωm to 2022.3 Ωm respectively. The respective layer thickness of first and second layer observed as 0.7 m to 41.5 m and 8.3 m to 65. 6 m. The average thickness of the first and second layers is 15.4 m and 22 m and high resistivity value observed in VES location 21 whereas low resistivity value observed in VES location 3. The second layer resistivity observed high in VES-23 and low in VES-3. The third layer resistivity value varied from 161.5 Ωm to 2500 Ωm. High resistivity value noticed in VES-21 and low value observed in VES-2.

The high resistivity of 250 Ωm indicated in the top layer is attributed by dry soil and kankar deposit and low resistivity could be red or black soil with low moisture. In second layer, resistivity indicated

up to 160 Ωm could be attributed by highly weathered gneiss and charnockite with saturation of water. The resistivity range 160 Ωm to 500 Ωm possibly indicated by poorly weathered gneiss and charnockite whereas high resistivity above 500 Ωm could be massive rocks. The low resistivity indicated in third layer attributed by fissured gneiss or semi weathered charnockite rock and high resistivity attributed by massive rocks. The interpreted resistivity layer is closely matched with the drilled hole lithology of the region. The lithology observed in bore hole (No.92442) to a depth of 43 m at Vellalankottai village of Kayathar block indicted that top soil thickness (0.5 m) followed by Kankar (0.5-6 m), Weathered Charnockite (6-8 m), Weathered and Jointed Charnockite (8-17 m) and Fissured Charnockite (17-43 m) [22].

The lineament pattern also plays an important role in groundwater potential and development. For good groundwater potential in hardrocks,the fracture systerm should be well connected. The secondary porosity in the form of fissures, fractures and joints in hard rocks provides higher permeability. Well connected fracture systems can be expected to form principal pathways for groundwater flow and mass transport. The study area has identified with two sets of lineament pattern in NW-SE and NE-SW directions. NW-SE pattern density and length of lineaments are observed comparatively high in the study area. NW-SE pattern of lineaments observed in western and southwestern part. In the eastern part of the study area observed with NE-SW pattern of lineament. The river also running on the lineament course of the study area.

In general, the first layer represents the top layer which is soil cover has shown a wide variation of geoelectrical response due to the soil condition and its nature. The subsequent second and third layers thickness and respective resistivity values are generally considered for locating groundwater potential zones. In general, the freshwater horizon encountered in hard rock terrain observed with a resistivity value range of 60 Ωm to 160 Ωm [23,24]. As the resistivity value obtained with a wide variation in the study area, it has been conveniently taken the

VES location No	Curve Type	Resistivity (Ωm)				Layer Thickness(m)		
		1	2	3	4	1	2	3
1	A	75.9	618.6			24.9		
2	K	56.2	297.9	**161.5**		5.5	22.8	
3	AK	**3.6**	**65.7**	1051.0	247.0	2.6	23.5	29.0
4	A	33.0	127.4	288.9		4.4	45.0	
5	A	72.8	309.0			8.3		
6	A	56.8	187.1	769.8		14.3	15.9	
7	K	122.1	702.9	352.0		18.3	23.5	
8	K	129.3	391.2	233.0		6.1	9.8	
9	K	23.0	363.5	617.4		**0.7**	**65.6**	
10	KH	25.1	469.0	142.0	446.1	1.2	27.8	12.5
11	A	92.5	515.8	1070.8		21.0	25.0	
12	AK	72.8	268.0	1898.5	627.6	7.9	**8.3**	46.4
13	K	83.4	593.8	154.7		36.1	49.0	
14	A	106.2	1163.8			**41.5**		
15	A	111.5	434.2			14.1		
16	A	31.6	223.4	752.0		19.0	26.3	
17	A	23.0	112.1	265.5		17.9	28.9	
18	A	71.0	592.3	1081.3		36.3	40.3	
19	A	42.0	122.2	663.3		14.0	12.0	
20	A	21.8	230.3			20.8		
21	A	**256.0**	1089.0	**2500.0**		10.0	30.0	
22	A	51.8	640.2	890.3		6.4	54.1	
23	A	89.7	**2022.3**			24.2		

Table 2: Interpreted Resistivity and Layer thickness.

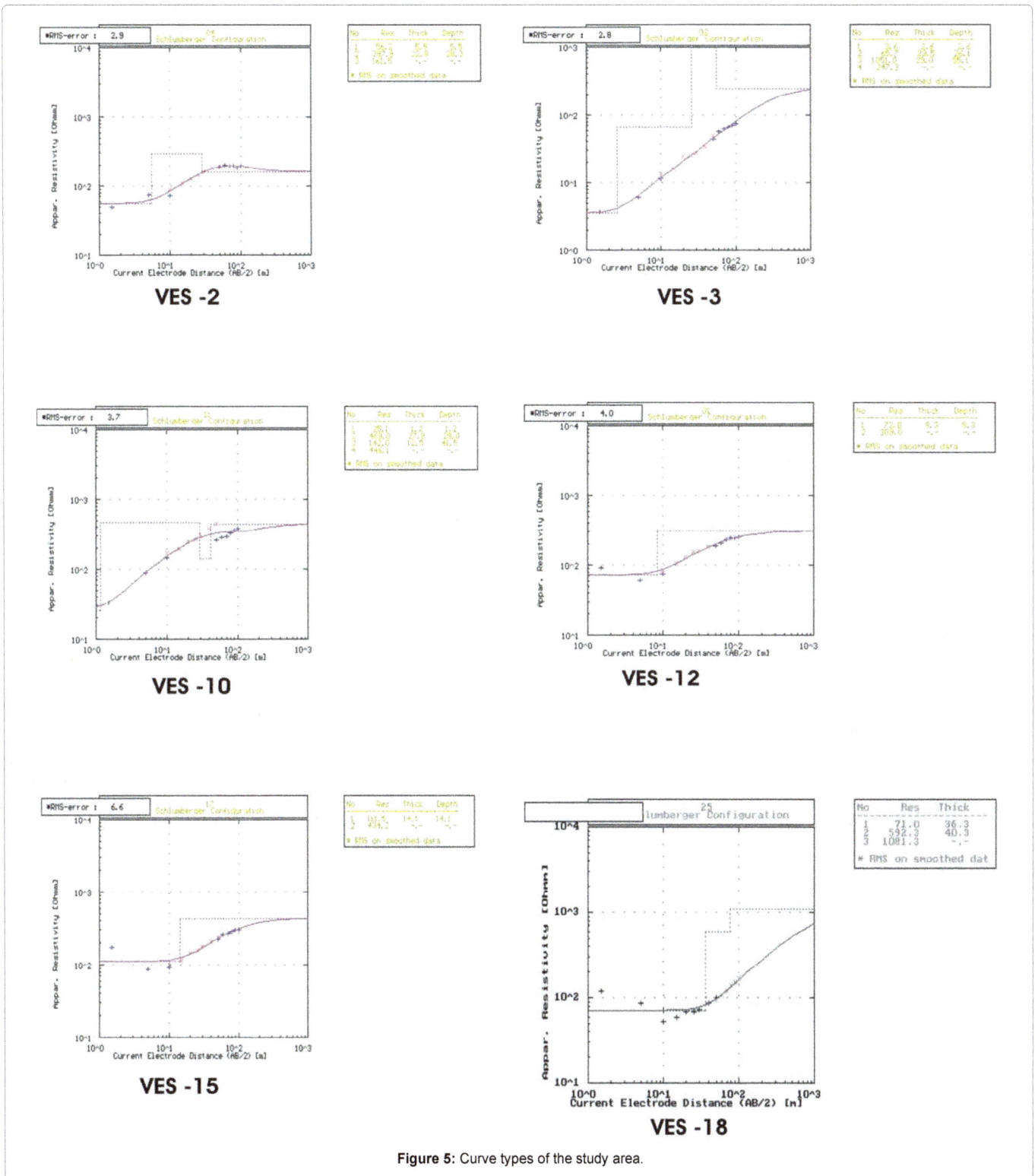

Figure 5: Curve types of the study area.

resistivity value below100 Ωm of second layer. The low resistivity with high second layer thickness zones is very limited in the present study area. Such promising zones were observed at some patches whereas high resistivity was observed in remaining part of the study area. When compare with the lineament density, it is observed that the lineaments not seems a potential aquifer in shallow depths.

Conclusion

The geoelectrical resistivity survey is a good and effective method for groundwater development and management. It is observed that the overall groundwater potential is very limited in the study area. The average second layer thickness found high in south and northwestern part. The groundwater potential is low in the study area, but the lineaments are present widely with shallow in depth and not with groundwater movement. Even the VES locations on the and near the lineament structure shows high resistivity and high second layer thickness. Hence, the lineaments are not much influencing at shallow depth levels but would have a chance for deeper depth levels. In certain locations low resistivity value observed with high second layer thickness. This high weathered and jointed zone are favourable for groundwater development. Remaining part of the study area is feasible for shallow aquifer developments. The high resistivity value noticed in certain locations is mainly due to the presence of high resistive minerals such as biotite and hornblende in gneissic rocks. Besides, the increased number of open and bore wells, over exploitation and transporting water to the nearby irrigation lands by pipe also could lead to lower the water table and poor groundwater potential. This study not only helpful for groundwater development but also to monitor and conserve the resources.

References

1. Patel AS (2004) Water resources Management and Peoples Participation. J App Hydrology 17: 1-7.

2. Al'pin LM (1950) In: Al'pin LM, Berdichevskii MN, Vedrintsev GA, Zagarmistr AM (Eds.) The Theory of Dipole Sounding. Dipole Methods for Measuring Earth Conductivity (1966). Consultants Bureau, New York 1-60.

3. Keller GY, Frischknecht EC (1966) Electrical methods in geophysical prospecting, Pergamon Press, London 517.

4. Battacharya PK, Patra HP (1968) Direct current geoelectric sounding, Elesvier, Amsterdam pp: 25-30.

5. Koefoed O (1979) Geosounding Principles: Resistivity Sounding Measurements. 1st Ed. Elsevier, Amsterdam.

6. Janarthana Raju N, Reddy RVK, Naidu PT (1996) Electrical resistivity surveys for groundwater in the Upper Gunjanaru catchment, Cuddapah district, AP. Jour eol Soc 37: 705-716.

7. VenkateswaraRao CH, NagamalleswaraRao B, Kasipathi C, Vasudev K, Satyanarayana KVV, et al. (2004) Application of geoelectrical method for interpretation of geohydrological conditions around Rajamundry, AP. J Geol Soc 63: 262-270.

8. Yadav GS, Singh PN, Srivastava KM (1997) Fast method of resistivity sounding for shallow groundwater investigations, Journal of Applied Geophysics 37: 45-52.

9. Jagadeeswara Rao P, Suryaprakasa Rao B, Jagannadha Rao M, HariKrishna P (2003) Geoeletrical Data Analysis to Demarcate Groundwater Pockets and Recharge Zone in Champavathi River Basin,Vizianagaram District, A.P. J Ind Geophysis Union 7: 105-113.

10. Kumar TJR, Balasubramanian A, Kumar RS, Dushiyanthan C, Thiruneelakandan B, et al. (2012) Vertical Electrical Sounding for Groundwater Prospecting in Uppodai of Tambaraparani River Tirunelveli and Thoothukudi Districts, Tamil Nadu, India. International Journal of Environmental Engineering and Management 3: 147-157.

11. Shenoy KN, Lokesh KN (2000) Electrical resistivity survey for Groundwater exploration in Udupi Municipal area. J App Hydrology 13: 30-35.

12. Mohan HSR (1984) A climatological assessment of the water resources of Tamil Nadu. Ind Jour Power & River Valley Developt 34: 58-63.

13. Abdulla NN, Paranthaman S (1983) Geology of parts of Tenkasi and Shencottah Taluks, Tirunelveli District, Tamil Nadu, Unpublished Report of G.S.I.

14. Abdulla NM (1981) Geology of parts of Nanguneri taluk, Tirunelveli District, Tamil Nadu, (Un.Pub.report of G.S.I).

15. Varadaraj N (1989) Groundwater Resources and developmental potential of Tirunelveli (Nellai Kattabomman) District, Tamil Nadu, CGWB, Southern region, Hyderabad.

16. www.igisindia.com

17. Velpen V (1988) RESIST'87 User's guide ITC, Department of Earth Resources Survey, Netherlands.

18. Balakrishna S, Ramanujachari KR (1979) Resistivity investigations in deccan trap regions. Geophys Res Bull 16: 31-40.

19. Balasubramanian A (1986) Hydrogeological Investigations in the Tambaraparani River basin, Tamilnadu, Unpublished PhD Thesis (University of Mysore) 348.

20. Kumar TJR (2006) Hydrogeological modeling of Chittar – Uppodai sub basin of Tambaraparani river basin, Tirunelveli District, Tamilnadu, India (Un published Ph.D Thesis), Annamalai University University 243.

21. Muthuraj D, Srinivas Y, Chandrasekar N (2010) Delineation of Groundwater Potential areas - a Case study from Tirunelveli District, Tamil Nadu, India. International Journal of Applied Environmental Sciences 5.

22. Public Works Department (PWD, 2005) A Profile of Thoothukudi District, Government of Tamil Nadu 114.

23. Balasubramanian A, Sharma KK, Sastri JCV (1985) Geoelectrical and hydrogeochemical evaluation of coastal aquifers of Tambraparni basin, Tamil Nadu. Geophys Res BulL 23: 203-209.

24. Melanchthon VJ (1988) Resistivity survey for mapping fresh water pockets. J Assoc Expl Geophys 9: 71-78.

4

Fold-thrust Style and Fluid Reservoir Potential of Eocene Sakesar Limestone: Sothern Surghar Range, Trans-Indus Ranges, North Pakistan

Alam I*

Atomic Energy Minerals Centre, Lahore

Abstract

The Surghar Range western extension of the Trans-Indus ranges constitutes the southeastern frontal fold-and-thrust belt of the Kohat Plateau. This structural province is comprised of various local to regional scale anticlines right from Serkia-Mitha Khattak to Kutki areas. The existing range-front anticlinal trend is well-built along the east-west trending segment of the Surghar Range. These anticlinal features reveal infantile tendency from east to west and unearthing the platform rock sequences ranging from Permian to Eocene which is unconformably overlain by the Mitha Khattak Formation equivalent facies to the Rawalpindi Group. This facies in turn has overlain by the fluvial sediments of Siwalik Group. Overall three major anticlines have been mapped from west to east as the Mitha Khattak, Makarwal and Malla Khel Anticline. Differential stratigraphic levels are exposed in cores of these anticlines which proved excellent prospect to be potential hydrocarbon reservoirs horizons. The Eocene Sakesar Limestone has been selected for details studies of his fractures and joints analysis. The Sakesar Limestone exposed along the range front making fraction of the frontal limbs of different anticlines. Various fractures networks and joints pattern has been observed in the Sakesar Limestone at different localities reveal high secondary porosity and permeability. Most of the secondary tectonically induced and primary diagenetic opening and ruptures planes are interconnected and tenders proficient conduit lattice for munificent circulation of fluids in the Sakesar horizon. Origin of fractures and joints growth is mainly associated to force folding in response to the compressional, transpressional and trans-tensional deformation being observed in the region. The studied anticlines reveal that they are the product of fault-bend and fault-propagation folding tender excellent structural fluid trapping philosophy. The range frontal flanks reveal that different level of strata thrust against the foredeep showing inconsistency in the subsurface level of basal detachment horizon which is too hopeful for the construction of structural traps at various levels. Blending of the structural style of the area with the sedimentary structural features of the Sakesar Limestone of Surghar Range urges that this structural province is significantly associated to make hydrocarbon reservoir potential at the stratigraphic level of Sakesar Limestone.

Keywords: Surghar range; Deformation style; Sakesar limestone; Reservoir potential; Tectonic; Diagenetic fractures

Introduction

Surghar Range is the outer most fold-and-thrust belt of the sub-Himalayas making the easternmost extension of the Tran-Indus ranges (TIR) bifurcated by the KaIabagh fault system from the western Salt Range of North Pakistan [1]. The range follows east-west structural trend along the sothern margin of the Kohat plateau and switches to north-south trend along the easternmost flank of Bannu Basin (Figure 1) [2]. Along the range front the non-outcropping Permian to Eocene rocks underneath the Kohat and Bannu Basin are exposed at surface. The range displays arcuate structural style in plan and exhibits distinct mountain forefront geometries along its map trace. It is characterized by south facing structures along its east-west trending segment. Whereas the north-south trending segment of the range is dominated east vergent fold-thrust assemblages. The whole range displays bidirectional structural trend, north-south toward its sothern terminus to Malla Khel and oriented east-west from Malla Khel to its eastern terminus up to junction of KaIabagh Fault Zone, making two broad segments. The north-south segment is composed of Siwalik sequence penetrated in progression by Eocene to Jurassic rocks from south to north in the proximity of Malla Khel village across the Baroch Nala section. The north-south oriented segment of the range is dominated by east facing structural geometries in additional to west vergent active back thrusting and tectonic wedging [3]. Previous work is mostly attributed to stratigraphy, economic geology and geological mapping of the Surghar Range [2,4,5] where as its outer eastern flanks of the north-south segment and southeast flanks of the east-west segments has been remained unaddressed since long. Sothern part of the north-south oriented segment is dominantly controlled by the uplifts of Siwalik sequence where Chinji Formation is exposed in the core of Qubul Khel

Anticline. Near Sirkai a broad anticline has been mapped in the Siwalik sequence where Mitha Khattak Formation and Sakesar Limestone are thrust in the hanging wall over the Dhok Pathan Formation in the footwall. The structural geometries of east-west oriented segment are characterized by south facing overturned concentric anticlinal folds with a prominent south vergent thrust fault. The Sakesar Limestone is exposed and mapped along three major anticlinal folds right from northwest of Sirkia Village to Baroch Nala. These anticlines have been designated from south to north as Mitha Khattak, Makarwal and Malla Khel anticlines. The frontal limbs of these folds are thrust over Siwaliks Group rocks toward Punjab Foredeep. This thrust fault is observed laterally extended along the foothills of the range from Sirkia to the eastern terminus of the Surghar Range. Transpressional tectonics is the significant structural style of the north-south trending segment while overturned concentric folding and thrust faulting is the dominant structural mechanism of the east-west trending segment of the range. Maturity in deformation style and variable stress regimes are observed from sothern plunging end to the eastern termination of the range all along the structural pathway in the form of tectonic progression, structural growth, maximum crustal shortening and unearthing of Paleozoic-Mesozoic rocks of the cover sequence. Maturity in the

*Corresponding author: Alam I, Atomic Energy Minerals Centre, Lahore, Pakistan, E-mail: iakhattak40@yahoo.com

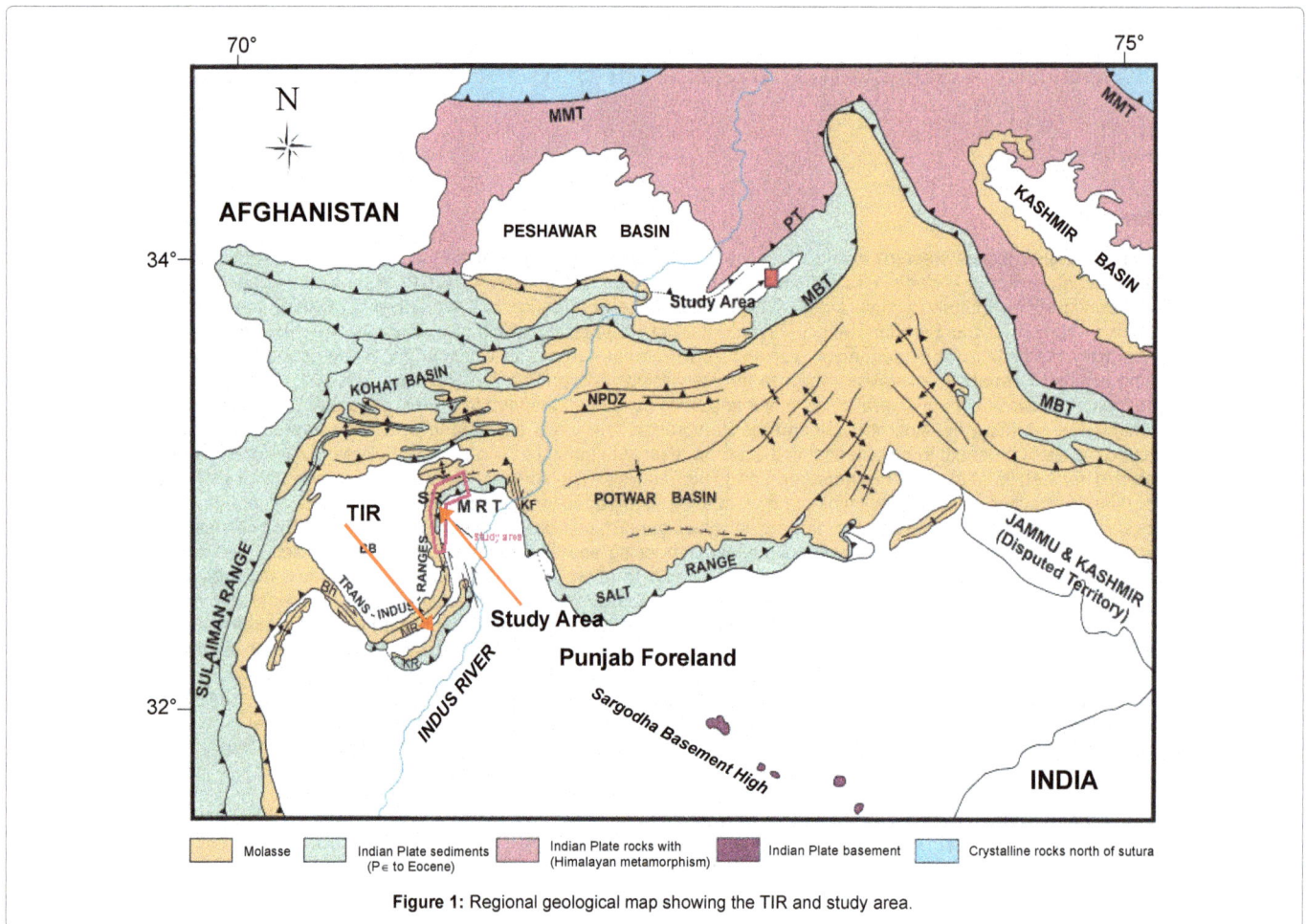

Figure 1: Regional geological map showing the TIR and study area.

tectonic phases and exhumation of the older strata from interior to surface is well predictable. Sakesar Limestone has been studied for their reservoir potential by considering the growth of natural fractures and joints network developed attributable to the tectonic and diagenatic processes. The fractures network and their distribution are important parameter and essential prerequisite for the potential hydrocarbon reservoir. Connected fractures have been observed which provide high permeability for the carbonate reservoirs and eventually helpful for the enhancement of yield.

Geological setting

Pakistan occupies the northwestern structural province of the subducting Indian lithospheric plate underneath the Eurasian Plate. This global tectonic event has produced compressional and transpressional tectonic elements since Eocene on the northern and northwestern fringes of Pakistan. Continual under thrusting of the Indian Plate since Cretaceous created the amazing elevated mountain ranges of Himalaya and a series of foreland fold-and-thrust belts as thick sheets of sedimentary origin and thrust over the Indian Craton [6].

The Trans-Indus extension of the Salt ranges is composed of several frontal ranges creates an "S" shaped double re-entrant and surrounds the Bannu Basin (Figure 1). These ranges symbolize the western fraction of the northwestern Himalayan foreland fold-thrust belt that produced by continual south-directed décollement-related thrusting of the Indian Plate crust during long-term collision between India and

Eurasia [7-10] crustal architecturing normally advanced southward with time space. The youngest and latest sothernmost fracture zone has transpired along the forefront thrust mechanism contiguous to the Trans-Indus ranges (Figure 2) [11,12]. The current study area is the easternmost extension of the TIR. Along the Surghar frontal fault Paleozoic to Cenozoic platform sequence is thrust southward over the undeformed Quaternary sediments of the Punjab Foredeep. The TIR characterized the foremost deformational front of the Kohat fold-thrust belt and Bannu Basin in North Pakistan. Consequently, the tectonic mode is generally observed thin-skinned for the outcropping structures.

Stratigraphic setting

During field studies, it has been observed that the eastern and southeastern flank of the Surghar Range is comprised of Permian to Eocene platform sediments unconformably overlain by Plio-Pleistocene fluvial sediments (Table 1). The platform sediments become thicker and more complete from west to east along the range. In western part of range, northeast of Malla Khel in Baroch Nala and east of Pannu in the Chichali Nala excellent sections ranging in age from Permian to Miocene are exposed. The stratigraphic successions were studied along these sections of the range as shown in Table 1.

In this area, the base of stratigraphic succession is occupied by the Permian Zaluch Group rocks of Wargal Limestone overlain by the Chhidru Formation. This succession is occupied by the Triassic sequence of Musa Khel Group rocks of Mianwali, Tredian and Kingriali

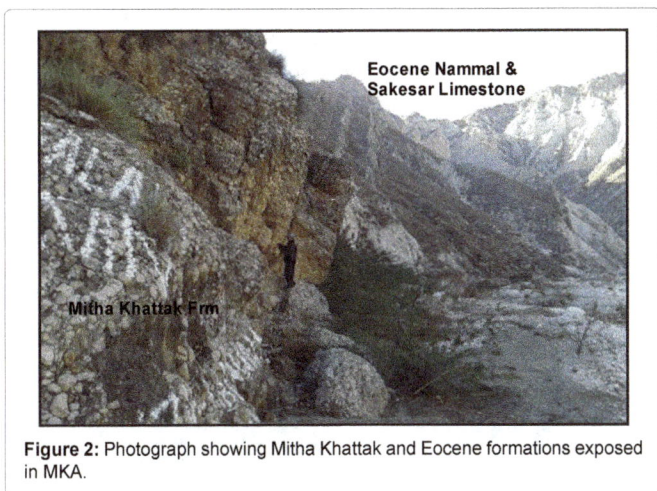

Figure 2: Photograph showing Mitha Khattak and Eocene formations exposed in MKA.

Era	Period	Epoch	Group	Formation
Cenozoic	Tertiary	Pliocene	Siwalik	Dhok Pathan
				Nagri
				Chinji
		Miocene	Equ. Rwp GP	Mitha Khattak
		Eocene	Chharat	Sakesar limestone
				Nammal
		Paleocene	Makarwal	Patala
				Lockhart Limestone
				Hangu
Mesozoic	Cretaceous	Early	Surghar	Lumshiwal
				Chichali
	Jurassic	Late	Baroch	Samana Suk
		Middle		Shinawari
		Early		Datta
	Triassic	Late	Musa Khel	Kingriali
		Middle		Tredian
		Early		Mianwali
Paleozoic	Permian	Late	Zaluch	Chhidru
				Wargal Limestone

Table 1: Stratigraphic framework of Surghar Range [5].

Dolomite overlying the Jurassic that include Datta, Shinawari and Samana Suk Formation. On top the Cretaceous sequence is mapped and comprised of Chichali and Lumshiwal Formation unconformably overlain by Paleocene sequence that comprised of Hangu, Lockhart and Patala formations. In turn the sequence is overlain by the Eocene Nammal Formation and Sakesar Limestone which is unconformably overlain by the Mitha Khattak Formation and Siwaliks Group rocks.

Methodology

Before performing fieldwork relevant literature was extensively reviewed to acquire knowledge regarding the geological setup of the study area. In the field I have studied the lithologic composition, primary and secondary structural features and stratigraphic position of the Sakesar Limestone in the exposed stratigraphic succession of the area. For the purpose conducted several stratigraphic/structural traverses across the range for the preparation of structural geological map of the area. Traverses were planned in the east-west and north-south directions in the western and eastern domains approximately right angle to the trend of the outcrops of the range. Besides studying the physical properties of the exposed Sakesar Limestone, detailed

structural data regarding strike and dip of bedding and, faults and attitudes of the fold axes were collected and correlated with each other in order to establish the pattern of the various structures.

Reservoir potential of eocene horizon

The following prominent fold structures have been mapped from south to northeast toward the frontal flanks of the Surghar Range exposing Eocene horizon comprised of Sakesar Limestone. Medium to thick beds of the Sakesar Limestone make the hanging wall ramp of the frontal thrust sheet against the Siwaliks sequence in the footwall toward the Punjab Foredeep. The Sakesar Limestone is developed throughout the Surghar Range is a hard prominent cliff-forming formation. The limestone horizon maintains a relatively uniform character all over the area. It consists of white to whitish gray, greenish gray to gray, nodular, medium to thick bedded limestone with alternate thin marl beds. Cherts concretion and nodules are frequently observed in middle and upper part of the limestone. The whole limestone is well fossileferous, highly fractured, jointed, moderate to rationally cavernous, and visible to measurable beds disjointing have been observed in the outcrop exposure. Random joints and fissures have been observed along the bedding planes making the horizon well porous and permeable. Interconnected joint net observed on the outcrop level. The morphology of the interconnectivity of the joints/fractures sets enhanced the rate of fluid flow inside the medium. Visually observed that the area falls in higher fracture density. Thickness of the Sakesar Limestone in the Landa Psha section is 128 meters whereas in the Makarwal section of the Surghar Range is 300 meters [13]. Thickness reported 220 meters in Chichali section and 600 meters at Makarwal area [5]. Physical characteristics including its lateral extension, thicknesses, primary and secondary connected and unconnected fractures fabricates its permeability and storage capacity of the formation is appropriate and in hand basic parameter for the reservoir potential of the horizon. Its lower contact with the Nammal Formation and upper contact with the Chinji Formation both comprised of thick shale beds making the seal horizons, observed conformable and unconformable respectively. The Sakesar Limestone evolved in open marine carbonate depositional environment [14] during early Eocene.

Mitha Khattak anticline (MKA)

This anticline is mapped north of Mitha Khattak village and comprised of rocks of the Siwaliks Group underlain by the Eocene Sakesar Limestone and Nammal Formation. The Eocene rocks are underlain by the Paleocene Patala Formation and Lockhart Limestone. Base of the Lockhart Limestone is not exposed in the core of MKA. This is an east facing and south plunging prominent structural of the area [15]. The eastern limb is steeply dipping to southeast to make the anticline is overturned.

The fluvial and Siwaliks Group rocks composed of Mitta Khattak and Chinji formations. The Mitta Khattak Formation dominantly consists of dark brown thick bedded sandstone at the top and 8 to 9 m thick, compact and well cemented cobble gravels conglomerate beds of coarsening upwards sequence observed at base of formation, representing unconformity. The Mitta Khattak Formation could be equivalent faciess of the Rawalpindi Group. Structural trend of the beds is N10~15°E and dipping at an angle of 75°~80°SE. Thick horizon of the Sakesar Limestone observed fractured and jointed making a potential reservoir format for the accumulation of hydrocarbon in this structure (Figure 3).

Joints data

Joints data has been collected on the eastern limb of the MKA

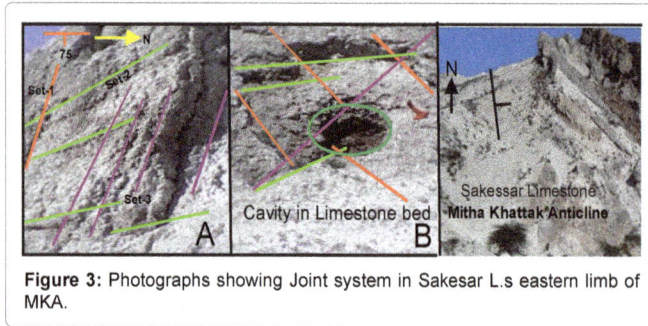

Figure 3: Photographs showing Joint system in Sakesar L.s eastern limb of MKA.

on the exposed beds of Sakesar Limestone observed highly fractured tectonically and diagenatically. Tectonically induced three different joints sets have been observed and acquired data of each parameter which is tabulated in Table 2.

The surface exposure of Sakesar Limestone observed highly fractured, jointed, vugy and showing nodulation. Visible and splitted bedding planes, longitudinal and transverse joints pattern, chicken wire net fractures and existence of elliptical to circular holes right angle to strike make the horizon encouraging reservoir for fluid circulation as well as potential storage compartment (Figure 4).

Karandi anticline (KA)

Rocks exposed on the eastern limb of KA are the Mitha Khattak and Chinji formations. The Eocene Nammal Formation and Sakesar Limestone are thrust eastward on the Mitha Khattak Formation in the hanging wall. The oldest formation exposed in core of the KA is the Cretaceous Lumshiwal Formation, where base of the Lumshiwal Formation is not exposed in the core. General trend of the strata is N5°~10°W and dipping steeply toward 80°SW. Tectonically induced Joints trend and other diagenetic apertures up to13 cm diameters, data has been acquired of the Sakesar Limestone to reveals its reservoir potential capability in the region. The obtained data is tabulated in Table 3.

The Sakesar Limestone is medium to thick bedded showing prominently bidirectional joint pattern with some visible circular to semicircular openings. The fracture density visually observed high at the outcrop level. The same density is expected to be existed at the subsurface level because these ruptures are the product of diagenetic as well as tectonic process. In view of, the Sakesar Limestone will be considered virtually as potential reservoir for the accumulation of hydrocarbon in the Surghar Range (Figure 5).

Malla Khel anticline (MKA)

The Malla Khel Anticline is located northwest of Malla Khel village. It is a prominent structural feature of the region. The frontal limb of the anticline is overturned. The back limb of the anticline gently dipping to northeast while its forelimb is asymmetrical to overturned and dipping at high angle ranging from 80°-85° northeast, the oldest Datta Formation is exposed in its core. In frontal flank Eocene strata are thrust in a hanging wall ramp over Siwaliks in the footwall ramp. Geometry of the anticline revealed as fault bend-fold (Figure 6).

General structural orientation of beds is N05°~10°E and steeply dipping to 80°~88°NW~NE. Secondary induced joints/fractures trend along with diagenatically produced openings data has been acquired along the frontal limb of anticline of the Sakesar Limestone to reveal its fluid reservoir potential in the Surghar Range. The collected data is tabulated in Table 4.

The surface exposure of the Sakesar Limestone observed remarkably fractured, jointed by penetrative strain with the creation of open bedding planes. Some major dissolution of the primary minerals observed in the form of cavities (Figure 7) which is responsible for the

S. No	Strike	Dip	Length m	Spacing m	Opening m	No of Joints
Set -1	N70°~80°E	50°~60°NW	01~3.0	0.5~1.0	0.001~0.002	25
Set -2	N20°~30°W	35°~40°SW	0.5~1.0	1.0~3.0	0.001~0.0015	27
Set-3	N30°~40°W	50°~65°SW	02~4.0	0.30~0.7	0.005~0.01	10

Table 2: Showing joints trend dimensional data of the Sakesar Limestone, MKA.

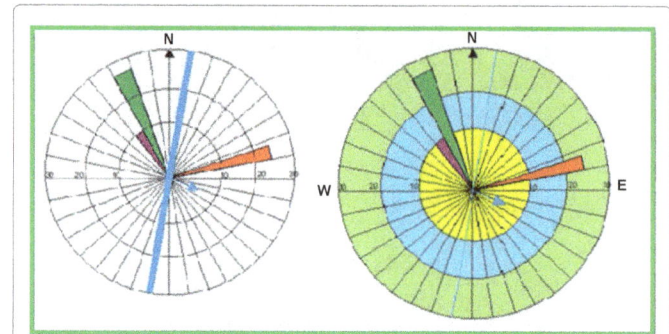

Figure 4: Joints rose diagram along with structural trend data eastern limb of MKA.

S. No	Strike	Dip	Length m	Spacing M	Opening M	No of Joints
Set -1	N68°~75°W	70°~80°SW	0.3~2.0	0.2~0.5	0.001~0.002	12
Set -2	N20°~25°E	50°~56°SE	0.5~4.0	1.0~3.0	0.01~0.15	15
Set-3	N55°~60°E	50°~65°NW	0.5~3.0	0.5~1.5	0.01~0.03	12

Table 3: Showing joints trend dimensional data of the Sakessar Limestone, MKA.

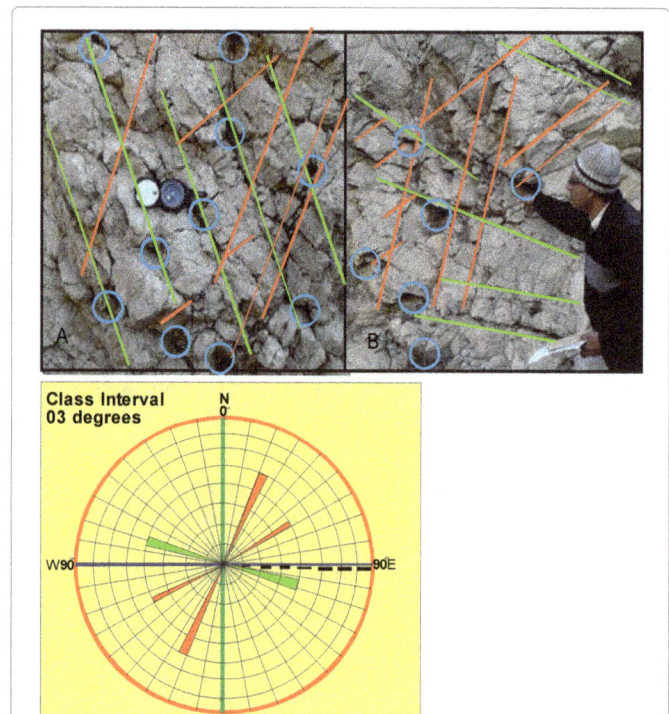

Figure 5: Showing fractured Sakesar Limestone with joints rose diagram of KA.

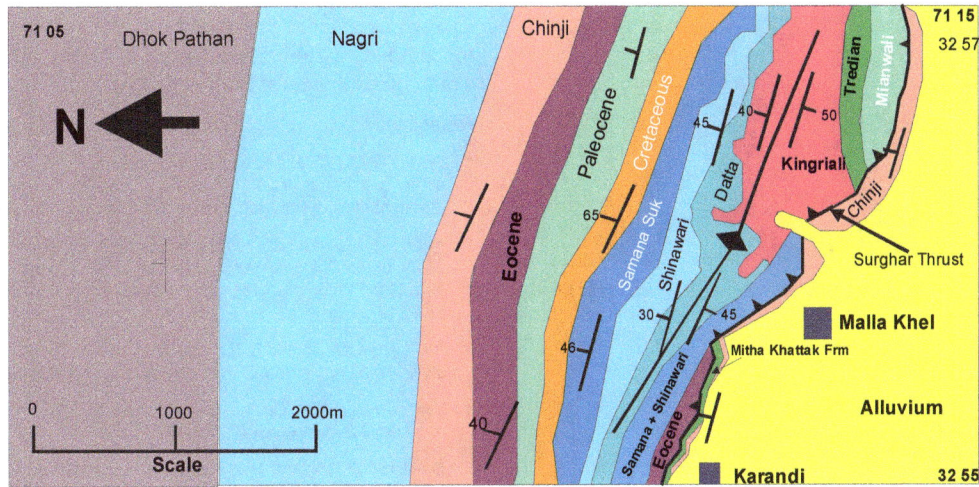

Figure 6: Geological map of the Karandi and Malla Khel areas of Surghar Range.

S. No	Strike	Dip	Length m	Spacing m	Opening m	No of Joints
Set -1	N40°~60°W	65°SW~Vert	0.2~1.0	0.5~ 2.0	0.001~0.0015	15
Set -2	N20°~25°E	30°~40°SE	0.5~2.0	0.5~1.5	0.002~0.004	18
Set-3	N60°~65°E	70°~75°NW	0.5~2.5	0.5~2.0	0.01~0.03	12

Table 4: Joints trend and diagenetic apertures in the Sakesar Limestone, MKA.

Figure 7: Photographs showing overturned Malla Khel Anticline, Surghar Range.

enchantment of permeability of the desired horizon. These fractures are observed interconnected with each other to make a net of conduits for the fluent flow of fluids in the studied horizon (Figure 8).

Discussion

Tectonics and structural deformations in the Surghar Range is younger proportional to the northern mountain belts of the inner Himalayas. This outermost frontal fold-and-thrust belt of the sub-Himalayas is comprised of latest tectonic and local to regional scale structural features well developed in the frontal flanks of the range. The mapped folds and faults are the product of compressional to transpressional tectonic regime. The Surghar Range is an arcuate feature and bounded all along its periphery by the combination of major thrust and local scale strike slip faults. The younger segment of the range is north-south oriented whereas the well-grown segment of the range is east-west oriented. The Sothern plunging terminus of the range is bounded by the Kundal strike-slip fault whereas the eastern terminus of the range is bounded by the Kalabagh fault system. The northwestern boarder is demarcated by the Karak Thrust fault and the southeastern perimeter is decoupled by the Surghar Range forethrust.

The in-between compartmentalized area is comprised of small to large sized force folds being the product of fault-bend and fault-propagation compressional phenomenon. The oldest Paleozoic sequence of Permian rocks is cropped out along the frontal thrust sheet and protrudes southeastward against the Punjab foredeep. In structural route from south to east along the range different stratigraphic horizons have been exhumed along the frontal ramp reveals detachment flux from the prime basal décollement. The Eocene strata towards south and Permian strata toward east along the range front have thrust over Siwaliks in the footwall ramp. The outcrop exposure of the rocks along the range front is remarkably fractured and jointed. The Sakesar Limestone of Eocene has been selected for detailed fractures analysis to show its hydrocarbon reservoir potential along the eastern and southeastern flanks of the range.

Conclusion

The structural defects have been observed in the Sakesar Limestone and found that they are the products being induced diagenatically and tectonically during depositional and post depositional phases respectively. The tectonically secondary induced fractures are

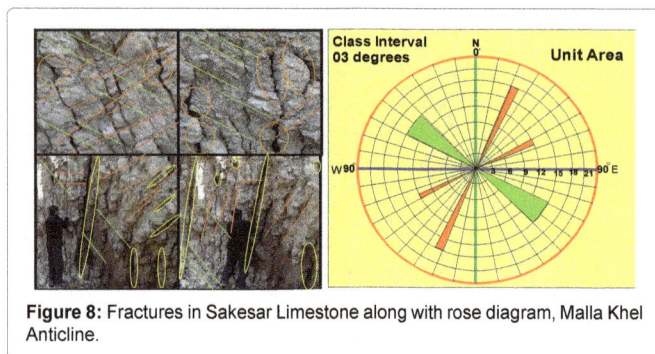

Figure 8: Fractures in Sakesar Limestone along with rose diagram, Malla Khel Anticline.

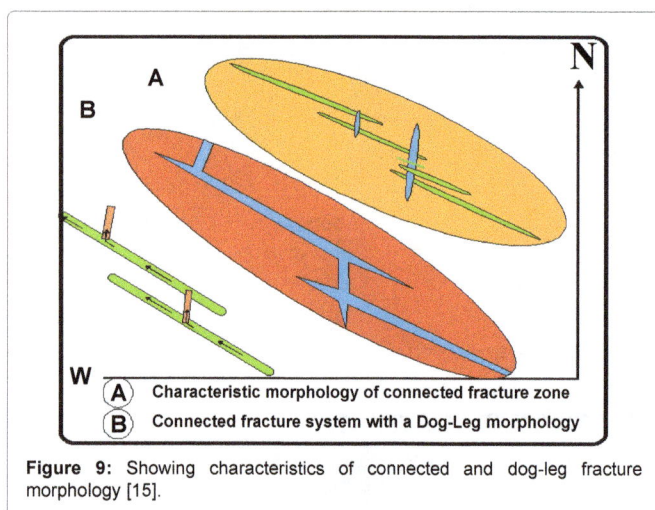

Figure 9: Showing characteristics of connected and dog-leg fracture morphology [15].

elongated reveal systematic and preferred orientations, joint spacing and joint openings/apertures. Three different joint sets have been observed at the three different observatory stations at three different anticlines along the range front shown high fracture densities. Some of the unsystematic/random joints and fractures have also been observed at each location raises the reservoir capability of the Sakesar horizon. The diagenetic apertures are generally shown on the photographs in the form circles, because they are more or less spherical in appearance. The sizes of sphere and their openings are significantly larger than the secondary tectonically induced fractures. The inner diameter walls of the spheres are uneven, rough and harsh and reveal the dissolution of some unstable minerals masses subsequent to their deposition.

The differences in fracture directions that might disclose differences in the regional tectonic stress patterns between the two time periods. Most of the joint patterns are observed interconnected with each other and give dog-leg morphology (Figure 9). Visually the fracture density in the Sakesar Limestone horizon observed greater in the east-west trending segment as compare to the north-south trending segment of the range. These fractured rocks facilitate the fluid storage capacity and transmissivity along the medium to enhance the reservoir quality of the Sakesar Limestone. That's why one of the most essential prerequisite for the hydrocarbon accumulation is in hand in the Surghar Range, Trans-Indus Ranges of the outer Himalayan orogenic province.

References

1. Powell CMA (1979) A speculative tectonic history of Pakistan and surroundings: some constraints from the Indian Ocean. In: Farah A, DeJong KA (eds.) Geodynamics of Pakistan. Geological Survey of Pakistan Quetta pp: 5-24.

2. Khan MJ, Opdyke ND (1993) Position of Paleo-Indus as revealed by the mag. Strati. of the Shinghar-Surghar Ranges, Pak. In: Shroder JF (Ed), Himalaya to sea: Geol, Geomorphology and the Quaternary Rutledge Press, London pp: 198-212.

3. Ali F, Khan IM, Ahmad S, Rehman G, Ali HT, et al. (2014) Range front structural style: An example from Surghar Range, north Pakistan. JHES 47: 193-204.

4. Wynne AB (1880) On the Trans-Indus extension of the Punjab salt Range. India Geol Survey Mem 17: 95.

5. Danilchik W, Shah MI (1987) Stratigraphy and coal resources of the Makarwal area, Trans-Indus Mountains, Mianwali District, Pakistan. US Geological Survey Prof Paper pp: 131: 38.

6. Kemal A (1991) Geology and new trends for hydrocarbon exploration in Pakistan, proceedings, International petroleum seminar Ministry of Petroleum & Natural Resources, Islamabad, Pakistan pp. 16-57.

7. Wells NA (1984) Marine and continental sedimentation in the early Cenozoic Kohat Basin and adjacent northwestern Indo-Pakistan: Ann Arbor, University of Michigan, Ph.D. dissertation 465.

8. Yeats RS, Hussain A (1987) Timing of Structural events in the Himalayan foothills of north-western Pakistan. Geological Society of America Bulletin 99: 161-175.

9. Smith HA, Chamberlain CP, Zeitler PK (1994) Timing and duration of Himalaya metamorphism within the Indian plate, Northwest Himalaya, Pakistan. Journal of Geol 102: 493-503.

10. Beck RA, Burbank DW, Sercombe WJ, Riley GW, Brandt JR, et al. (1995) Stratigraphy evidence for an early collision between Northwest India and Asia. Nature 373: 55-58.

11. Khan MJ, Opdyke ND, Kheli RAKT (1988) Magnetic Stratigraphy of the Siwalik Group, Bhittani, Marwat and Khasor ranges, north-western Pakistan and the timing of neocene tectonics of the Trans-Indus. J Geophy Res 93: 11773-11790.

12. Alam I (2008) Structural and Stratigraic framework of the Marwat-Khisor ranges, N-W.F.P, Pakistan. Unpublished Ph.D. Thesis submitted to NCEG, Univ of Pesh, Pak.

13. Ali A (2010) Structural analysis of the TIR: Implication for the hydrocarbon pot. of the NW Himalayan, Pak. Unpublished Ph.D. Thesis, submitted to NCEG, Univ. of Pesh. Pak.

14. Tectostrat (1992) Trans-Indus and Salt Range Study Report.

15. Jamison WR (1998) Quantitative evaluation of fractures on Monkshood Anticline, a detachment fold in the foothills of western Canada. AAPG B 81: 1110-1132.

5

Interpreting Seismic Profiles in terms of Structure and Stratigraphy with Implications for Hydrocarbons Accumulation, an Example from Lower Indus Basin Pakistan

Majid K[1,2*], Shahid N[3,] Munawar S[2,4] and Muhammad H[1,2]

[1]Institute of Geology and Geophysics, Chinese Academy of Sciences, 100029, Beijing, China
[2]University of Chinese Academy of Sciences, 100049, Beijing, China
[3]National Centre of Excellence in Geology, University of Peshawar, 25000, Khyber Pakhtunkhwa, Pakistan
[4]Shanghai Astronomical Observatory, Chinese Academy of Sciences, 200030, Shanghai, P.R. China

Abstract

Structural delineation and Stratigraphic evaluation of proven geological basins are of prime importance for hydrocarbon exploration and exploitation. The current study aims to map the subsurface geology in a part of Lower Indus basin of Pakistan in terms of structure and stratigraphy using seismic data with some borehole information. In this paper, conventional integrated geophysical technique has been used to analyze the seismic data to characterize reservoir formations. To serve this purpose, 2D seismic data in SEG-Y format was used along with velocity and well logs information. Seismic profiles are interpreted and then transformed into 2D and 3D (Time and Depth domain) contour maps, which is the representation of true subsurface geology. Three promising closures at shallow time have been identified on structural contour maps which are indicative of possible prospects. A closure, covering approximately 1 Sq Km area over Eocene aged Sui Main Limestone and further two over Cretaceous aged Lower Goru formation covering areas of 500 and 400 Sq km along shot point 260 and 380-390, respectively. Conjugate normal fault system among the stratigraphic layers existing in the study area can smoothly provide the hydrocarbon trapping mechanism. In order to further confirm the probable future prospects and support the study, modeling of P and S wave Impedance have also been done. The advance integrated study of AVO modeling and numerical rock physics analysis may be helpful in providing further insights into current research.

Keywords: Impedance; Compressional and shear waves; Structural interpretation; Lower indus basin; Pakistan

Introduction

The study area having coordinates 2652'50"N and 68°55'60" E lies in Lower Indus Basin of Pakistan which is one of the proven geological provinces as well as contributing in the oil and gas requirements of the country. The study area has a prominent position in hydrocarbon prospecting with Sargodha high, Indian Shield and marginal zone of Indian Plate in north, east and west respectively [1], furthermore, this field is situated in Sindh Province of Pakistan, discovered in 1991 and came in regular production after successful drilling of three wells in 1995. The production from this field on regular basis is in the form of gas with condensate while it is bounded by nearby famous gas fields namely Miano and Sawan. Major basins in Pakistan are result of to regional tectonic activity about 200 M.y. ago, continuous rifting gave birth to formation of Lower Indus basin, which was further subdivided into two parts because of rotation of Indian plate about 55 M.y. ago. In the original 1992 Field Development Plan the core area, which is the eastern part of the field Lower Goru was considered more prospective as the reservoir [2].

Stratigraphic analysis comprises of interpreted seismic section where the different lithological layers exhibit as genetically related sedimentary succession. Structural interpretation involves identifying proper geological structures for probable accumulation of hydrocarbon, mainly tectonics play vital role in structural styles of a geological entity. Tectonic setting usually governs the structural network and associated features. The Lower Indus basin of Pakistan has been resulted due to large scale extensional tectonic forces that resulted into normal faulting and associated horst and graben geometry. Structural traps include the faults, anticlines and duplex etc. [3]. One of the most common ways of identifying seismic reflections is to compare a seismic section with another section to find the regularity in different horizon of the area to be investigated. To grasp strong command on seismic interpretation

and structure delineation, synthetic seismograms are frequently used to identify the reflectors [4]. For new prospects in any area, seismic data interpretation for hydrocarbon traps is not sufficient, further detail study like Petrophysical analysis, reservoir characterization, rock physics analysis and seismic modeling is required. According to Robinson and Coruh [5], in order to get detail information of the subsurface, velocity modeling is indeed essential. However, if the geophysical data is sparse, then structural and stratigraphical interpretation is the most suitable method to extract more information regarding petroleum system. In general, velocity increases with depth as density and overburden pressure increases, velocity in the subsurface varies in both laterally and vertically. Vertical variations are due to lithological changes of layering and increasing pressure due to increasing depth [6]. Lateral variations are due to slow changes in density and elastic properties due to changes in lithology or physical properties [6]. Meanwhile acoustic impedance provides useful information about the lithological successions as well as variation in different rock properties [7,8]. Acoustic impedance variations are directly related to the lithological variations and hydrocarbons content in a reservoir formation. A 60 fold 2D seismic data with 6 seconds of record length was acquired in 1989 by LASMO oil Pakistan limited in SEG-Y format, the seismic data for this study was provided in post stack migrated (PSTM) format for structural

---

*Corresponding author: Majid K, Institute of Geology and Geophysics, Chinese Academy of Sciences, 100029, Beijing, China
E-mail: majidktk.000@gmail.com

and stratigraphical interpretation. In this study, we have interpreted the data of 2D seismic lines for better understanding of structure and stratigraphy in a part of Lower Indus basin Pakistan. The horizons of interest are defined on the basis of their seismic character and continuity as well as wells formation tops. Three seismic horizons were interpreted in the study area as Eocene aged Sui Main Limestone; Cretaceous aged Upper Goru and Lower Goru formations. For further insights, P and S waves' acoustic impedance were numerically computed and modeled to locate probable hydrocarbon bearing prospects in the study area.

Geological setting

Based on the sedimentation history and structural style, the Indus Basin is divided into three segments namely Upper, Central and Southern Indus basins [9,10]. Relatively high areas from shoreline of Jacobabad-Khairpur and Mari- Kandhkot highs distributed the lower portion of Indus basin in central and southern basins mutually known as Sukker Rift.

The study area is situated on the eastern and southeastern flank of the regional north-south– trending Jacobabad-Khairpur high as shown in Figure 1, which is an important factor in the formation structural traps in the study area [11], Kadanwari [2], Sawan and Tajjal gas fields. The Jacobabad-Khairpur has been developed by domal uplifting during the early Cretaceous, and later on along deep seated faults in the Late Cretaceous and Paleocene ages. The three tectonic events mainly responsible for the structural configuration of the study area are: First is the late Cretaceous uplift and erosion, Second event is a late Paleocene

right-lateral wrenching, and third one is the late Tertiary to Holocene uplift/inversion of the Khairpur high [2]. The Khairpur high can be characterized by a high geothermal gradient of up-to 4.8°C/100 m (328 ft) [10]. The study area has been evolved through above mentioned tectonic events and as a result the stratigraphic successions from Pre-Camb-Mesozoic and Paleogene have been deformed by more than one episode [9]. The variations can be seen in stratigraphic succession of Lower Indus basin from east to west, as no succession older than Late Triassic have been drilled in the Southern Indus basin [11]. The stratigraphic package of the Jacobabad-Khairpur High includes Mesozoic,Tertiary and Quarternary lithologies as shown in Figure 2. The major unconfirmity occurs between base Permian and base Tertiary and the presence of Jurassic rocks in the area show deposition in early rifting stage. During early collision in Paleocene Ranikot, clastics were deposited followed by Eocene carbonates including the important reservoir unit Sui Main Limestone in the study area. The Sui Main Limestone was deposited on a shallow water carbonate platform with sporadic influx of clastic materials. This unit is widely distributed on a Jaccobabad - Khairpur High and which could be considered as good hydrocarbon reservoirs. Medium to coarse-grained sandstones in a shallow-marine setting constitute the main reservoir (Lower Goru Formation) in the field [10]. The sandstone of the Cretaceous Lower Goru is the most productive reservoir rock unit in the study area and nearby oil and gas fields. The lower part of this member has informally been divided, from bottom to top, into A, B, and C intervals, respectively [11].

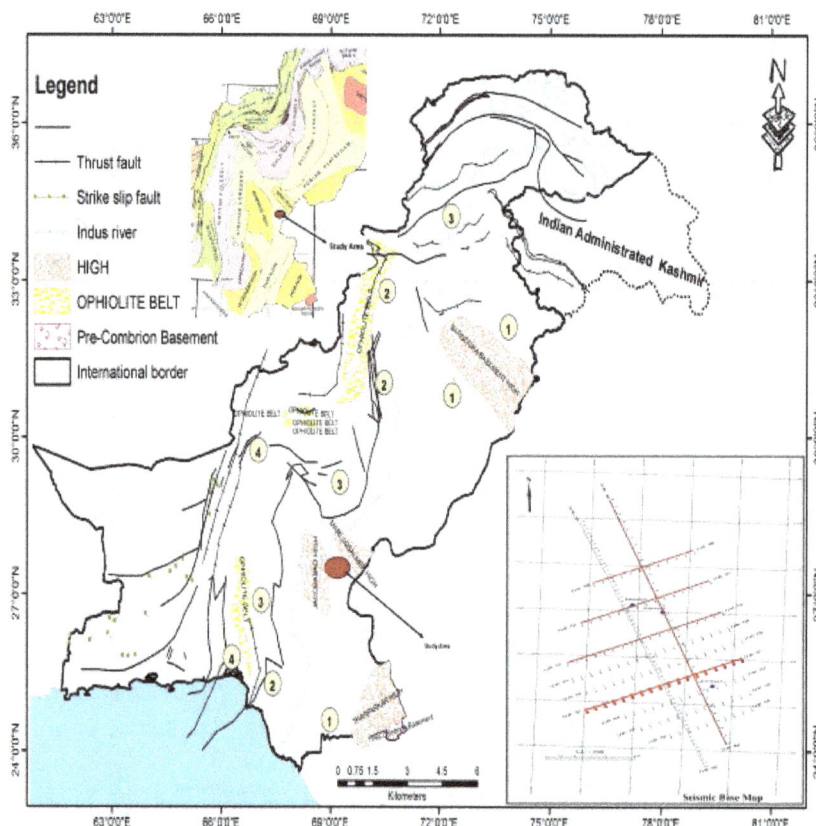

Figure 1: Geological and Tectonic setting of Lower Indus Basin Pakistan [1]. The Base map in the bottom right corner shows the seismic profiles. The study area is highlighted by circle.

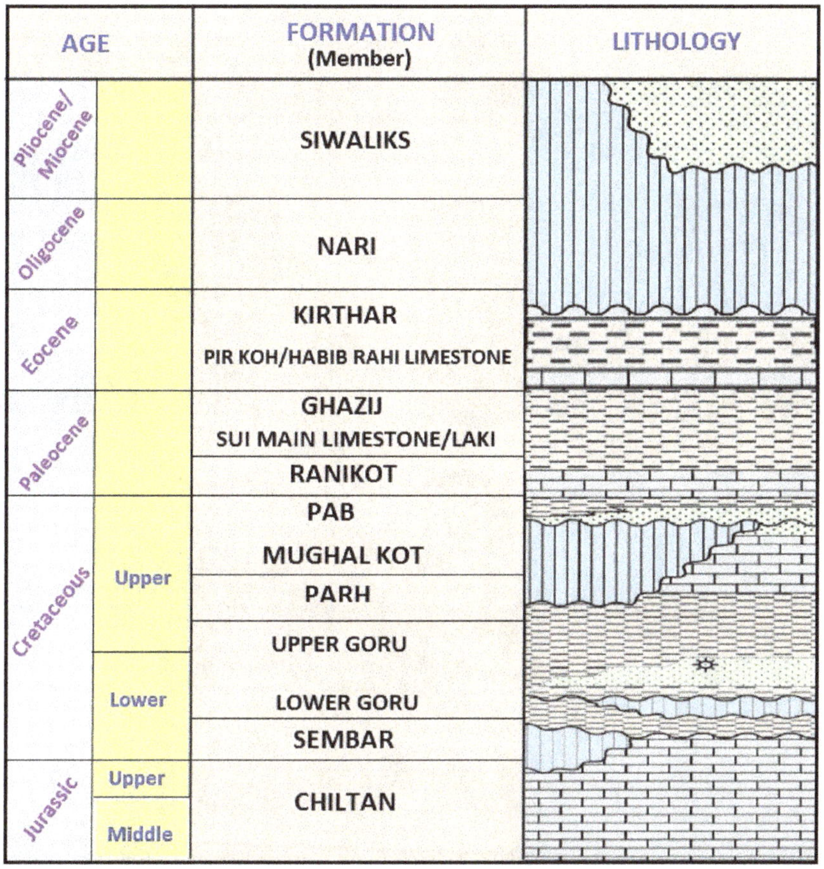

Figure 2: The generalized stratigraphy of Lower Indus basin Pakistan. The formations shown are mainly from the present study area.

Petroleum prospects

The Lower Cretaceous formation (Lower Goru) has consecutive layers of sand and shale. These sands act as reservoir rock units with rapidly varying reservoir characteristics within few kilometers. The foremost influence on these two processes is the degree of sands supply. This formation is the most proven reservoir rock unit throughout the entire basin; it is producing oil and gas in many fields located in Lower Indus basin. The Eocene aged Sui Main Limestone is also found productive in the vicinity gas fields. The shales of Cretaceous and Eocene act as source rock units in the study area [12].

Seismic coverage

The study area is an old exploration concession known as Old Tajjal Concession situated in the Lower Indus basin. The initial 2D seismic data in investigated area was acquired 6 to 7 kilometers along a horizontal profile in the year 1988. The seismic profiles are generally east west trending with some lines ties in north south direction. The prospective areas were subsequently in-filled with more data; the current study area is one of them, in which the central area was considered more prospective than western area. In the year 1989, a NE–SW orientated, 2 kilometers spaced seismic grid, with a shot point interval of 50 m and 60-fold coverage was acquired. This was extended in 1990 to the west, but with a 25 m shot point interval and 120-fold coverage. The seismic source used was vibroseis. The area is generally covered with sand dunes of Thar Desert [2].

Materials and Methods

Seismic interpretation is the processes of transforming seismic

data into geological section to get information about structure and stratigraphy. It possibly will find out the general information about an area, locate prospects for drilling new exploratory wells, or guide development for already discovered field. According to Badley [13], there are two main approaches for interpreting seismic data, one is stratigraphic analysis and another is structural analysis. For seismic data interpretation, the identification of seismic reflection packages has been done using seismic sections as shown in Figures 3a-3c, depths from formation tops and average thicknesses from available wells, interval and average velocities during the processing of seismic data, geological and seismic characteristics of different lithologies. In stratigraphic interpretation, lateral continuity, variation in sedimentary deposits and different episodes of sedimentation affected by tectonic activities and recognition of time depositional units is studied.

In this study, we investigated new prospects for fast track implementation of oil and gas exploration activities in parts of Lower Indus basin of Pakistan. Regional geology, tectonics and stratigraphy were reviewed as part of the objectives in order to properly identify the structural and stratigraphical traps which were then confirmed by compressional and shear wave's velocity and Impedance modeling. On the basis of structural and stratigraphic trends of the formations and geological history, migration pathways are also predicted along deeper horizon on seismic section as shown in Figure 4. The Iso-velocity map hich commonly shows velocity variations along horizontal and vertical directions over a specific profile was constructed using velocity information obtained from seismic sections.

The primary objective of seismic interpretation is therefore to

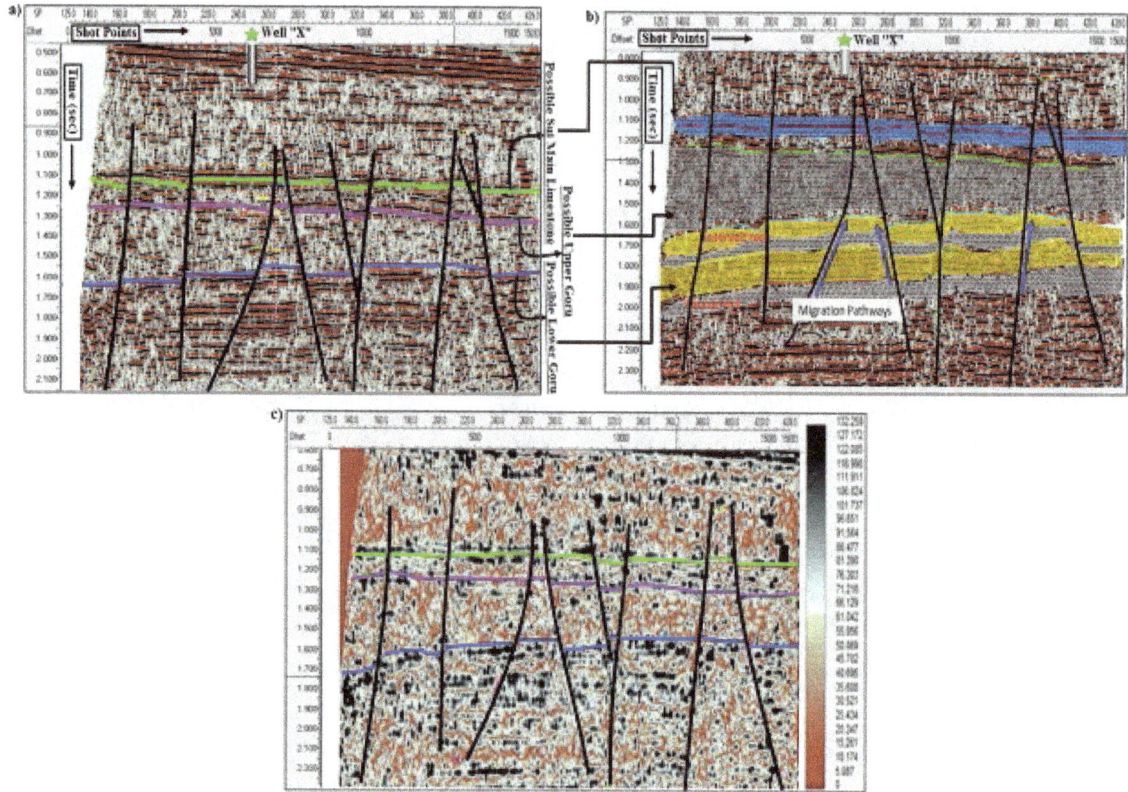

Figure 3: (a) Interpreted seismic lines showing normal faulting with horst and graben geometry (b) lithologies filled and predicted hydrocarbon migration path (c) Trace Envelope map of seismic line "X".

prepare contour seismic maps showing the two way time to a reflector as picked on the seismic sections. This time (isochron) map must be converted to depth (isodepth) map through the seismic time-depth conversion process. Contour maps are the most accurate representative of subsurface geology; consequently interpreted seismic sections along with average velocity information were used to generate 2D and 3D contour maps of formations of interest in the study area. The contour maps of Lower Goru formation and Sui Main Limestone are shown in the Figures 5 and 6 respectively.

Trace Envelope as attributes analysis has the ability to enhance strength, amplitude and energy of the reflector while increasing the basic properties of seismic reflectors. Trace Envelope was initially developed for the oil industry by Nigel Anstey in the 70's in order to identify "bright spots" related to hydrocarbon accumulations. Trace envelope can be computed from complex trace which was first introduced by Taner [14]; the complex trace can be defined as;

$$CT(t) = T(t) + t H(t) \tag{1}$$

Where;

$CT(t)$ = Complex trace

$T(t)$ = Seismic trace

$H(t)$ = Hilbert's transform of T (t)

$H(t)$ is a 90° phase shift of T(t)

The Trace Envelope is calculated from complex trace by the following formula;

$$E(t) = SQRT\{T^2(t) + H^2(t)\} \tag{2}$$

Trace envelop is helpful in identifying the subsurface discontinuities, lithological variations, changes in deposition system and sequence boundaries. In order to differentiate the top and bottom of various geological units or formations in the study area the trace envelope was designed using seismic lines.

An empirical relationship was established for computation of acoustic impedance by Gardner et al. [15] using density and velocity information. This relationship is widely used for brine saturated sedimentary rocks and is given in Eq (3);

$$\rho = 0.23V_p^{.25} \tag{3}$$

$$V_s = \sqrt{\frac{\mu}{\rho}} \tag{4}$$

Where;

ρ =Density (g/cm³)

V_p =P-wave (compressional) velocity (ft/s)

μ =Shear Modulus

Consequently, the acoustic impedance can be computed by using the compressional and shear wave velocity information, the empherical relation between velocity and density is given by;

$$Z = Vp,s\rho \tag{5}$$

Z=Acoustic Impedance

Root Mean Square (RMS) velocity read from velocity panels of the seismic data provided were used to derive different velocities using Dix equations to construct compressional and shear wave's models, density sections and impedance sections. The impedance sections for P and S waves were created after numerical solution for available velocity data as shown in Eq (5). To achieve this target, SMT Kingdom 8.4 and Geosoft softwares were used in integration with other Geological and Geophysical softwares.

Results and Summary

Structural and stratigraphical interpretation of a geological entity is important for expediting hydrocarbons exploration. In order to understand the subsurface geology and structural trends for possible hydrocarbon prospects in the study area, seismic data with some well information were used on an interactive workstation. Distinguishable results have been achieved by using 5 seconds two way travel time data to explore the hydrocarbon potential of the geological structures a part of Lower Indus basin Pakistan. The structural interpretation of the seismic data is presented in Figure 3. Based on the well to seismic ties three horizons were identified in the seismic sections. These three horizons were named Possible Sui Main Limestone, Possible Upper Goru and Possible Lower Goru formation. The fault analysis was carried out on dip lines because it provides clearer picture of the fault pattern. Figure 3a shows an interpreted seismic line oriented NE-SW direction. The green, purple and blue colors represent Possible Sui Main Limestone, Possible Upper Goru and Possible Lower Goru formation. Faults are marked as black in order to differentiate between horizon and fault as it is discontinuity in seismic reflectors. Since the study area is an extensional regime disturbed by drifting with some rotational component therefore the interpretation results show that the area has normal nature of faulting with some wrench components. The associated geometry is horst and graben, horst is geologically uplifted part of the

normal fault while graben is the down ward portion of the fault which is bounded by two normal faults. Graben is relatively deep portion, where chances of hydrocarbon accumulations are less. Each horizon/ formation show horst and grabens geometry, since hydrocarbons always move from high potential or high Pressure area to low potential or low pressure area therefore the probability of hydrocarbons accumulation is mainly possible on "horst" rather than graben. Pressure in Horsts structure is comparatively low as compared to grabens because of the fact that horsts are shallow portions as compared to grabens in entire structure. The traps in the study are mainly stratigraphic traps. Sui Main Limestone (SML) and Lower Goru formation have strong reflection on the seismic section. A number of normal faults can be seen on the seismic section. SML formation is not much disturbed by faults on the western sides of the seismic section. But if we go deeper in the seismic section the throw of the faults increases. A very minor horst structure at Sui Main Limestone (SML) level can be seen on the seismic section between Shot Points 380 to 395 at approximately 1.138 seconds. Figure 3b shows lithologies filled in respective formations; it is close to real situations in the subsurface, predicted path for migration of hydrocarbons is shown with blue color arrows. Once the hydrocarbons are matured within the source formation (Shales of Lower Indus basin: probably Sembar formation), they are migrated towards the Lower Goru formation (reservoir formation) through these predicted paths. Generally the primary migration takes place along the fault or fractured planes or pores within rock units. Envelope map for the known seismic line is shown in the Figure 3c. Trace envelope rose with ultimate energy horizon on seismic section and often associated with huge hydrocarbon accumulation in the form of bright spots. It is affected by the slope, rather than envelope magnitude, hence indicate sharp interfaces.

Figure 4 shows the Iso-velocity contour map of the study area. The Iso-velocity contour map represents the lateral as well as vertical

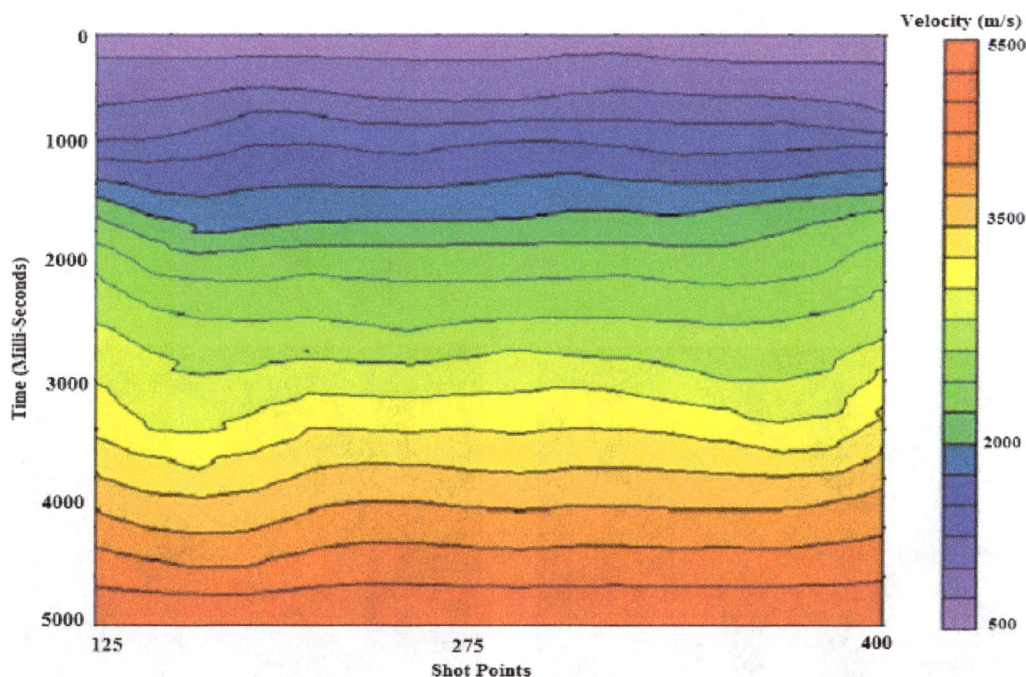

Figure 4: Iso velocity contour map with shot points distributed on horizontal axis while TWT (milli-seconds) on vertical axis. Blue color shows low velocities and red color shows high velocities.

variations encountered in the study area. In practice, these variations are resulted due to changes in the rock properties and discontinuities

affected by tectonic force and over burden pressure. This map shows the push-up and pull-up velocities. These velocities are representative of

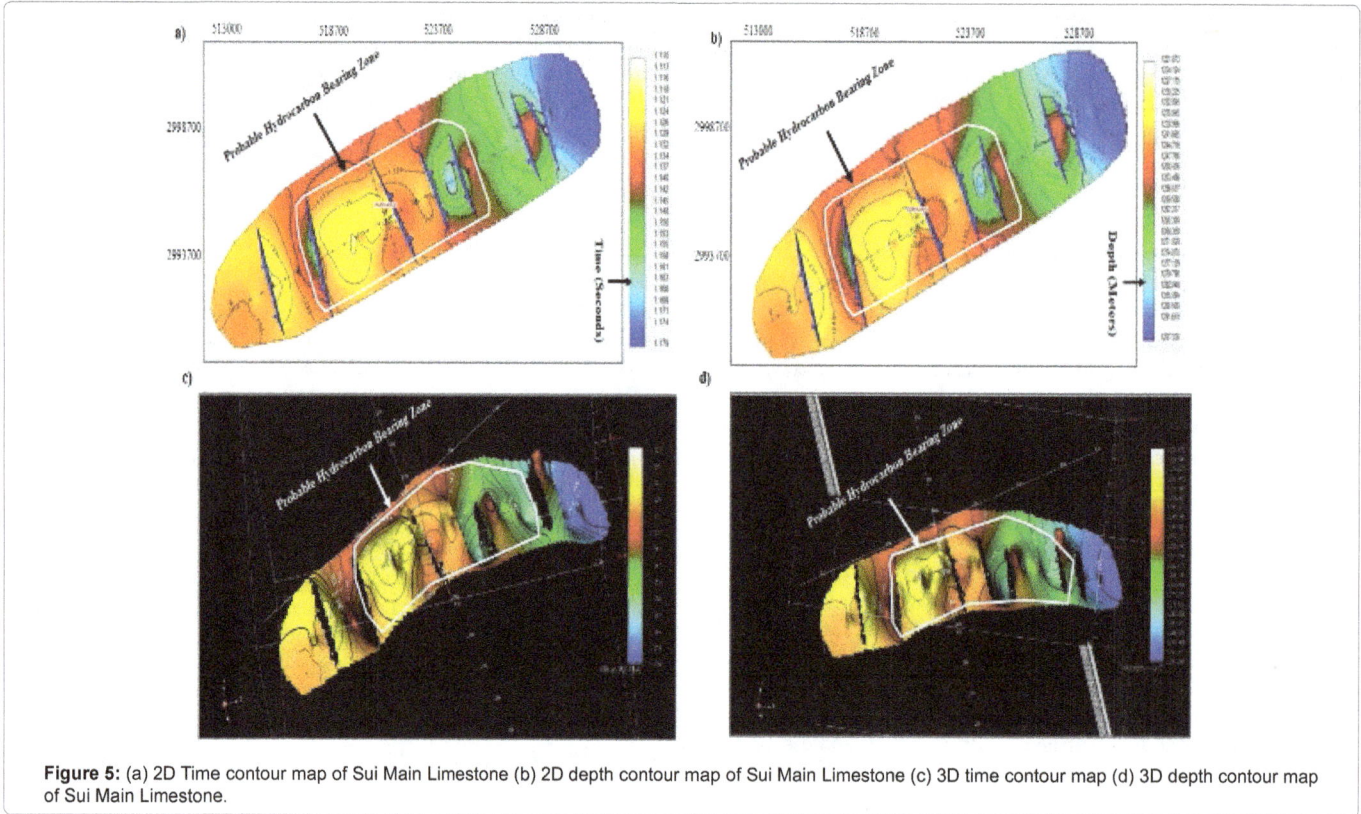

Figure 5: (a) 2D Time contour map of Sui Main Limestone (b) 2D depth contour map of Sui Main Limestone (c) 3D time contour map (d) 3D depth contour map of Sui Main Limestone.

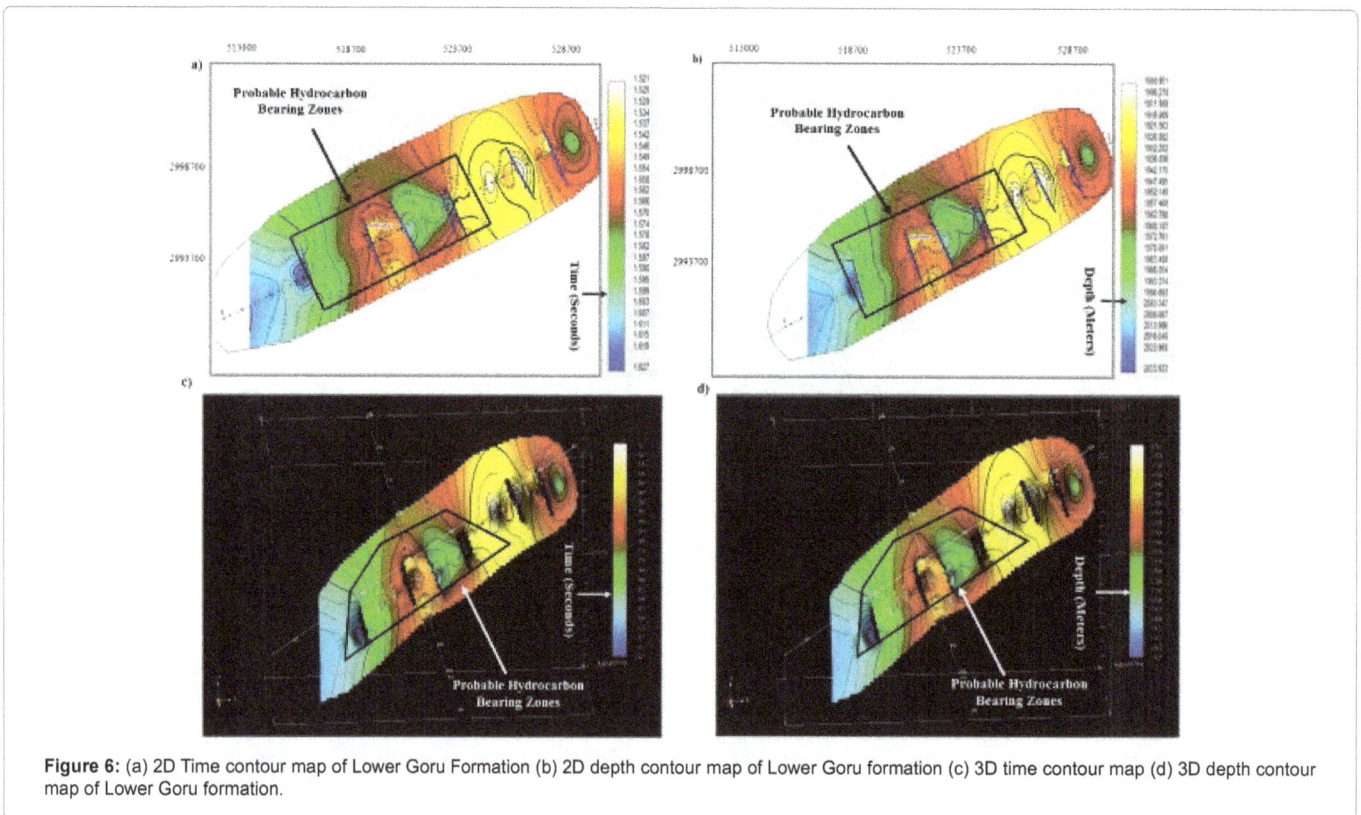

Figure 6: (a) 2D Time contour map of Lower Goru Formation (b) 2D depth contour map of Lower Goru formation (c) 3D time contour map (d) 3D depth contour map of Lower Goru formation.

variation of average velocities at certain time along different vibrating points. The purple color represents low velocities associated with shallow subsurface unconsolidated sediments while the red color represents higher velocities and deeper portions of the subsurface, it follows the general trend of velocity increase with increase in depth. The Iso-velocity map is important for current study in order to observe anomalies associated with structural disturbances in the study area.

Contours represent the lines joining the same elevation so they are important tool of seismic data interpretation. Contours represent the picture of the subsurface formations. It is generally applied to basin wide study because data in time or depth domain of all the seismic lines are used to contour a single formation. Time and depth contour maps of Sui Main Limestone and Lower Goru in 2D and 3D are shown in Figures 5 and 6.

Figure 5 represents the time and depth contour maps of Sui

Main Limestone, the yellow color shows the shallowest part (horst) and blue color shows deeper parts (graben). A closure of light yellow contour having 0.05 sec interval at shallow time of 1.125 sec covers approximately area of 1 square kilometer. Well "X" is drilled at the extreme eastern boundary on the horst structure. Contour value lies between 1.110 sec-1.179 sec, complete structure of the area has been covered by this interval of two way travel time. The highlighted area in the same figure may be the probable zone of hydrocarbon accumulation, because it provides a good stratigraphic and structural trap. The time contour map of Sui Main Limestone was converted into depth domain using the information of average velocity in the study area. The contour map of Sui Main Limestone is shown in depth domain in Figure 5b, the depth ranges from 1221 to 1297 meters and geometry is same as the time domain. Figures 5c and 5d show 3D representation of the time and depth contour maps of Sui Main Limestone.

Contour maps of Lower Goru formation are shown in Figure 6. The

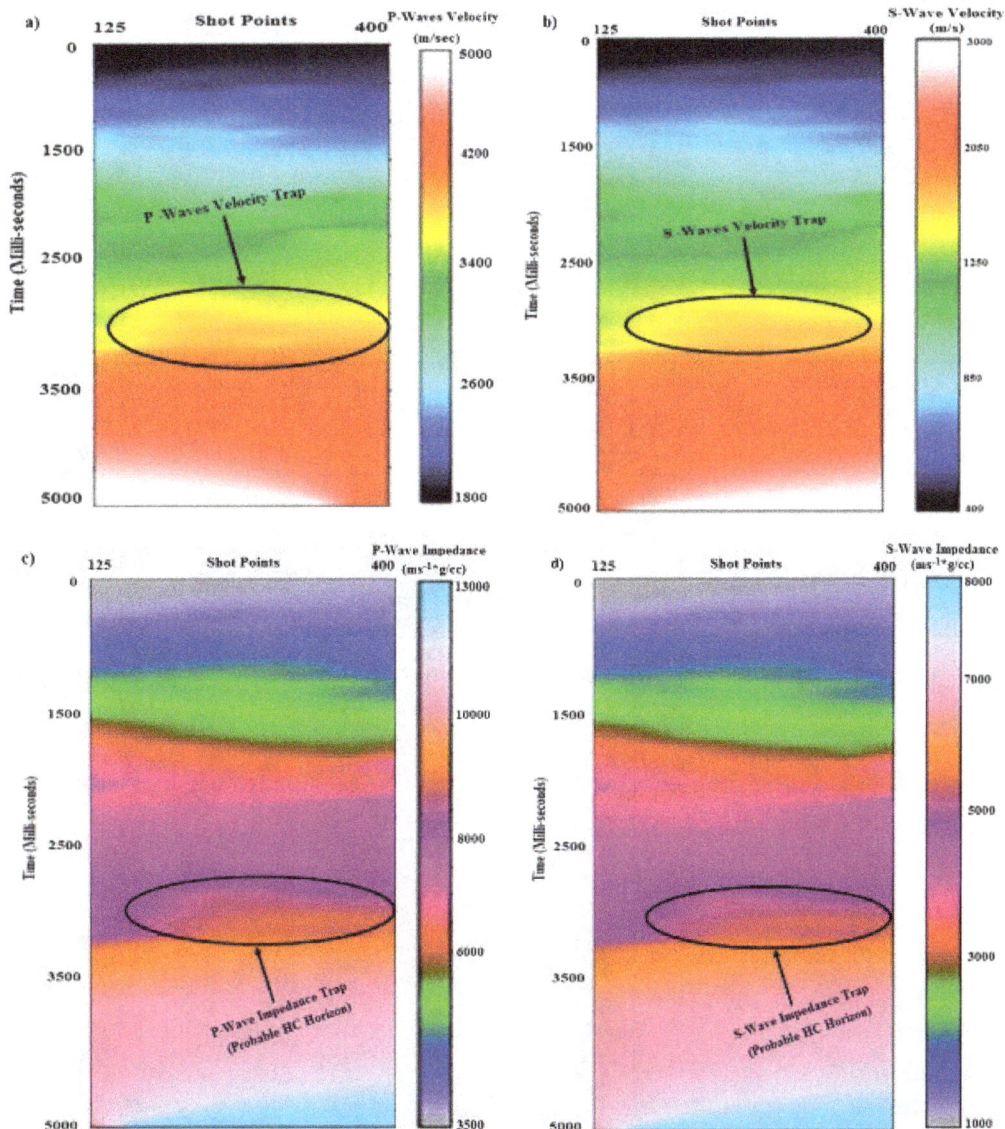

Figure 7: Models of Compressional and Shear waves Acoustic Impedance for seismic profile "A". (a) Compressional wave velocity model having a possible trap between 2500 to 3500 milli-seconds (b) Shear wave model (c) Compressional wave Acoustic Impedance with impedance trap between 2500 to 3500 milli-seconds (d) Shear wave acoustic Impedance.

contour mapping of 0.05 seconds contour interval shows a closure of 500 sq km at Shot Point # 260 of two horst structures with approximately 0.1 seconds vertical throw and in terms of depth it is approximately 15 meters. A small lead is also present at Shot Point 380 – 390 of closure 400 sq km bounded by fault towards western side of the Lower Goru formation. This could be very good promising zone for accumulation of hydrocarbons and can be regarded as future prospects. 3D contour maps of Lower Goru formation is also shown in the Figures 6c and 6d. The maps clearly depict that area has conjugate normal fault system with extensive horst and graben geometry.

Figures 7a-7d shows the velocity modeling of the simple body wave's propagation and acoustic impedance sections of a seismic profile "A". The interval velocity information was used to investigate the variations of compressional and shear wave and their response to different formations. Velocities generally increase with depth, tight sand show sharp increase while shales generally show blunt variations in seismic velocities. In seismic, velocity information play vital role in providing the subsurface details. Figures 7a and 7b show a velocity trap between 2500 to 3500 milli-seconds, throughout the section the velocity of both P and S waves uniformly increases with depth but the highlighted zone shows a trap of intermixed velocity which depicts a

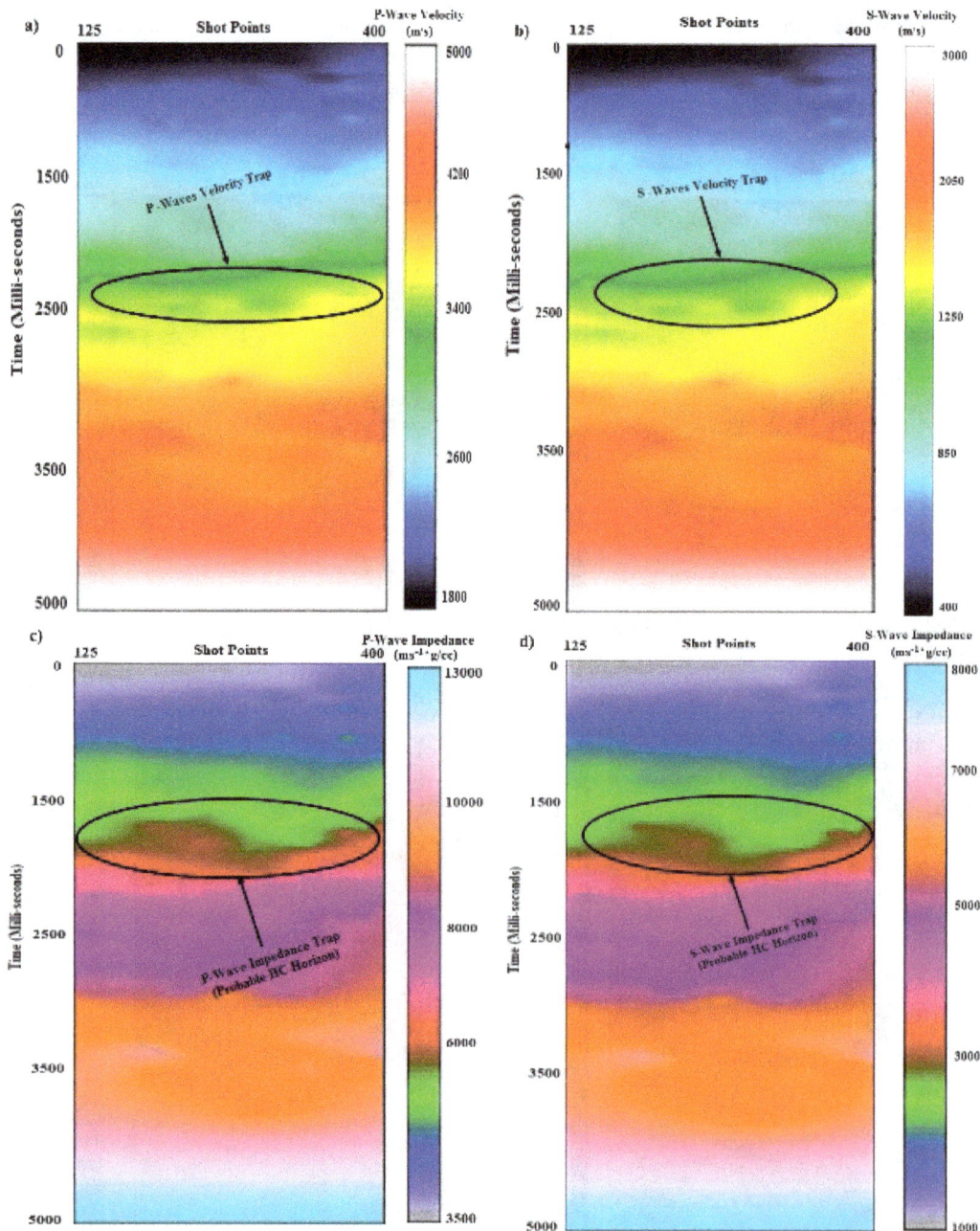

Figure 8: Models of Compressional and Shear waves Acoustic Impedance for seismic profile "B". (a) Compressional wave velocity model having a possible trap between 1500 to 2500 milli-seconds (b) Shear wave model (c) Compressional wave Acoustic Impedance with impedance trap between 1500 to 2500 milli-seconds (d) Shear wave acoustic Impedance.

transition or presence of gas sand inside the formation. The trap was confirmed with acoustic impedance model for the same profile as well. In Figures 7c and 7d the sky and pink color represents zones of high acoustic impedance while blue color represents low acoustic impedance contrast. The zone between 2500 to 3500 milli-seconds may be the possible future prospect. The same

Zone was marked in contour maps of Lower Goru Formation (Main Reservoir rock in study area); both the structural and stratigraphical interpretation and seismic wave's propagation models confirm an anomalous zone. Figures 8a-8d shows the compressional and shear wave propagation model and impedance sections for the seismic profile "B". According to Castagna [16], the trapping of seismic waves in the subsurface always depicts presence of fluid inside the rock formation. The same anomaly was seen close to 2500 milli-seconds as marked by the circle in Figures 8a and 8b, the same anomaly was seen in the impedance sections of the same profile. The zone may be considered promising for future exploration.

Conclusion

The structural and stratigraphical based seismic interpretation of possible hydrocarbon bearing formations in a part of Lower Indus Basin has been done using integrated approach of reflection seismic profiles, geology and acoustic impedance modeling. Three horizons have been marked on the basis of available information of well "X". The structural analysis of seismic profiles A and B shows conjugate normal fault system which are responsible for hydrocarbon entrapment in field. A closure of contour at shallow time of 1.125 sec on Eocene aged Sui Main Limestone covers an area of approximately 1 Sq Km, while for Lower Goru formation a closure of 500 Sq Km at Shot Point 260 of two horst structures with approximately 0.1 sec vertical throw and in terms of depth it is approximately 15 meters and another closure is of 400 Sq Km between Shot Point 380-390. We have concluded that about 50 M.y ago tectonic deformation in the study area has resulted variety of structural and stratigraphical traps. Based on our findings, the Lower Indus basin can be regarded as unique for hydrocarbon accumulations. The compressional and shear waves and impedance modeling show the same anomalous zone as depicted by seismic data interpretation. The acoustic waves modeling investigation shows the prospective to include the appropriate information of subsurface geological and geophysical understandings. We are interested in lengthening this approach for further studies with good quality data in order to facilitate the researchers as well as industry professionals for expediting hydrocarbons exploration and exploitation in Lower Indus basin of Pakistan. This study, however, can provide additional information for precise well placement in further exploration and production of oil and gas. Within the limits of the available data, it is recommended that further studies should include integration of velocity (check shot) and biostratigraphic data of all the wells. This will provide more reliable data for interpretation of the depositional environments.

Acknowledgment

We express great gratitude to Dr. Gulraiz Akhtar and Dr. Aamir Ali of Department of Earth Sciences of Quaid-i-Azam University, Islamabad and Dr. Khaista Rehman NCEG, Peshawar, Pakistan. We are thankful to Directorate General of Petroleum Concessions, Ministry of Petroleum and Natural Resources, Islamabad Pakistan for providing the public domain data set.

References

1. Kazmi A, Rana R (1982) Geology and Tectonic of Pakistan. Graphic publishers Karachi Pakistan Geological Survey of Pakistan.

2. Ahmad N, Chaudhry N (2002) Kadanwari gas field Pakistan: A disappointment turns into an attractive development opportunity. Petroleum Geoscience 8: 307-316.

3. Sheriff R (1999) Encyclopedic Dictionary of Applied Geophysics.

4. Coffeen JA (1984) Seismic Exploration Fundamentals. PennWell Publication Company.

5. Robinson ES, Coruh C (1988) Basic Exploration Geophysics.

6. Lillie R (1991) Whole Earth Geophysics: An Introductory Textbook for Geologists and Geophysicists Prentice Hall.

7. Lavergne M (1975) Pseudo diagraphies-de vitesse en offshore profound Geophysics Prospecting. 23: 695-711.

8. Lindseth RO (1976) Seislog process uses seismic reflection traces. Oil and Gas Journal 74: 67-71.

9. Ahmad N, Mateen J, Shehzad KCh, Mehmood N, Arif F (2012) Shale Gas Potential of Lower Cretaceous Sembar Formation in Middle and Lower Indus Basin, Pakistan. Pakistan Journal of Hydrocarbon Research. 22: 51-62.

10. Berger A, Gier S, Krois P (2009) Porosity-preserving chlorite cements in shallow-marine volcaniclastic sandstones: Evidence from Cretaceous sandstones of the Sawan gas field Pakistan. AAPG Bulletin 93: 595-615.

11. Krois P, Mahmood T, Milan G (1998) Miano field, Pakistan, a case history of model driven exploration: Proceedings of the Pakistan Petroleum Convention. Pakistan Association of Petroleum Geologists, Islamabad p: 111-131.

12. Kadri IB (1995) Petroleum Geology of Pakistan. Pakistan Petroleum Limited.

13. Badley ME (1985) Practical seismic interpretation. Boston International Human Resource Development Corporation p: 266.

14. Taner MT, Koehler F, Sheriff RE (1979) Complex seismic trace analysis. Geophysics 44: 1041-1063.

15. Gardner GHF, Gardner LW, Gregory AR (1974) Formation velocity and density-the diagnostic basics for stratigraphic traps. Geophysics 39: 770-780.

16. Castagna JP, Batzle ML, Eastwood RL (1985) Relationships between compressional wave and shear-wave velocities in clastic silicate rocks. Geophysics 50: 571-581.

6

Isolated Paleomagnetic Component and its Dating from the Malha Formation, Southwestern Sinai, Egypt

Khashaba A[1]*, Soliman SA[2], Takla EM[1], Farouk S[2] and Mostafa R[2]

[1]*National Research Institute of Astronomy and Geophysics, Egypt*
[2]*Egyptian Petroleum Research Institute, Egypt*

Abstract

The age of Malha Formation that located at Gebel Sarbut el-Gamal in the southwestern Sinai is a controversial point, due to the absence of index fossils. Therefore, 125 oriented core samples have been collected for paleomagnetic investigation to determine the age assignments of this formation. Rock magnetic experiments revealed that, the hematite is the main magnetic carrier within the studied rocks. Analysis of the demagnetization data isolated a single primary component of magnetization with D=351°, I=36.2° with α95=3.7° and the corresponding pole lies at lat.=77.8°N and long.=257.4°E. The calculated paleomagnetic pole has been found to be primary and enforced, suggesting the age of Malha Formation to be (Aptian-Albian). For the reliability of the obtained pole, a comparison with the published corresponding poles of Cretaceous age in Egypt was done.

Keywords: Rock blasting; Peak particle velocity; Open pit mine; Ground vibration; Air blast

Introduction

Many authors have attained different conclusions about the age of Malha Formation; Jurassic based upon the presence of Middle Jurassic miospores; Neocomian based upon well preserved palynomorph assemblages, which consist mainly of spores and pollen grains in the kaolin-bearing Malha Formation, and Aptian-Albian based upon the stratigraphic relationships [1-9]. In addition, the studied ferruginous sandstone is considered by the Geological Survey of Egypt [10] and Moustafa [11] on their maps as Triassic Qisaib Formation. On the contrary, the Egyptian General Petroleum Corporation and Conoco [12] drew these successions as Malha Formation of Aptian-Albian age (Figure 1). The present study tries to solve the problem in age controversy of these successions by a paleomagnetic investigation to determine the correct age assignments.

Lithology

Nubian sandstone ranges in age from the Cambrian to Upper Cretaceous eras below the first major marine transgressive during the Upper Campanian Duwi Formation [7]. It consists of mainly continental sandstones with thin beds of marine limestones, and marls. The Nubia sandstone now is classified into different rock units, where the Aptian-Alpian in Western Desert occurs within upper part of Sabaya Formation and lower part of Maghrabi Formation based upon palynology styudies [13]. The Malha and the Risan Aneiza Formations are part of the Nubian Sandstone group that rests unconformably on the basement rock units [14].

The term "Malha Formation" was first introduced by Abdallah et al. [5] for the multicolored and fluvio-marine sandstone with interbeds of clay at Wadi Malha, at the western side of the Gulf of Suez. According to Said [15], the thickness of the Malha Formation in the north Eastern Desert varies greatly from one place to another, Formation varies from 40 to 100 m in West Sinai, to 250 m in East Sinai, and to more than 1000 m in the subsurface of North Sinai.

This formation unconformably overlies different rock units, according to the paleotopographic and geologic settings. It overlies the Lower Paleozoic Naqus Formation in East Sinai, whereas in West Sinai and the Gulf of Suez, it unconformably overlies either the Carboniferous, as in the present study, or the Jurassic deposits. The

Malha Formation underlies unconformably the Galala Formation and the contact between them is marked by the transition zone from varicolored sandstone to pale grey siltstone of the Galala Formation.

Sampling

Samples have been collected from 6 sites of different lithologies at the surface outcrops of Malha Formation at Gebel Sarbut el-Gamal at latitude 29°08'15"N and longitude 33°13'25"E (Figure 2). The number of samples differs from one site to another, ranging between 3 and 5 hand blocks. Hand block samples have been cut into

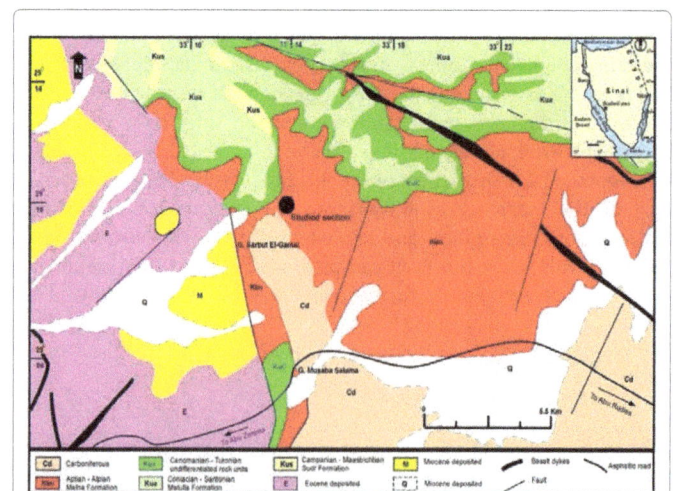

Figure 1: Location and Geological map of the study area (modified after the Egyptian General Petroleum Corporation and Conoco [12]).

*Corresponding author: Khashaba A, Ph.D, National Research Institute of Astronomy and Geophysics, Egypt, E-mail: khashaba80@yahoo.com

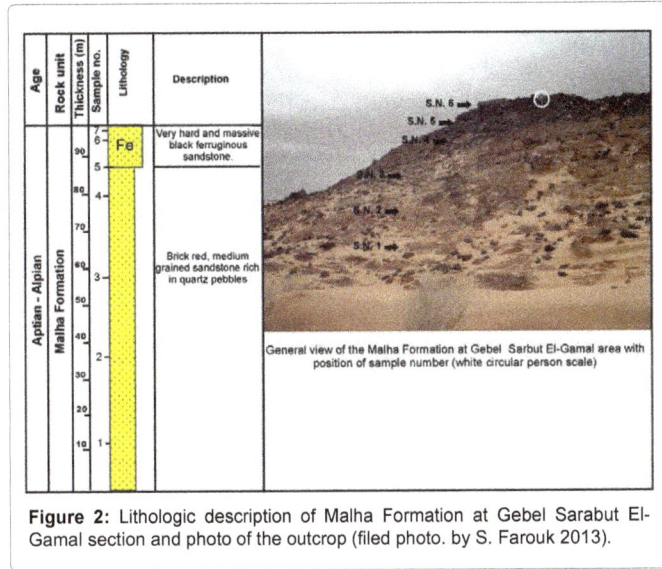

Figure 2: Lithologic description of Malha Formation at Gebel Sarabut El-Gamal section and photo of the outcrop (filed photo. by S. Farouk 2013).

total of 125 oriented core specimens using a drilling machine at the Paleomagnetism Laboratory of the National Research Institute of Astronomy and Geophysics (NRIAG), Helwan, Egypt. Samples were oriented using magnetic compass.

Methodology

Several rock magnetic experiments, as the Thermomagnetic (Js-T) curves, hysteresis loops, Isothermal Remanent Magnetization (IRM) acquisition curves and the back-field coercivity curves, using MicroMag, were done, in order to identify the main carriers of remnant magnetization.

The natural remanent magnetization (NRM) was measured in two different laboratories using 2G-SQUID (in the Paleomagnetic Laboratory at the Institute of Geophysics, Warsaw, Poland) and the JR-6 spinner magnetometer (in the Paleomagnetism Laboratory at NRIAG). Demagnetization steps were done, using both alternating field (AF) and thermal (Th) techniques. The AF- demagnetization was carried out using a magnetically shielded Degausser attached with the SQUID, with 5-10 mT steps, up to 120 mT. Thermal demagnetization, on the other hand, was done using non-magnetic thermal demagnetizer of Magnetic Measurements (MMTD80). During the thermal demagnetization, the samples were heated in steps of 50°C - up to 350°C and then the steps reduced to be 15°C up to 700°C in most cases.

The magnetic susceptibility was measured before the demagnetization process and also after the selected heating steps, using Bartington Susceptibility meter (MS2B), in order to monitor any mineralogical alteration in the specimen during thermal demagnetization.

Analysis of the demagnetization data was carried out visually; using the stereographic and orthogonal projections and statistically; using the Principal Component Analysis (PCA) for each specimen to determine the characteristic magnetic component(s) [16,17]. The obtained components, from each site, were gathered and the respective site-mean direction was calculated. The overall mean directions, as well as the virtual geomagnetic pole (VGP) position were computed using Fisher statistics for each locality [18]. Data processing was done using Rema soft program.

Experimental Results

Rock magnetic experiments showed that, hematite is the main magnetic mineral in almost all the studied specimens:

Figure 3 represents an example for the IRM curves. As shown, the magnetization in the studied specimen did not reach the saturation state until 1000 mT, indicating the presence of hard magnetic mineral(s). The same specimen was subjected to back field curves experiment, in order to identify the coercivity of remanence. As we can see in Figure 4, the existed magnetic mineral is of high coercivity value, of about 400 mT. This hard magnetic mineral of high coercivity could be hematite and/or goethite. Hysteresis loop curve in Figure 5 shows that, no saturation state was reached, indicating the presence of hematite as the main magnetic carrier. Studying the magnetization against temperature (thermomagnetic curves) Figure 6 reveals that,

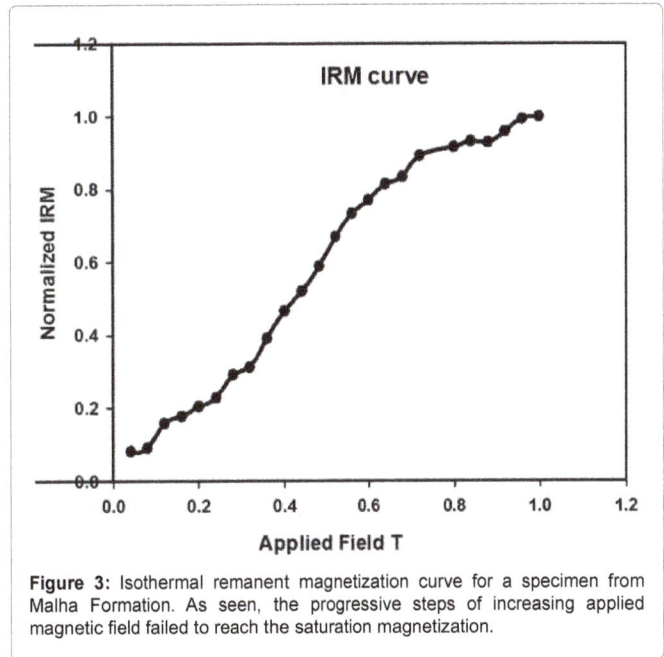

Figure 3: Isothermal remanent magnetization curve for a specimen from Malha Formation. As seen, the progressive steps of increasing applied magnetic field failed to reach the saturation magnetization.

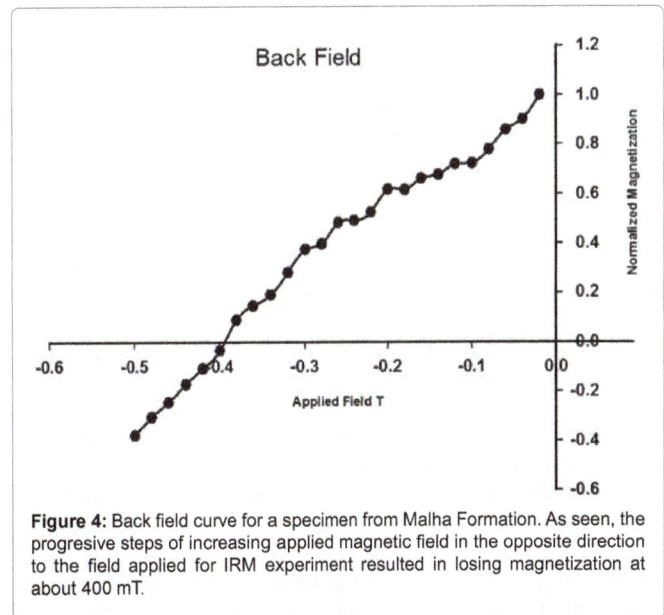

Figure 4: Back field curve for a specimen from Malha Formation. As seen, the progresive steps of increasing applied magnetic field in the opposite direction to the field applied for IRM experiment resulted in losing magnetization at about 400 mT.

Hysteresis Loop

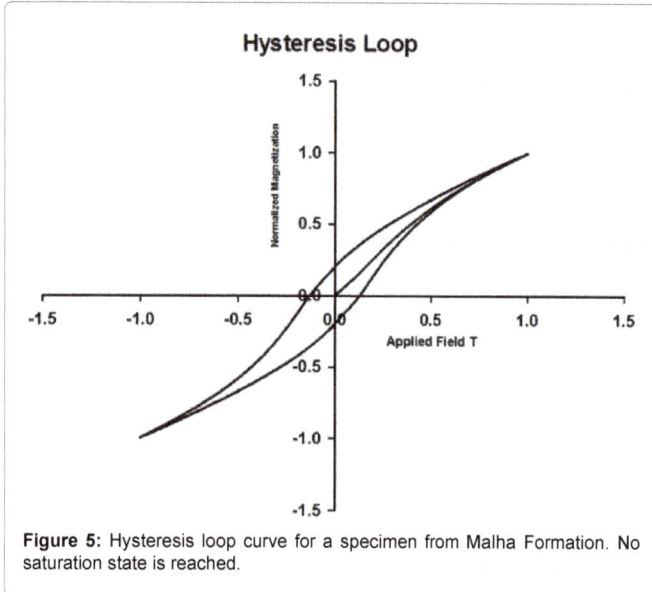

Figure 5: Hysteresis loop curve for a specimen from Malha Formation. No saturation state is reached.

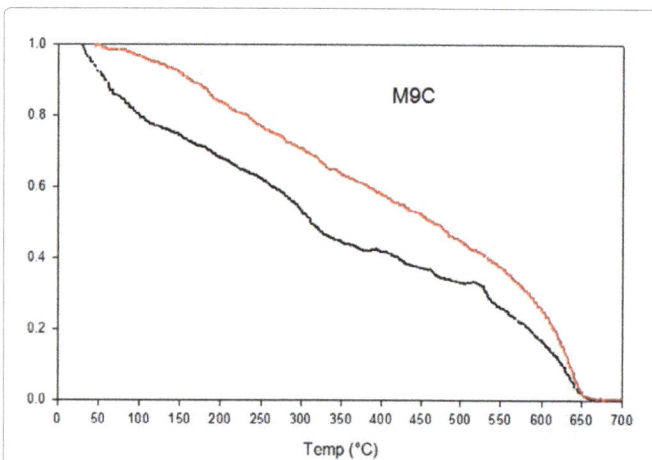

Figure 6: Thermomagnetic curve for a specimen from Malha Formation. The specimen lost magnetization at about 650°C, where the Y axis represents normalized magnetization.

the Curie the temperature value of the existed magnetic mineral is about 650°C. Such a high value of Curie temperature characterizes the hematite mineral. This mineral could carry a stable magnetization within the investigated specimens [20-25].

Demagnetization process relied mainly on the thermal demagnetization, as the AF demagnetization failed to isolate the characteristic magnetic component, due to the presence of hematite, as the main magnetic mineral. As expected in the study of sedimentary rocks, the intensity of remanence was low of around 2.0×10^{-3} A/m. Also, the values of magnetic susceptibility for the studied specimens were low around 5×10^{-6} SI units. As shown in Figure 7, the intensity of NRM decreased gradually during successive demagnetization steps till around 650°C, where about 90% of the remanence was removed. Zijderveld diagram showed that a straight line pointed toward the origin, indicating single component of stable magnetization. Also, an abrupt change in the susceptibility values was noticed, indicating the change in the mineralogy during demagnetization steps (Figure 7). The site-mean values of the obtained components that resulted from the demagnetization process of the studied specimens were calculated and

tabulated in Table 1. In the Geologic Time Scale Figure 8; the base of Aptian age is found to be reverse magnetization and this agree with results of the present work shown in Table 1 where the component of the site (M1, M2 and M3) are (-ve) at the base of Malha Formation and the upper three sites are of (+ve) components indicating that the Malha Formation started deposition at the very beginning of Aptian age [19].

The paleopole position obtained in this study, agrees with the

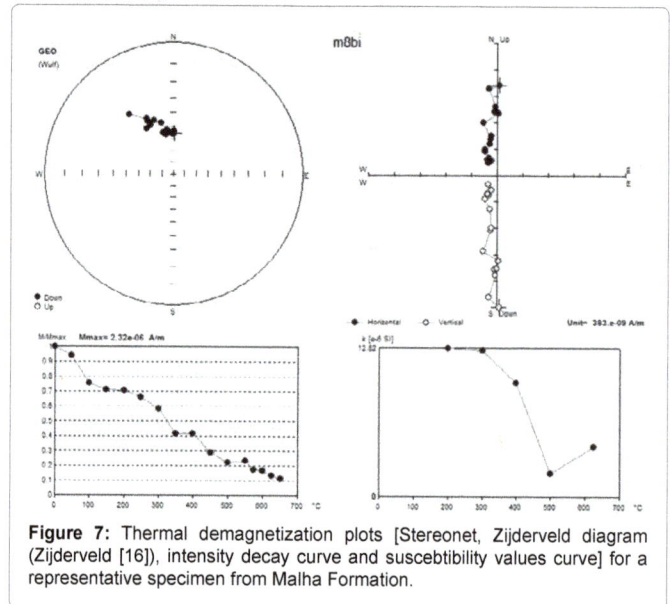

Figure 7: Thermal demagnetization plots [Stereonet, Zijderveld diagram (Zijderveld [16]), intensity decay curve and suscebtibility values curve] for a representative specimen from Malha Formation.

Site No	N	D (°)	I (°)	α95 (°)	k	Lat. (°N)	Long. (°E)
M1	11	170.3	-39	7.0	44.09	78.7	266.4
M2	10	173.7	-40.2	9.0	29.56	81.6	256.9
M3	14	172.6	-36.3	9.0	20.42	78.8	251.7
M4	12	358.9	41.5	7.2	37.33	84.6	224.1
M5	8	339	29.9	15.9	13.08	66.7	273.7
M6	10	348.5	26.6	7.7	40.79	71.5	250.9
Mean	65	351	36.2	3.7	23.84	77.8	257.4

Table 1: Paleomagnetic results of the Malha Formation.

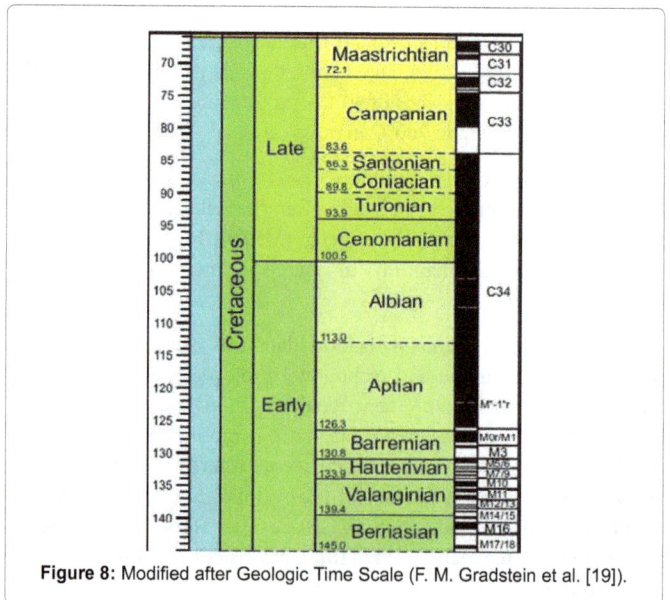

Figure 8: Modified after Geologic Time Scale (F. M. Gradstein et al. [19]).

Cretaceous paleomagnetic poles obtained in Egypt (Table 2 and Figure 9) [26-29].

Discussion and Conclusions

The present study is an attempt to throw some light on the age of the ferruginous sandstone successions, which is a controversial point in the study area. Our conclusion, based upon paleomagnetic investigations agrees with the Egyptian General Petroleum Corporation and Conoco [12], and contradicts with the term Triassic Qisaib Formation, which introduced in other studies by different authors [10,11]. The presence

Site No	Rock unit, locations	Age Ma	Lat. (°N)	Long. (E)	Reference
1	Ekma chalk, Sinai	66-98	40	226	Kafafy & Abdeldayem
2	Qabaliat Sandstone, Sinai	91-97	50	227	Kafafy & Abdeldayem
3	Nazzazat Tronian, Sinai	88-90	50	229	Kafafy & Abdeldayem
4	Qusier Trachytes	63-92	63	252	Ressetar, et al. [25]
5	East El Owienat volcanic	65-97	68	296	Hussain & Aziz [22]
6	El Kahfa Ring complex	74-85	61-	238	Abd El-Aal
7	Abu khrug Ring complex	87-91	59	266	Ressetar, et al. [25]
8	E. Aswan Nubian Sand Stone	45-135	75	203	El-Skazly & Krs [26]
9	Wadi Natash Nubian S.S	45-135	82	223	Schult et al. [20,27]
10	Aswan N.S.S.	Cretaceous	80	227	Schult et al. [20]
11	Aswan Iron &sandston	Cretaceous	75	2.03	El Shazly & Krs [26]
12	Abu Khruqe-El kahfa intrusion	72-98	65	249	Ressetar et al. [25]
13	Wadi Natash Volcanic	86-100	69	258	Schultet al. [27]
14	Qena, N.S.S.	70-145	76	265	Hassain et al. [21]
15	Wadi Natash S.S & volcanic	78-111	64	218	El Shazly & Krs [26]
16	Idfu-Marsa Alam N.S.S	Cretaceous	80	252	Schult, et al. [20]
17	Wadi Natash Intrusion	78-111	76	228	Ressetarel. Al [25]
18	East Owienat N.S.S	Cretaceous	77	258	Hussain & Aziz [22]
19	Abu Rawash sediments	~80	61	230	Kafafy el al.
20	N.s.s. G El Minisherah	Cretaceous	84	288	Ibrahim [28]
21	N.S.S. G. ElHalal	Cretaceous	78	288	Ibrahim [28]
22	N.S.S. Central Eastern Desert	Cretaceous	74	244	El-Hemaly et al. [23]
23	Gifata Sediments	Cretaceous	81	224	Saradeth et al. [29]
24	East El Oweinat syenite	Cretaceous	68	269	Hussain and Aziz
25	Phosphate and sandstone at Gabel Gifata	66-75	82	271	Abd El –All [4]
26	El-Naga Ring Complex	~140	69	268	Abd El-All [4]
27	Nubia S.S. W. Desert	113-124	78	294	El-Shayeb et al. [24]
28	Malha Formation	120-127	77.8	257.4	Present work

Table 2: Selected Cretaceous paleomagnetic poles of Egypt, with the pole of the present work.

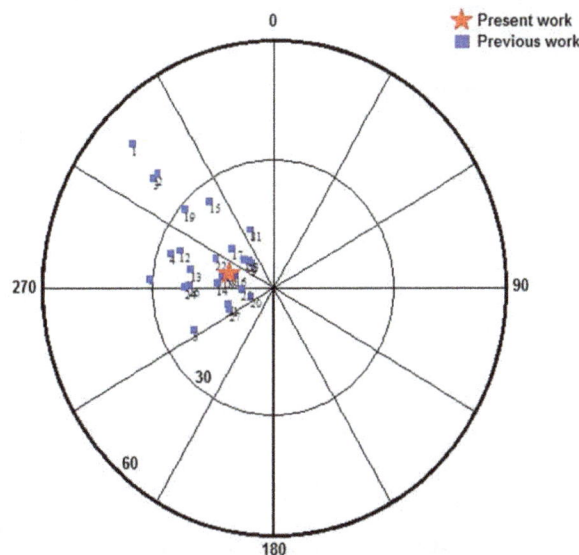

Figure 9: Previous obtained Cretaceous poles of Egypt with the pole of the present study.

of anti-parallel directions is an evidence for the primary origin of the resultant components, as well as the reversal of the geomagnetic field during the deposition of Malha Formation (Table 1 and Figure 8). To the knowledge of the authors; the present work is the first paleomagnetic study of Malha Formation in Sinai, so a comparison with relevant studies on Nubia Sandstone in other places in Egypt was done in Table 2 and Figure 9, such as: Schult et al. [20] studied some Nubia Sandstone rocks in Aswan, they obtained VGP lies at Lat.=80°N and Long.=227°E [Table 2, pole no. 10], Hassain et al. [21] carried out a paleomagnetic study of the Nubia Sandstone at Qena, They found that the VGP lies at Lat.=76°S and Long.=265°E [Table 2, pole no. 14], Hussein & Aziz [22] studied the paleomagnetism and magnetic mineralogy of Nubia Sandstone in East Owienat, The pole position obtained from that study lies at Lat.=77°N and Long.=258°E [Table 2, pole no. 18], in their paleomagnetic studies on Nubia Sandstone at Eastern Desest; El-Hemaly et al. [23] showed that the VGP lies at Lat.=74°N and Long.=244°E [Table 2, pole no. 22] and El-Shayeb et al. [24] studied the paleomagnetism of some Nubia Sandstone formations in the Western Desert and obtained pole position at Lat.=78°S and Long.=294°E [Table 2, pole no. 27]. The paleopole position obtained in this study, agrees with the Cretaceous paleomagnetic poles obtained in Egypt (Table 2 and Figure 9). The results show obviously the presence of a single and stable component as revealed from the demagnetization process [25-29].

References

1. Sultan IZ (1985) On the presence of Middle Jurassic miospores at Gebel El Iseila, west central Sinai, Egypt. Bulletin of Faculty of Science 25: 26-40.

2. Aboul Ela NM, Abdel Gawad GI, Saber SC (1990) Jurassic microflora of Gabal El Minshera, northern Sinai, Egypt. N Jb Geol 3: 129-140.

3. Kora M, El Beialy S (1989) Early Cretaceous palynomorphs from Gebel MusabaSalama area, southwestern Sinai, Egypt. Review of Palaeobotany and Palynology 58: 129-138.

4. Abd El-All EM (2004) Paleomagnetism and Rock Magnetismof El-Naga Ring Complex South Eastern Desert, Egypt. NRIAG J of Geophys 3: 1-25.

5. Abdallah AM, Adindani A, Fahmy N (1963) Stratigraphy of Lower Mesozoic rocks, western side of Gulf of Suez. Egyptian geological survey special papers 1: 23-27.

6. Abu-Zied RH (2007) Palaeoenvironmental significance of Early Cretaceous foraminifera from northern Sinai, Egypt. J cretres 28: 765-784.

7. Issawi B, Francis MH, Youssef EA, Osman RA (2009) The Phanerozoic Geology of Egypt: A Geodynamic Approach, Ministry of Petroleum, the Egyptian Mineral Resources Authority, Cairo, Egypt.

8. El Beialy SY, Martin JH, Atfy HS (2010) Palynology of the Mid-Cretaceous Malha and Galala formations, Gebel El Minshera, North Sinai, Egypt. Palaios 25: 517-526.

9. Abd-Elshafy E, Abd El-Azeam SA (2010) Paleogeographic relation of the Egyptian Northern Galala with the Tethys during the Cretaceous Period. J cretres 31: 291-303.

10. Geological Survey of Egypt (1994) Geological Map of Sinai, Egypt.

11. Moustafa AR (2004) Exploratory notes for the geologic maps of the eastern side of the Suez rift (western Sinai Peninsula), Egypt.

12. Egyptian General Petroleum Corporation and Conoco (1987) (Scale 1 : 500,000, NH 36 NE South Sinai).

13. Schrank E, Mohmoud MS (1996) Palynology (pollen, spores and dinoflagellates) and Cretaceous stratigraphy of the Dakhla Oasis, central Egypt. Journal of African Earth Sciences 26: 167-I93.

14. Shata AA (1982) Hydrogeology of the Great Nubian Sandstone basin, Egypt. Q J Eng Geol 15: 127-133.

15. Said R (1990) The Geology of Egypt. Balkema Rotterdam: 1-721.

16. Zijderveld J (1967) "A.C. demagnetization of Rocks: analysis of results". In: Methods in paleomagnetism. Elsevier Amsterdam, New York.

17. Kirschvink J (1980) "The least squares line and plane and analysis of "Paleomagnetic data". Geophys J Astr Soc 62: 699-718.

18. Fisher RA (1953) "Dispersion on a sphere". Proc R Soc London Ser A 217: 295-305.

19. Gradstein FM, Ogg JG, Schmitz M, Ogg G (2012) Geologic Time Scale, published by Elsevier, and the website of the Geologic TimeScale Foundation.

20. Schult A, Soffel H, Hussain AG (1978) Paleomagnetism of Cretaceous Nubian Sandstone. J Geophy 44: 333-340.

21. Hussain AG, Schult A, Fahim M (1976) Magnetization of the Nubian Sandstone in Aswan area, Qena-Safaga and Idfu-Mersa Alam districts. Helwan Observ Bull No 133: 1-13.

22. Hussain AG, Aziz Y (1983) Paleomagnetism of Mesozoic and Tertiary rocks from East El Weinat area south Egypt. J Geophysics Res 88: 3523-3529.

23. El-Hemaly IA, Odah HH, Abd El-All EM (2004) Magnetization of the Upper Cretaceous Nubian Sandstone, Eastern Desert, Egypt. Delta Journal of Science 288: 87-97.

24. El-Shayeb H, El-Hemaly IA, Abdel Aal EM, Saleh A, Khashaba A, et al. (2013) Magnetization of three Nubia Sandstone formations from Central Western Desert of Egypt. J nrjag 2: 77-87.

25. Ressetar R, Nairn AEM, Monrad JR (1981) Two phases of Cretaceous-Tertiary magmatism in the Eastern Desert of Egypt: paleomagnetic, chemical and K-Ar evidence. Tectonphysics 73: 169-193.

26. El-Shazly EM, Krs M (1973) Palaeogeography and Palaeomagnetism of the Nubian sandstone, Eastern Desert of Egypt. Geol Rundsch 62: 212-225.

27. Schult A, Hussain AG, Soffel H (1981) Paleomagnetism of upper Cretaceous volcanics and Nubian Sandstones of Wadi Natash, SE Egypt and implications for the polar wander path for Africa in the Mesozoic. J Geophy 50: 16-22.

28. Ibrahim EH (1993) Paleomagnetism of some Phanerozoic rocks from the Nile Valley, Egypt. Ph. D Thesis, Fac Sci, Mansoura Univ, Egypt.

29. Saradeth S, Soffel HC, Horn P, Muller-Sohnius D, Schult A (1989) Upper Proterezoic and Phanerozoic pole positions and potassium-argon (K-Ar) ages from the East Sahara craton. Geophys J 97: 209-221.

7

Seismic Refraction Method to Study Subsoil Structure

Fkirin MA[1], Badawy S[1] and El deery MF[2]*

[1]*Electronic Engineering, Department of Industrial and Control Engineering, Menoufia University, Egypt*
[2]*Hanwha Chemical Research and Development Center, Tukh, Egypt*

Abstract

Seismic waves are used in many fields. It can be applied on determine subsoil structure, and materials. In this work use seismograph for the measurement seismic wave propagation velocity in the real geological (subsoil layers) medium. These investigations are applied in petroleum research institute Egypt. Predication results of subsoil layers thickness and seismic wave's velocity analysis are obtained. The seismograph recorded receiving sample data from geophones and by "SeisImager" software is extract the final seismogram. The seismograph is dependent on Snell law for wave's propagation. The obtain result of P and S waves of this work are the same as P and S waves references. P waves are shake ground in the direction they are propagating (longitudinal waves), and S waves are shake perpendicularly or transverse to the direction of propagation.

Keywords: Seismic waves; Seismograph; Usage; Data processing; Layers; Layers samples analysis

Introduction

The propose work are determine subsoil layers and measure P and S-waves velocity. The propose applied experimental work is implemented in petroleum research institute zone Egypt, the experimental location between latitudes 30-02-42.13N and longitudes 31-20-27.66E as shown in Figure 1. The seismic wave's propagations are Unique; they are used in many fields. Features of the seismic reflection and refraction waves applicable to many fields [1,2] these fields are subsoil structure, carbonate layer, gas reservoir, and groundwater. Seismic wave generated under the action of the short pulse is a complex wave that consists of the following components as shown in Figure 1. These components are longitudinal compressive P-wave, Transverse S-wave, and Rayleigh-Surface R-wave. Longitudinal P-waves and transverse S-waves are known as the body waves. Body waves are propagating through the medium by means of the hemispherical wave front. The type of the component being considered depends on the source of vibrations. Rayleigh wave which is propagated radially and has the cylinder-like wave front. It is appears simultaneously with the body waves. Displacement of the ground is the vertical direction.

System model

The seismology descript as block diagram as shown in Figure 2. All block has main function to extract final seismogram. In short, we are talking about all function block.

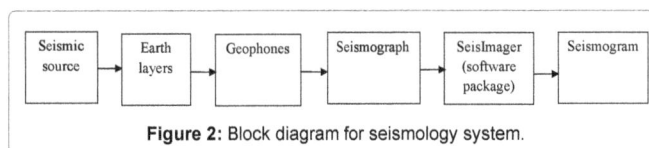

Figure 1: Propagation of the surface and body waves (A is the point of the excitation) [1]. The seismic source locates at first geophone. The distance from first geophone to last geophone is 46 m. Seismic source is manual hummer as in first block. Geophones are traditional geophones as in third block.

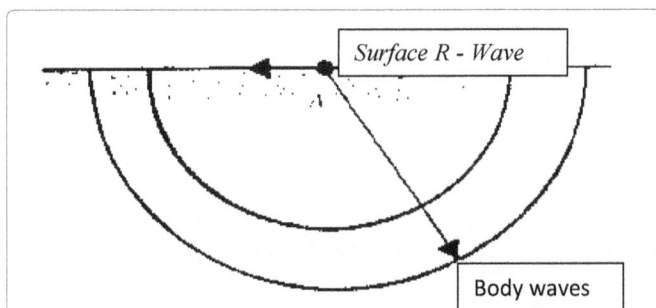

Figure 2: Block diagram for seismology system.

First block is seismic sources

Seismic sources are vibration sources. It generates sound waves. These waves penetrate the earth layer. The seismic sources must be strong vibrator. There are many types of seismic sources. Seismic sources on land and seismic sources on water, but the most often used sources for land surveys are vibrator trucks and for marine surveys are air guns. In this investigation used manual hummer. It is weight 5 Kg. From Newton second law F=ma, It is expressed in Newton (N) or Kg m/s^2 where m is mass and a is acceleration of mass in x direction. This acceleration from personal thumps the earth layers. For three dimension F=m (ax+ay+az).

Second block is earth layers

All material in subsoil layers make layer. These layers have different thickness. Seismic waves penetrate the subsoil layers has low frequency, high amplitude and high wavelength. The P and S waves equation as following [2,3].

$$V_p = \sqrt{\frac{K + \frac{4}{3}\mu}{ñ}}$$

$$V_s = \sqrt{\frac{\mu}{\rho}}$$

Where elastic constants are k=bulk modulus, μ=shear modulus and ρ=density of material (subsoil layers). The seismic waves speed (V_p and V_s) changed from layer to another dependent to density and porous

---

***Corresponding author:** El deery MF, Senior Engineer, Hanwha Chemical Research and Development Center, Tukh, Egypt
E-mail: engmohamed.eldeery@gmail.com

of the rock layers, some reading value of V_p and V_s in sandstone and carbonate. Obtain P and S waves by applied these equations.

An applied a force to real object causes some deformation of the object i.e. change of its shape. If the deformation is negligibly small this body considers a rigid body. The rigid body retains a fixed shape under all conditions of applied forces (original shape). If the deformations are not negligible, we have to consider the ability of an object to undergo the deformation i.e. its elasticity, viscosity or plasticity [4]. The refraction or angular deviations that sound rays (seismic pulse) undergoes when passing from one material to another depends upon the ratio of the transmission velocities of the two materials. The fundamental law that describes the refraction of sound rays is Snell's Law and this together with the phenomenon of critical incidence is the physical foundation of seismic refraction surveys p-waves [5]. Snell's Law and critical incidence are shown in Figure 3A, velocity V_1 underlain by a medium with a higher velocity V_2 incidence are shown in Figure 3A.

The layers thickness by applied Snell's [5] as shown in Figure 3, $\frac{V1}{V2} = \frac{\sin \alpha}{\sin \beta}$ at $\beta = 90$ this is critical angle then the rays totally refracted

$$X = AB + BC + CD \quad V1 = \frac{Z1}{T1}, \quad V2 = \frac{Z2}{T2}.$$

$$Ttotall = \frac{AB + CD}{V1} + \frac{BC}{V2},$$

At Boundary between the two layers then the equation become $\frac{V1}{V2} = \sin \alpha$ from Figure 3B Z_1 and Z_2 are thickness of layers 1 and 2. AB from source to layer 2 then the rays totally refraction from B to C the rays reflection from C to D. AB=CD and

$$AB + CD = \frac{2Z1}{\cos\alpha}, \quad BC = X - 2CD.$$
$$CD = Z1\tan\alpha$$
$$T_{total} = 2\left(\frac{Z1}{V1\cos\alpha} + \frac{X - Z1\tan\alpha}{V2}\right) + \frac{X}{V2},$$

$$T_{total} = 2Z\left(\frac{1}{V1\cos\alpha} + \frac{\sin\alpha}{V2}\right) + \frac{X}{V2},$$

$$T_{total} = 2Z1\left(\frac{V2 - V1\cos\alpha\tan\alpha}{V1V2\cos\alpha}\right) + \frac{X}{V2}$$

For $\sin\alpha = \frac{V1}{V2}$, then

$$T_{total} = 2Z1V1\left(\frac{\frac{1}{\sin\alpha} - \sin\alpha}{V1V2\sin\alpha\cos\alpha}\right) + \frac{X}{V2},$$

Then

$$T_{total} = \frac{2Z1\cos\alpha}{V1} + \frac{X}{V2}$$

$$Z1 = \frac{TiV1}{2\cos\alpha}$$

As shown in Figure 4 multi layers but in our study we applied on two layers.

Third block is conventional geophones

Geophones are instrument to sense and measure velocity seismic waves. There are two types of geophone based on application, on land or on water, it is called hydrophone. Most of the geophones are based on the principle of moving coil as shown in figure. The Figure 5 shows geophone structure. It is composed of voice coil suspended inside permanent magnetic. The voice coil is suspended by spring and it is free movement. The voice coil can move up and down inside the magnetic field and produce a voltage Vg. Total spread geophones in this study as shown in Figure 6. Geophones are distributed in horizontal line. Reflected and refraction signals are received by geophones and transmitted from geophone to seismograph by a spread cable. The total geophones distance from first to last geophone (Figure 6) should be 3 to 5 times the depth of interest (depth recommended of seismograph).

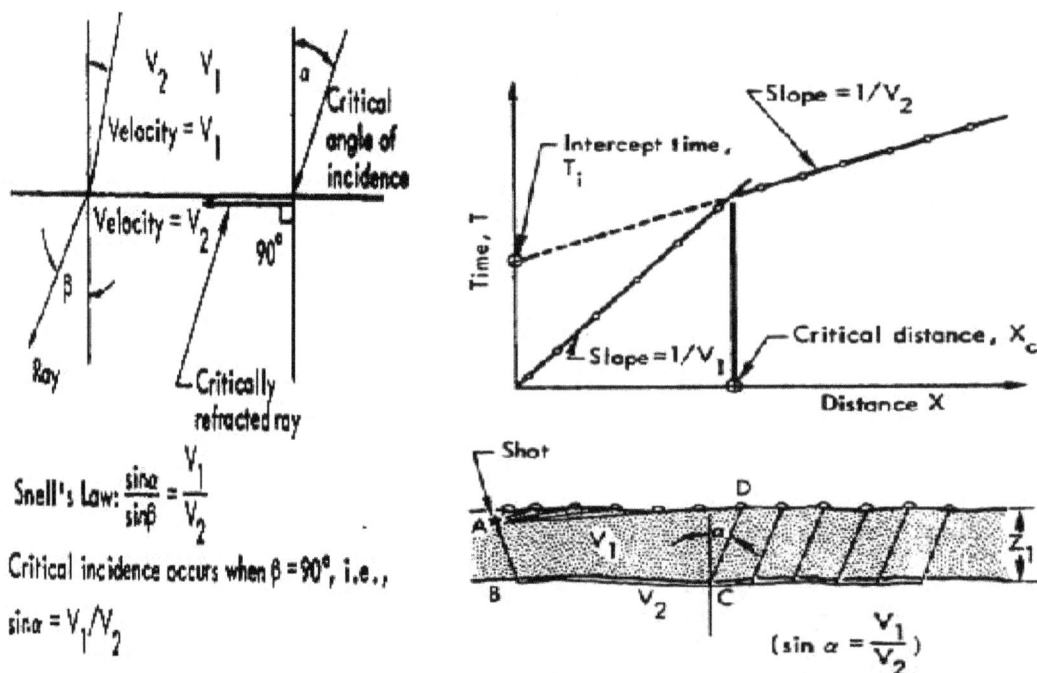

Figure 3: A) Shows the Snell's law reflect and refract rays. B) Shows the rays at critical angle.

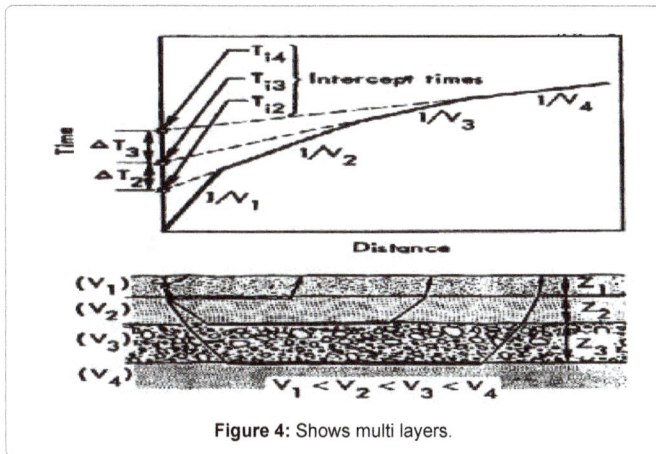

Figure 4: Shows multi layers.

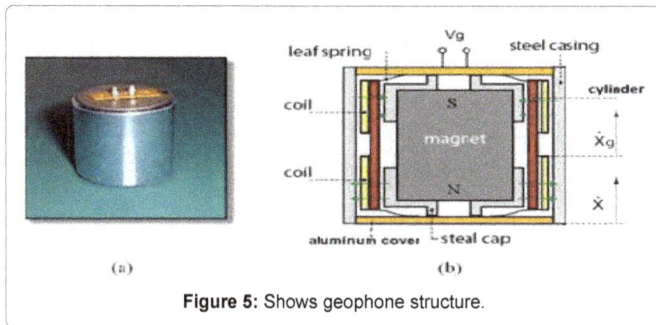

Figure 5: Shows geophone structure.

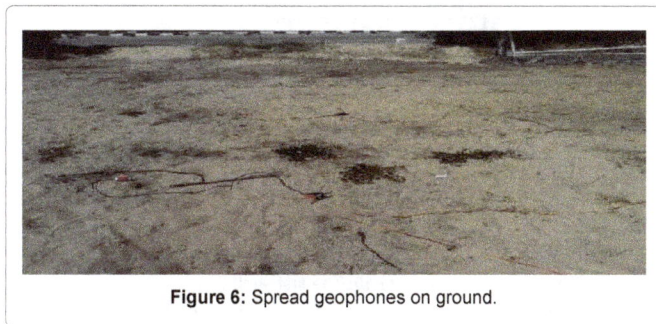

Figure 6: Spread geophones on ground.

In this experiment is 46 m. Seismic source at a minimum there should be two shots located at either began and end of line. It is best practice to also have one center shot (S3) so in our experiment we made three shot (S1, S2 and S3) [5].

Fourth block is seismograph

Seismograph generally consists of sensors (geophones), a low-pass anti-aliasing filter, analog to digital (A/D) converter, and a recorder. Modern digital seismographs are complicated by the extensive electronic circuitry involved [6]. Seismograph used in this experiment as shown in Figure 7.

Seismograph model

OYO McSeis-SX24, the seismography McSEIS-SX is a portable and it have a 24 channel for a 24 geophones to refraction exploration downhole P-S velocity logging and crosshole seismic for engineering and construction. This is also used as a data acquisition system for multi-channel surface wave analysis. The system is compact, light in weight to transport and do the job with a smaller 12VDC battery

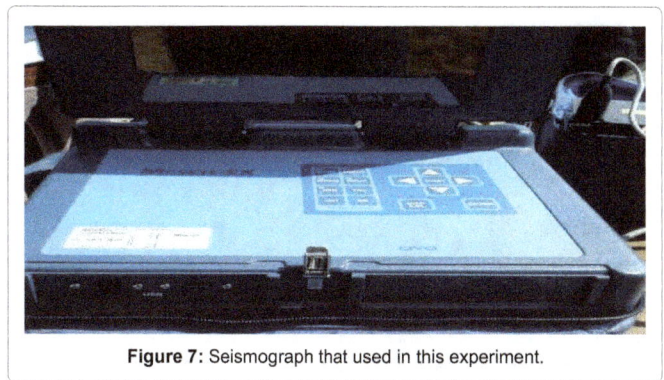

Figure 7: Seismograph that used in this experiment.

anywhere. It's based on the Windows XP SP2 professional with XGA/TFT colour display, hard disk drive, USB2.0 ports with higher quality and reliable field performance [7].

The seismic waves transmitted from seismic source impulses subsoil layers, and received by geophones. The impulses transmit from seismic source to subsoil layers from point to another under the influence of an external impulse. Particles of matter in the subsoil layers move from seismic source positions toward each other and collide, and thereby transmit mechanical motion subsoil layers from one point to another. Particles of matter in the subsoil layers begin to vibrate in the direction of seismic wave propagation. The elastic waves propagate and transmit mechanical energy from one point to another as shown in Figure 8. The seismograph represented an oscilloscope in Figure 8. Times elapsed from sending to receiving a seismic wave depend on depths of studied structures and velocities of propagation of seismic waves in medium.

The seismograph setting as following below [7]

1. Sample interval: 0.125 to 0.25 mill second (over-sampling is fine) in our experiment is 200 micro second.

2. Data length: 0.25 to 8 k (should be long enough to capture distant arrivals) in our experiment is 4 k.

3. Stacking as needed to increase signal to noise ratio, 5 to 10 times in our experiment is auto (automatic).

4. Delay: -10 ms allows the first break on the near geophones to be more easily viewed in our experiment is zero.

5. Acquisition filters acquisition filters are not recommended because effect is irreversible; should be carefully applied to filter signal you are certain you will never want such as 60 Hz power line noise.

6. Preamp gains highest setting in our experiment are gain 1: set and gain 2: set.

7. Display gains: Fixed gain (same gain over time for a given trace, but variable from trace to trace; traces far from the source will need a higher gain setting than those that are near).

The seismograph settings above are set and now the seismograph is already to receive. To initiate mechanical force explosives hammer. Successful application of seismic methods is based on the fact that subsoil layers have different elastic properties and density that directly depend on their lithological composition [8,9]. Analysis of thus changed properties of seismic waves therefore allows determining the tectonic structure of the subsoil layers, lithological composition of subsoil layers strata, and in favorable circumstances also directly locating reservoirs of oil and gas.

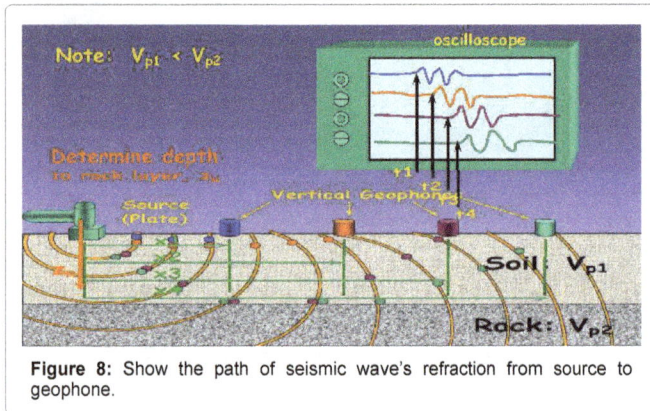

Figure 8: Show the path of seismic wave's refraction from source to geophone.

Analyze waveform file of the first shot as shown in Figure 9. Quality is little prefirst break noise, the first breaks are obvious. When there is one refractions break in slope indicates there are another layers. The crossover distance is Break in slope at 5 traces [8].

The first shot, the recoded and display in seismograph is shown in Figure 10. The figure show data collected by 24 geophones. The deviation from geophone to other represented the distance between geophone to another. The geophones signal as stairway left direction. The first geophone is the first signals arrive to seismograph, and so on to last geophone is the last signals arrive to seismograph.

The second shot seismic source located between geophone number 12 and 13 (at middle distance=24 m). In this case the signals from all geophones as triangle shape the head of this triangle at middle. The first geophones reading seismic signal at geophones number 12 and 13, and so on to geophones number 1 and 24, Figure 9 shows the geophones signal.

From Figure 11 it is very clarification the shot at middle distance. Because the time for arriving signals for all two geophones symmetric is constant, so there is symmetric shown in figure.

Third shot in third shot the signals are at variance in first shot signals. The seismic source at last geophone (geophone number 24), so the Figure 10 is at variance in Figure 12. The geophone number 24 is the first reading signals and so on until geophone number one. The geophones signal as stairway right direction. Third shot as shown in Figure 10.

Fifth block is software package (seisImager/2D) [10]. The data that collected in seismograph export to P.C. the program make process on this data. The first processing on data, Set the display gain so the first breaks are clearly visible.

Determine the first arriving time. The data consists of a travel time T (sec) for each geophone location X (m). Data used to plot travel time curve. It is use to determine the first arriving time as shown in Figure 13. In this experiment determine by manual technique.

Results and Discussion

After determine first arriving time; use time traveling curves to identify subsoil layers. This technique applied by the GRM method according to Palmer (Generalized Reciprocal Method, Palmer, 1981 and 1991). Dromocrones of longitudinal waves are obtaining by analyzing first arrivals of elastic waves (Figure 13).

If we read the geophones data and plot this data we find there is break down, this break down indicate the wave path through another

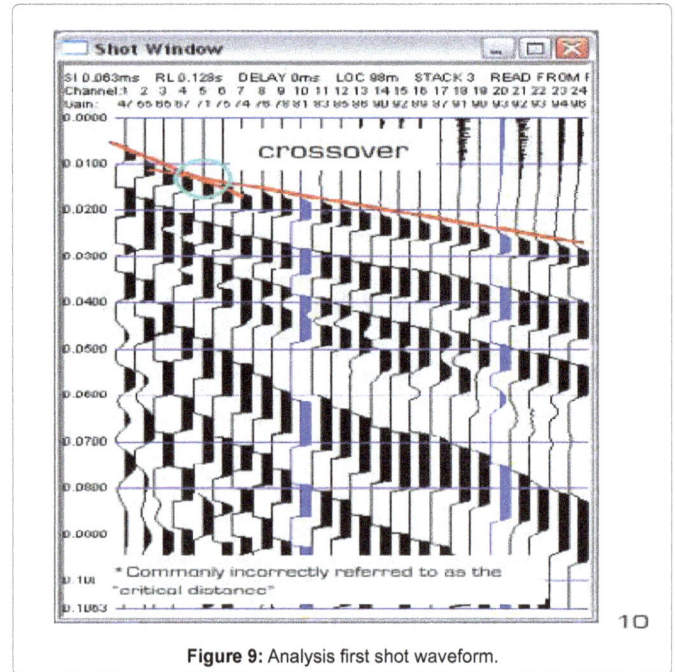

Figure 9: Analysis first shot waveform.

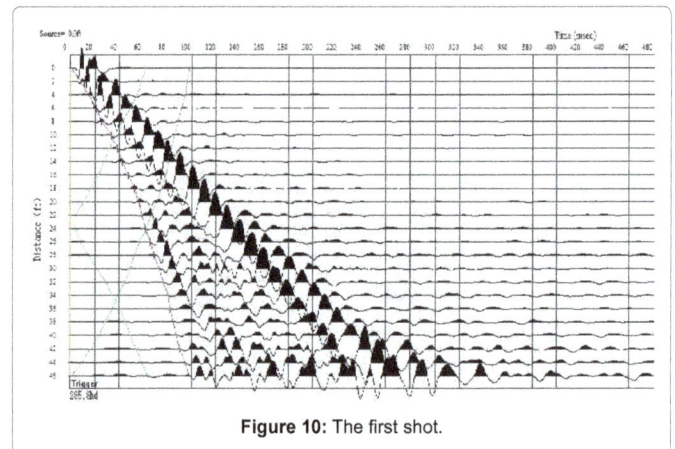

Figure 10: The first shot.

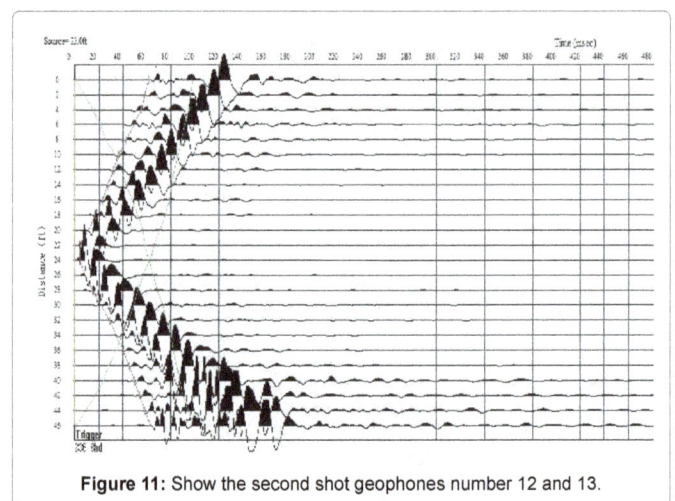

Figure 11: Show the second shot geophones number 12 and 13.

layer as shown in Figure 14 and by using the generalized reciprocal method (GRM) is a technique for delineating undulating refractors at

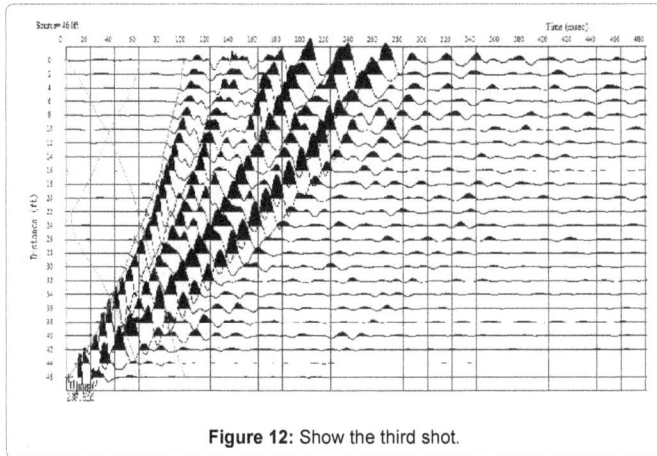

Figure 12: Show the third shot.

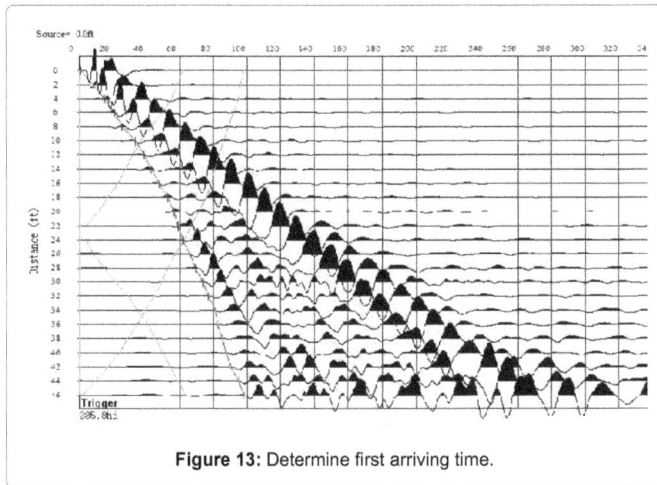

Figure 13: Determine first arriving time.

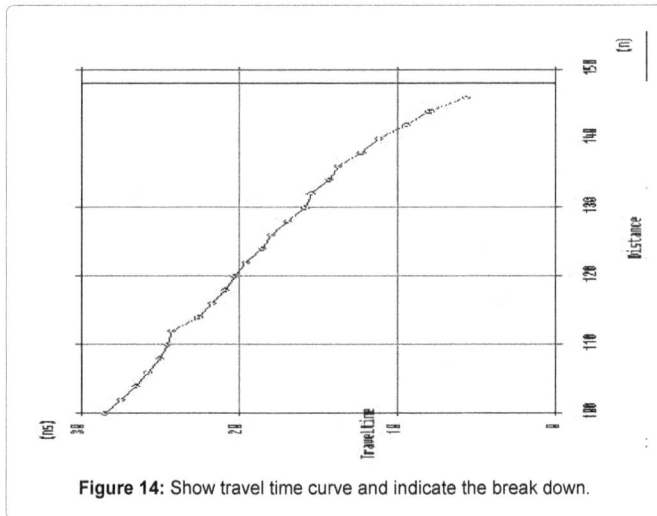

Figure 14: Show travel time curve and indicate the break down.

as producing gross smoothing of irregular refractor topography. The depth conversion factor is relatively insensitive to dip angles up to about 20 degrees. As a result, depth calculations to an undulating refractor are particularly convenient even when the overlying strata have velocity gradients. The GRM provides a means of recognizing and accommodating undetected layers provided an optimum XY value can be recovered from the travel time data. The presence of undetected layers can be inferred when the observed optimum XY value differs from the XY value calculated from the computed depth section. The undetected layers can be accommodated by using an average velocity based on the optimum XY value. This average velocity permits accurate depth calculations with commonly encountered velocity contrasts. Reciprocal times are about equal (good quality check). The forward and reverse shots are symmetrical about the center indicating there is little or no dip on refractor. The Generalized Reciprocal Method (GRM) as shown in Figure 15. The GRM method is necessary because dependent on absolute time to inverts from time depth (ms) to depth (m) this method is very sensitive to thickness of slow superficial soil layers [12]. All profiles were interpreted as a two layers configuration. The first layer with velocity of 300-400 m/s corresponds to the superficial soils. The next layer 550-850 m/s to further near surface of mixes sand, these results refer to Table 1 are considered the same results. The same velocities for all layers. Time term inversion gives a quick solution for 2 to 3 layer cases with evident breaks in slope. After we determine the travel time curve the program plot the layers module result. The layers module result plot the detecting layers and velocity in all layers as shown in Figure 16.

Sixth block is seismogram. From the figure show there are two layers the first layer have thickness almost 3 m. It is as surface layer,

any depth from in line seismic refraction data consisting of forward and reverse travel times [11]. The travel times are used in refractor velocity analysis and time depth calculations.

This results of refractor velocity analysis being the simplest and the time depths showing the most detail. In contrast conventional reciprocal method which has XY equal to zero is especially prone to produce numerous fictitious refractor velocity changes. As well

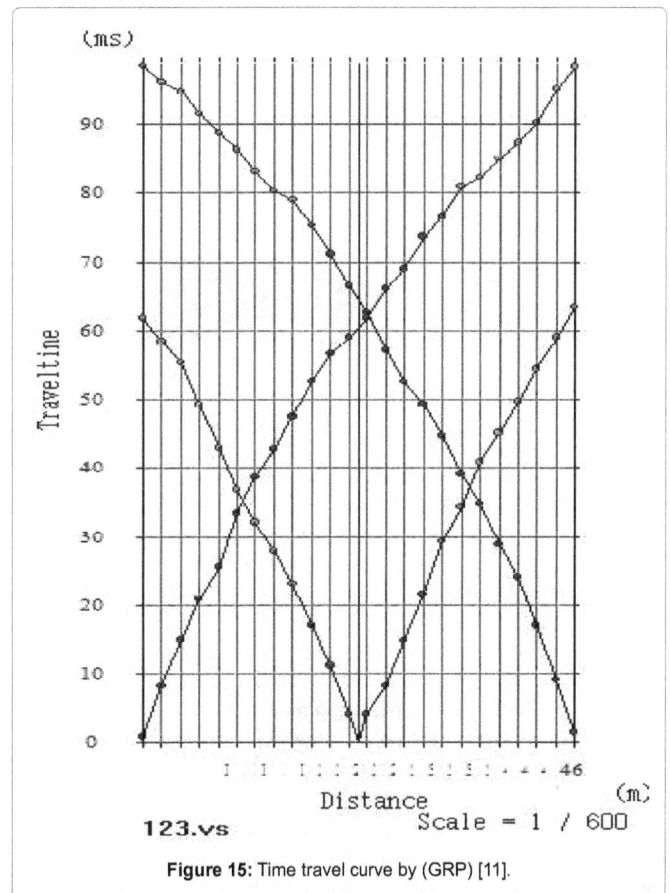

Figure 15: Time travel curve by (GRP) [11].

Material	Velocity Vp m/s
1- Unconsolidated materials	
Sand dry	200-1000
Sand (water-saturated)	1500-2000
Clay	1000-2500
Glacial till (water-saturated)	1500-2500
Permafrost	3500-4000
2-Sedimentary rocks	
Sandstones	2000-6000
Tertiary sandstone	2000-2500
Pennant sandstone (Carboniferous)	4000-4500
Cambrian quartzite	5500-6000
Limestone	2500-6000
Cretaceous chalk	2000-2500
Jurassic oolites and bioclastic limestones	3000-4000
Carboniferous limestone	5000-5500
Dolomites	2500-6000
Salt	4500-5000
Anhydrite	4500-6500
Gypsum	2500-3500
3-Igneous/Metamorphic rocks	
Granite	5500-6000
Gabbro	6500-7000
Ultramafic rocks	7500-8000
Serpentinite	5500-6500
4-Pore fluids	
Air	300
Water	1400-1500
Ice	3400
Petroleum	1300-1400
5-Other material	
Steel	6100
Iron	5800
Aluminum	6600
Concrete	3600

Table 1: Compressional wave velocities in Earth materials Vp [14,15].

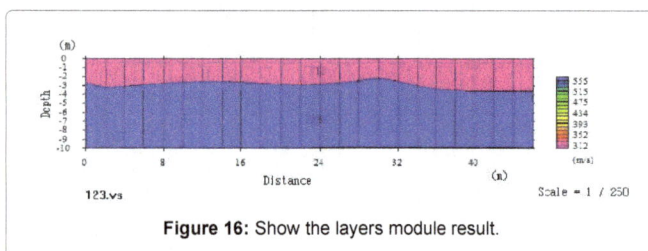

Figure 16: Show the layers module result.

and P-waves velocity in this layer is 312 m/s. the second layer have thickness almost 7 m, and P-waves velocity in this layer is 558 m/s [13]. The velocity the subsoil layers give indication about the subsoil layers formations contain, such as sandstone, carbonate [14], gas and oil reservoir and other material. For the Table 1 we have same reading of P-waves velocity for many formations material, the P and S waves velocity (Table 1) are the same velocity in our investigations. S-waves measure by planting horizontal geophone in seismograph studies. It calculated by S-waves equation, it equal 0.7 P-waves velocity [15-17].

Conclusion

Seismology is curtailment as system model. This system model consists of blocks diagram. All blocks have main function. Seismic refraction techniques used in exploration for minerals, subsoil layers petroleum, and subsoil layers water. This study obtained by seismograph. Seismograph is dependent on refraction P-waves. In this work the predicted layers are determined the first layer velocity (p-wave) is 300-400 m/s corresponds to the superficial soils, and velocity in second layer 550-850 m/s to further near surface of mixes sand. P-waves velocity increase with depth increase, Due to increase porosity, density and wetness layers. The seismograph results indicate that the seismic wave's velocities are equal seismic waves velocity obtains by seismic waves model.

References

1. Vlastimir DP, Zoran SV (1998) Measurement of the Seismic Waves Propagation Velocity in the Real Medium. Scientific J Facta Univers 1: 63-73.

2. Brian HR, Tad S (2007) The relationship between dry rock bulk modulus and porosity-An empirical study. CREWES Research Report 19.

3. Ross JA, Jenniferm MJ, Hans JR, Sergio S (2009) Elasticity measurements on minerals: a review. Eur J Mineral 21: 525-550.

4. Introduction to Theoretical Seismology (2006) Bratislava.

5. Bruce BR (1973) Seismic Refraction Exploration for Engineering Site Investigations. Springfield, Virginia 22151.

6. Won-Young K (1995) Instrument Responses of Digital Seismographs at Borovoye, Kazakhstan by Inversion of Transient Calibration Pulses. Lamont-Doherty Earth Observatory of Columbia University Palisades, NY 10964.

7. McSEIS-SX, 24ch Portable Engineering Seismograph.

8. Salman ZK, Al-Awsi MDh (2014) Investigation of Subsurface Structures by Using Seismic Refraction Method, Northwestern of Al-Anbar Governorate / Iraq 10: 112-121.

9. Don WS, Richard DM (1990) Seismic Reflection Methods Applied to Engineering, Environmental, and Groundwater Problems.

10. SeisImager Manual Version 3.0 (2003) PickWin v. 2.84, Plotrefa v. 2.66.

11. Derecke P (1981) An Introduction to the generalized reciprocal method of seismic refraction interpretation. Geophysics 46: 1508-1518.

12. Aziman M, Hazreek ZAM, Azhar ATS, Haimi DS (2016) Compressive and Shear Wave Velocity Profiles using Seismic Refraction Technique. J Phys Conf Ser 710.

13. Horrent C, Brouyere S, Demanet D, Michiels T, Jongmans D (1997) Seismic refraction in bernburg, Germany: application of generalized reciprocal and phantoming method. Geo Soc Lon Eng Geo Spec Pub 12: 407-411.

14. Mohamed AK, Andreas W (2015) Study on P-wave and S-wave velocity in dry and wet sandstones of Tushka region. Egypt J Pet 24: 1-11.

15. Salman ZK, AL-Khersan EH, AL-kashan AZ (2006) P and S-Waves evaluation for engineering site investigation at a hostel complex inside Basrah University, Southern Iraq 24: 27-37.

16. Uyanik O (2010) Compressional and shear-wave velocity measurements in unconsolidated top-soil and comparison of the results. Int J Phys Sci 5: 1034-1039.

17. Williams RA, Stephenson WJ, Odum JK, Worley DM (2005) P- and S-wave Seismic Reflection and Refraction Measurements at CCOC.

8

Roadside Air Pollutants along Elected Roads in Nairobi City, Kenya

Shilenje ZW[1*], Thiong'o K[1], Ongoma V[2,3], Philip SO[1] Nguru P[4] and Ondimu K[4]

[1]Kenya Meteorological Department, Nairobi, Kenya
[2]College of Atmospheric Science, Nanjing University of Information Science and Technology, Nanjing, Jiangsu, P.R. China
[3]Department of Meteorology, South Eastern Kenya University, Kitui – Kenya
[4]National Environment Management Authority, P.O. Box 67839-00200, Nairobi – Kenya

Abstract

This paper presents a statistical analysis of air quality monitoring in Nairobi city, at three major roads and Industrial Area, a site closer to the main industrial activities. The study was carried out using different gas analyzers and samplers. From the statistical analysis it was found that, there were extremely high values of black carbon which went beyond the upper limit of the instruments (50,000 ŋg/m³) during the day on Ladhis road. Nakumatt Junction site recorded extreme values of Black carbon (14,008 ŋg/m³) in the evening hours, while at Pangani Roundabout site, the diurnal mean value was extreme (14,446.5 ŋg/m³) for the period. None of the four sites exceeded the WHO 24 h limit for both PM_{10} (50 µg/m³) and $PM_{2.5}$ (25 µg/m³). The 24 h mean values of PM_{10} in the three sites also did not exceed the ambient air quality tolerance Kenyan limit of 100 µg/Nm³ and 150 µg/Nm³ in industrial area.

The diurnal mean of SO_2 over the four sites was generally low with the highest amount of 1.08 ppb recorded at Pangani Roundabout. This amount is far much below the diurnal WHO and Kenyan limit of 10 ppb and 48 ppb respectively. The global background concentration of carbon monoxide ranges between 0.05-0.12 ppm. The mean 24 h amount of CO in all the sites was above the background concentration, with Pangani Roundabout recording the highest amount of 1.73 ppm. The eight h means for ozone in all the sites were below WHO limit of 51 ppb with the highest amount of 20.2 ppb recorded in industrial area.

Keywords: Ambient air quality; Air pollutants; Ecosystem; Amelioration

Introduction

With the rapid accumulation of vehicular flow on Nairobi roads and the City's economic development, the City's environment is placed under increasing pressure due to air pollutants [1-3]. Public and private transportation sector is important from an environmental and ecosystem point of view; it uses energy, particularly fossils fuels which are a source of air pollutants when ingested and processed. The sector has long been among the main topics of environmental attention in the industrialized countries and in the recent past the issue of air pollution has taken priority in developing countries such as China, Indonesia, and parts of West Africa [2-5]. Knippertz et al. [4] describes the work undertaken by the Dynamics-aerosol-chemistry-cloud interactions in the West Africa project (www.dacciwa.eu) that is set to investigate the influence of anthropogenic and natural emissions on the atmospheric composition over South West Africa and to assess their impact on human and ecosystem health and agricultural productivity.

Road transport emissions load ambient air with attendant pollutants. Air pollution also comes from production of goods and services, heating of houses, and rising particulates from dusty roads and pavements. Time spend on travelling on the roads are long due to traffics build up and jams. For instance, according to Boman and Thynell [2] the average travel time to work in Nairobi is estimated at about an h in traffic, but could have since gone up. Ambient air pollution is closely related to the health of people [6,7]. Air pollution too, has impacts on other aspects of the economy such as agriculture, and contributes to greenhouse gas emissions. Typical values for the vehicle share in total air pollution range from about 40% to 90% for carbon monoxide, hydrocarbons and nitrogen oxides and are somewhat lower for fine particulate matter. Nairobi City is an important part of Kenya's economy and a regional hub on many fronts and has been developing rapidly in all aspects including population density that stands at about 4 million persons [8]. This means that the city holds about 10% of the total population of Kenya. With a population density of about 4000 persons per square kilometer, the city's air pollution levels, atmospheric and environmental issues affect many persons.

Vehicular traffic is an important source of particulate pollution in cities of the developing world, where vehicle population grows fast [3], coupled with a lack of an effective public transport policy, land use planning, resulting in harmful levels of pollutants in the air near major roads. Other factors that aggravate air pollution levels in the city include; uncontrolled or lack a proper urban integrated development plan, allowing into city of transit vehicles that would otherwise not need enter the city, exponential rise in the number of private cars and lack of a credible alternative rail transport system.

Technical support for air pollution control and air quality amelioration must be provided. The dynamic changes in air quality, especially due to vehicular flow, must be systematically monitored and analyzed, especially due increasing urban population, less regulated traffic activities, mechanical state of vehicles on the roads. This study specially examines and assesses the state of emissions on the road corridor under consideration; Ngong Road, Landhis Road, Pangani intersection and the area around industries in industrial area. In recent years, research on atmospheric environmental change focused on compounds of Nitrogen (NO_2, NO, NOx, and NH_4), compounds

*Corresponding author: Shilenje ZW, Kenya Meteorological Department, P. O. Box 30259 – 00100, Nairobi, Kenya, E-mail: zablonweku@yahoo.com

of sulphur (SO_2, H_2S), carbon dioxide, carbon monoxide and total suspended particulate matters. Although flux and concentration measurements, modelling and assessment techniques have been developed to estimate the emissions there is still a need to continuously monitor emissions from specific sources such as road networks. Several organizations have established pollutants limits for atmospheric pollutants concentrations considered as harmful for air quality; these include WHO the European Union and the national agencies such as the US Environmental Protection Agency.

The WHO, 2012 guidelines set atmospheric concentration limits for particulate matter (PM), Ozone (O_3), NO_2, and SO_2. In an assessment of 26 cities, 24 had an annual average of PM_{10} over the level set. The European Commission's Air Quality Standards cover these four pollutants and set standards of atmospheric concentrations for lead (Pb) and carbon monoxide (CO). These represent most atmospheric pollutants in the urban environment.

According to WHO [9] an estimated 4.3 million people a year die prematurely from illness attributable to air pollution caused by the inefficient use of solid fuels. Among these deaths, 12% are due to pneumonia, 34% from stroke, 26% from ischemic heart disease, 22% from COPD, and 6% from lung cancer [10]. Previous findings regarding the association between air pollution and preeclampsia/eclampsia, mostly conducted in developed countries, are limited and have been inconsistent [11-13]. Several recent studies have reported positive associations between preeclampsia and air pollutants including nitrogen oxides (NOx), NO, NO_2, CO, O_3, PM <2.5 μm in aerodynamic diameter ($PM_{2.5}$), and particulate matter <10 μm in aerodynamic diameter (PM_{10}) [11] but others have reported no association with $PM_{2.5}$ or CO [13] or inconclusive findings for PM_{10} and NO_2. Two studies have reported positive associations between gestational hypertension (a risk factor and early symptom of preeclampsia) and air pollutants (PM_{10} and $PM_{2.5}$ [14] and NO_2 and PM_{10}). A study in Spain also observed an increased risk of preeclampsia associated with exposure to fine particulate air pollution [15].

Data and Methodology

Nairobi area

This study was carried out in Nairobi city (Figure 1) and the air samples were collected from the selected sites. The sites were designed to be at busy intersections or roundabouts (Ngong road, Landhies road and Pangani Roundabout) and a site in Industrial Area. The selection was done after a feasibility assessment that mapped out busy and potential monitoring sites considering witnessed traffic jams, possible elevated emissions and assessing the general prevailing meteorological conditions that influences pollution levels. Nairobi is Kenya's capital city is located within 1°9'S, 1°28'S and 36°4'E, 37°10'E and covers an area of 684 km². The climate is warm and tropical occasionally becoming cool and cloudy June through July yearly. Over the past decades Nairobi has undergone significant land use and land use change transformation largely occasioned by city expansion, industrialization, mushrooming high rise residential estates and a rising population curve. The current road expansion network, therefore, is a necessity to pass, fast and efficiently, the resulting attendant traffic for human, vehicular, goods and services. Such recently expanded roads include; Thika Road, Southern and Northern by-passes and the ongoing Outering Road.

Data and methodology

The data was collected every one minute using various gas analyzers from Kenya Meteorological Department (KMD). The analyzers are shown in Table 1. The meteorological parameters were measured by an automated weather observing stations.

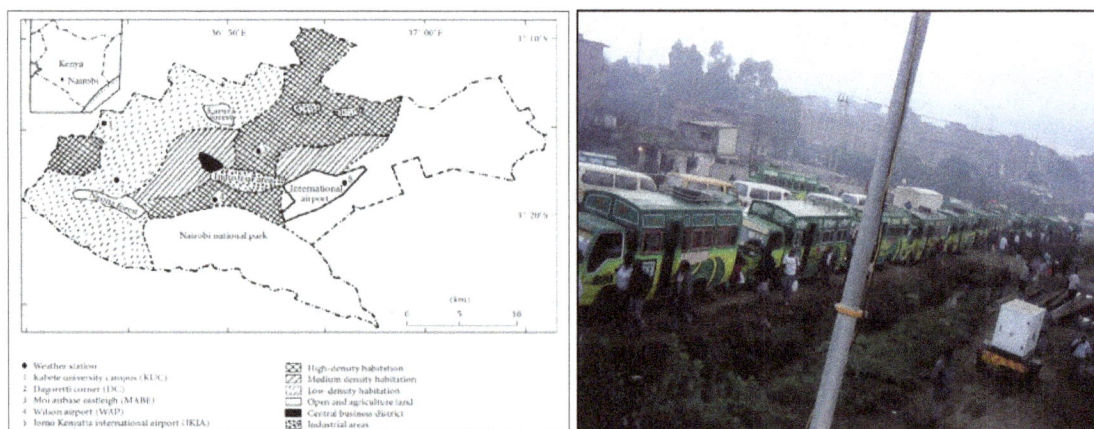

Figure 1: Map of buildup of traffic along outering road. The map is adopted from [20].

Instrument model	Parameter
EcotechSerinus 51	Hydrogen Sulphide and sulphur dioxide
EcotechSerinus 44	Nitrogen Oxides and Ammonia
EcotechSerinus 10	Surface ozone
EcotechSerinus 30	Carbon Monoxide
Ecotech EC 9820 Series	Carbon dioxide
Environmental Dust Monitor Model 180	PM_{10} and $PM_{2.5}$
Aethalometer	Black Carbon
Automated weather observing station	Ambient and screen temperatures, Solar radiation, precipitation, Atmospheric pressure, relative humidity, wind speed and direction

Table 1: Summary of instrumentation used in the study.

An aethalometer is a foremost instrument for the real-time measurement of optically-absorbing black or elemental carbon aerosol particles. It has been applied by other researchers in different settings [16,17]; it has also been used in extended applications in the area of public health and epidemiological studies, by allowing for real-time measurements of carbon particulate concentrations. Aethalometers provide fully automated data collection operation. The sample is collected as a spot on a roll of quartz fiber filter tape and performs a continuous optical analysis, while the sample is collecting, during this process, the tape does not move. The tape only moves forward when the spot has reached a certain density. Aethalometer draws the air sample through the inlet port, typically at a flow rate of a few liters per minute, using a small internal pump. The flow rate is monitored by an internal mass flow meter and is stabilized electronically to the set point value entered in software [17].

The optical method in the aethalometer is a measurement of the attenuation of a beam of light transmitted through the sample when collected on the fibrous filter. When calculated as expressed in equation 1, this quantity is linearly proportional to the amount of BC in the filter deposit;

$$ATN = 100 * In \left(\frac{I_o}{I} \right) \quad (1)$$

Where I_0 is the intensity of light transmitted through the original filter, I as the intensity of light transmitted through the portion of the filter on which the aerosol deposit is collected and ATN is optical attenuation while the factor 100 is for numerical convenience. This measurement is affected by the wavelength of the light with which it is made, provided that the particle size is somewhat smaller than the wavelength. The absorption of light by a broad band absorber such as graphitic carbon is inversely proportional to the wavelength of the light used. Thus, for a given mass of black carbon [BC], the optical attenuation at a fixed wavelength (λ) may be written as given in equation 2.

$$ATN (\lambda) = \sigma(1/\lambda)* [BC] \quad (2)$$

Where (BC) is the mass of black carbon, and $\sigma (1/\lambda)$ is the optical absorption cross-section that is wavelength dependent, and which is referred to as specific attenuation.

Environmental dust monitor

The environmental dust monitor model 180 was used for measuring PM_{10} and $PM_{2.5}$. The ambient-air, to be analyzed, is drawn into the monitor via an internal volume-controlled pump at a rate of 1.2 liters/minute. The sample passes through the measuring cell, past the laser diode detector and is collected onto a filter. The pump also generates the necessary clean sheath air, which is filtered and passes through the sheath air regulator back in to optical chamber. This is to ensure that no dust contamination with the laser-optic assembly. The sample flow is 1.2 l/min. Then a fine dust filter removes all the particles from the sample air. A membrane pump sucks the clean air through a valve, a protection filter, an orifice and a three way valve to the sample outlet. The sample flow is controlled by a flow controller which monitors the pressure drop over the orifice. A part of the cleaned air is used to supply the measuring chamber with rinsing air to keep the optic and the measuring chambers clean [18]. This clean air is also used during the functional self-test to calibrate the system for zero particles.

Nitrogen oxides were measured with EcotechSerinus 44 analyzer. The measurement of the $NO/NO_2/NOX/NH_3$ is performed via the gas phase chemiluminescence method. Sample air is drawn into the reaction cell via three separate (alternating) channels the NO, NOX and NX. The NOX channel travels through a delay coil enabling the same sample of air to be sampled for NO, NO_2 and NOX. The NOX channel passes through an NO_2 to NO converter, NO_2 is converted to NO. The analyzer also draws in air through an external converter (NX channel) which converts NH_3 into NO (and some NO into NO_2). This is then passed through the molyconverter to convert any NO_2 into NO. The other ecotech series instruments are described in detail in the operating manual GC 50003/2013.

Results and Discussion

The 24 h mean for all parameters sampled are presented in Table 2, while the morning and evening mean are show in Table 3 and the average meteorological parameters that were measured for all sites are presented in Table 4. From the findings the PM_{10} and $PM_{2.5}$ levels were found to be less than the recommended WHO guidelines, the levels were higher in the evenings and morning hours when the motor vehicle movement is higher. As Boman and Thynell [2] observed there

Pollutants	PM_{10}	$PM_{2.5}$	BC	SO_2	NO	NO_2	O_3	CO	CO_2
Units	ug/m³	ug/m³	ng/m³	ppb	ppb	ppb	ppb	ppm	ppm
Nakumat Junction	10.9	7.26	6474.66	0.46	10.44	9.21	7.39	0.61	368.94
Landhis Road	21.58	14.56	26115.54	0.82	54.89	17.41	4.11	0.97	368.74
Pangani	19.17	10.79	14446.53	1.08	36.99	15.73	5.64	1.72	385.51
Industrial Area	15.93	12.31	5996.32	0.78	4.53	8.37	10.34	0.57	379.51

Table 2: 24 h mean of pollutants.

Pollutants	PM_{10}	$PM_{2.5}$	BC	SO_2	NO	NO_2	O_3	CO	CO_2
Units	ug/m³	ug/m³	ng/m³	Ppb	ppb	ppb	ppb	ppm	ppm
Nakumat Junction (Morning)	22.0	15.01	14008.13	0.25	15.18	8.24	4.41	0.84	387.02
(Evening)	20.3	11.52	9835.37	0.53	20.25	15.34	1.84	1.06	377.54
Landhis Road (Morning)	16	14.84	29855.8	0.69	67.48	20.13	2.16	1.16	385.06
(Evening)	32.3	23.02	33778.66	1.65	90.42	23.28	2.09	1.33	358.62
Pangani (Morning)	16.96	14.75	29855.84	0.57	0.84	67.4	2.14	1.13	385.05
(Evening)	32.25	22.94	33778.58	1.57	1.15	90.41	2.21	1.32	358.53
Industrial Area (Morning)	33.21	24.93	12007.57	1.13	19.45	13.74	3.13	0.75	403.65
(Evening)	25.82	19.65	8702.74	1.5	4.07	12.35	3.66	0.99	378.72

Table 3: Morning/evening peaks of pollutants.

is a significant difference between roadside concentrations and urban background concentrations. The Boman study finds that 24 h $PM_{2.5}$ background concentration in the city center site as 22 µg/m³. This compares well with the present study that finds a city center site (Ladhis road) recording an average of 21.58 µg/m³ daily. This study found the each concentrations of $PM_{2.5}$ at the sampling sites varied with highest levels observed at the site within the CBD.

Nakumat junction site

Figure 2 depicts 24 h mean of PM_{10}, $PM_{2.5}$ and black carbon at

Nakumatt Junction near Dagoretti Corner. The wind direction was mainly easterly (Figure 3) with mean speed of 2.4 m/s as indicted in Table 3. The wind regime compares well to fin dings by Ongoma et al. [19]. Figure 2 shows two distinct peaks, one in the morning and the other in the evening for the PM_{10}, $PM_{2.5}$ and BC. These peaks can be attributed to vehicular emission during the morning and evening rush hours. The mean morning peak concentrations were 20.3, 11.5 µg/m³, 9835.3 ŋg/m³ for PM_{10}, $PM_{2.5}$ and black carbon respectively, while the evening peaks were 22.0, 15.0 µg/m³, 14008.1 ŋg/m³ for PM_{10}, $PM_{2.5}$ and black carbon respectively. It is important to note that the

Parameters	Wind Speed	WD	Temp	RH	AP	Precip	SR
	m/s	degrees	°C	%	hPa	mm/hr	W/m²
Nakumat Junction	2.364	97.173	18.3	72.49	825.74	0.016	350.064
Landhis Road	2.169	92.946	19.19	78.26	839.52	0.091	302.19
Pangani	1.894	107.228	19.61	72.12	839.12	0.000	362.017
Industrial Area	3.566	129.614	22.21	61.84	838.9	0.517	380.187

Table 4: Average weather parameters (24 h mean).

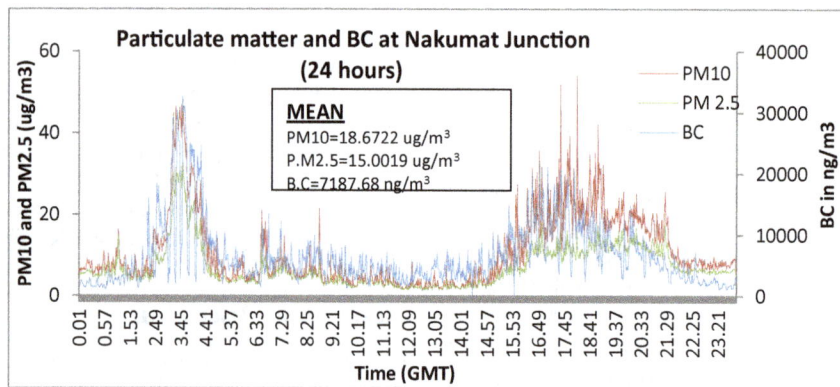

Figure 2: Diurnal variation particulate matter.

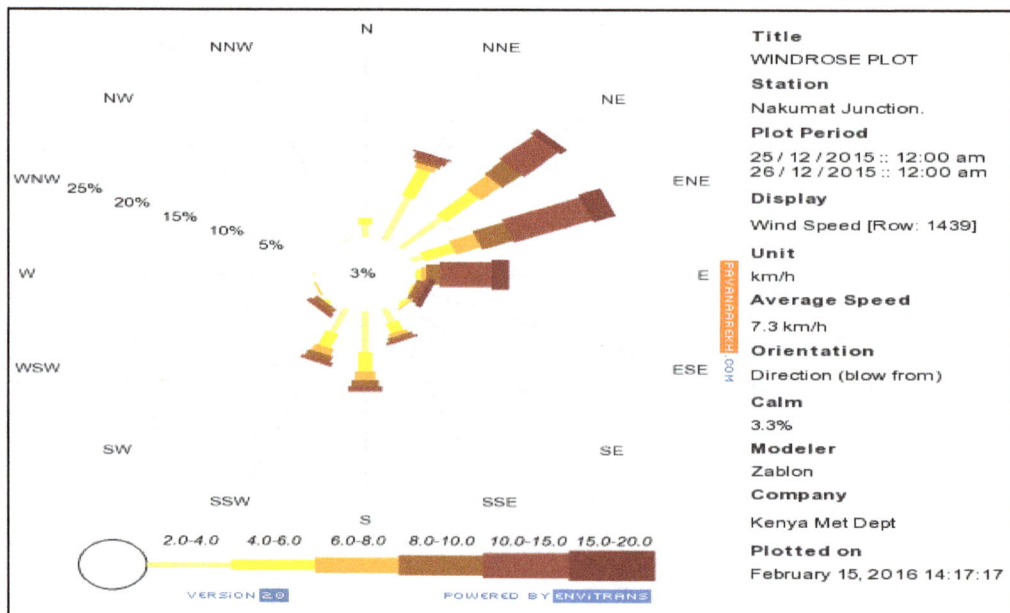

Figure 3: Nakumat junction windrose and frequency.

measurements were taken around Christmas and end year holidays when the volume of vehicles on the road was low. According to World Health Organization (WHO) Air Quality Guideline (AQG) of 2005, the levels for $PM_{2.5}$ and PM_{10} for 24 h duration are 25 and 50 µg/m³ respectively. For this site the levels were 7.26 and 10.90 µg/m³ for $PM_{2.5}$ and PM_{10} respectively consequently not exceeding WHO limits. Black carbon levels were noted to be very high, where by the 24 h mean was 6474.7 ŋg/m³ while the morning and evening means were 14,008 and 9,835 ŋg/m³ respectively (Table 5).

Figure 4 shows diurnal distribution NO_2 and nitric oxide (NO) at Junction Site. The 24 h mean are 10.4 and 9.2 ppb for NO and NO_2 respectively. There are two pronounced peaks in the morning and evening. The morning and evening peaks for NO and NO_2 are 15.1 and 20.2, and 8.2 and 15.3 respectively. The peaks are associated with the morning and the evening rush hours since motor vehicles account for over 50% of the total NO_2 generated. The highest hourly concentration of NO_2 was 17.5 ppb, far much below the WHO limit of 102 ppb.

The 24 h concentrations of SO_2, CO_2 and CO are shown in Figure 5. They also depict same characteristic as the other pollutant whereby, they have two peaks in the morning and evening and it is attributed to the build-up in traffic at those times. These three parameters are

generated from fossil fuels among other sources. Consequently, the two peaks in concentration realized in the morning and evening.

The diurnal amount of SO_2 is generally low averaging at 0.46 ppb. This is far much below the WHO 10 ppb for 24 h, while that of carbon dioxide was 368.94 ppm, which is also below the global annual average of 402 ppm. However, ambient guidelines for CO_2 do not exist. The global annual mean concentration of CO_2 in the atmosphere has increased by more than 40% since the start of the Industrial Revolution, from 280 ppm in the mid-18th century to 402 ppm as of 2016. The present concentration is the highest in at least the past 800,000 years and likely the highest in the past 20 million years. The increase has been caused by anthropogenic sources, particularly the burning of fossil fuels and deforestation. Global background concentrations of carbon monoxide range between 0.06 and 0.14 mg/m³ (0.05-0.12 ppm). The mean 24 h amount of CO was 0.62 ppm which is slightly above the background concentration. In the streets, the carbon monoxide concentration varies greatly according to the distance from the traffic; it is also influenced by topography and weather conditions. In general, the concentration is highest at the leeward side of the street, and there is a sharp decline in the concentration from pavement to rooftop level. Figure 6 depicts the diurnal variation of ozone at NJ. The figure generally shows low amount in the early morning and at night with the peak realized during the day. The peak in midday is due to the fact that surface ozone is produced by photochemical oxidation of CO, CH_4 and non-methane volatile organic carbons (NMVOCs) in the presence of NOx. The eight h mean was 10.3 ppb which is below the WHO mean of 51 ppb.

Results for ladhis road

Figure 7 depicts generally extremely high values of BC during the day which went beyond the upper limit of the instruments (50,000

Site	Ozone 8 Hour Mean (PPB)
Nakumatt Junction	10.3
Landhies	7.8
Pangani	10.1
Industrial Area	20.2

Table 5: Ozone 8 h mean values.

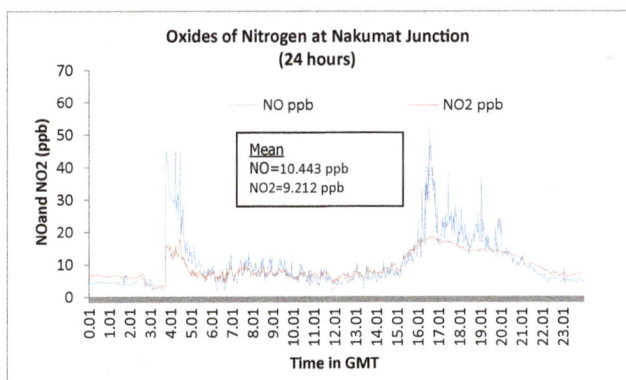

Figure 4: Oxides of nitrogen.

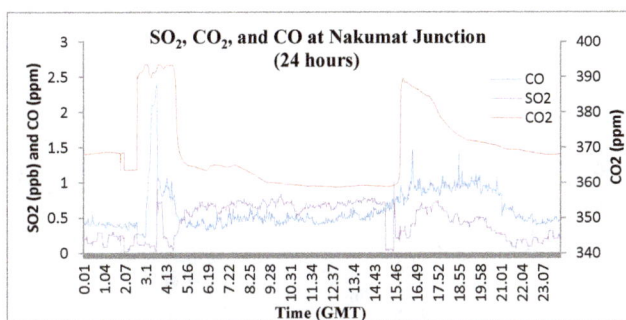

Figure 6: Diurnal variation of ozone.

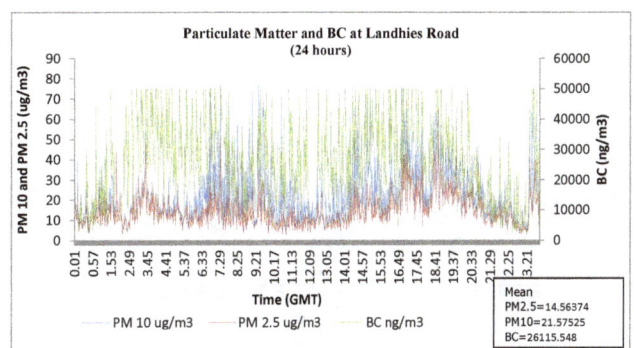

Figure 5: Diurnal SO_2, CO_2 and CO at Nakumat Junction (The mean for SO_2, CO_2 and CO is 0.4630 ppb, 368.9391 ppm and 0.6174 ppm respectively).

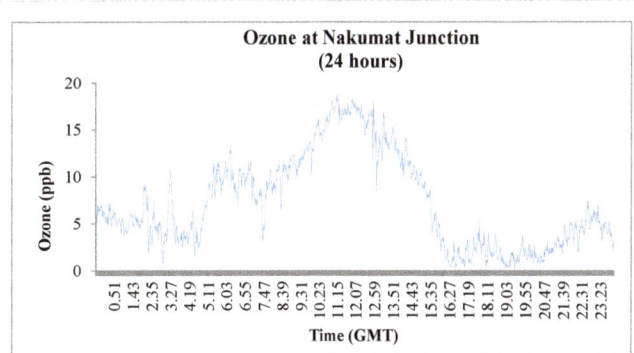

Figure 7: Diurnal variation of PM.

ŋg/m³). The 24 h mean were 26,115.5 ŋg/m³, with the mid-morning and evening peaks at 29,855.8 and 33,778.6 ŋg/m³ respectively. These are high values as indicated in Table 4 which have negative impact on human health. The PM$_{2.5}$ and PM$_{10}$ 24 h mean was 14.6 and 21.6 µg/m³ respectively consequently not exceeding WHO limits. Observation at Ladhis road shows the wind direction was mainly northeasterly (Figure 8) with mean speed of 2.2 m/s as indicted in Table 5. The diurnal distribution of both NO$_2$ and nitric oxide (NO) are shown in Figure 9. The 24 h mean are 54.9 and 17.4 ppb for NO and NO$_2$ respectively (Table 4). This mean is lower than the Kenyan tolerance limit of 100 ppb. There are two pronounced peaks in the morning and evening. The morning and evening peaks for NO and NO$_2$ are 67.4 and 90.4 and 20.1 and 23.2 respectively as indicated in Table 5. The peaks are associated with the morning and the evening rush h. The highest hourly concentration of NO$_2$ was 44.4 ppb, which is far much below the WHO limit of 102 ppb and Kenyan limit of 200 ppb.

Figure 10 depicts the mean 24 h concentration of SO$_2$ and CO. These pollutants are generated from fossil fuels among other sources. Consequently, the peaks realized in the morning and evening, were attributed to the build-up in traffic. The diurnal amount of SO$_2$ is generally low averaging at 0.82 ppb. This is far much below the WHO 10 ppb for 24 h and Kenyan limit of 48 ppb. Global background concentrations of carbon monoxide range between 0.06 and 0.14 mg/m³ (0.05-0.12 ppm). The mean 24 h amount of CO was 0.97 ppm which is way above the background concentration. Figure 11 depicts the diurnal variation of ozone. The figure generally shows low amount in the early morning and at night with the peak realized during the day.

The peak in midday is due to the fact that surface ozone is produced by photochemical oxidation of CO, CH$_4$ and non-methane volatile organic carbons (NMVOCs) in the presence of NO$_2$. The eight h mean was 7.8 ppb which is below the WHO mean of 51 ppb.

Results for pangani round about site

The mean wind speed was 1.9 m/s with predominant wind direction of 107.2° as indicated in Table 5 and Figure 12. The mean temperature was 19.6° with no rainfall recorded. Figure 12 shows the diurnal mean of particulate matter in Pangani Roundabout. There is no discernible pattern. However, high values were observed for the three pollutants between 1.49 and 19.49 GMT with the remaining time indicating low values. The 24 h mean of PM$_{2.5}$ and PM$_{10}$ were 10.8 and 19.2 ug/m³ respectively which are both below WHO and Kenyan limit. The black carbon mean was 14,446.5 ng/m³ which is extremely high (Figure 13). Figure 14 shows diurnal distribution of both NO$_2$ and nitric oxide (NO). The 24 h mean are 37.0 and 15.7 ppb for NO and NO$_2$ respectively. The mean 24 h of NO$_2$ is below Kenyan limit of 100 ppb. There are two pronounced peaks in the morning and evening with sustained relatively high values during the day. The morning and evening peaks for NO$_2$ and NO are 17.9 and 24.1 and 67.4 and 90.5 ppb respectively. The peaks are associated with the morning and the evening rush h. The highest hourly concentration of NO$_2$ was 52.2 ppb, which is below both WHO and Kenyan limits of 102 and 200 ppb respectively.

Figure 15 depicts the mean 24 h concentration of SO$_2$ and CO. At this site, there are no pronounced peaks in the morning and evening.

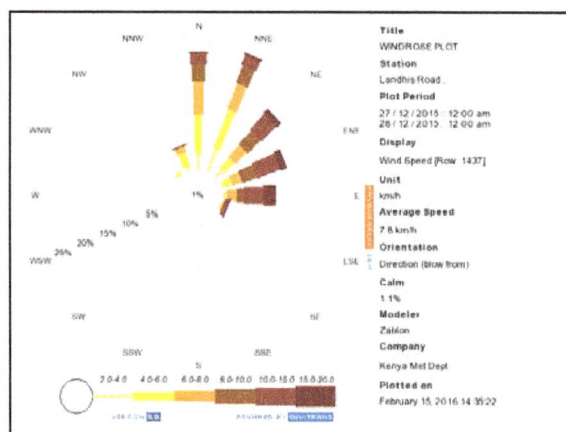

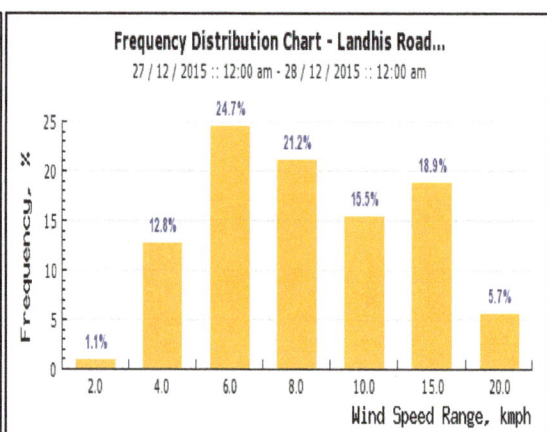

Figure 8: Landhis road windrose and frequencies.

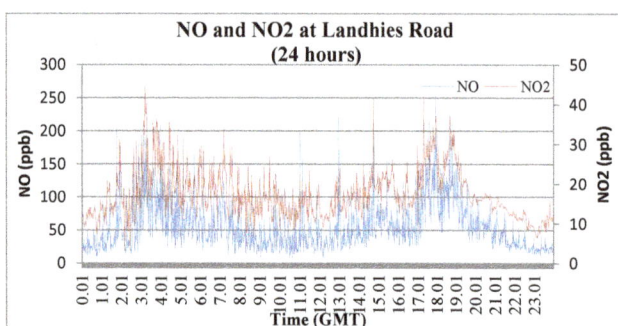

Figure 9: Diurnal variation of oxides of Nitrogen.

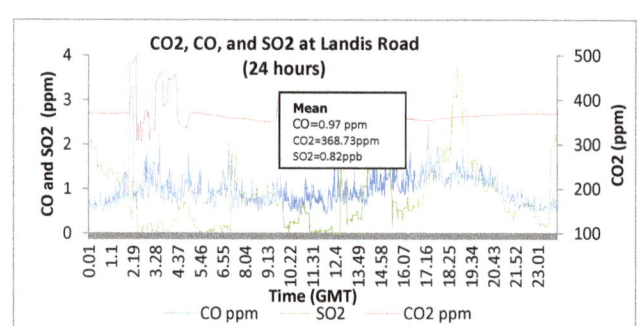

Figure 10: Diurnal SO$_2$, CO$_2$ and CO at Ladhis road.

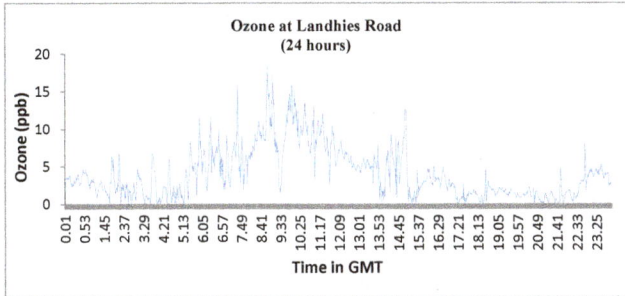

Figure 11: Diurnal variation of ozone.

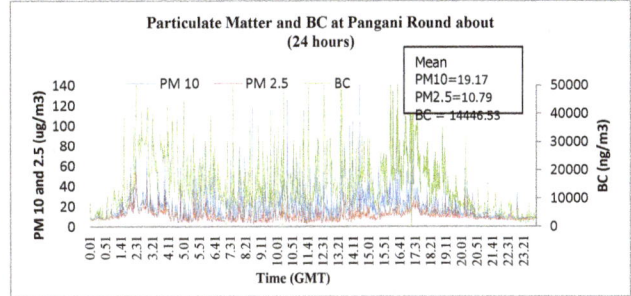

Figure 12: Diurnal variation of particulate matter.

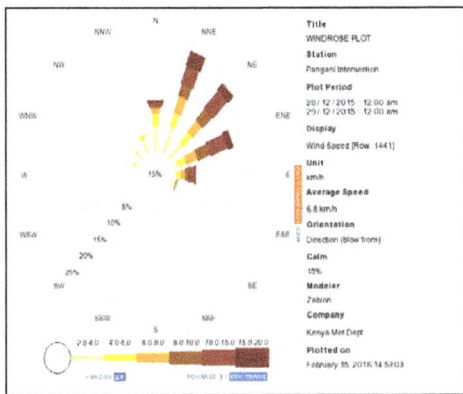

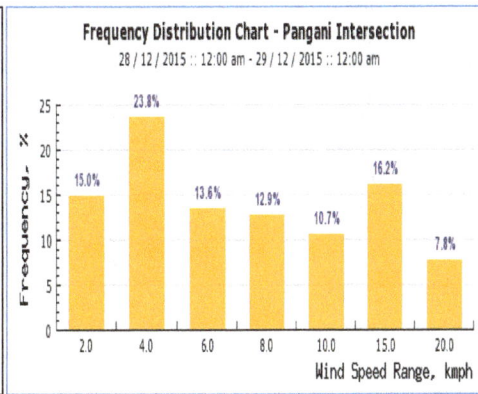

Figure 13: Wind rose and frequency for Pangani site.

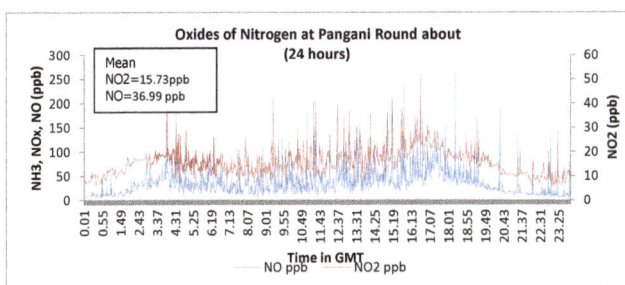

Figure 14: Diurnal variation of Oxides of nitrogen.

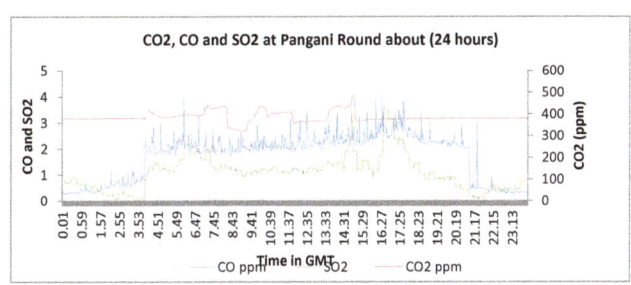

Figure 15: SO$_2$ and CO at Pangani Round about.

However, high values are dominant during the day with low values experienced at night. This can be attributed to the high numbers of vehicles during the day. The diurnal amount of SO$_2$ is generally low averaging at 1.08 ppb. This is far much below the WHO 10 ppb for 24 h. Global background concentrations of carbon monoxide range between 0.06 and 0.14 mg/m^3 (0.05-0.12 ppm). The mean 24 h amount of CO was 1.73 ppm which is above the background concentration.

Figure 16 depicts the diurnal variation of ozone. The figure generally shows low amount in the early morning and at night with the peak realized during the day. The peak in midday is due to the fact that surface ozone is produced by photochemical oxidation during the day. The eight h mean was 10.1 ppb which is below the WHO mean of 51 ppb.

Results for industrial area region

The wind direction was mainly Southeasterly (Figure 17) with mean

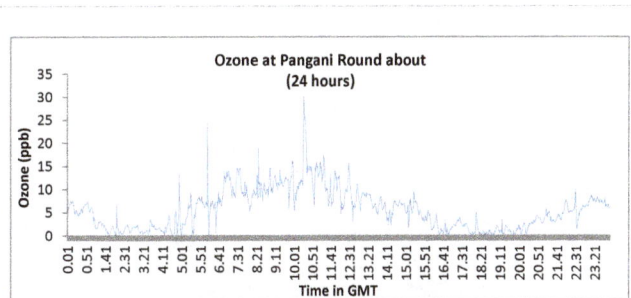

Figure 16: Diurnal variation of ozone at Pangani round about.

wind speed and direction of 3.6 m/s and 129.6° as indicted in Table 5. A drizzle of 0.52 mm/h was experienced during the monitoring duration.

Figure 18, shows low concentration of the pollutants between 8.00 AM and 6.30 PM. The rest of the time, which is generally during the night, indicates high values. The observed variation can be associated with low and high rate of dispersion at night (stable atmospheric conditions) and daytime (unstable atmospheric conditions) respectively. With both low temperature and wind speed experienced at night, the dispersion rate of pollutant is low due to less atmospheric mixing and therefore higher concentrations around surface level. The converse is true with both high temperature and higher wind speeds during the day. The 24 h means for $PM_{2.5}$ and PM_{10} were 12.3 and 15.9 μg/m^3 respectively consequently not exceeding WHO limits, while that of black carbon is 5,996.3 ŋg/m^3 which is very high as indicated in Table 5.

Figure 19 shows diurnal distribution of both NO_2 and nitric oxide (NO). It indicates two peaks, one in the morning and the other in the evening. The morning is more pronounced especially for NO while the evening peaks are broader for both parameters. The 24 h mean are 4.5 and 8.4 ppb for NO and NO_2 respectively (Table 4). The morning and evening peaks for both NO and NO_2 are 19.5 and 4.1 and 13.7 and 12.3 respectively (Table 5). NO and NO_2 are mainly produced by fossil fuel with motor vehicles accounting for over 50% of the total NO_2 generated. The peaks can therefore be attributed partly to the buildup of traffic near the monitoring site. The low concentration in the day is due to unstable atmospheric conditions which increase the dispersion rate of pollutants. The highest hourly concentration of NO_2 was 18.9 ppb, far much below the WHO limit of 102 ppb.

Figure 20 shows the diurnal concentration of SO_2, CO_2 and CO.

The pollutants are generated from fossil fuels among other sources. Consequently, the two peaks in concentration realized in the morning and evening were attributed partly to the build-up in traffic from the nearby roads. The high values observed during the night may be attributed to stable atmospheric conditions. Consequently, the dispersion rate of pollutant is low due to less atmospheric mixing and therefore higher concentrations around surface level. The diurnal amount of SO_2 is generally low averaging at 0.78 ppb. This is far much below the WHO 10 ppb for 24 h and Kenyan limit of 48 ppb. Global background concentrations of carbon monoxide range between 0.06 and 0.14 mg/m^3 (0.05-0.12 ppm). The mean 24 h amount of CO was 0.57 ppm which is slightly above the background concentration.

Figure 21 depicts the diurnal variation of ozone. The figure generally shows low amount in the early morning and at night with high amount of ozone throughout the day. The peak in the day is due to the fact that surface ozone is produced by photochemical oxidation of CO, CH_4 and non-methane volatile organic carbons (NMVOCs) in the presence of NOx. The high values of ozone show that there is high concentration of ozone precursors in this site. The eight h mean was 20.2 ppb which is below the WHO mean of 51 ppb. However, this is the highest recorded amount of the four sites (Figure 4).

Conclusion and Recommendation

At Nakumatt Junction site particulate matter depicted two distinct peaks, one in the morning and the other in the evening. The peaks in pollutant concentration were attributed to vehicular emission during the rush hours. However, in Landhies road and Pangani Roundabout,

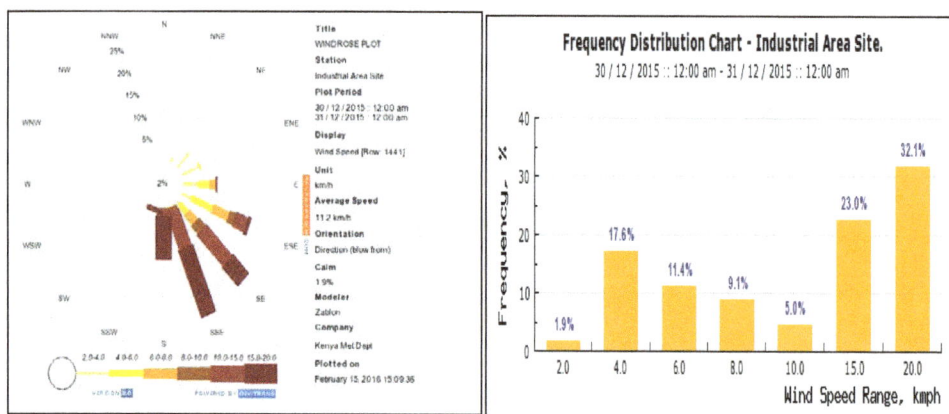

Figure 17: Wind rose and frequency for Industrial area site.

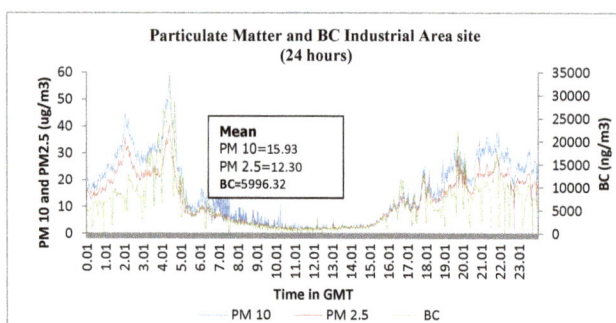

Figure 18: Diurnal variation of particulate matter and black carbon.

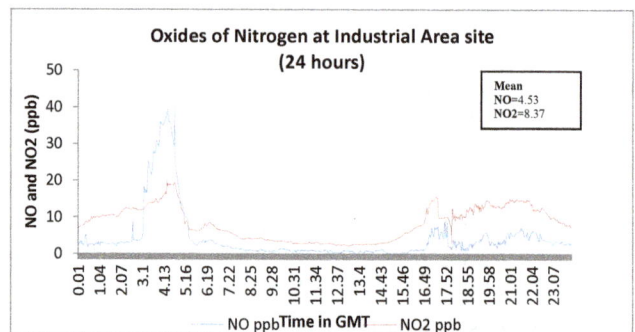

Figure 19: Diurnal distribution of both NO_2 and nitric oxide (NO).

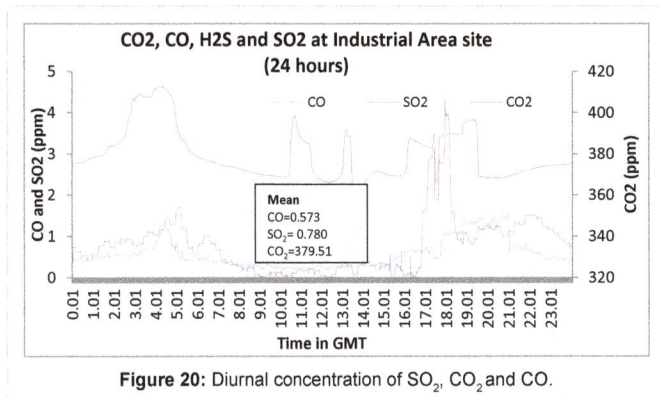

Figure 20: Diurnal concentration of SO_2, CO_2 and CO.

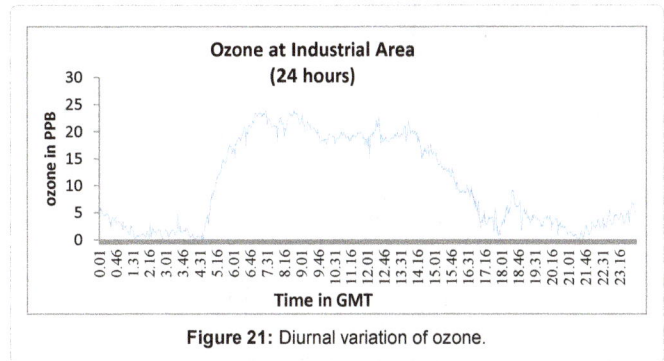

Figure 21: Diurnal variation of ozone.

the peaks were not discernible due to the high volume of vehicles throughout the day. Industrial Area site particulate matter showed high and low concentration during the night and day respectively The observed trends was associated with low and high rate of dispersion at night (stable atmospheric conditions) and daytime (unstable atmospheric conditions) respectively.

None of the four sites exceeded the WHO 24 h limit for both PM10 (50 μg/m³) and PM2.5 (25 μg/m³). The 24 h mean of PM10 in the three sites also did not exceed the ambient air quality tolerance Kenyan limit of 100 μg/Nm³ and 150 μg/Nm³ in Industrial Area. Extremely high values of black carbon which went beyond the upper limit of the instruments (50,000 ng/m³) were observed during the day in Landhies road. Nakumatt Junction recorded extreme values of Black carbon (14,008 ng/m³) in the evening peaks while Pangani Roundabout, the diurnal mean value was extreme (14,446.5 ng/m³). In all the monitoring sites, Oxide of nitrogen showed two pronounced peaks, one in the morning and the other in the evening. NO and NO_2 are mainly produced by fossil fuel with motor vehicles accounting for over 50% of the total NO_2 generated. The peaks are therefore associated with the morning and the evening rush h even in Industrial Area due to the nearby roads.

None of the site exceeded the Kenyan 24 h mean ambient air quality tolerance limit of NO_2 (100 ppb). Pangani Roundabout recorded the highest hourly concentration of NO_2 (52.2 ppb) which is far much below the WHO and Kenyan limits of 102 and 200 ppb respectively.

The SO_2 and CO exhibited two peaks in concentration realized in the morning and evening. This trend was attributed to the build-up in traffic since the two parameters are generated from fossil fuels among other sources. The diurnal mean of SO_2 over the four sites was generally low with the highest amount of 1.08 ppb recorded at Pangani Roundabout. This amount is far much below both WHO and Kenyan limit of 10 and 48 ppb respectively. The global background concentration of carbon monoxide ranges between 0.06 and 0.14 mg/m³ (0.05-0.12 ppm). The mean 24 h amount of CO in all the sites was above the background concentration with Pangani Roundabout recording the highest amount of 1.73 ppm.

The diurnal variation of Ozone in Nakumatt Junction, Landhies road and Pangani Roundabout showed low amount in the early morning and at night with the peak realized during the day. The peak in midday is due to the fact that surface ozone is produced by photochemical oxidation of CO, CH_4 and non-methane volatile organic carbons (NMVOCs) in the presence of NOx. The eight h means for the three sites were below WHO mean of 51 ppb. In Industrial Area, low amount of ozone were realized in early morning and at night with high amount observed throughout the day. The high values of ozone show

that there is high concentration of ozone precursors in this site. The eight h mean was 20.2 ppb which is below the WHO mean of 51 ppb. However, this is the highest recorded amount of ozone in the four sites.

Kenya seeks to be industrialized by the year 2030, at a time when world over, policymakers and the general public are concerned with the degradation of air quality, especially in urban centers. She has to come up with adequate strategies of tackling air pollution, which has direct health impacts upon her increasing population. Therefore, concerted efforts have to be made to find a sustainable balance between industry, human health and environmental protection. The study recommends that;

- Further monitoring of air pollution to be conducted along major roads in Nairobi.

- Develop an atlas of air pollution levels in major cities in Kenya

- Enhancement of ad hoc air pollution monitoring in different counties in order to profile pollution levels within the country.

Acknowledgments

The authors are grateful to National Environmental Management Authority (NEMA) for financial support and Kenya Meteorological Department (KMD) for technical assistance.

References

1. Shilenje ZW, Thiong'o K, Ondimu KI, Nguru PM, Nguyo JK, et al. (2015) Ambient Air Quality Monitoring and Audit over Athi River Township, Kenya. Int J Scien Res Environ Sci 3: 291-301.

2. Boman J, Thynell M (2013) Bad air quality is equal to bad life quality minus the influence of transportation on air quality in the city of Nairobi, conference proceedings, 15th-18th July, 13th World Conference on Transport and Research, Rio, Canada.

3. Kinney PL, Gichuru MG, Volavka-Close N, Ngo N, Ndiba PK, et al. (2011) Traffic Impacts on PM2.5 Air Quality in Nairobi, Kenya. Environ Sci Pol 14: 369-378.

4. Knippertz P, Evans MJ, Field PR, Fink AH, Liousse C, et al. (2015) The possible role of local air pollution in climate change in West Africa. Nat Clim Chan 5: 815-822.

5. Paeth H, Born K, Girmes R, Podzun R, Jacob D (2009) Regional climate change in tropical and northern Africa due to greenhouse forcing and land use changes. J Clim 22: 122-132.

6. WHO (2012) Diesel Engine Exhaust Carcinogenic. In: Cancer IAfRo (Ed.), World Health Organization.

7. Harrison RM, Yin J (2000) Particulate matter in the atmosphere: which particle properties are important for its effects on health? Sci Total Environ 249: 85-101.

8. Kenya National Bureau of Statistics (KNBS) (2009) Kenya Population Census, Ministry of State for Planning, National Development and Vision 2030, Government Print Press, Nairobi.

9. WHO (2013) Outdoor Air Pollution a Leading Environmental Cause of Cancer Deaths. In: Cancer IAfRo (Edn.), World Health Organization.

10. Agrawal S, Yamamoto S (2015) Effect of Indoor air pollution from biomass and solid fuel combustion on symptoms of preeclampsia/eclampsia in Indian women. Indoor air 25: 341-352.

11. Lee PC, Roberts JM, Catov JM, Talbott EO, Ritz B (2013) First trimester exposure to ambient air pollution, pregnancy complications and adverse birth outcomes in Allegheny County, PA. Matern Child Health J 17: 545-555.

12. Pereira G, Haggar F, Shand AW, Bower C, Cook A, et al. (2013) Association between pre-eclampsia and locally derived traffic-related air pollution: a retrospective cohort study. J Epidemiol Community Health 67: 147-152.

13. Rudra CB, Williams MA, Sheppard L, Koenig JQ, Schiff MA (2011) Ambient carbon monoxide and fine particulate matter in relation to preeclampsia and preterm delivery in western Washington State. Environ Health perspect 119: 886-892.

14. Vinikoor-Imler LC, Gray SC, Edwards SE, Miranda ML (2012) The effects of exposure to particulate matter and neighbourhood deprivation on gestational hypertension. Paediatr Perinat Epidemiol 26: 91-100.

15. Dadvand P, Parker J, Bell ML, Bonzini M, Brauer M, et al. (2015) Maternal exposure to particulate air pollution and term birth weight: a multi-country evaluation of effect and heterogeneity. Environ Health Perspect 121: 267-373.

16. Dons E, Temmerman P, Van Poppel M, Bellemans T, Wets G, et al. (2013) Street characteristics and traffic factors determining road users' exposure to black carbon. Sci Total Environ 447: 72-79.

17. Hansen ADA (2005) Theaethalometer manual. Magee Scientific, Berkeley, California, USA.

18. GRIMM Aerosol Technik (2006) Environmental Dust Monitor 180 User Manual, Ainring, Germany.

19. Ongoma V, Muthama JN, Gitau W (2013) Evaluation of Urbanization Influences on Urban Winds of Kenyan Cities. Ethiopian J Environ Stud Manage 6: 223-231.

20. Makokha GL, Shisanya CA (2010) Trends in mean annual minimum and maximum near surface temperature in Nairobi City, Kenya. Advan Meteorol.

9

The Correlation of North Magnetic Dip Pole Motion and Seismic Activity

Williams B*

Retired, Gillette, Wyoming, United States

Abstract

Viterito states in his paper that increasing seismic activity for the globe's high geothermal flux areas (HGFA) is correlated with the average global temperature from 1979 to 2015 with a correlation factor r of 0.785 and that this explains 62% of the variation in the earth's surface temperature. This makes the geothermal activity the most significant element in the change in temperature of the earth at this time.

Knowing that this is the cause of a major concern for so many and a critical piece to understanding our earth and its processes, the mechanisms surrounding the seismic activity was investigated. It was found that there is a high positive correlation (r=0.935) between the speed of the North Magnetic Dip Pole motion and the seismic activity during this time period.

Keywords: Seismic activity; Geothermal forcing; Magnetic poles

Introduction

Viterito [1] studied the interaction between the temperature of the globe and seismic data to address a point of difference within the scientific community as to assigning the proper percentages of temperature change to the known factors that add heat to the globe. In the past the majority of the heat was attributed to the addition of CO_2 to the atmosphere due to anthropogenic activity. He derived that in actuality 62% of the heating could be attributed to geothermal heating induced by introducing heat to the oceans through seismic activity in areas of the spreading tectonic plates in the oceans. This raises a question of if there is a correlation to any activity below the surface of the earth which could be the cause of this increased activity or being affected by this activity. One of the major unusual activities since 1860 has been the motion of the North Magnetic Dip Pole (NMDP).

This paper looks at the relationship between this geological activity and the motion of the NMDP. That is, the locations where inclination is 90 degrees, i.e. where the field is perpendicular to Earth's surface; the Black dots in Figure 1. This is not the geo-magnetic poles which define a theoretical dipole arrangement that is the sum of all the magnetic anomalies due to the earth's complex magnetic field. The geo-magnetic field is shown in red dots.

Materials and Methods

The geological activity values were taken from the paper by Viterito [1]. The position of the NMDP was retrieved from NOAA [2] geomagnetic data as was the South Pole [3] position. Using the latitude and longitude values, the distance between each year's positions was calculated as well as the distance from the rotational pole using first order true spherical distance calculations. From these values the rate of change for the pole was calculated on a yearly basis for both the absolute distance moved and the distance moved relative to the North Rotational Pole (NRP).

Results

Referring to Figure 2, from 1591 until 1974 the NMDP was moving at an average of around 6.9 Km per year with a low of 0.34 Km/yr and a high of 16.4 Km/yr. And from 1975 until 1996 the North Magnetic Pole was moving at an average of around 12 Km/yr with a low of 11.8 Km/yr and a high of 17 Km/yr. The position of the NMDP (Black Dots) as it moved is shown in Figure 1.

Then inexplicably, in 1996 the rate of travel doubled from its 1995

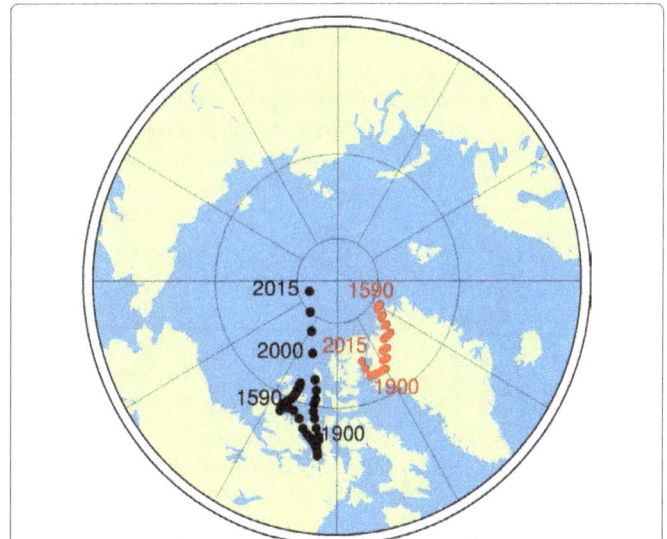

Figure 1: Earths North Magnetic Poles 4. Black=Magnetic Pole | Red=Geomagnetic Pole

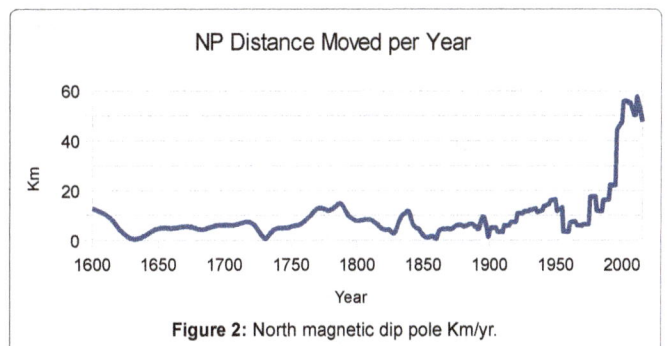

Figure 2: North magnetic dip pole Km/yr.

***Corresponding author:** Williams B, Retired, P.O.Box 3938, Gillette, Wyoming 82717, United States, E-mail: bbwilliams@bresnan.net

value to 44.1 Km/yr. In addition its average has been 3 times the 1975-1995 average. This rate of motion coincides with the change in geological activity with a correlation of 93.45%. This indicates an extremely close correlation and a strong probability of causal interaction [4]. The high degree of correlation between the NMDP speed and the geological activity can be seen in Figure 3. By looking at Figure 3 it would seem that the geological activity began to rise in 1994 and then the poles speed increasing a year later in 1995, but a fit for a one year offset is only 91.27% correlation, 2.2% less than a one to one correlation.

This would indicate that we cannot tell which causes which, at least not without further analysis or more finely separated data, i.e. on a monthly basis or even finer [5]. But there is an indicator in the data that says it is more likely that the geological activity is driving the poles motion and not the other way around.

When you look at the graphs in Figure 4 and 5 you will notice that at low geological activity the speed acts in a quantum way. In other words the Pole speed is grouped into 4 distinct speeds at 12 Km/yr, 16.25 Km/yr, 17.5 Km/yr and 22.2 Km/yr. The Geological activity at 12, 16.25 and 17.5 Km/yr varies from 175 to 325 and at 22.2 Km/yr it varies from 175 to over 600.

The actual data shows the NMDP speed was 22.2 Km/yr ± 0.1 Km/yr in the years 1991 to 1995, inclusive. The geological activity was

around 200 for the first 2 years then jumped to 361 and then to around 600 then 627 before the speed broke away in 1996 to 44.1 Km/yr. All of this type of activity would indicate some kind of a strong inertial or 'frictional' holding mechanism which prevented the magnetic pole from changing position until there was sufficient geothermal activity to overcome the holding mechanism and allow a speed change.

From a dynamo theory of the magnetic poles [6] we assume that the poles are set up following something similar to the Glatzmaier-Roberts model, which simulates convection and magnetic field generation in a fluid outer core surrounding a solid inner core. And in this case one of the driving forces would be the temperature and motion of the mantle. Geological activity would affect several factors in this model including the temperature of local areas of the mantle which would affect the conductivity of the material in these areas as well as the viscosity of the materials, both of which will have an effect on the resulting magnetic field.

One possible scenario is that the cooling effect of the release of heat from the mantle into the oceans would increase localized areas viscosity, which would tend to hold it in place, acting as a frictional force. But, its cooling would also cause its resistance to decrease, causing the regenerating fields in this area to grow stronger while other areas grow weaker. The increased electron flow would then increase the

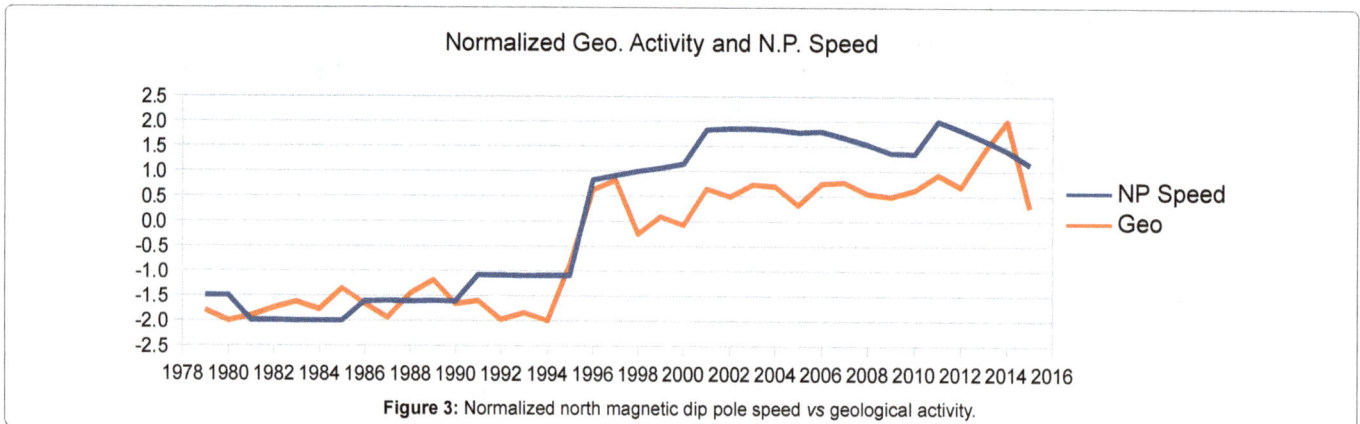

Figure 3: Normalized north magnetic dip pole speed *vs* geological activity.

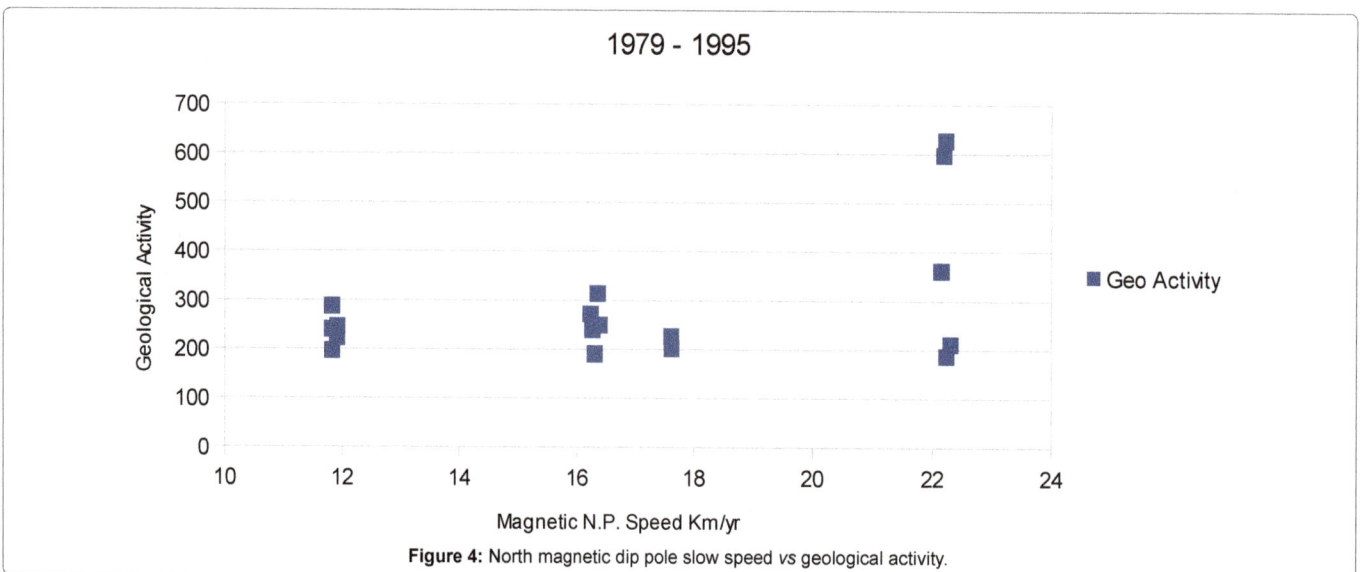

Figure 4: North magnetic dip pole slow speed *vs* geological activity.

resulting motor action until a tipping point is reached. In this case when the cooling is causing the movement to be over 22 km/year the driving force of the motor action overcomes the drag from increased viscosity and the pole acts more in a linear manner. That is not to say this will be a universal condition under all circumstances, just that it is the way the present configuration is responding.

Figure 6 shows the 88.24% correlation between the distances of the NMDP to the NRP with the geological activity increasing as the NMDP gets closer to the NRP. There are several physical processes which can explain this. If a significant component, such as the rate of relative motion caused by the mantle's rotation relative to the core, then one would expect that the maximum effect would be when the rotating members are aligned with the rotational axis, any offset from direct alignment would reduce the relative B X V velocity. Where B is the magnetic field strength, V is the relative velocity between the material cutting the B field and X is the vector multiplication operator. Since V is a velocity, speed with a vector direction, the absolute value of the magnitude of the induced currents will be a function of the sine of the angle between the motion and field direction. Motion in this case does not have to be relative motion between the mantle and the core. It can be just the direction of the magnetic current within the mantle.

It could also be that under the existing conditions the magnetic field is being driven towards the NRP because of where the geological activity is occurring. In the geological past and future, because the continental plate boundaries were not in the same place and will not stay in the same place, similar heat dissipation may produce dissimilar results due to where the heat loss is occurring. All of these considerations and many more must be taken into consideration and analyzed before definitive answers can be arrived at. But just knowing that these particular characteristics exist is the first step in determining their cause and their ultimate effects on both the atomic and macroscopic scales.

The amount of work necessary to derive these answers is not within the scope of this paper and will require many man hours and many people to solve. The references in [6] and [5] is an indication of the complexity of the earth's magnetic field and the long hard work that has gone into understanding what we know at present and the amount of work it will take to understand it more fully.

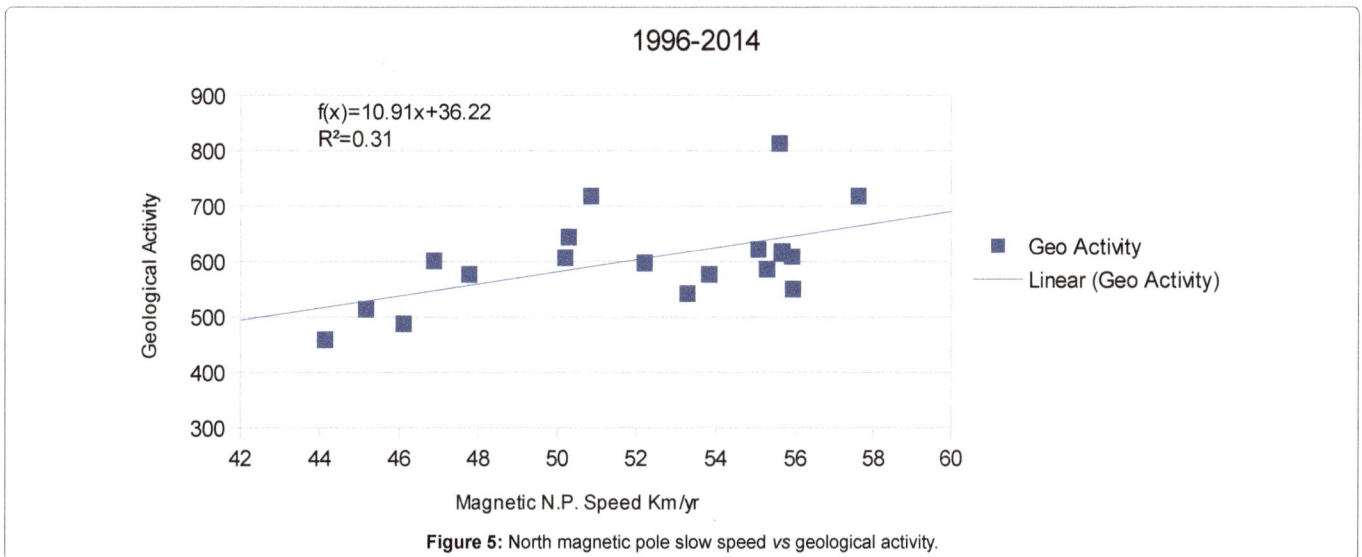

Figure 5: North magnetic pole slow speed *vs* geological activity.

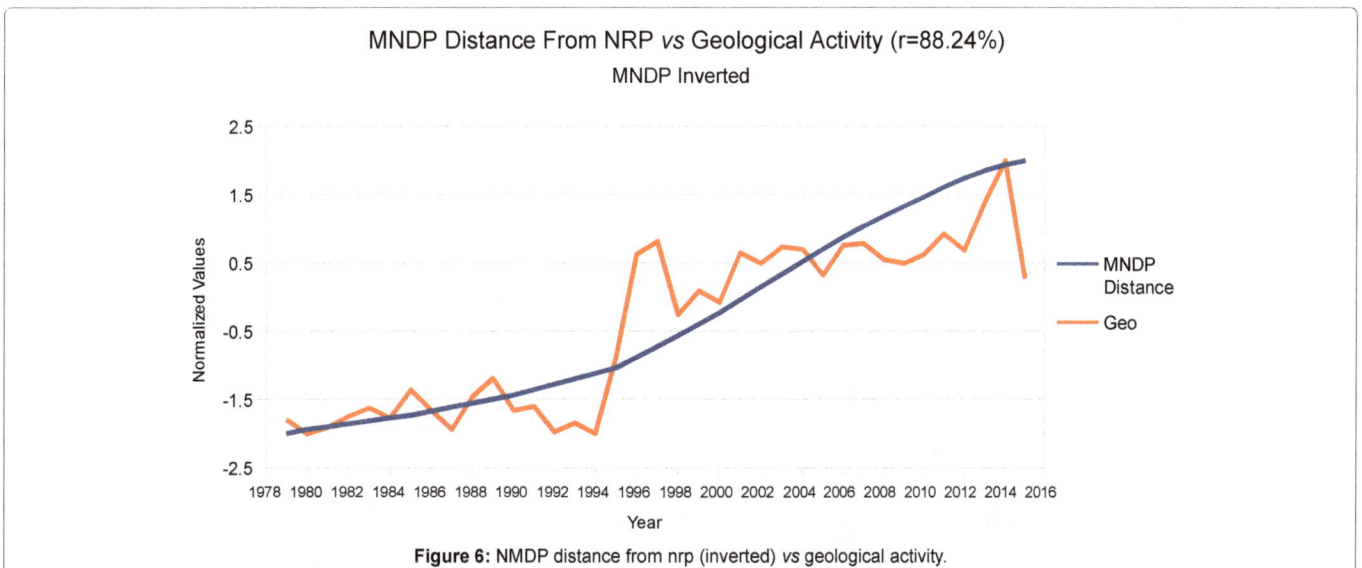

Figure 6: NMDP distance from nrp (inverted) *vs* geological activity.

Acknowledgment

This research was independently conducted, and was not funded by any corporation, government agency, or non-profit organization.

References

1. Viterito A (2016) The Correlation of Seismic Activity and Recent Global Warming. J Earth Sci Clim Change 7: 345.

2. http://www.ngdc.noaa.gov/geomag/data/poles/NP.xy

3. http://www.ngdc.noaa.gov/geomag/data/poles/SP.xy

4. http://www.gfz-potsdam.de/typo3temp/pics/npole_01_a5d35fc f3c.png

5. Olson P (2016) Mantle control of the geodynamo: Consequences of top-down regulation. Geochem Geophys Geosyst 17: 1935-1956.

6. Merrill RT, McFadden PL (1995) Dynamo theory and paleomagnetism. J Geophys Res 100: 317-326.

10

Volumetric Assessment through 3D Geostatic Model for Abu Roash "G" Reservoir in Amana Field-East Abu Gharadig Basin-Western Desert-Egypt

Abu-Hashish MF[1]* and Ahmed Said[2]

[1]*Geology Department, Faculty of Science, Menoufiya University, Egypt*
[2]*Qarun Petroleum Company, Cairo*

Abstract

The main objective of constructing a geostatic model is to determine the 3-D geometry of the reservoir rock and assess its hydrocarbon volumetrics. To achieve this goal, the Amana oil field is taken as a real example. An integrated methodological approach was applied starting with data collection and quality control and then followed by interpretation of the available geological, geophysical and petrophysical data. The field area is located in the most eastern trough of the Abu Gharadig basin in the East Bahariya Concession in the Western Desert of Egypt. The source rock in this area is the Upper Jurassic Khatatba Formation that was deposited in a continental to inner-middle shelf environment. The reservoir rock is Abu Roash "G" sand, one of the seven lithologic members of the Abu Roash Formation.

Interpretation of seismic data together with well logs revealed the presence of a horst block, acting as a good structural trap, meanwhile the Abu Roash "F" carbonate and Abu Roash "G" shale are the seal rocks. Well data have shown that the reservoir rock in the Amana field is the Abu Roash "G" Member, which comprises three sand lithologic zones; namely the upper, meddle and lower zones. Of these zones, the middle one is the most attractive and has the best reservoir quality. The shale content in this sand is 8% compared to 13% and 26% in the upper and lower zones. In addition, the net to gross thickness ratio is less than 18%, more than 35%, and about 10% in the upper, middle, and lower zones, respectively.

Data analysis also indicates that the upper and lower zones are appreciably water wet. However, the middle sand zone appears to be prospective. The net pay thickness in this zone varies between 10 and 32 ft, porosity 19-22%, water saturation 18-40% and average permeability 40 md. Based on the geostatic model of the Amana field reservoir, it is concluded that this area is a positive prospect in the most eastern part of the Abu Gharadig basin. The volume of oil (STOLIIP) in Amana reservoir is calculated as 10 million barrels, with an initial recoverable oil of 1.4 million barrels.

Keywords: Geostatic model; Petrophysics; Seismic interpretation; Western desert abu gharadig basin; Egypt

Introduction

Abu Gharadig basin is considered the most petroliferous basin in Western Desert as far as hydrocarbon production and potential. It is a deep E-W trending asymmetric graben, has dimensions of 300 km E-W and 60 km N-S, with an area of about 17500 km², and basement at depths over 10 km. Its structure has been recognized as a major rift basin in which there are numerous localized highs that in NE-SW oriented plunging anticlines that are believed to be fault-controlled folding [1]. The northern margin of this basin is marked by a major border fault zone which up-throws the basement to about 10,000 feet forming Sharib-Sheiba ridge, and the southern boundary is called Sitra platform [2]. Amana oil field is located in the most eastern trough of the Abu Gharadig basin (Figure 1), exactly in East Bahariya concession that exists in Mubarak sub-basin.

Tectonic framework and structural settings

Abu Gharadig Basin may have begun life as a pull-apart basin between two right-lateral wrench faults, its development began during the Jurassic and Cretaceous and the tectonic activity reached a peak during the Upper Cretaceous to Eocene interval [3]. It is subdivided into three structural units from east to west; Mubarak sub-basin, Abu Gharadig basin and Qattara depression including Mubarak High, Abu Gharadig Anticline and Mid basin Arch (Figure 2).

Litho-stratigraphy

The study area is related to the unstable shelf which covers the northern belt of Egypt, which is characterized by a thick stratigraphic succession that ranges, in age, from Pre-Cambrian Basement to Holocene time, varying in lithology (Figure 3). Abu Roash Formation is characterized by a cyclic alteration of deltaic flood plain sandstones, coastal sandstones and shales, and shallow marine shales and limestones [4]. The formation has been divided into seven units (members) "A" to "G". Units "B", "D" and "F" are mainly carbonates, units "A", "C", "E" and "G" contains variable amounts of detrital materials [5]. Abu Roash "G" shale acts as a very good top and lateral seal while Abu Roash "F" carbonate acts as a lateral seal, especially in the central part Amana field , where the major fault throw exceeds 300 ft. The pay sand zone in Amana field is involved within Abu Roash "G" Mbr. which is composed mainly of marine shale intercalated by carbonate streaks, with three cycles of sand facies considered as the main reservoirs in Abu Gharadig basin, and named (from top to base) as Upper sand zone, Middle sand zone and Lower sand zone) (Figure 4).

Methodology and Workflow

In order to perform a reliable 3D Geostatic Model, it is necessary to quality control and manages all types of the data, starting from seismic

---

*Corresponding author: Abu-Hashish MF, Geology Department, Faculty of Science, Menoufiya University, Egypt, E-mail: mfarouk64@gmail.com

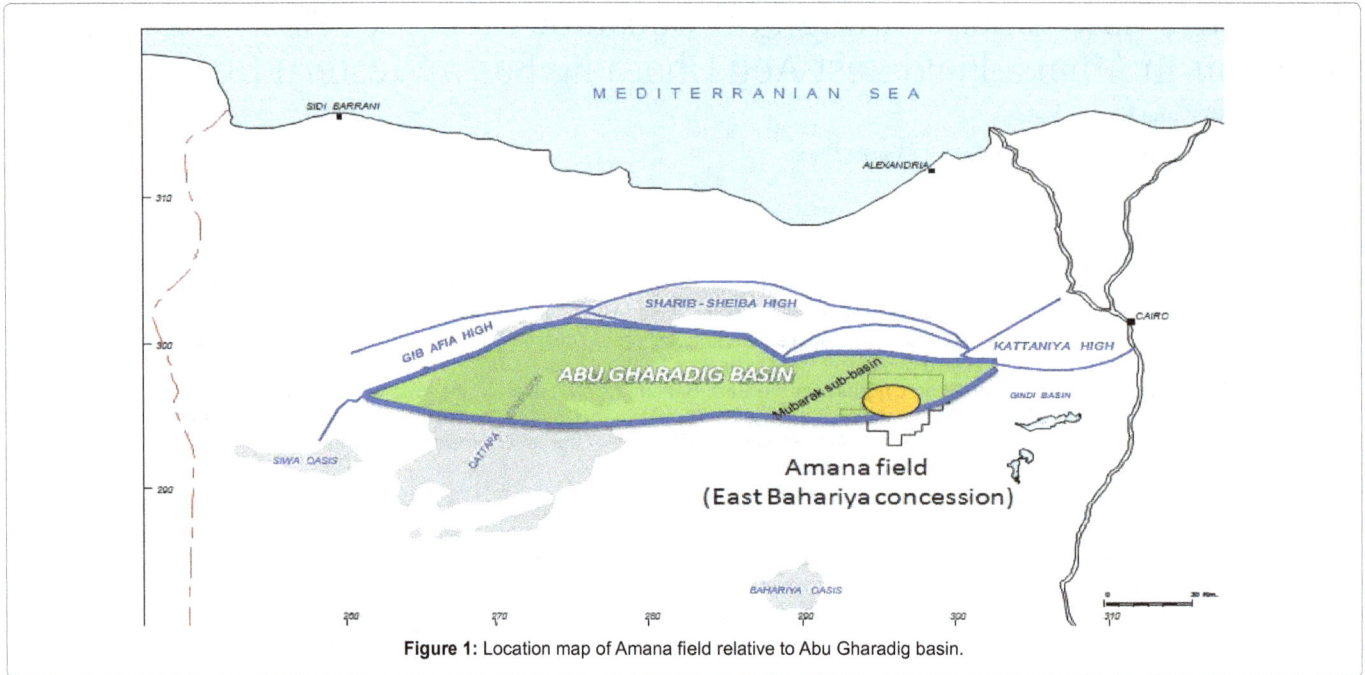

Figure 1: Location map of Amana field relative to Abu Gharadig basin.

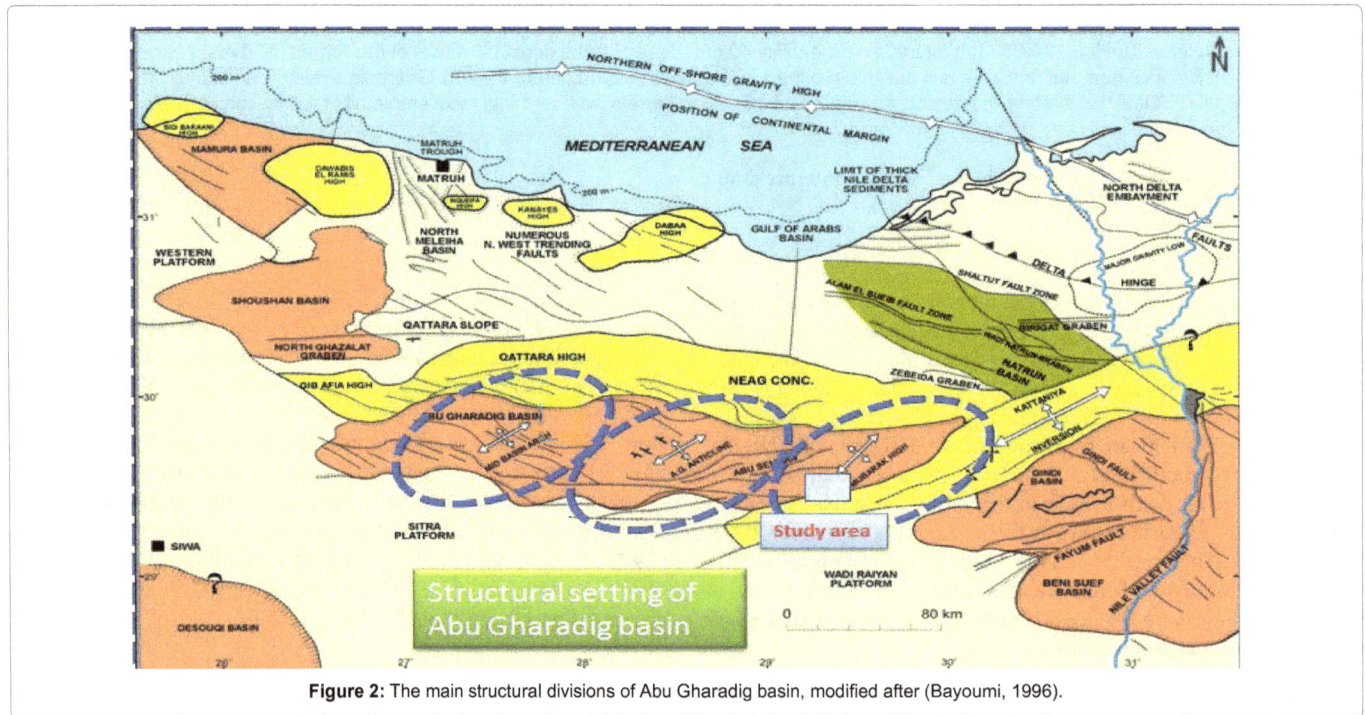

Figure 2: The main structural divisions of Abu Gharadig basin, modified after (Bayoumi, 1996).

data, well logs and well test results. Conducting the Model is starting through seismic data to be interpreted explaining the areal extent of the interpreted surfaces and the regime of the faults running in the area, and then suppose the horizontal gridding in which the model spreads the values, with determination of the trends of the modeled area. After that, a high light should be focused on the reservoir bed by Zoning stage, frequently, the tying with well data and well correlation should be processed to help in facies modeling layering the reservoir bed for more accuracy. The facies and petrophysical properties are distributed (i.e. interpolated and extrapolated) to assign a value for each cell in volume calculation stage as shown in Figure 5.

Data collection and quality control

In order to acquire a reliable 3D geostatic model, it is necessary to use all available data related to the study area. The data should be quality checked, and environmental corrected, in addition to removing the spikes or any unreliable readings. Almost all seismic data are enhanced and qualified for interpretation after passing through the stage of processing. Some complicated corrections should have been conducted on the well logs to be suitable for log analysis. These corrections are varied in certainty levels and have various types of mechanism. The most common log corrections that have been run are:

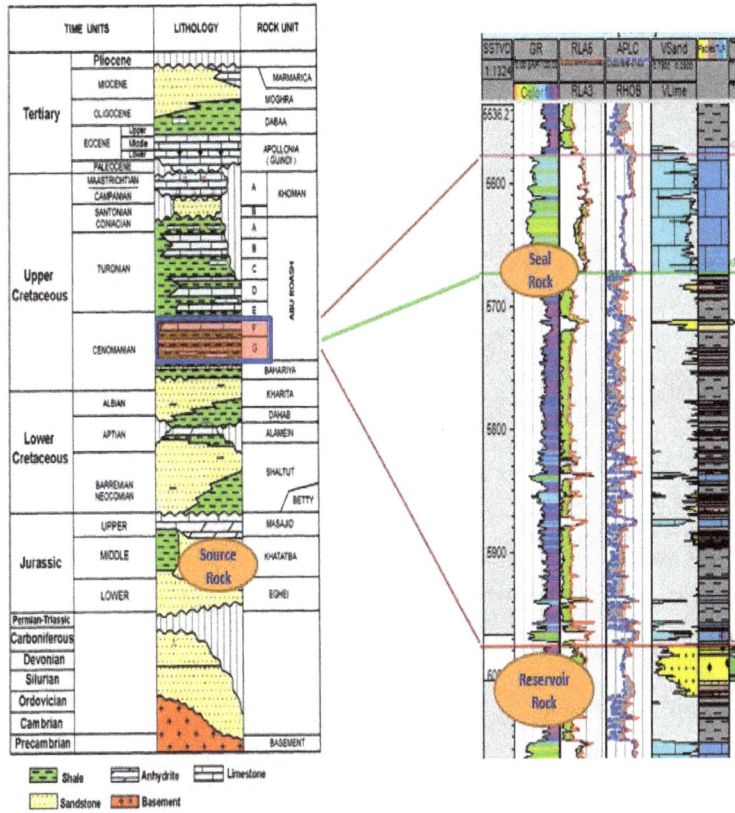

Figure 3: Generalized lithostartigraphic column in Abu Gharadig basin with high light on reservoir and seal beds in Amana field [14].

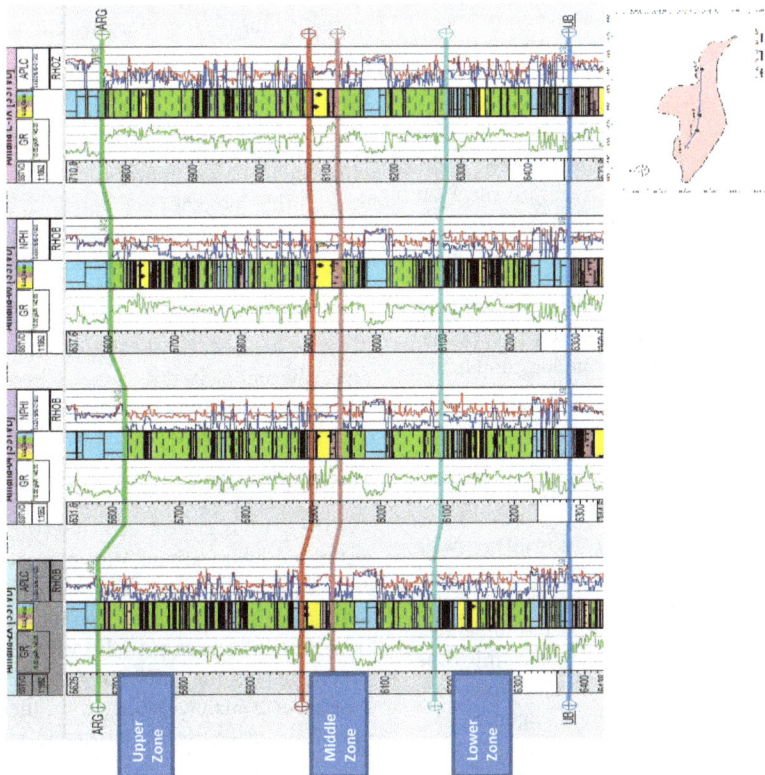

Figure 4: Stratigraphic correlation for Abu Roash "G" Member through Amana wells with key map.

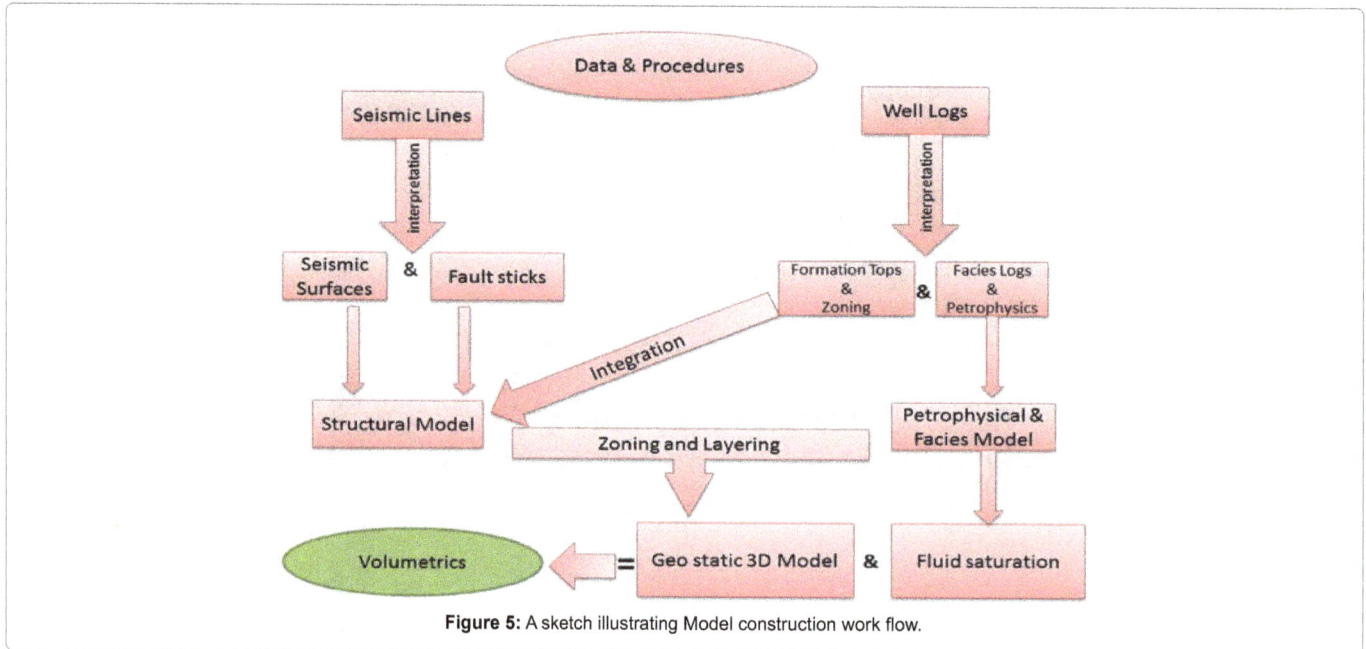

Figure 5: A sketch illustrating Model construction work flow.

Temperature and pressure effect removal.

Mud weight, salinity and mud resistivity calibration.

Gamma Ray log corrections for mineral and organic matter effect, borehole washout, and mud weight.

Resistivity logs' corrections for mud salinity, borehole size, and mud resistivity.

Neutron, Density and sonic logs' corrections for borehole size, drilling fluids and formation fluids, mud cake thickness.

Well log analysis

In order to conduct the 3D geostatic model, it is necessary to analyze the well logs to obtain the most interesting properties needed for spatial distribution. After the logging acquisition has been done, petrophysicist will start the correction and the quality control on the data, and then will interpret and evaluate the logs resulting in a traditional petrophysical evaluation. This petrophysical evaluation is the most reliable stage in the static reservoir study, since it generates a series of vertical profiles, for each well in the reservoir, describing the main properties of the reservoir pore system [6], such as Vshale, porosity, fluid saturation, estimated permeability from logs, and net to gross ratio.

Shale volume

The estimation of the volume of shale and clay minerals within the reservoir interval has an essential impact on the effective porosity [7]. In this study gamma Ray logs (wireline or LWD) are the main indicator for determining the volume of shale for Abu Roash "G" reservoir. And it was found that the middle zone is considered a clean sand zone with average volume of shale less than 8%, and it is clearly recommended to use Archie's equation in determining water saturation for this zone while the upper zone has an average volume of shale more than 13% and the lower zone shows 23% average volume of shale (Figure 6).

Total porosity

Determination of porosity can generally be considered the least complex stage in the petrophysical interpretation but the most important one because it defines the quantity of the hydrocarbons present in the reservoir [8]. The most frequently used methods in determining porosity are those based on the interpretation of well logs. Actually, none of the logging tools measure porosity directly, but the interpretation is carried out by indirect measurements [6], even neutron technique, which detects the hydrogen resides in the pore spaces (as a function of water and/or hydrocarbon), and the logging unit used this hydrogen index in order to simply deduce the formation porosity, thus the total porosity can be directly read from the neutron display.

Effective porosity

Not all the fluid presented in the rock is movable, because of the existence of isolated pores (not connected together) and wettability, thus during determining porosity, it should be the effective one regardless the immobile fluid. The effective porosity is determined clearly by using total porosity and volume of shale from this formula:

$$\Phi_{eff} = \Phi_{total} * (1\text{-}Vsh)$$

Abu Roash "G" Mbr. has a wide variation in effective porosity, especially among its three zones, where the Upper zone has an average effective porosity of 11%, Middle zone has 21% and the Lower zone displays about 14% Φ_{eff} (Figure 6).

Water saturation

The porous system of a reservoir rock is filled fluids, typically water and hydrocarbons with percent depending upon the chemical and physical properties of the rock and the fluids as well as the interaction between rock and fluid (rock wettability). The water saturation influences not only the volume of hydrocarbons, but also the productivity of the wells. For the purposes of reservoir studies, water saturation is mainly predicted on the basis of log interpretation for open holes using the famous Archie equation for clean formation [9].

$$Sw = \sqrt{\frac{F*Rw}{Rt}}$$

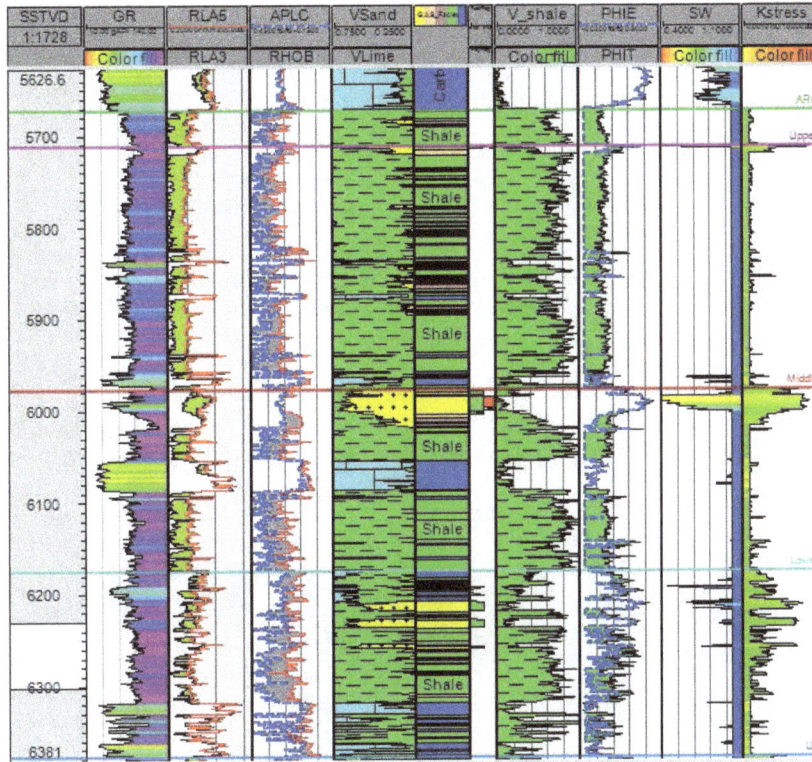

Figure 6: Log interpretation in Amana field.

Where: **Sw** ..... is water Saturation

Rw ..... is Formation water resistivity

Rt ...... is true resistivity

F...... is the formation resistivity factor

Calculating water saturation for Abu Roash "G" Mbr. has indicated that the whole member is a water bearing zone S_w is more than 90%, even the Upper and Lower Zones, except the Middle zone which is subsequently considered as a hydrocarbon bearing zone with average water saturation of 30% (Figure 6).

Permeability

The productivity of the wells and reservoir ability to feed drainage areas are the function of the permeability which is the most difficult parameter to be determined because of the extreme spatial variability among wells. The measurement of permeability derives from Darcy's law in case of available core data, but if the core data is absent, it will be estimated from the well logs by Timur equation [10].

Timur permeability: Timur permeability is given by equation

$$K = \left(\frac{100 * \Phi e^{2.25}}{Swirr} \right)^2$$

Estimation of permeability for Abu Roash "G" Mbr. showed that the permeability average of the Middle zone is about 100 mD and more, while all the rest of the member is ranging from 1 to 40 mD (Table 1).

Net to gross ratio (N/G)

For more detailed description of reservoir, the gross and net thickness should be calculated. Some geologists depended on the porosity cut-off to distinguish between the gross and net thickness, and others are depending on permeability cut-off, but a more rigorous approach to net thickness determination is on the basis of a detailed analysis of the rock properties [11,12]. The net to gross is a relation compiling the gross thickness and net thickness in one function obtained from dividing the net thickness by the gross thickness as a decimal fraction or percentage. The N/G ratio was calculated for the three zones in Abu Roash "G" Mbr. as shown in the figure below, illustrating that the N/G ratio increasing in the Middle zone in the central part of the study area (Figure 7).

Structural modeling

Structural model is the initial point of the reservoir model construction. It is considered as the backbone of the reservoir model. The structural model for the reservoir is basically conducted through two major procedures; mapping the structural geometry of the top of reservoir, and defining the set of faults running through the reservoir. These two procedures are ideally performed through the stage of seismic interpretation, where this stage involves two modules of interpretation; Horizon interpretation which is acquired on the horizons of interest, and Fault interpretation for the faults affect the reservoir [13,14]. In Horizon interpretation, all seismic lines and section used for covering all the area of interest, while in Fault interpretation, the cross lines (the lines that are perpendicular to the major fault trend) should be selected at the first in order to clearly pick the faults with their real throw and trend. Interpretation of the seismic lines in Amana field explains the structural setting affected the reservoir and formed the hydrocarbon trap in the area. As shown in the figure below, Amana oil field displays a horst block that is considered a good structural trap for accumulating the hydrocarbon in it.

Well Name	Total Porosity	V _ Shale	Eff. Porosity	Water Saturation	Net Pay	Estimated Permeability
	Φt (%)	(%)	Φeff (%)	Sw (%)	(Ft)	K (mD)
Amana-1X						
Middle zone	24	8-Feb	22	18	32`	100 - 250
Lower zone	15	20 - 33	14	52	wet	0.5 - 4
Amana-2						
Middle zone	22	8-Jun	22	28	28`	110 - 150
Lower zone	20	23-Oct	16	50	wet	20 - 30
Amana E-1X						
Middle zone	23	7-Apr		21	30	20`	100 - 300
Lower zone	21	20 - 30		17	55	wet	Feb-40
Amana-3						
Middle zone	22	16-Jan	19	41	10`	50 - 150
Lower zone	19	20-Dec	15	90	wet	30 - 50

Table 1: Average petrophysical properties for Abu Roash "G" Reservoir in Amana Field.

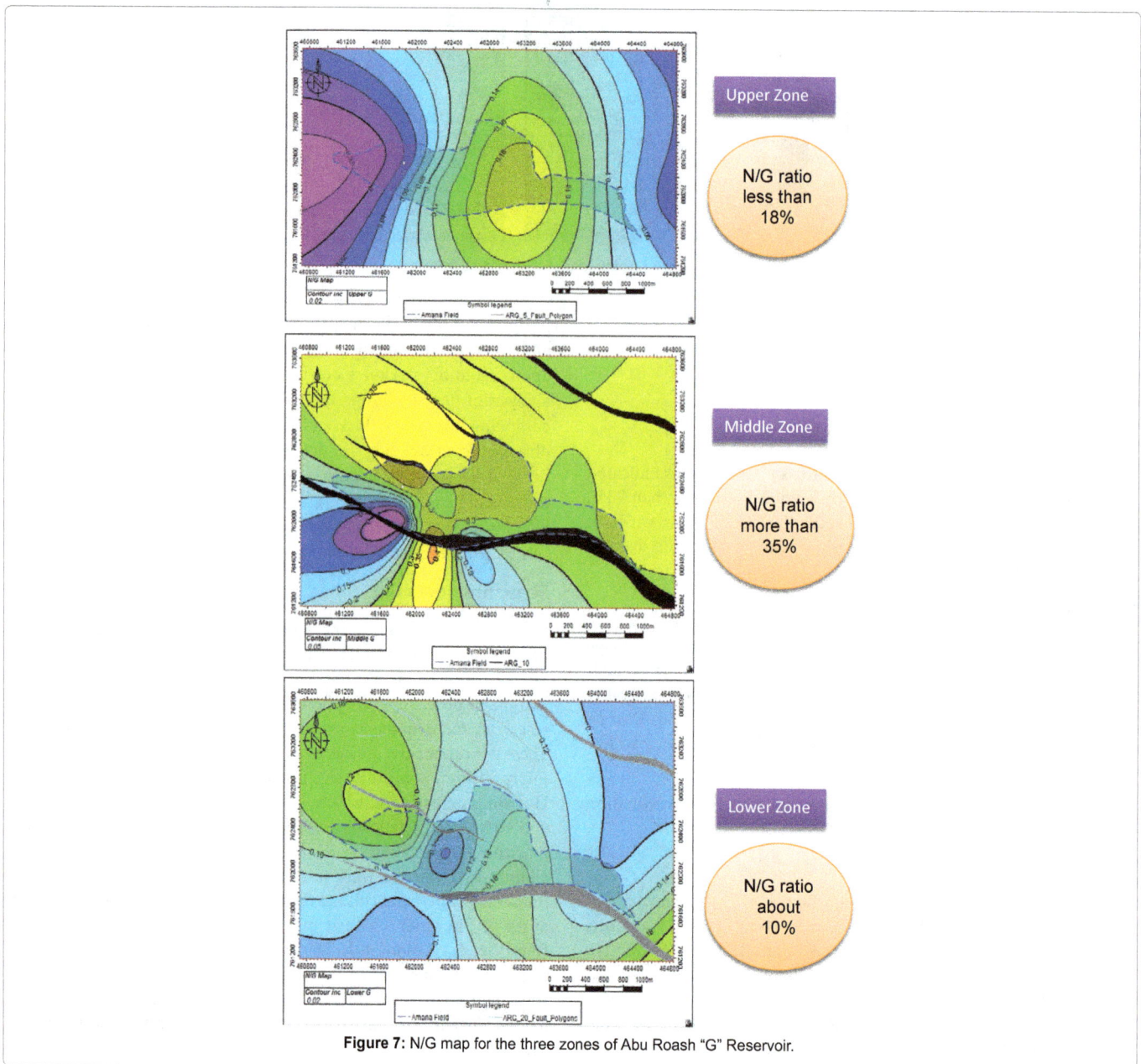

Figure 7: N/G map for the three zones of Abu Roash "G" Reservoir.

The direct results for seismic interpretation are summarized in the lateral explanation of the fault system affecting the reservoir as shown in Figure 8, in which the tectonic setting shows a flower structure that indicates a wrench fault system resulting in a horst structure, and mapping the structural geometry of the top of reservoir that displays NW-SE trending normal faults as shown in Figure 9.

The structural model of the reservoir is the result of compiling the output of Horizon interpretation and Fault interpretation, thus the indirect results for seismic interpretation are the basic inputs for constructing structural model, such as fault sticks (*the fault interpretation in lateral view*), also defined as sets of line data

that represent the fault plane, and seismic surfaces (*the horizon interpretation*), so that an interpretation has been carried out on Abu Roash "F" "G" and Upper Bahariya surfaces (Figure 10).

The structural model is carried out through four steps as follow.

Fault model: This step converts the fault sticks (*the output from seismic interpretation*) into key Pillars which are easily editable in 3D grid. This step is considered as a fine tuning for the faults to be more fitting in the structural model as the key Pillars connecting two faults, are used for extending a short one, making cross faults, or dividing one to two faults (Figure 11).

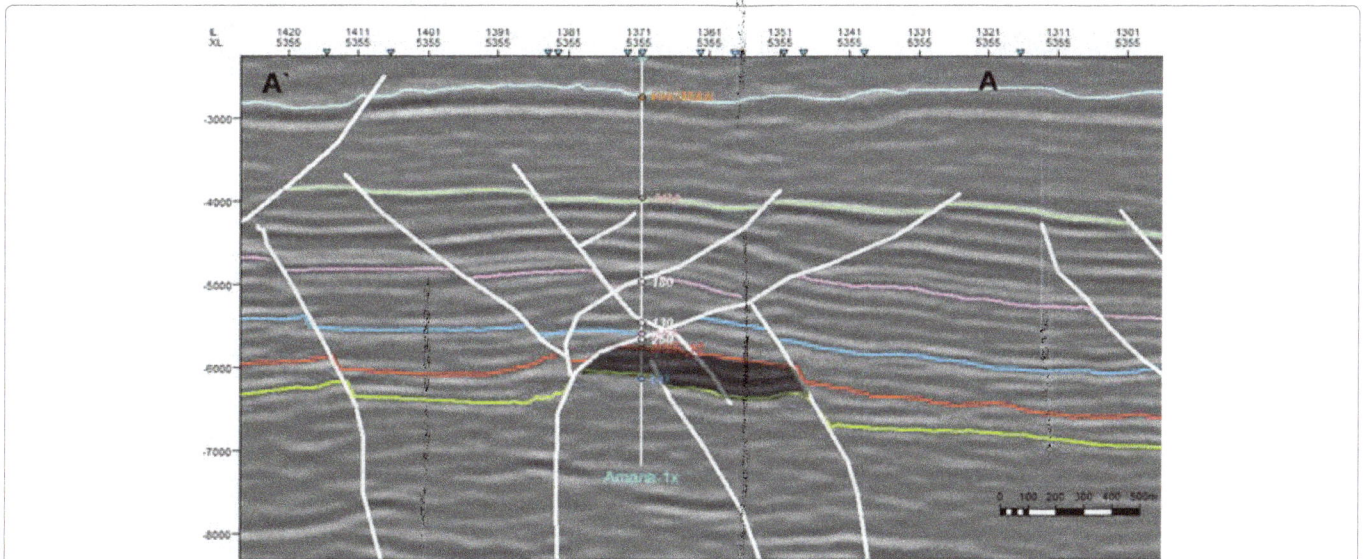

Figure 8: A Seismic X-line showing the structural setting in Amana block.

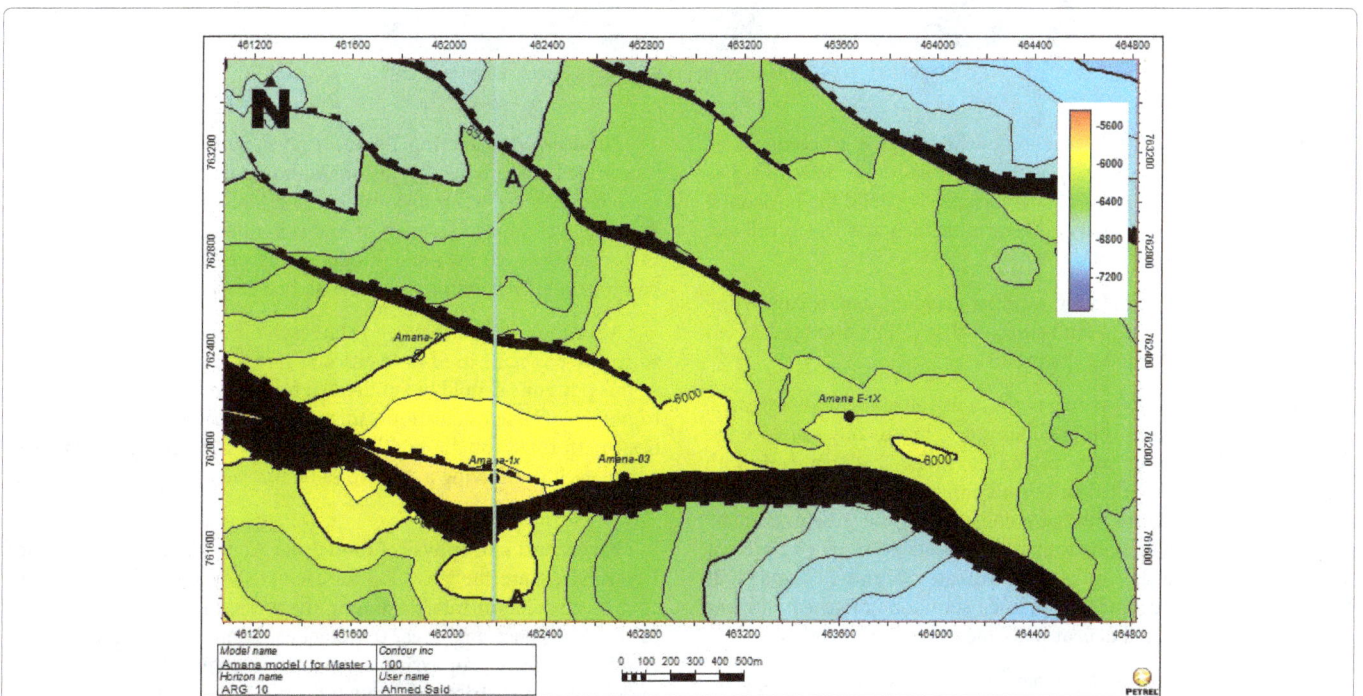

Figure 9: Structural map on top Abu Roash "G" Middle reservoir in Amana field.

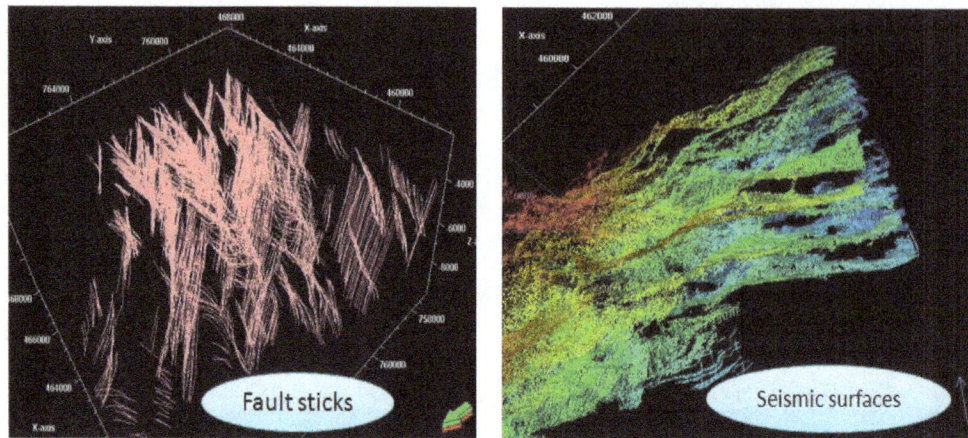

Figure 10: Showing seismic data as input for constructing the structural model.

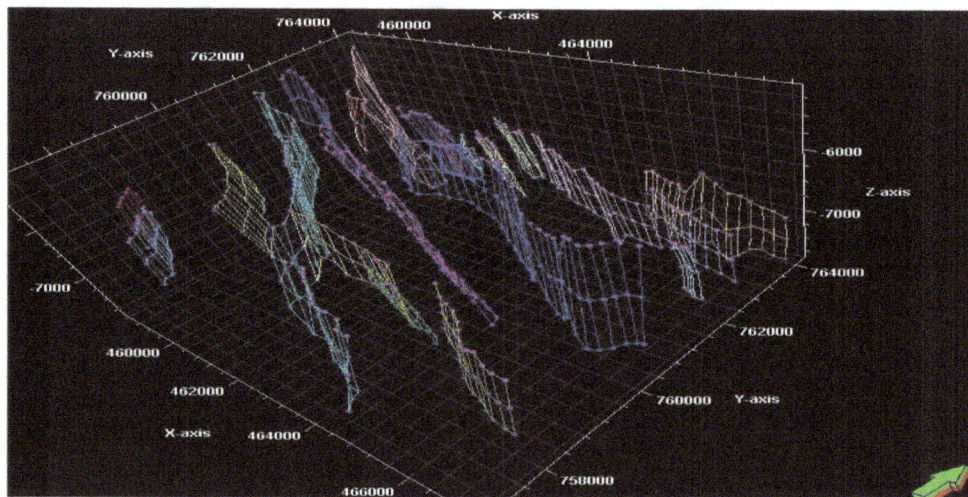

Figure 11: 3D window showing the Key Pillars in fault model step.

Pillar gridding: The construction of the Model is proceeded after that with Pillar Gridding step in which the trending is determined as **I, J,** and **K** directions (Figure 12). In this step we build the horizontal gridding of cells in the Model and determine their size which will aid, after that, in volume calculation.

Another important function of Pillar gridding step, is converting the faults in key Pillars into fault surfaces that will be displayed as cell walls in the final static model (Figure 13).

Making horizons: Regarding to the faults, all of them were treated, edited and smoothed in the previous stage, but in this stage we are dealing with the seismic surfaces and 2D grids. This step deals with the seismic surfaces resulted in seismic interpretation and uses them in order to make horizons fitting for constructing the structural model (Figure 14). The resulted horizons from this stage are celled horizons in which each cell takes a value of coordinates (**X** and **Y**) and a value of each property. The resulted horizons should be matched with the formation tops from well data at wells' locations.

The making horizons stage may be considered the final step for structural model, but in most cases, the resulted horizons are for major surfaces only, thus it is recommended to be zoned and detailed.

Zonation and layering: Because only major seismic surfaces can be interpreted, and frequently, just major horizons will be involved in the model, so it is necessary to consider the minor surfaces and sub-zones. By using well tops of formations and sub-zones, it is possible to make sub-zones between the major modeled surfaces for high lightening the pay zone involved within the reservoir bed as shown in Figure 15.

After the zoning process for the reservoir in order to focus on pay zone, the pay zone itself has vertically varied values of the properties, so this pay zone should be subdivided into a number of small layers in to obtain reliable calculations, this process is called layering and had been performed for the middle zone reservoir resulting in ten layers where each one having a reliable value for each property as shown in Figure 16.

Facies and petrophysical modeling: After Make zones stage and Layering, we use the facies log for each well and distribute the facies over all the area of interest among the wells, to predict the lithology in the non-penetrated area, this stage is named facies Modeling and is followed by Petrophysical Modeling in which the Petrophysical parameters are distributed through the pay facies (Figure 17).

Estimating properties for undrilled well locations: The main

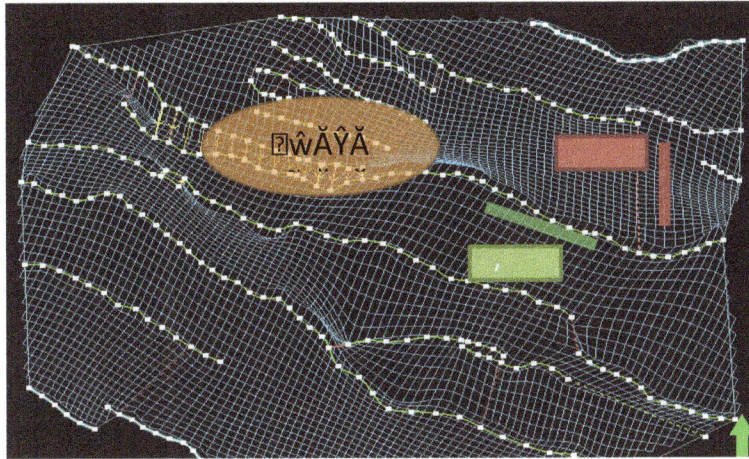

Figure 12: The I and J trends with horizontal Pillar gridding in Amana field.

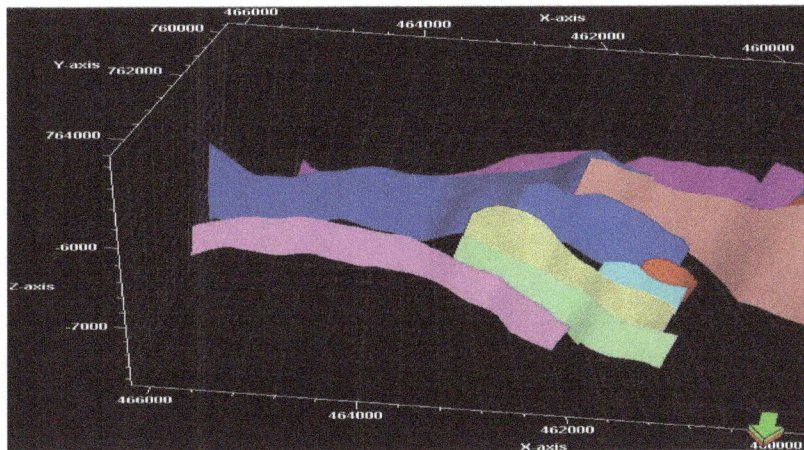

Figure 13: Showing the fault surfaces converted from Key Pillars.

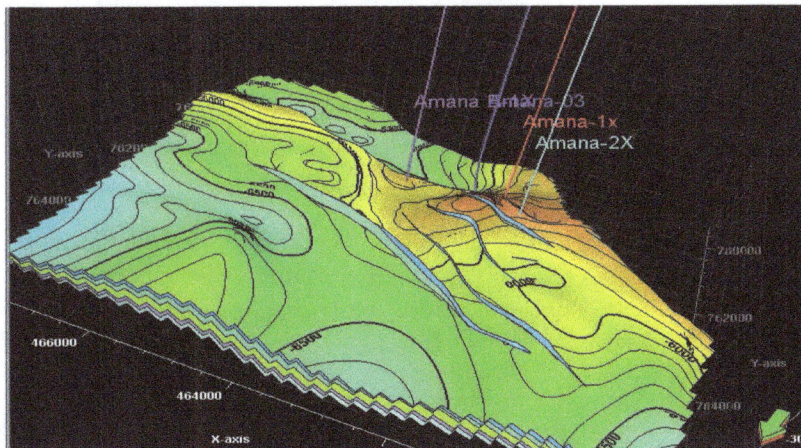

Figure 14: 3D modeled surfaces after making horizons stage.

target behind this work is to estimate facies and Petrophysical properties for the undrilled locations in order to drill new wells for increasing productivity and enhancing the field recovery. Figure 18 shows the area of interest (green area) above the oil water contact where the probability of oil presence is of a value, so it is recommended to drill more wells within the green area lying (far from the blue one), as the Geographic system, between:-

Latitudes: 29°33'20"N and 29°33'48"N

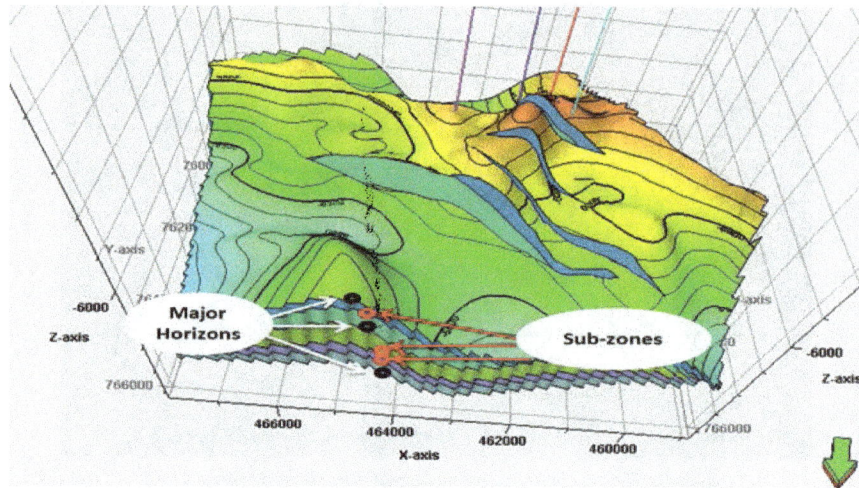

Figure 15: 3D zonation between the major horizons.

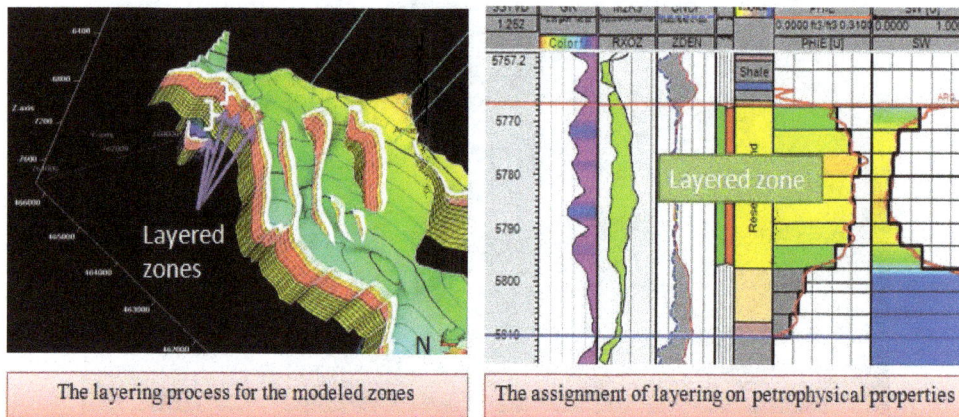

| The layering process for the modeled zones | The assignment of layering on petrophysical properties |

Figure 16: The layering process impact on vertical distribution of the data.

Longitudes: 29°25'17"E and 29°26`01"E

and as the Metric system, it lies between:

X: 462000 m E and 463200 m E

Y: 761800 m N and 762600 m N

Reserves and volumetric calculations: The final step (for completing this work) is volume calculation, where the well logs and seismic surfaces are used indirectly for calculating area, pay thickness, porosity, and fluid saturation in order to determine the OOIP and Reserves. In the 3D Modeling, as a result of facies distribution and calculating Effective Porosity and Water Saturation within the area of interest in the Model, the volume of oil in place could be calculated easily, and by adding the recovery factor, the recoverable reserve is determined.

The area of interest (above the oil water contact) is defined in a function of trap geometry delineated from the top by top seal and from the base by fluid contact as shown in Figure 18.

The net pay thickness is calculated from the well logs, petrophysical properties such as porosity and water saturation are averaged, thus, the application of recoverable reserve equation is achieved as follows [15]:

OOIIP = 7758 * A * H * Φ * So

Where:

OOIIP: Original Oil Initially in Place (reservoir barrel).

7758: conversion from (**acre.ft**) volume to barrel (**bbl**) volume.

A: the areal extent of reservoir (**acre**).

H: the net pay thickness (**feet**).

Φ: Porosity (**%**).

So: Oil saturation (1 - Sw) (**%**).

RR = OOIIP * FVF * RF

where:

RR: Recoverable Oil (stock tank barrel).

FVF: Formation Volume Factor (1/Bo). Bo is reservoir barrel multiplied by stock tank barrel (VR/VS).

RF: Recovery Factor (%).

The calculated volume of hydrocarbons in Abu Roash "G" member in Amana Field is displayed in Table 2.

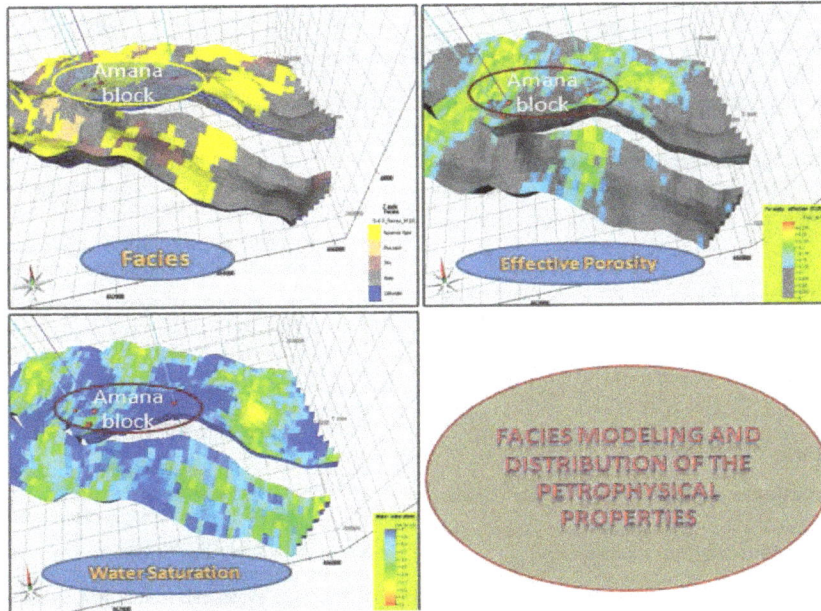

Figure 17: Spatial distribution through Amana block.

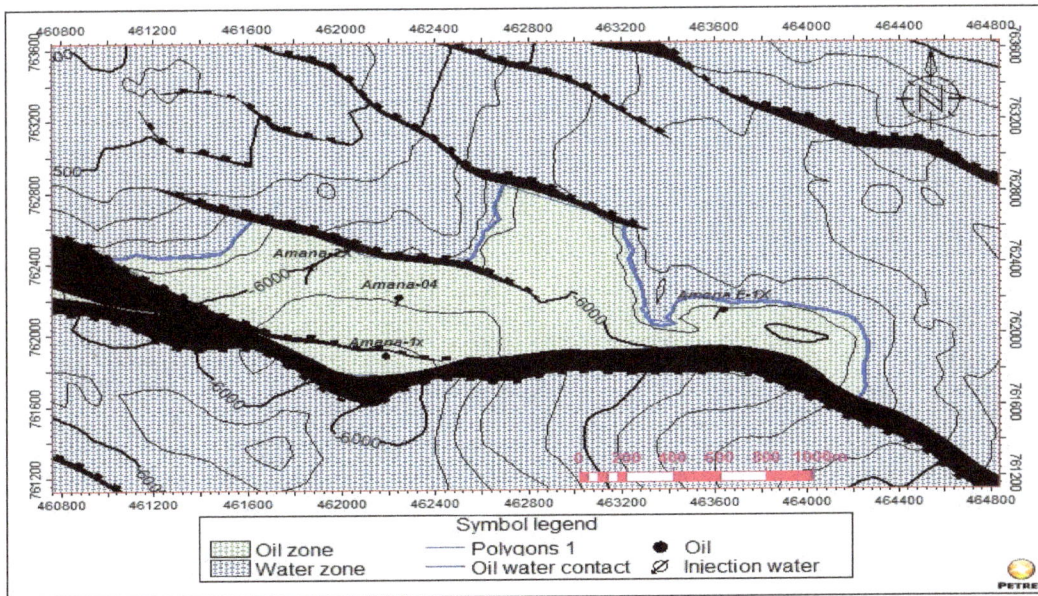

Figure 18: Promise area in Amana block.

Conclusions

The study area (Amana field) represents a positive prospect included in the most eastern part of Abu Gharadig basin where the proven oil that produced from Amana wells proved the presence of petroleum system in the area.

The seismic interpretation for the reservoir horizon and fault system in the area indicated that there is a structural petroleum trap (Horst style) that could be able to accumulate the oil in it.

Well logging interpretation was very useful in order to determine the reservoir and its lithology by constructing the facies log that

resulting in the existence of reservoir sandstone in middle zone in Abu Roash "G" Member which shows an average porosity more than 22% and permeability more than 100 mD.

Seismic interpretation integrated with well log analysis, indicats that Abu Roash "G" shales are typical top seal and lateral with Abu Roash "F" carbonate for the present reservoir and explained that the main structural features in Amana field appear to have had their maximum of development during the Late Cretaceous (base Khoman deposition), thus the timing of oil expulsion and primary migration (that had commenced during the Campanian age) relative to the structure trap formation should be favorable.

Case	Net volume[acre.ft]	Pore volume[RB]	HCPV oil[RB]	STOIIP [STB]	Recoverable oil[STB]
Middle Zone	12331	14,513,846	10,159,692	9,152,875	1,372,931

Table 2: The volume calculation results.

The reservoir areal extent and vertical thickness, in addition to, petrophysical properties are recorded in the 3D Model, so the volume of the original oil in place and the recoverable reserves have been easily calculated showing about 10 million barrels STOIIP and 1.3 million barrels initial recoverable oil without former of any stage of EOR(enhancing oil recovery).

3D petrophysical modelling yielded an area of promise in order to increase Amana field productivity, this area lies in the central northern part in the field, so it is recommended to drill more producer wells in this part and drill water injector wells in the most northern edge.

References

1. Schlumberger (1995) Well Evaluation Conference Egypt.

2. Enayet O (2002) Geology and petroleum potentiality of western desert- Egypt.

3. Meshref WM, Beleity A, Hammouda H, Kamel M (1988) Tectonic evaluation of the Abu Gharadig basin, AAPG Mediterranean Basins Conference.

4. Shora MM (2012) Sandstone as a reservoir rock.

5. Qarun Petroleum Company, Amana NW-1X well recommendation.

6. Krygowski D (2003) Guide to petrophysical interpretation, Austin, Texas, USA.

7. Alberty M (1992) Standard interpretation. In: Thompson DM, Woods AM (eds.) Development geology reference manual: AAPG Methods in Exploration Series 10: 180-185.

8. Cosentino L (2005) Oil field characteristics and relevant studies. In Exploration, Production and transport. Encyclopedia of hydrocarbons.

9. Archie GE (1942) The electrical resistivity log as an aid in determining some reservoir characteristics. American Institute of Mining, Metallurgical, and Petroleum Engineers. Transactions 146: 54-62.

10. Timur A (1968) An investigation of permeability, porosity, and residual water saturation relationships for sandstone reservoirs 9.

11. Gaynor GC, Sneider RM (1992) Effective pay determination. In: Thompson DM, Woods AM (eds.) Development geology reference manual, AAPG Methods in Exploration Series 10: 286-288.

12. Worthington P, Cosentino L (2005) The role of cut-off in integrated reservoir studies, In: Proceedings of the Society of Petroleum Engineers annual technical conference and exhibition, Denver (CO), 5-8 October, SPE 84387.

13. Bayoumi T (1996) The influence of interaction of depositional environment and synsedimentary tectonics on the development of some Late Cretaceous source rocks, Abu El-Gharadig basin, Western Desert, Egypt.

14. Mahsoub M, Abul-Nasr R, Boukhary M, Abdel-Aal H, Faris M (2012) Bio- and Sequence Stratigraphy of Upper Cretaceous - Paleogene rocks, East Bahariya Concession, Western Desert, Egypt. Geologia Croatica 65.

15. Murtha JA, Peterson SK (2001) Another Look at Layered Prospects. Presented at the SPE Annual Technical Conference and Exhibition, New Orleans, Louisiana, 30 September-3 October 2001. SPE-71416-MS.

11

Longitudinal Dependence and Seasonal Effect on Equatorial Electrojet Using MAGDAS Data

Ibrahim Khashaba A* and Essam Ghamry

National Research Institute of Astronomy and Geophysics, Geomagnetism, Egypt

Abstract

EE-index (EDst, EUEL), has been used to study the longitudinal dependence and seasonal variation of the equatorial electrojet (EEJ). The EUEL data eliminates many sources of disturbances by subtracting the median value of horizontal component (H) and the E Dst from H component data. EUEL data at a chain of stations along the dip equator have been analyzed to provide a detailed study on the equatorial electrojet. Data from eight stations (ANC, ILR, AAB, TIR, LKW, BCL, DAV and YAP stations) have been used for a period of three years (2009, 2010 and 2011). The longitudinal dependence has been studied for each year, a very good agreement between each year results has been found. This study shows that the magnetic signature of the EEJ is stronger in South America with a maximum at about longitude 77°W; and weaker in Asia, with a minimum in India, between longitudes 70°E and 90°E. The seasonal variations of the equatorial electrojet have been studied by both the whole data set (disturbed and quiet days) and quiet days' data. It has been proved that there is a semiannual variation in the equatorial electrojet with equinoctial maxima.

Keywords: EEJ-MAGDAS; Equatorial electrojet; Ground magnetic stations; EE index

Introduction

Egedal [1,2] discovered an electric current that flows in a narrow zone of approximately 600 km in width above the magnetic dip equator. This intense electric current, which in daytime flows in an eastward direction was named the "equatorial electrojet" by Chapman [3]. The EEJ represents a rather large enhancement of the diurnal variation in the horizontal or surface component of the geomagnetic field at and in the vicinity of the dip equator. The enhanced current was explained as being because of an abnormally large electrical conductivity [4,5]. Features of EEJ have been described for longitude regions of 75°W (Forbush and Casaverde 1961), 15°–19°E, 75°E, 5°W [6-12].Most of the first studies were carried out to explain the generating mechanism of EEJ [5,13]. Since the 1970s, some theories and physical models of the ionospheric dynamo have been developed in order to explain the mechanism of the EEJ flow and its main features (day to day, seasonal variability, counter-electrojet, electrodynamic processes of coupling with global scale current systems, etc.,) [14-19]. Another approach to simulate the EEJ has been done through the analysis of EEJ magnetic effects assuming simple current configurations. These configurations are the line current the thin-band current with different modes of latitudinal dependence: the uniform and parabolic and the "fourth degree" current distribution as well as the thick current distribution incorporating latitude and height dependence [3,6,7,20].

In the present study, the EEJ has been estimated using simultaneous ground based geomagnetic data recorded at 8 stations from MAGnetic Data Acquisition System (MAGDAS) at different longitudes (Figure 1 and Table 1). Also, the data of the three quietest days in each month during three years (2009, 2010 and 2011) has been analyzed. The obtained results have been used to study the longitudinal dependence and the seasonal variation of EEJ.

Data Analysis

At first, the three years (2009, 2010 and 2011) for 8 stations have been analyzed (Table 1). Here we deal with the whole data then focus on the three quietest days in each month for the whole period (Table 2).

The data used in the present work has been taken from the International Center for Space Weather Science and Education ICSWSE, Kyushu University, Japan. The data are in three forms H, ER, and EUEL data; where:

1. H component data: Variation in the North-South component at a certain magnetic station.

2. ER data (ΔH): The median value of the H component data, which is determined for the period from the start time of the observation to the end time, is subtracted from the H component data for each station.

3. EUEL data: Is calculated by subtracting the EDst index from the ER data.

The EE-index (EDst (Equatorial Disturbance in storm time), EU (index for equatorial electrojet), and EL (index for counter electrojet)), is proposed by Uozumi [21] to monitor temporal and long-term variations of the equatorial electrojet by using the MAGDAS/CPMN real-time data.

In the present study we deal with EUEL data in most cases.

Figures 2 and 3 represent the EUEL data for all magnetic observatories under investigation. The gaps that appear in most plots are due to problems in the measuring instrument.

Estimation of the EEJ

In this part, the EEJ has been estimated by measuring the maximum value of the EUEL for each quiet day at 12 Local Time at each station. Tables 3-5 represent the values of EEJ in the three quietest days each month during years 2009, 2010 and 2011 for the equatorial stations ANC, ILR, AAB, TIR, LKW, BCL, DAV and YAP.

*Corresponding author: Ibrahim Khashaba A, researcher, National Research Institute of Astronomy and Geophysics, Geomagnetism, Egypt
E-mail: khashaba80@yahoo.com

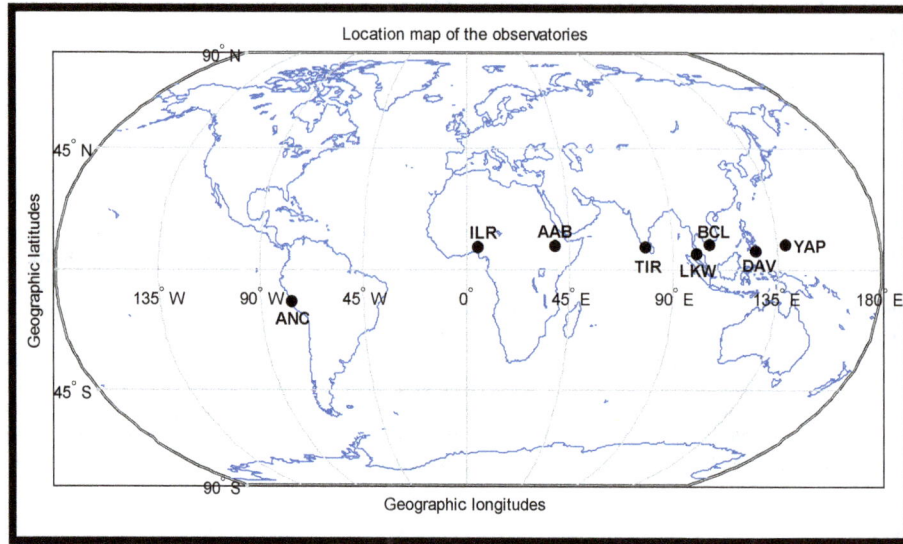

Figure 1: Location map of the geomagnetic stations used in the present work.

Stations inside the equatorial region						
Abbrev.	Station name	Country	GG Lat.	GG Long.	GM Lat.	GM Long.
ILR	Ilorin	Nigeria	8.50	4.68	-1.82	76.80
AAB	Adis Ababa	Ethiopia	9.04	38.77	0.18	110.47
TIR	Tirunelveli	India	8.70	77.80	0.21	149.30
LKW	Langkawi	Malaysia	6.30	99.78	-2.32	171.29
BCL	Bac Lieu	Vietnam	9.32	105.71	-0.66	177.96
DAV	Davao	Philippine	7.00	125.40	-1.02	196.54
YAP	Yap Island	FSM	9.50	138.08	1.49	209.06
ANC	Ancon	Peru	-11.77	-77.15	0.77	354.33

Table 1: Stations under investigation.

Months	Quiet days in 2009			Quiet days in 2010			Quiet days in 2011		
January	12	22	23	07	09	17	05	23	30
February	02	08	17	20	21	27	03	09	27
March	02	07	09	21	22	23	15	16	26
April	02	04	23	10	18	26	10	26	27
May	12	25	27	23	24	27	08	12	20
June	01	12	17	08	12	20	03	28	29
July	17	18	19	10	17	18	16	27	28
August	15	16	24	21	22	30	18	19	31
September	23	24	29	11	12	30	01	19	23
October	10	14	20	01	02	14	22	28	29
November	06	23	29	06	19	26	09	14	19
December	01	03	04	10	11	22	16	26	27

Table 2: The three international quietest days.

Longitudinal dependence during years 2009, 2010 and 2011

The longitudinal dependence of the EEJ is estimated through surface measurements of EUEL along the dip-equator. We use available magnetic data recorded at different longitudinal sectors.

In order to reduce the variables that may affect the strength of the EEJ, we will analyze simultaneous records from available stations for each year. The EEJ is given by the mean value of the conjoint days of EUEL.

In year 2009, there were only three conjoint days (22/01/2009, 02/02/2009 and 17/02/2009) between seven stations (ANC, ILR, AAB, TIR, LKW, DAV and YAP), while in year 2010 there were eleven conjoint days (22/08/2010, 30/09/2010, 01/10/2010, 02/10/2010, 14/10/2010, 06/11/2010, 19/11/2010, 26/11/2010, 10/12/2010 11/12/2010 and 22/12/2010) between seven stations (ANC, ILR, AAB, TIR, BCL, DAV and YAP). In year 2011, there were five conjoint days (26/04/2011, 27/04/2011, 08/05/2011, 12/05/2011 and 20/05/2011) between seven stations (ANC, ILR, AAB, TIR, LKW, DAV and YAP).

As shown in Figure 4, there is a very good agreement between the trends of the equatorial electrojet strength in the three studied years.

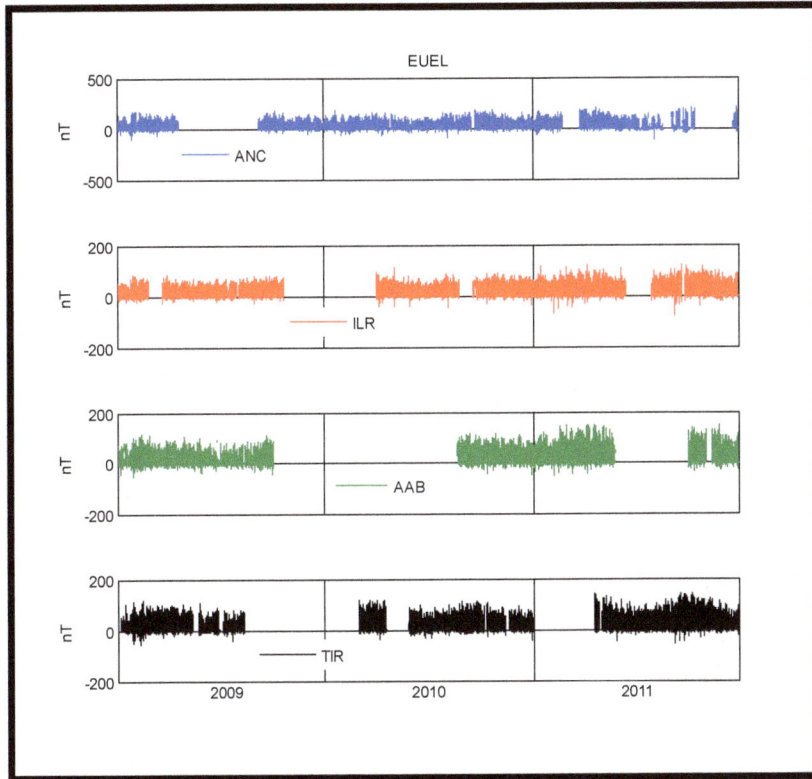

Figure 2: EUEL data for ANC, ILR, AAB and TIR stations during Years 2009, 2010 and 2011.

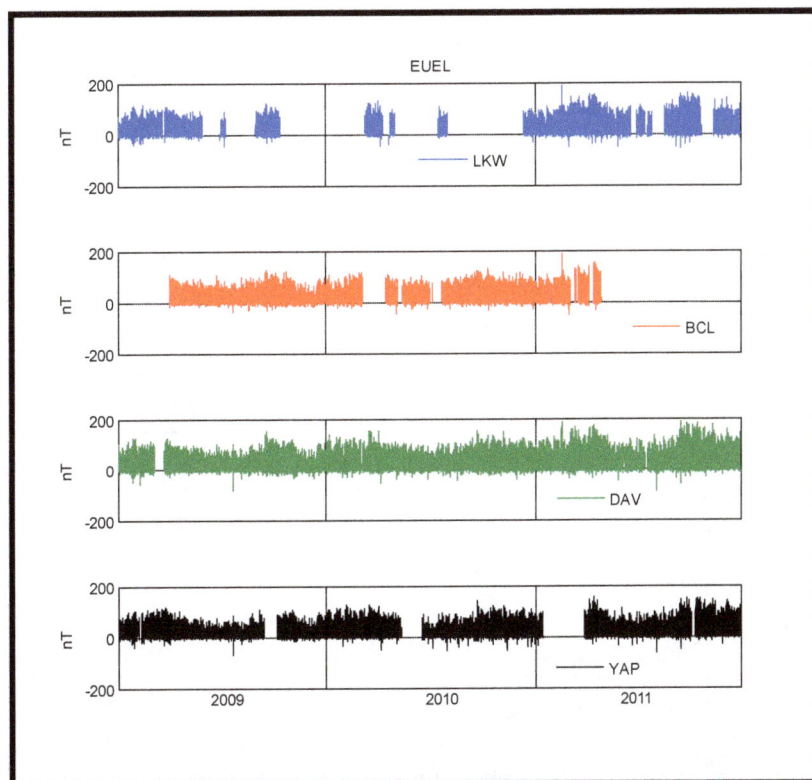

Figure 3: EUEL data for LKW, BCL, DAV and YAP stations during Years 2009, 2010 and 2011.

Dates	ANC	ILR	AAB	TIR	LKW	BCL	DAV	YAP
12/01/2009	128.4	41.7	--	39.5	43.7	--	55.6	51.9
22/01/2009	55.3	21.8	53.9	66	46.7	--	84.6	67.1
23/01/2009	55.5	32	--	53.7	64.3	--	82.7	66.5
02/02/2009	170	51.8	67.5	69	84.5	--	89.8	74.7
08/02/2009	93.5	59.6	92.9	68.6	51	--	90.1	--
17/02/2009	118.3	72	82.4	68.9	65.7	--	74.5	74.6
02/03/2009	132.1	--	59.1	70.4	91	--	94.2	96.4
07/03/2009	76.9	--	55.9	60.4	66	--	--	65.6
09/03/2009	132.9	--	69.1	77.5	68.8	--	--	105
02/04/2009	77.4	50.5	--	76.2	97.7	107.2	113.6	98
04/04/2009	93.3	38.3	88.3	--	91.3	93.2	92.1	66.5
23/04/2009	--	39.4	--	63.1	54	65.7	82	75.1
12/05/2009	--	43.2	49.6	65.7	76	76.1	83.9	59.5
25/05/2009	--	61.7	100.4	--	93.2	92.5	85.4	79.2
27/05/2009	--	54.6	74.3	77.1	74.8	76.8	70.6	60.2
01/06/2009	--	58.5	--	41.1	--	51.2	35.3	28.6
12/06/2009	--	52.4	81.8	57	--	59.4	48.9	34.7
17/06/2009	--	42.6	35.9	48.8	--	61.3	59.7	46
17/07/2009	--	38.8	54.2	61.9	--	53.6	58.9	51.1
18/07/2009	--	51.8	55.7	45.4	--	42.6	39.1	31.5
19/07/2009	--	38.7	35	44.4	--	54.9	47.9	40.9
15/08/2009	--	56.9	74.5	--	--	94.7	73.2	47.6
16/08/2009	--	45.2	48	--	--	62.7	58	35.9
24/08/2009	--	45.3	--	--	--	78.7	66.4	53.2
23/09/2009	160.2	66.6	--	--	80.8	85.7	94.1	--
24/09/2009	159.5	53.2	83.7	--	89.9	94	107.1	--
29/09/2009	147.9	57.9	69.2	--	75.1	71.1	64.1	--
10/10/2009	108.7	58.4	--	--	76.4	82.5	98	82.2
14/10/2009	99.9	39.5	--	--	56.3	61.2	73.3	61.7
20/10/2009	137.5	67.5	--	--	--	115.1	108.1	81
06/11/2009	115.6	--	--	--	--	87.1	97.2	83.1
23/11/2009	83.5	--	--	--	--	58	53	45.9
29/11/2009	104	--	--	--	--	39.9	45	48
01/12/2009	88.9	--	--	--	--	23.7	24.8	39.3
03/12/2009	128.4	--	--	--	--	--	24.2	22.7
04/12/2009	111.7	--	--	--	--	--	51.7	41.4

Table 3: Estimated EEJ in 2009.

Dates	ANC	ILR	AAB	TIR	LKW	BCL	DAV	YAP
07/01/2010	114.4	--	--	--	--	105.8	138.7	106.2
09/01/2010	80.4	--	--	--	--	--	81.7	78.4
17/01/2010	131.5	--	--	--	--	48.5	69.4	60.3
20/02/2010	115.4	--	--	--	--	101.6	123	102.5
21/02/2010	78.5	--	--	--	--	87.8	104	92.4
27/02/2010	122.1	--	--	--	--	86.7	102.7	91.6
21/03/2010	119.6	--	--	87.3	75.2	--	97.9	93.6
22/03/2010	109.1	--	--	87	80.7	--	104.1	101.8
23/03/2010	106.4	--	--	74.3	76.5	--	87.2	86.8
10/04/2010	72.5	60.2	--	78	--	--	95.1	98
18/04/2010	123.9	75.9	--	--	--	--	90.4	74.5
26/04/2010	--	64.4	--	--	98.9	93.9	98.4	95.9
23/05/2010	59.7	41.1	--	--	--	77	83	--
24/05/2010	127.2	51.7	--	--	--	54.3	71.8	--
27/05/2010	90.7	54.1	--	72.2	--	71.8	75.1	--
08/06/2010	57.9	40.2	--	71.8	--	83,7	78.8	--
12/06/2010	87.8	53.8	--	66.8	--	68.6	74.1	--
20/06/2010	71.1	30.1	--	43.8	--	64.5	36.6	22.6
10/07/2010	76.7	52.4	--	53.7	--	--	61.4	35.7
17/07/2010	82.6	44	--	69	82.3	--	71.6	53.4
18/07/2010	104.9	34.4	--	66.5	90.7	--	81.9	57.8

21/08/2010	99.9	48.7	58.7	50.9	--	--	86.4	62.3
22/08/2010	69.3	44.6	41.9	66.9	--	71.6	72.2	56.6
30/08/2010	138.5	--	75.8	70.5	--	63.2	64.6	50
11/09/2010	156.3	--	103.3	102.6	--	103	101.1	72.1
12/09/2010	155.8	--	72.6	91.3	--	94.2	91.5	71
30/09/2010	134.9	53.7	95	99.5	--	107	118.2	102.5
01/10/2010	136	55.3	44.1	67.6	--	86.3	105.3	93.6
02/10/2010	160	43.8	67.1	74.1	--	89.5	86	74
14/10/2010	120.8	59.8	71.4	64.6	--	80.2	99.4	84.3
06/11/2010	109.2	70.2	99.2	69.9	--	92.5	116.2	91.7
19/11/2010	121.9	46.1	69.9	74.9	--	76.7	86.5	80.3
26/11/2010	93.9	62.2	95.2	64.9	--	89.2	99.3	92.7
10/12/2010	110.7	80.1	99.3	76.8	--	105.7	125.1	110.6
11/12/2010	104.4	59.5	53.4	28.6	47.6	50.2	68.4	71.6
22/12/2010	79.9	61.1	87.3	78.4	95.9	97	89.6	80.9

Table 4: Estimated EEJ in 2010.

	ANC	ILR	AAB	TIR	LKW	BCL	DAV	YAP
05/01/2011	84.1	57.8	66.6	--	58.7	52.5	81.2	--
23/01/2011	131.9	61.5	69.3	--	98.1	95	114.1	--
30/01/2011	99.7	34.3	62.1	--	72.1	71	80.3	--
03/02/2011	83.2	60	89.7	--	53.4	47.5	66.9	--
09/02/2011	133.1	70.6	79.5	--	91.3	94	113.6	--
27/02/2011	--	43.4	77.3	--	71.7	73.4	89.5	--
15/03/2011	--	53.6	82.8	--	121.7	123.9	149.5	--
16/03/2011	--	63.8	99.1	--	111.1	--	110.2	--
26/03/2011	169	78.6	118.2	--	--	106.2	98.7	--
10/04/2011	129.1	79.3	131	--	116.2	--	117.7	95.7
26/04/2011	110.2	75.7	93.7	109.2	119.3	117.4	129.3	105.2
27/04/2011	155.3	77.3	104.9	106.9	114.6	--	135	106.5
08/05/2011	111.3	61.3	77	82.2	93	--	106.3	80.3
12/05/2011	106.2	72.8	90.1	80.3	75.5	--	95	73.8
20/05/2011	129.7	71.3	117.3	88.6	94	--	91.9	70.2
03/06/2011	75.2	57.6	--	56.7	75.5	--	74.8	65.8
28/06/2011	102.9	--	--	74.3	--	--	74.6	29.9
29/06/2011	128.1	--	--	88.2	--	--	86	69.7
16/07/2011	91.6	--	--	75.3	--	--	--	59.7
27/07/2011	--	--	--	74.9	77.3	--	68.5	59.4
28/07/2011	--	--	--	67.3	85.3	--	99.9	76.8
18/08/2011	--	62.1	--	67.5	--	--	123.6	110.8
19/08/2011	--	73.6	--	85.6	107.5	--	93.6	69.4
31/08/2011	--	55.5	--	60.5	74	--	80.1	74.3
01/09/2011	--	72.9	--	--	111.8	--	119.7	97.7
19/09/2011	177.8	60.1	--	141.3	143.3	--	159.5	127.5
23/09/2011	--	--	--	95.5	118.6	--	146.7	116.5
22/10/2011	--	91.8	112	123.7	141	--	121.6	104.8
28/10/2011	--	83	120	116.3	--	--	156.7	126.2
29/10/2011	--	80.1	106.4	106.4	--	--	162	138.8
09/11/2011	--	84.7	--	--	--	--	106.6	110.2
14/11/2011	--	94.9	--	106.8	--	--	177.3	145.2
19/11/2011	--	53.3	45.7	52	81.6	--	97.6	79.3
16/12/2011	--	50.5	75.5	54.6	76.9	--	115.7	112.8
26/12/2011	--	49	72.5	67	90.9	--	119.6	109
27/12/2012	153.9	81.7	82	--	80	--	94	100.6

Table 5: Estimated EEJ in 2011.

Mean longitudinal dependence

The longitudinal dependence of the EEJ is expressed by a numerical spline function estimated by calculating the mean value of the three years for each station.

Figure 5 shows that the magnetic signature of the EEJ is stronger in South America with a maximum at about 77°W. It is followed by a minimum in West Africa at about 4°E. The EEJ magnetic signature is weak in Asia-except Philippine, with a minimum in India, between 70°E

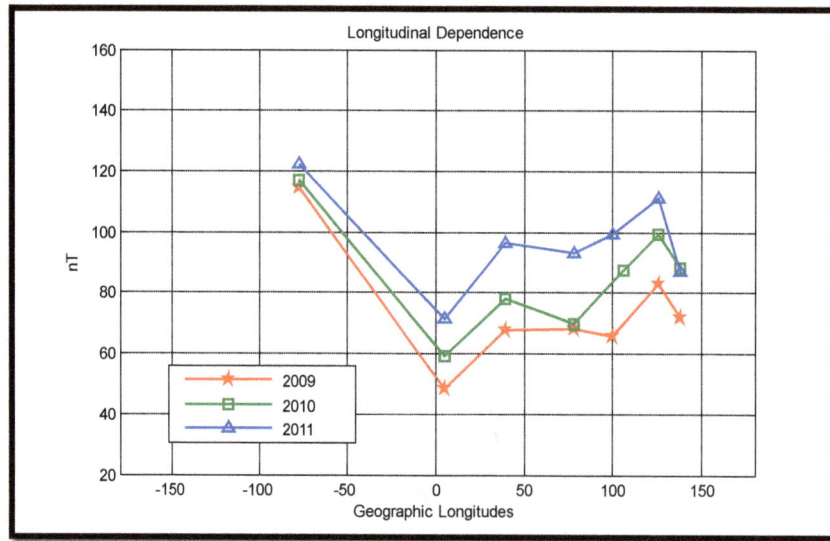

Figure 4: Longitudinal dependence of EEJ in years 2009, 2010 and 2011.

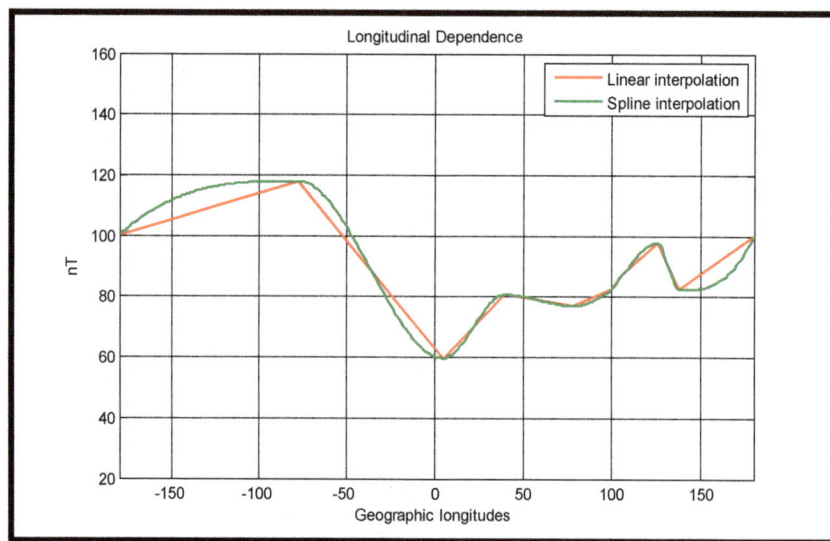

Figure 5: Mean longitudinal dependence.

and 90°E. A secondary maximum is at about 125°E. These longitude variations of the EEJ magnetic effect roughly follow variations of the inverse main field (1/B) at the dip equator.

Seasonal variation using whole data set

Figure 6 shows 30 day running means of the daily EUEL values for DAV station. These reveal clearly the presence of semiannual variation of the EEJ that maximizes during equinoctial months.

The seasonal variation of equatorial electrojet at DAV station reveals two cycles a year, each cycle reach maximum at equinoctial months. In year 2009 the EEJ started increasing until reaches maximum in March then decreased till midyear, the next cycle started from midyear then reached maximum in September and again started decreasing till end of the year. The same trend reoccurred in year 2010. In year 2011 the EEJ reached maximum in April and October.

Seasonal variation using quiet days

In this section we study the stations that almost have no or at least few missing days each year. These are, in 2009 DAV and YAP stations, in 2010 ANC station, DAV station and YAP station, in 2011 ILR station, LKW station and DAV station.

The number of day in the year is plotted against its EUEL value then a 4th degree polynomial fitting is performed for each station.

Figures 7-14 show clearly the semiannual variation of the equatorial electrojet.

Discussion and Conclusion

In the present study geomagnetic data from a chain of stations along the dip equator have been analyzed to provide detailed study in this region on the equatorial electrojet. Data from eight stations (ANC,

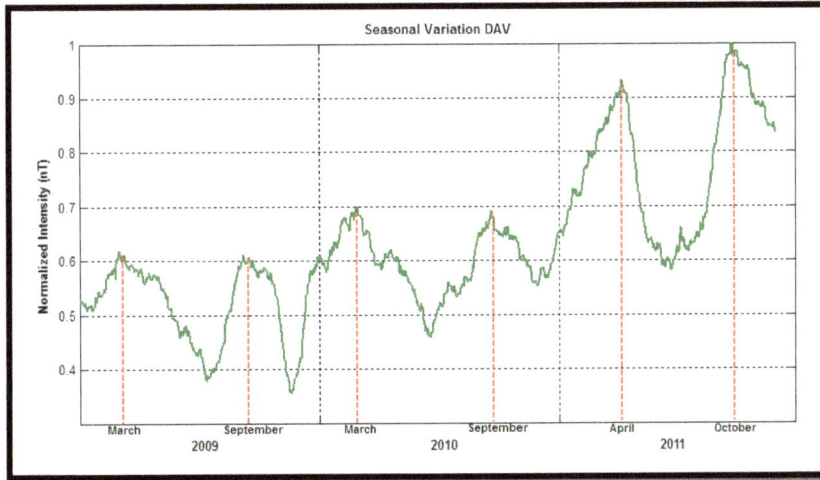

Figure 6: Seasonal variation at DAV (30 days running means of daily EUEL data), two cycles of current are clearly obvious with maximum at equinoctial months.

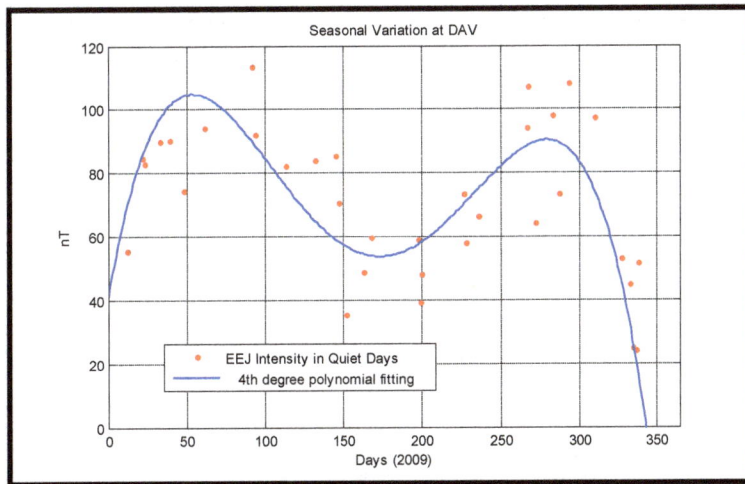

Figure 7: Seasonal variation at DAV in 2009, the EUEL values in quiet days are represented by red dots, the blue line represents a 4th degree polynomial fitting in order to view the trend of EUEL values, there is two cycles of current reaching their maximum during spring and autumn (equinoctial months).

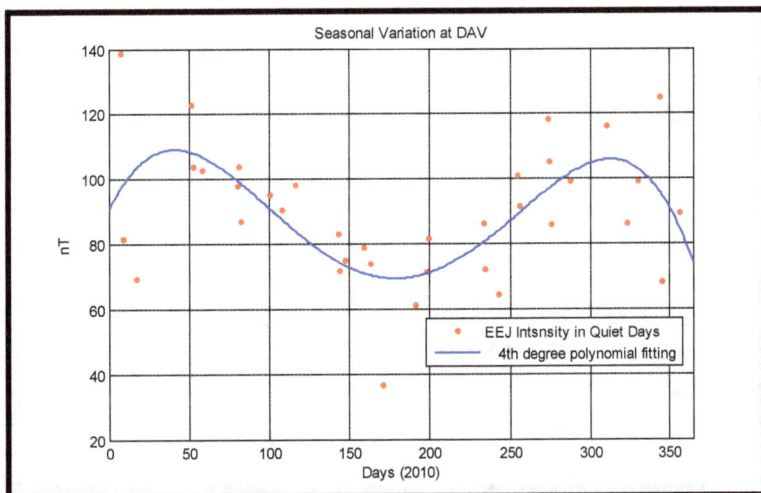

Figure 8: Seasonal variation at DAV in 2010.

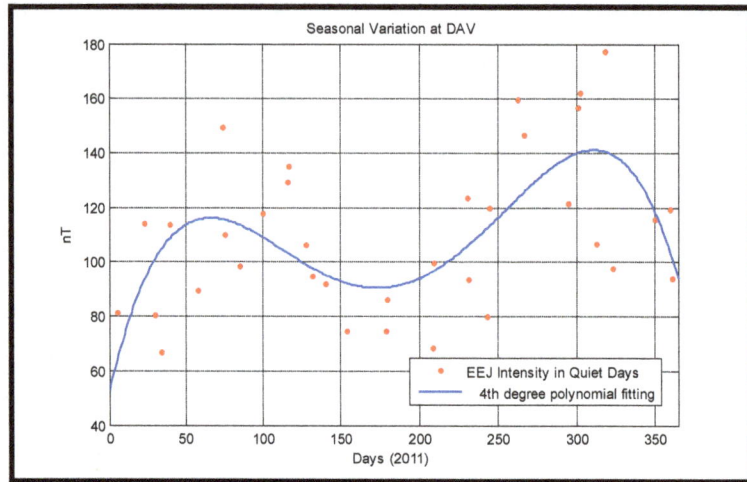

Figure 9: Seasonal variation at DAV in 2011.

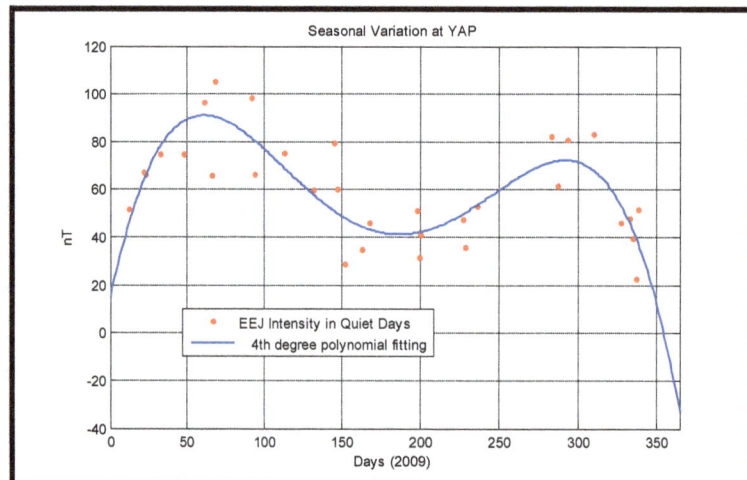

Figure 10: Seasonal variation at YAP in 2009.

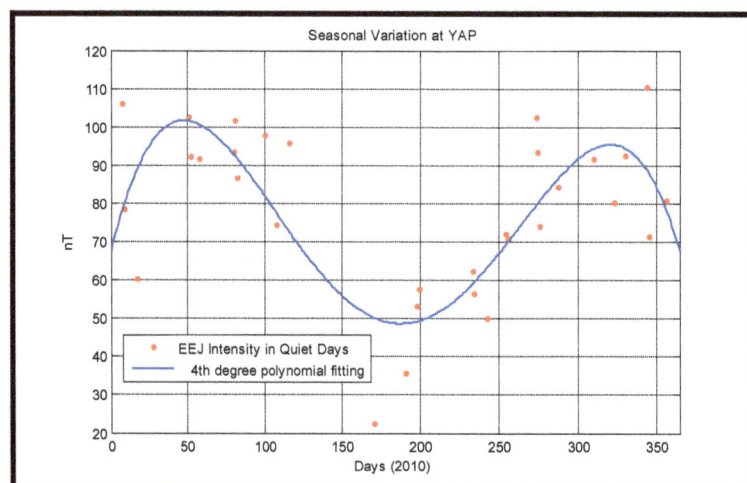

Figure 11: Seasonal variation at YAP in 2010.

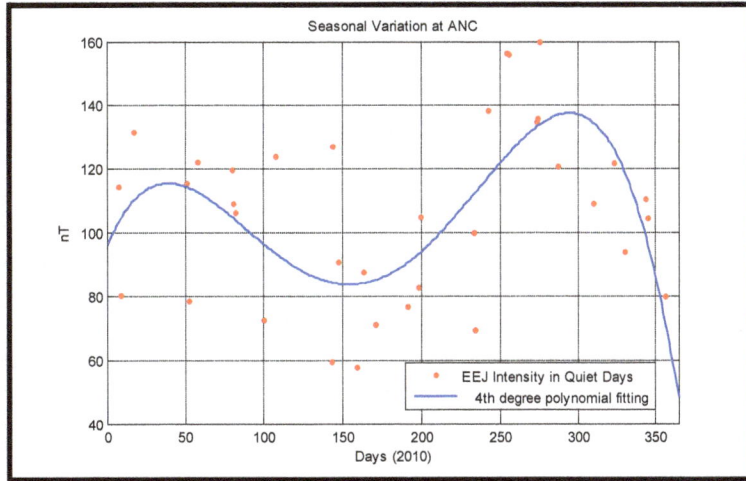

Figure 12: Seasonal variation at ANC in 2010.

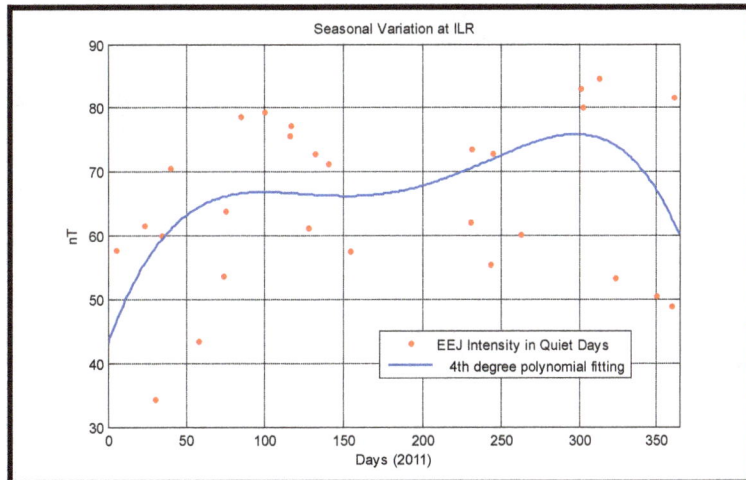

Figure 13: Seasonal variation at ILR in 2011.

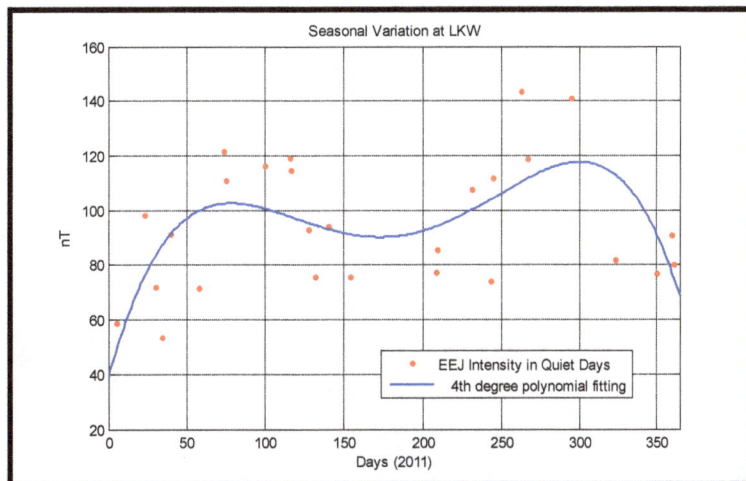

Figure 14: Seasonal variation at LKW in 2011.

ILR, AAB, TIR, LKW, BCL, DAV and YAP) have been used. Most of the analyzed data are EUEL data for a period of three years (2009-2011).

The strength of EEJ has is found to vary both from day to day for a single station and from station to another. The maximum value of EEJ in the period of study at ANC is 177.8 nT and the minimum is 55.3 nT, ILR maximum 94.9 nT and minimum 21.8 nT, AAB maximum 131 nT and minimum 35 nT, TIR maximum 141.3 nT and minimum 28.6 nT, LKW maximum 143.3 nT and minimum 43.7 nT, BCL maximum 123.9 nT and minimum 23.7 nT, DAV maximum 177.3 nT and minimum 24.2 nT, and YAP maximum 145.2 nT and minimum 22.6 nT.

As expected there is a local time dependence of EEJ. EEJ-related magnetic effects in the daily variations of the horizontal component appear at about 6 LT, reach a maximum near local noon and vanish after 18 LT.

Simultaneous surface magnetic records from eight stations in different longitude sectors have been used to study the longitudinal dependence of EEJ. Three years have been analyzed separately. There is a very good agreement between the results of the three years, which makes the obtained results as one of the reliable results to discuss the longitudinal dependence of EEJ. The intensity of EEJ is found to be stronger in South America with a maximum at about 77°W. It is followed by a minimum in West Africa at about 4°E. The EEJ magnetic signature is relatively weak in Asia with a minimum in India, between 70° and 90°E. A secondary maximum is at about 125°E. These longitude variations of the EEJ magnetic effect roughly follow variations of the inverse main field (1/B) at the dip equator.

The seasonal variations of the equatorial electrojet have been studied by both the whole data set (disturbed and quiet days) and quiet days' data. It has been proved that there is a semiannual variation in the equatorial electrojet with equinoctial maxima.

References

1. Egedal J (1947) "The Magnetic Diurnal Variation of the Horizontal Force Near the MagneticEquator". Terrestrial Magnetism and Atmospheric Electricity 52: 449-451.

2. Egedal J (1948) "Daily Variation of the Horizontal Magnetic Force at the Magnetic Equator". Nature 161: 443-444.

3. Chapman S (1951) "Some Phenomena of the Upper Atmosphere". Proceedings of the Physical Society of London 64: 833-843.

4. Hirono M (1952) A theory of diurnal magnetic variations in equatorial regions and conductivity of the ionospheric E region. J Geomagn Geoelec 4: 7-21.

5. Baker WG, Martyn DF (1953) Electric currents in the ionosphere.I. The conductivity. Phil Trans Roy Soc 246: 281-294.

6. Forbush S, Casaverde M (1961) Equatorial electrojet in Peru. Carnegie Institution of Washington, Washington.

7. Fambitakoye O, Mayaud PN (1976) Equatorial electrojet and regular daily variations of SR-I. A determination of the equatorial electrojet parameters. J Atmos Terr Phys 38: 1-17.

8. Fambitakoye O, Mayaud PN (1976) Equatorial electrojet and regular daily variations SR-II. The centre of the equatorial electrojet. J Atmos Terr Phys 38: 19-26.

9. Arora BR, Mahashabde MV, Kalra R (1993) Indian IEEY geomagnetic observational program and some preliminary results. Rev Brazil Geofis 11: 365-385.

10. Rastogi RG (1999) Ionospheric current system in Indo-Russian longitude sector. Science and Culture 65: 269-282.

11. Doumouya V, Vassal J, Cohen Y, Fambitakoye O, Menvielle M (1998) Equatorial electrojet at African longitudes: First results from magnetic measurements. Ann Geophys 16: 658-676.

12. Doumouya V, Cohen Y, Arora BR, Yumoto K (2003) Local time and longitude dependence of the equatorial electrojet magnetic effects. Journal of Atmospheric and Solar-Terrestrial Physics 65: 1265-1282.

13. Chapman S, Bartels J (1940) Geomagnetism. Oxford University Press, United Kingdom.

14. Sugiura M, Cain JC (1966) A model equatorial electrojet. J of Geophysical Research 71: 1869-1877.

15. Untiedt J (1967) A model of the equatorial electrojet involving meridional current. J of Geophysical Research 72: 5799-5810.

16. Sugiura M, Poros DJ (1969) An improved model equatorial electrojet with meridional current system. J of Geophysical Research 74: 4025-4034.

17. Richmond AD (1973) Equatorial electrojet-I. Development of a model including winds and electric field. Journal of Atmospheric and Terrestrial Physics 35: 1083-1103.

18. Stening RJ (1985) Modeling the equatorial electrojet. Journal of Geophysical Research 90: 1705-1719.

19. Reddy CA (1989) The equatorial electrojet. Pure and Applied Geophysiscs 131: 485-508.

20. Onwumechili CA (1967) Geomagnetic variations in the equatorial zone. Physics of Geomagnetic Phenomena-I. Academic Press, New York, London.

21. Uozumi T (2008) A new index to monitor temporal and long-term variations of the equatorial electrojet by MAGDAS/CPMN real-time data. EE-Index Earth Planets Space 60: 785-790.

12

Continental Drift and Plate Tectonics vis-a-vis Earth's Expansion: Probing the Missing Links for Understanding the Total Earth System

Sen S*

Council of Scientific and Industrial Research, Hemantika, O-26, Patuli, Kolkata, India

Abstract

Hypotheses like continental drift, plate tectonics or Earth's expansion should not be considered as viable in view of solid and rigid state of mantle which, in contrast, would resist such phenomena. Based on Hilgenberg's model of Earth's expansion (1933), the author elucidates that before expansion when the Earth was small and devoid of oceans, its mantle must have been sufficiently fluid owing to association of ocean-forming water. Further, matching thickness of fluid outer core and extent of radial expansion of the Earth strongly support that the outer core was opened up as a void geosphere owing to planetary expansion. At the deep interior of the planet, owing to occurrence of a void or pseudo-fluid geosphere separating basaltic mantle and solid iron core, an additional force of reverse gravity would develop acting in opposite direction of normal inwardly directed force of gravity. This postulation leads us to consider that in the deep interior of the planet an upwardly directed force of gravitational attraction would act in a predominant manner, thereby sustaining sufficiently low temperature and pressure condition and magnetic nature of the inner core which completely agrees with observed features of terrestrial magnetism.

Over the Earth's surface the crustal layer was fragmented due to expansion while through the expansion cracks widespread incidences of magma emission occurred forming rudimentary ocean basins. With further expansion, these basins were expanded and filled up with water that degassed from the mantle associated with the process of magma emission while owing to desiccation, the mantle itself eventually turned into a rigid body. Before expansion of the planet when the iron core and the mantle were juxtaposed to each other, due to external magnetic influence the magnetic iron core was deflected causing major change in polar and equatorial disposition of the planet. Subsequently in younger geological period when due to expansion a major void geosphere was opened up between the iron core and mantle, external magnetic influences caused the magnetic core to execute smooth revolutions giving rise to new magnetic phenomena like pole reversal and polar wandering, documented over the planet's younger strata. It may be noted that while due to expansion, the continental fragments would tend to move away from one another, owing to rotation of the planet along its axis of rotation some continental fragments came closer to each other or even collided to form mountain ranges.

Keywords: Geospheres; Plate tectonics; Silicate rocks; Celestial body

Introduction

More than four hundred years Ortelius [1] prepared the first authentic atlas of the world which showed occurrence of parallel shorelines between Africa and South America across the Atlantic Ocean. This remarkable feature led the pioneer cartographer to consider existence of these continents in a conjoint manner in the past. With further improvement of the atlas occurrence of parallel shore lines between distantly placed continents became conspicuous and in due course many Earth scientists attempted to adjust the relevant continental fragments.

With progress in geology, a new branch termed global tectonics was developed for elucidation of Earth's structure. Further advancement in a consorted manner by geologists and geophysicists established presence of three thick geospheres in the interior of the planet, two of which are solid, namely the 2867 km thick basaltic mantle occurring below the granitic crust of 33 km and the deepest or inner-most part of the planet termed inner core of 1391 Km radial thickness and constituted of solid iron with some nickel [2-5]. These two solid geospheres are separated by a 2080 km thick fluid geosphere, termed Outer Core which too has been considered in the prevalent concept to be composed of iron although in fluid state. Based on the observation of increase of temperature with depth recorded in the crustal surface and mantle, the prevalent concept conjectures that values of temperature and pressure steadily increase up to the center of the core attaining to nearly 4000°C and 3.5 million atmospheres, respectively.

In course of time various concepts on Earth sciences were developed amongst which two most extensively discussed views are based on unaltered dimension of the of Earth throughout the past geological periods. These views are continental drift, vehemently debated since 1912 when the concept was put forward by Wegener [6-8], though not for the first time, and the concept of plate tectonics-which in reality is a modified version of the drift theory. The concept of plate tectonics [9], developed in the sixties, considers existence of several plates or continents that emerge at certain places and are destroyed elsewhere, thereby, in consequence keeping the dimension of the planet unaltered.

Besides these two concepts another idea that endorses altered dimension of the Earth is expansion of the Earth and, of late, many scientists are seriously considering it as a plausible model for elucidating global features. The view of expansion for defining the remarkable parallelism between the distantly placed continental coastlines was first

*Corresponding author: Sen S, Retired Scientist, Council of Scientific and Industrial Research, Hemantika, O-26, Patuli, Kolkata, 700094, India
E-mail: ssennagpur82@yahoo.com

hinted out by s Bacon [10] when he used the Latin word 'exporrecti', meaning expansion for the fit. Amongst many who attempted to adjust continental fragments without altering the dimension of the globe, the name of Alfred Wegener is well-known although much before him several scientists, such as the pioneer Pellegrini of France [11], shown in Figure 1, American geologists Taylor [12] and Baker [13], as well as, the South African Alexander du Toit [14] attempted continental fitting. On the other hand, those who advocated Earth's expansion, the names of Yarkovskii [15] of Russia, Mantovani [16,17] of Italy, Hilgenberg of Germany [18] and Carey [19-21] of Australia-who spent almost his entire life persuading studies on expansion, deserve special mention. Of late, many Earth scientists are intrigued with the concept of Earth expansion theory although – since a possible cause of the phenomenon has not been put forward – presently the concept has yet to be duly acknowledged or accepted.

Continental Drift, Plate Tectonics and Earth Expansion

Wegener's view of continental drift [6], despite some initial support was criticized by several Earth scientists, including the renowned British Geophysicist Harold Jeffrey's who differed mainly because of solid and rigid state of the mantle, as well as, inappropriate fitting of continents even where parallelism of distant shorelines appeared to be distinct. The author considers that both the views, namely, that of Wegener and also of Jeffrey's are partially correct indicating that there must be a missing link in the relevant model of tectonics that needs to be fathomed out. The author in fact argues that in case of rigid and solid state of the mantle, as firmly established by seismic studies, all these three major global features would not be possible since for manifestation of such mobilistic phenomena it is essential that the mantle must have to be sufficiently fluid.

Wegener's view of continental drift, shown in Figure 2, assumes that initially the continents were joined together forming a super-continent called Pangaea surrounded by a large ocean termed Panthalassa. The concept fails to deliver a satisfactory explanation for fragmentation of the huge super-continent. Both continental drift and plate tectonics cannot explain formation of oceans and emergence of water over the Earth's surface as well as how the long elongated features known as mid-oceanic ridges were formed occurring mid-way between in the oceans and the continental shores.

The concept of presence of convection current in the mantle was mooted by the distinguished Scottish geologist Homes [22,23], who nevertheless categorically warned that his ideas were "purely

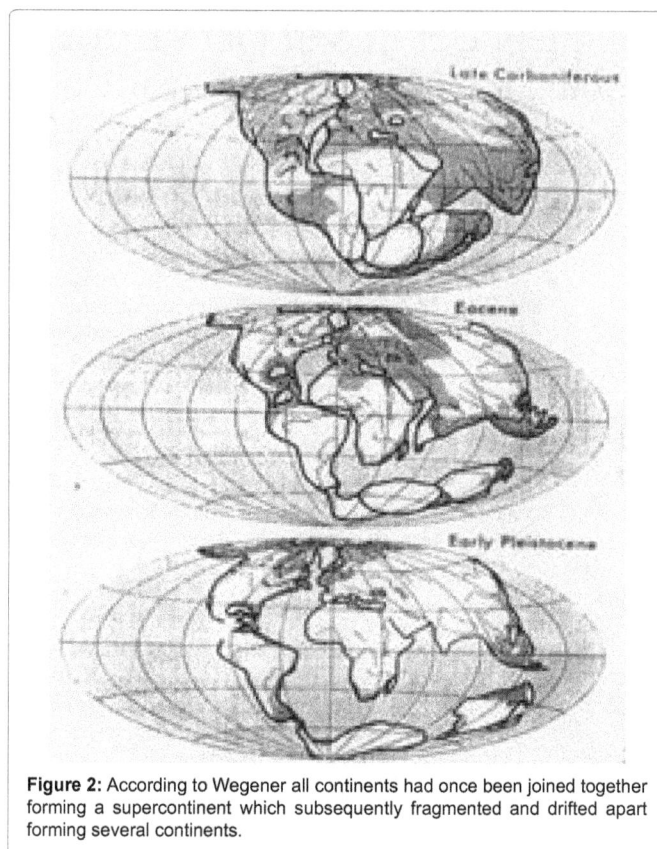

Figure 2: According to Wegener all continents had once been joined together forming a supercontinent which subsequently fragmented and drifted apart forming several continents.

speculative". Holmes' suggestion that the continents are carried out by flow of the mantle caused by convecting current was accepted by a large number of supporters of continental drift or its modified version plate tectonics. Apparently the main reason for acceptance of the vague concept was – as the author considers – owing to failure to fathom out a plausible cause for the much cherished concept that views unchanged global dimension throughout the geological periods. The author points out again that as the mantle is a rigid and solid, firmly confirmed by seismic waves, occurrence of convecting current with in the mantle is absurd. Another major absurd fantasy on mantle convection is that the speed of the process is extremely slow. How such slow moving convection can drift continents to great distance and, also, cause development of geomagnetism in the core? Harold Jeffreys' point of inappropriate fitting of continents in case where parallelism of distant shorelines appear to be distinct is indeed an expected feature since the dimension of the Earth was kept unaltered. That perfect fitting of the continents is feasible in a globe of reduced dimension is a definite confirmation that originally the planet's dimension was much smaller. Based on the model of ocean-less condensed Earth obtained by snug-fit of landmasses in a globe reduced to two-third of its present radial length, shown in Figure 3 [18], the author [24,25] has conceived that since the primordial small Earth was devoid of oceans, the ocean-forming water, was originally associated with the mantle. Consequently the original mantle was sufficiently fluid because of incorporation of ocean-forming water under ultrahigh pressure – a view conceived based on studies of hydrothermal systems of silicate rocks at elevated temperature and pressure, confirming lowering of melting point of silicate rocks under ultrahigh pressure condition [26]. Hence, it can be reasonably conceived that the small Earth at the unexpanded stage could sustain widespread global expansion in response to

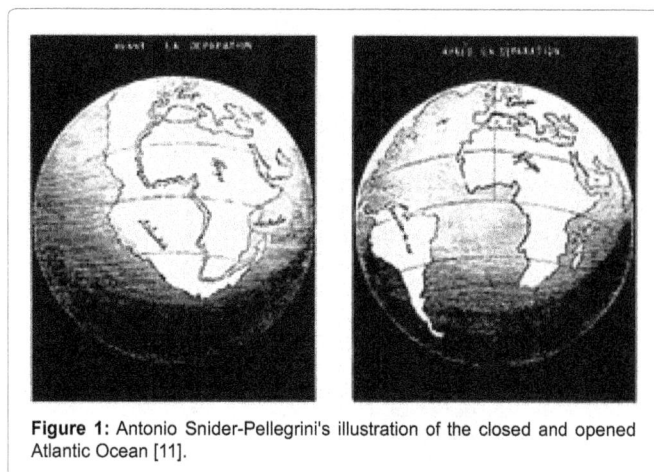

Figure 1: Antonio Snider-Pellegrini's illustration of the closed and opened Atlantic Ocean [11].

Figure 3: Otto Hilgenberg of Germany, put forward earth expansion theory independently in 1933 with models showing that if we could reduce the radius of the Earth to two-third of its present thickness all the continental blocks could be perfectly fitted in the small and ocean-less globe.

an external force of suitable magnitude.

A stable and static condition of the Earth with a vast quantum of ocean water already present, as conceived in plate tectonics and other concepts, on the other hand, leaves no scope for considering a sufficiently fluid or mobile condition of the mantle under which plate movements could be contemplated.

Causes of the Earth's Expansion

Based on the evidence of the gravitational pull over the Earth, induced by the Moon causing periodic tides, it can be conceived that the Earth's expansion too is significantly related to the Moon's gravitational pull, especially affecting the planet's semi-fluid mantle. In the primordial pre-expansion stage of the planet when its mantle was significantly fluid, the effect of such extra-terrestrial gravitational attraction was obviously much more pronounced than what it is today over the rigid terrestrial body.

To account for a fairly uniform pattern of expansion of the Earth with identical bulge around its axis of rotation and uniform pattern of disposition of the internal geospheres, a fairly slow rate of expansion, as well as existence of the Earth in a stable, condensed and consolidated state during the pre-expansion stage has been contemplated. It appears that the extra-terrestrial body which was initially situated at a far greater distance during the Precambrian and earlier periods, came nearer to the Earth in a smooth and gradual manner and was captured by the larger planet Earth and thus increasing the magnitude of the periodic gravitational pull over the terrestrial surface. In all probability it can be concluded that it was owing to the Moon's gravitational pull accomplished in a periodic manner the Earth was expanded, especially affecting the planet's semi-fluid mantle.

Impact of Magnetic and Gravitational Influences of the Extra-terrestrial Body on Earth

It was due to approach of the extra-terrestrial body - the Moon; the Earth's surface, as well as, internal features were conspicuously affected, cumulatively causing major reorientation of various global structures. Geological data [4] show that originally equator of the Earth was oriented at right angles to its present disposition. Nearly 500 million years ago this orientation started to change in a steady and gradual manner and, in a span of nearly 250 years, the equatorial disposition of the planet was deflected or tilted by nearly 90° from its original disposition. This change in the axis of rotation and equatorial disposition of

the Earth can be explained by considering that the sources of the terrestrial magnetic field, which was firmly attached within the Earth's body up to the Permian, was tilted or deflected by the magnetic influence of a similar dipolar extra-terrestrial magnetic object. It appears that a planetary body with similar characteristics approached the Earth in a slow and steady manner, causing to shift the terrestrial magnetic poles which in turn changed the Earth's disposition in space. With the advent of this extra-terrestrial magnetic body or the Moon, the equator and the axis of rotation of the Earth were tilted which must have initiated during the Precambrian period and continued up to the Middle Permian, bringing about a rotation of the Earth's axis by 90° (Figure 4).

It can be conceived that as the magnetic celestial body approached nearer to the Earth, the mutual force of gravitational attraction between the two also enhanced. By the time the Earth experienced a tilt of 90° by the magnetic influence of the celestial body, the gravitational pull exerted by the same celestial body, in all probability the satellite Moon, caused a rupture along the original mantle-core conjunction of the planet Earth along which, in consequence, a void zone was subsequently developed. Towards the beginning of the Mesozoic era this void zone encircling the solid iron core was sufficiently enlarged so that the latter could execute free and secular rotation within it in response to external magnetic forces.

Discussion

Based on information obtained from various sources,

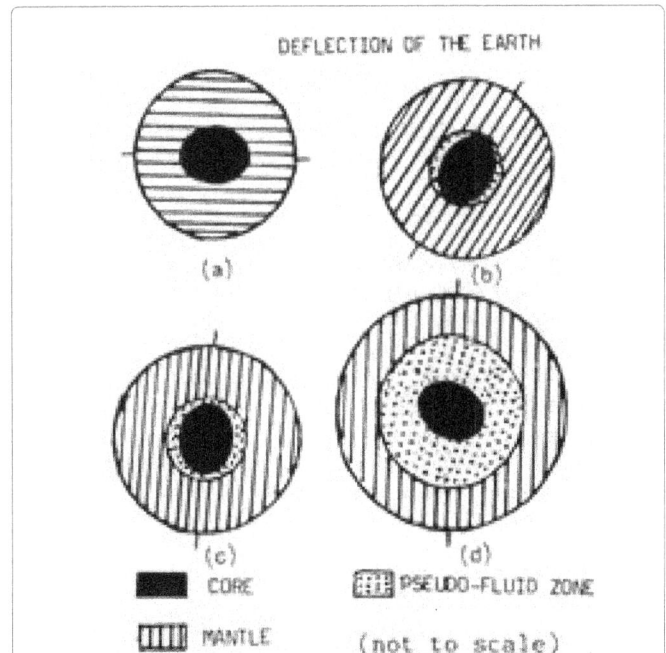

Figure 4: Before expansion of the planet, the solid iron core and semi-fluid mantle were juxtaposed to each other (Figure a). Due to expansion the original core-mantle conjunction was ruptured (Figure b) and along the ruptured surface a void zone or pseudo-fluid geosphere was developed (Figure c). Hence, due to occurrence of two solid geospheres separated by a virtually void zone, in the deep interior, the phenomenon of reverse gravity would prevail, generating low temperature and low pressure zones in the Earth's core and deeper parts. After development of the thick pseudo-fluid geosphere (Figure d), the iron-core remained within it in a suspended condition. Hence, at this stage in response to extra-terrestrial magnetic influences, polar wandering, pole reversal and west-ward drift could take place (not to scale).

followed by relevant updated discussion, it can be concluded that the primordial Earth was considerably small, ocean-less and covered with a sialic super-continent. Since the small Earth was devoid of oceans, it is reasonable to presume that the ocean-forming water was originally associated with its mantle, thereby turning the geosphere adequately fluid and especially suitable for expansion. Based on evidence of the Moon's gravitational pull over the fluid bodies on Earth, it can further be deduced that such external force or pull was responsible for expansion of the Earth, predominantly affecting the planet's primordial semi-fluid mantle.

The gravitational pull exerted by the extra-terrestrial body, obviously the Moon, over the semi-fluid mantle of the Earth which was rotating or spinning on an axis of rotation, added by the satellite's movement around the planet caused uniform distribution of the pertinent force on Earth. Consequently, the expansion of the Earth was manifested in a uniform manner without causing any uneven protrusion or irregular rising over the relevant portion of the planet. Due to such external pull the following remarkable alterations over the granitic crust, as well as, in the deep interior of the planet took place.

- The planet's dimension was enhanced concomitantly increasing the horizontal space between the adjacent fragmented sialic crusts. The newly developed space was continuously filled up with semi-fluid mantle to form floor of the rudimentary ocean bodies.

- The individual fragmented sialic blocks, completely encircled by oceans, formed continents. Formation of both oceans and continents can be linked with upliftment of the semi-fluid mantle, bring about development of additional horizontal space between the separated broken fragments of the super-continent. This appears to look as if the relevant broken continental fragments - which were once joined together - moved horizontally away from one another. Strictly speaking it was an apparent look caused due to increase of space owing to upliftment of the mantle.

- The expansion cracks along which several oceans grew, although horizontally remained fixed in their original position, due to expansion were uplifted or experienced vertical lift, thereby, giving an apparent impression of shifting of positions towards the middle part of the newly developed ocean bodies to form mid-oceanic ridges. In reality, these fissures continued to remain stationary or fixed in their original position but the oceans gradually grew up between the continents due to upward rise of the mantle over which continents and oceans and mid-oceanic ridges rest.

- There is yet another prominent force caused by rotation of the planet around its axis of rotation which is most effective around the Earth's equatorial plane but least so around the Polar Regions. The continents which remained stationary in regard to horizontal movement, though uplifted as discussed earlier, also experienced drift – which is horizontal positional shift – caused due to rotation of the planet. Several major features over the crust, including mountain belts, have been formed owing to this force.

- As discussed earlier, owing to expansion widespread disgorge of molten magma took place spreading on both sides of the expansion cracks to form floors of ocean basins. Extrusion of molten magma is invariably associated with expulsion of

extensive quantum of volatiles containing large amount of water vapour. The water that continuously released from the mantle due to degassing in tandem filled up the ocean basins, at the same time turning the geosphere solid and rigid in a gradual manner due to depletion of water. It is known that the melting point of a silicate rock would be lowered owing to presence of adequate amount of water. The mantle - which is a silicate rock – would, therefore, acquire sufficient fluid or semi-fluid characteristics owing to presence of ocean-forming water and absence of it would turn the medium solid and rigid.

- The same silicate medium – the mantle – due to depletion of water caused by widespread degassing, would gradually turn into a solid and rigid body, representing its present state when further expansion of the planet would be virtually stopped.

- It has been explained that due to expansion caused by extra-terrestrial pull, the Earth's semi-fluid mantle was swelled up. This in turn caused fragmentation of the sialic super-continent to form continents surrounded by oceans over the crustal surface. In the interior of the planet upward rise of the semi-fluid mantle initiated splitting the core-mantle contact along which with further upliftment of the mantle a void zone was gradually developed. Due to influx of particles and fine materials and volatiles from the adjoining walls, mainly from the rocky mantle, the void zone would turn into a thick pseudo-fluid or low density zone, separating the mantle from the core (Figure 5).

- Under such set up of occurrence of two thick geospheres separated by a thick zone pseudo-fluid or low density zone, both the geospheres would exert gravitational pull on each other. Consequently the trend of gravitation in the mantle would follow normal downward direction, while in the core due to the pull exerted by the mantle, an upward trend of gravitational pull would result giving rise to the phenomenon of reverse gravity in the deep interior of the planet, including the core.

- Such trend of resultant force of gravity in reverse direction in the iron core would cause low temperature-pressure condition at the core - a state congenial for sustenance of magnetic properties of iron. Hence, it is conceivable that the Earth's inner core, constituted of iron with some nickel, from where magnetic lines of force emit, is indeed a huge dipolar magnet. The prevalent concept conjectures very high temperature-pressure condition at the core - a state under which magnetic properties of magnet would be destroyed.

- Before expansion, the Earth was small and at that stage its magnetic iron core and semi-fluid mantle were juxtaposed to each other. Hence at that pre-expansion stage of the planet owing to magnetic influence of an external planetary object, such as - like pole of Earth facing like pole of the external object -would bring about major changes in the Earth's disposition in space in a slow and steady manner which in turn would cause remarkable alteration in the Earth's climatic zones (Figure 4).

- The fluid geosphere, termed in the prevalent concept as outer core is composed of iron in fluid state. The present treatise views that this geosphere - the thickness of which matches with extent of expansion of the planet – has been opened up owing to expansion of the planet. Earlier observers have not noticed this significant clue neither any have put forward any reason why such an enigmatic fluid zone could emerge in the deep interior of the planet (Figure 5).

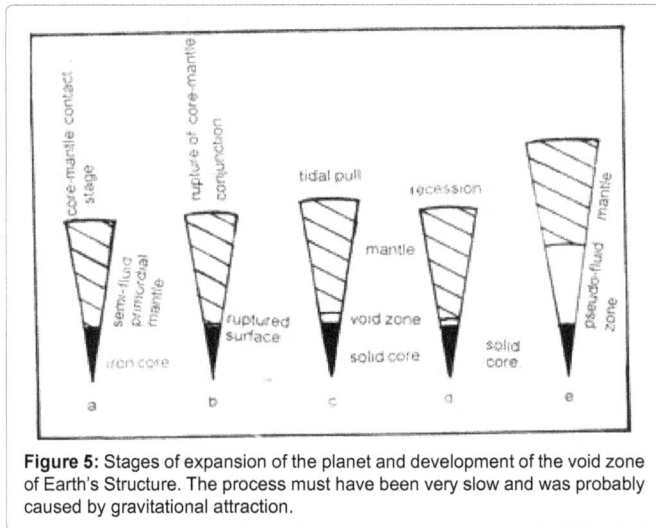

Figure 5: Stages of expansion of the planet and development of the void zone of Earth's Structure. The process must have been very slow and was probably caused by gravitational attraction.

- After sufficient expansion of the planet, when a thick pseudo-fluid geosphere was developed which completely encompassed the magnetic inner core, the latter could execute free and secular oscillation within the former in response to magnetic influence of an extra-terrestrial planetary object – presumed to be the Moon. Such oscillations of the magnetic core gave rise to the phenomena of polar wandering, pole reversals and west ward drift which have been taken place several times during younger geological periods as recorded in various iron bearing sedimentary rocks (Figure 4).

Conclusion

Earth expansion theory is one of the oldest concepts on Earth system sciences supported and developed over the years by a number of scientists belonging to a wide spectrum of disciplines, such as, geology, cartography, paleontology, astrophysics, physics, including Nobel Laureate physicist Dirac [27]. Detail updated information on certain crucial aspects of global features reveal that initially the planet Earth was considerably small and covered with a thin silicate crust. Below the crust three major geospheres occur which, in sequential order, are made of semi-fluid mantle of basic composition, followed downward by a virtually void zone –opened up owing to expansion of the Earth that encompassed the inner-most region of the planet composed of magnetic iron with some nickel from where magnetic lines magnetic force emerge. Before expansion of the Earth, when the pseudo-fluid zone was not opened up, its magnetic core iron and semi-fluid mantle were juxtaposed to each other. At that stage owing to influence of an external magnetic object – possibly the Moon which was approaching Earth – the planet was deflected in a steady manner thereby causing major alteration in its climatic feature that are precisely documented in the crustal rocks of the planet. Because of presence of a void-like zone between core and mantle, in addition to normal inwardly directed force of gravitation, a reversely directed gravitational force would occur in deep interior, including the core, in consequence of which the temperature – pressure condition of the core would be sufficiently low and congenial for maintenance of magnetic properties of the core. After opening up of the pseudo-fluid zone that encompassed the solid magnetic core, the latter could execute smooth rotation or revolution in response to external magnetic influence which have been accurately recorded on several sedimentary rocks containing iron. The mantle which was originally of semi-fluid turned into solid and rigid body owing to extensive degassing of volatiles containing water from it that filled up the ocean basins.

References

1. Abraham O (1596) Thesaurus Geographicus (in Latin) (3rd edition), Antwerp: Plantin, OCLC 214324616.

2. Harold J (1925) The Earth and its Origin, History and Physical Constitution, Cambridge University Press, Cambridge 51: 64-65.

3. Harold J (1970) The Earth, 5th Edition, Cambridge University Press, Cambridge 21: 541.

4. Robert D, Batten RL (1971) Evolution of the Earth, McGraw-Hill, New York p.649.

5. Sen S (2007) Earth-The Planet Extraordinary, Allied Publishers, 1/13-14, Asaf Ali Road, New Delhi-2, India p.232.

6. Alfred W (1912) The emergence of large molds the earth's crust (continents and oceans), on geophysikalischerGrundlage. Petermann Geographical Releases 63: 185-195, 253-256, 305-309.

7. Wegener A (1929) The origin of continents and oceans (4th edition), Braunschweig: Friedrich Vieweg & SohnAkt Ges.

8. Wegener A (1929/1966) The origin of continents and oceans. Courier Dover Publications, ISBN 0-486-61708-4.

9. McKenzie DP, Parker RL (1967) The North Pacific: An example of tectonics in a sphere. Nature 216: 1276-1280.

10. Bacon F (1620) Novam Oraganum Scientiarun, London.

11. Snider-Pellegrini A (1858) La Création et ses mystères dévoilés, Paris: Frank and Dentu.

12. Taylor FB (1910) Bearing of the tertiary mountain belt on the origin of the earth's plan. GSA Bulletin 21: 179-226.

13. Baker HB (1911) Origin of the Moon. Detriot Free Press, 23 April Features Section 7.

14. du Toit A (1926, 1937) Our Wandering Continents. Oliver and Boyd, Edinburg p.361.

15. Yarkovskii IO (1888) Hypothsecritiéquqs de la gravitation Universal de connexion avec la formation of elements chimiquees p.134.

16. Mantovani R (1889) Les fractures de l'écorceterrestreet la théorie de Laplace. Bull Soc Sc Et Arts Réunion 41-53.

17. Mantovani R (1909) L'Antarctide: The m'instruis. La science pourtous 38: 595-597.

18. Hilgenberg OC (1933) Vom Wachsenden Erdball, Berlin: Giessmann & Bartsch.

19. Carey SW (1958) The tectonic approach to continental drift. In: Carey SW Continental Drift-A symposium. Hobart: Univ. of Tasmania. 177-363.

20. Carey SW (1961) Palaeomagnetic evidence relevant to a change in the earth's radius. Nature 190: 36.

21. Carey SW (1976) The Expanding Earth. Elsevier, Amsterdam p.488.

22. Arthur H (1919) Radioactivity and Earth's Movement. Trans Geol Soc Glasgow 18: 559-606.

23. Arthur H (1944) Principles of Physical Geology (1st edition). Edinburgh: Thomas Nelson & Sons.

24. Sen S (1984) Unified global tectonics-a new qualitative approach in earth sciences. J Min Met Fuels 32: 20-22.

25. Sen S (2003) Unified global tectonics: structure and dynamics of the total earth system. J Min Met Fuels 51: 351-355.

26. Roy R, Tuttle OF (1961) In: Ahrens LH, Rankama K, Runcorn SK (eds.) Investigation under hydrothermal conditions. Phys Chem Ear 1: 138-180.

27. Dirac PAM (1938) A new basis of cosmology. Proc Roy Soc London 165: 199-209

13

Risk Assessment and Remedial Solutions of Coastal Flooding: Case Study of Hammam Lif Coastline, Northern Tunisia

Abir B[1]*, Mohamed B[2], Samir M[3] and Chokri Y[1]

[1]Laboratory of Sedimentary Dynamic and Environment, National Engineering School of Sfax, Street of Soukra km 4, 3038 Sfax, Tunisia
[2]Geotechnical Company Thynasondage and Geotechnical Engineer, Tunisia
[3]High Institute of Technological Studies, Sfax, BP46, Sfax, 3041, Tunisia

Abstract

This study focused on the coastal area of Hammam Lif, an urban zone – about 20 kilometers to the south of the capital city Tunis - threatened by flooding caused by rainwater and water generated by the wave run up on the shoreline area. In this study we evaluated the potential flood threat using in the calculation of runoff flow rate in the area of Hammam Lif caused firstly by rain and partly by the wave run up. The study, also, tried to highlight the wrong conceptual characteristics of the urban area infrastructure in Hammam Lif. In order to be able to face the potential flood threats caused by the rain in Hammam Lif, some adequate protective solutions were proposed and other soft protective solutions were suggested to face the wave run originating at the sea.

Keywords: Coastal flooding; Rain; Wave run up; Urban coastal zone; Beach nourishment; Artificial reef submerged formed by geotextile; Borrow site

Abbreviations: SC_1: Core Drilling 1; SC_2: Core Drilling 2; SC_3: Core Drilling 3; EI_{11}: Intact Sample Number 1 of the Core Sampling Number 1; EI_{12}: Intact Sample Number 1 of the Core Sampling Number 2; EI_{21}: Intact Sample Number 2 of the Core Sampling Number 1; EI_{31}: Intact Sample Number 3 of the Core Sampling Number 1; ER1d: Overhauled Sample Number 1 taken from the Carry Sand Drijet; ER2d: Overhauled Sample Number 2 taken from the Carry Sand Drijet; ER3d: Overhauled Sample Number 3 taken from the Carry Sand Drijet; ER1b: Overhauled Sample Number 1 taken from the Carry Sand Borj Hfaieth; ER2b: Overhauled Sample Number 2 taken from the Carry Sand Borj HJaieth; ER3b: Overhauled Sample Number 3 taken from the Carry Sand Borj Hfaieth; R: Wave Run Up

Context of the Study

It seems obvious from previous studies that the Tunisian coastal zone is subject to a risk of coastal flooding, mainly during storm periods, where the premium sea level can reach 1.13 m NGT for a 50-year old return period. The northern coastline, especially the beaches of the southern area of Greater Tunis, for the coasts that are relatively characterized by low altitude, the highly urbanized Rades, Ezzahra and Hammam Lif are threatened by flooding caused firstly by waves that can exceed the protective dikes (the action of wave run up) and secondly by resulting rain water, which in its greater percentage, flows through a highly urbanized area [1]. However, this area is rather a waterproof urban zone which sanitation networks are mostly old inadequate and easily over flown. Such a situation promotes the high chances of the flooding of the Wadi Méliane, the nearest to the study area [2].

This study focused on the evaluation of the flooding potential of the Hammam Lif coastline. This study is, therefore, made up of the component: Risk assessment of coastal flooding on the coastline of Hammam Lif and proposition of some protective solutions.

Flooding was studied relying on three simulations. The first highlighted only the rain component and its effect on the urban zone whereas the second dealt with the wave run up (the waves overcoming the breakwater to reach the area of habitats).

Two types of solutions were figured out: either solution by adding adequate coastal defense structures such as breakwaters, other rock fill dams, coastal protection cobs and or others through strengthening the already existing maritime structures against the ascending waves [3]. The choice of the solution obviously depends on several parameters such as the impact of the strength of the wave on the existing wall, the wave quality (beachcomber or not) and the crossing flow rate.

Location of the study area

The Beach of Hammam Lif is located in the central part and in the Gulf of Tunis NE / SW axis, between the restaurant called « la SIRENE » and the Municipal Stadium. The beach has a length of 1540 m and a width of 45 m (Figure 1).

The restricted area of study of the rain component and the wave run up component is 125,000 m^2 thus 12.5 Ha (Figure 2).

Materials and Methods

Study of the rain component: runoff flow rate calculation

Our calculations were carried out according to Eq. (1):

Runoff flow rate:

$$Q=K \times A \times W \times Pc \tag{1}$$

With K: conversion factor, A: Area, I: Intensity, PC: runoff coefficient.

Flow estimation by using transformation method of rainfall in runoff: the rational method in hydraulics: The oldest method of estimating peak flow from rainfall is called rational method (Figure 3). The rain is supposed with a constant intensity over rain intensity (ip) a time t=tc and flood volume is proportional to the rain volume. The answer is a flow hydrograph triangular of duration is 2TC and peak

*Corresponding author: Abir B, Laboratory of Sedimentary Dynamic and Environment, National Engineering School of Sfax, Street of Soukra km 4, 3038 Sfax, Tunisia, E-mail: abir.baklouti@gmail.com

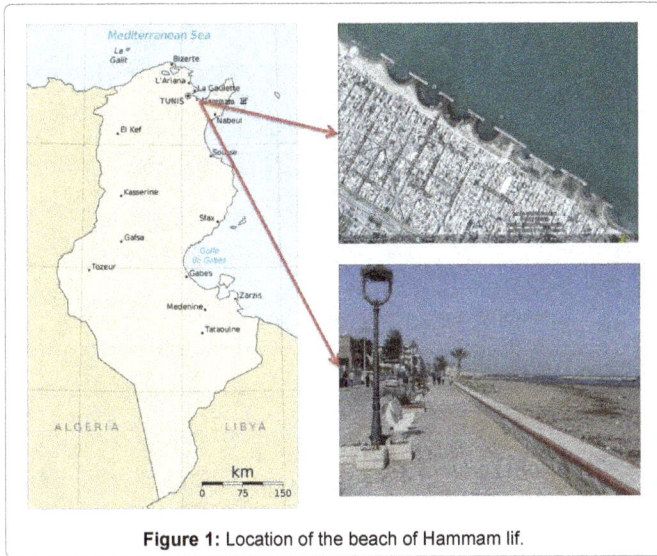

Figure 1: Location of the beach of Hammam lif.

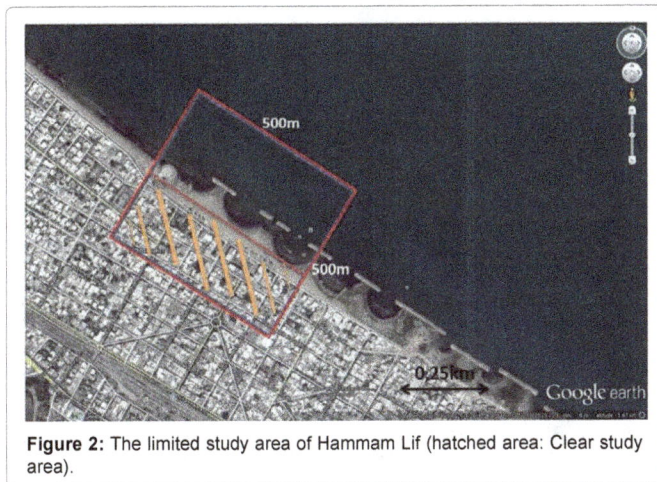

Figure 2: The limited study area of Hammam Lif (hatched area: Clear study area).

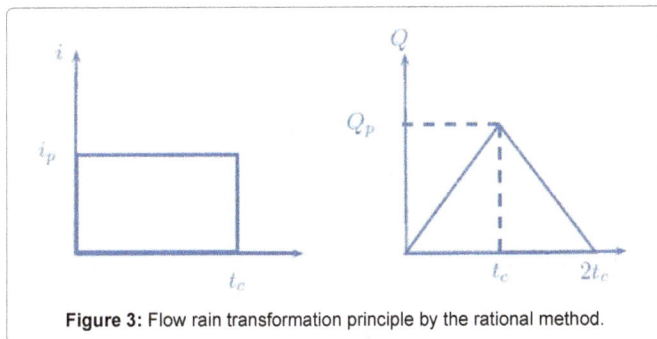

Figure 3: Flow rain transformation principle by the rational method.

flow Qp. The rain volume Vp=tcipS. The flood volume is (see Eq. (2)):

$$Vc = 2 \times \frac{1}{2} Qptc \qquad (2)$$

With S is the surface of the watershed. It is assumed that the coefficient of proportionality is C ($0 < C \leq 1$), also known as peak runoff coefficient. S from the equality Vc=CVp such that Eq. (3):

$$Qp = Cip\,S \qquad (3)$$

Note that ip is usually expressed in mm/h while Qp is in m^3/s. For the previous formula to be expressed be in these units, it could be modified as follows in Eq. (4):

$$Qp = \frac{CipS}{3,6} \; (m^3/s) \qquad (4)$$

Study of the component wave run up

For the calculation of wave run up of the coast of Hammam Lif, field trips to the study area were organized during the period between 01st March 2012 and 1st March 2013 (for a monthly output average) 13 follow up months of the wave of Hammam Lif in different weather conditions suitable for the calculation of the wave run up.

The choice of days depended on the weather conditions in the site. In fact, to achieve good measures we had to choose the days following the storms in the time when the sea is in a high tide state [4]. Our study area is limited to the case of sea bright water (wet part of the beach) for the calculation of maximum run up of the wave in the Swach area.

Calculation of the wave runs up from measurement of the sea foreshores and the corresponding tide: The dividing line is that between the dry sands (irregular surface made up of dry sands) and the wet sands (smooth surface). This boundary is a precise and easily identifiable limit of the foreshore to calculate the level reached by the sea waves.

On the beach of Hammam Lif, the levels achieved by the sea foreshores were measured along the profile only. Sea foreshores were measured between 1st March 2012 and 1st March 2013 [5]. The surveys were geo referenced and connected to the Tunisian general leveling system (NGT) using GPS (Figure 4).

To deduce the run up values from the altitude of the sea foreshores, it was necessary to obtain observed tidal values. The run up values observed for each survey were, in fact, obtained by subtracting the maximum rise of the tide observed from the sea foreshores altitude (Figure 5).

For the range of Hammam Lif, the observed tidal data were obtained from the Tunis _Carthage station. The sea level observed is measured by the tide gauge and corrected at the study site.

Muller's method for the calculation of wave runs up: Relying on 200 laboratory experiments, Müller calculated the lift height (run up)

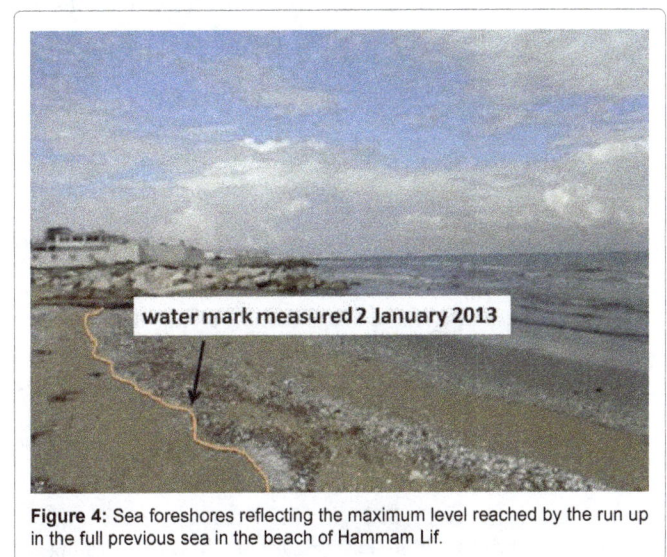

Figure 4: Sea foreshores reflecting the maximum level reached by the run up in the full previous sea in the beach of Hammam Lif.

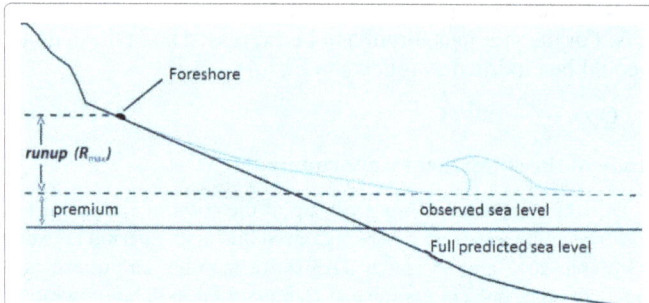

Figure 5: Diagram summarizing the principles of the method A used in this study. Each observed run up value is obtained by subtracting the level of sea observed from the sea foreshores altitude.

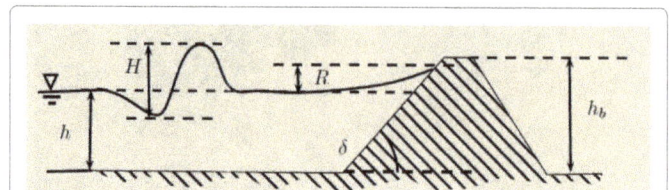

Figure 6: A wave run up against a breakwater.

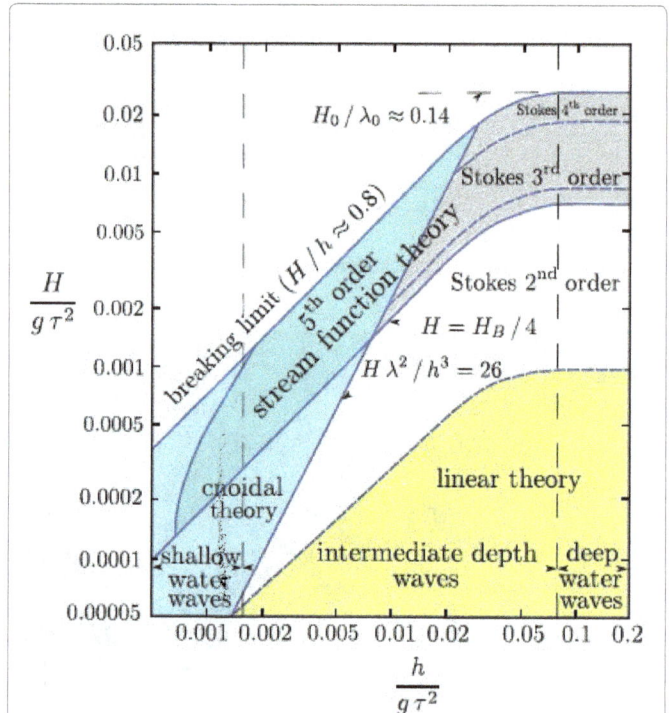

Figure 7: Range of validity of the various theories based on the wave height H of the water height h, the time $\tau = \lambda / c$. The light blue area is the area of conoïdal waves. The yellow area is the Airy theory. The blue area is the Stokes wave theory. According to a classification proposed by Le Méhauté.

R of a pulse wave along an obstacle (such as that facing a dam) [6] as follows in the Eq. (5):

$$R = 1.25h\left(\frac{\pi}{2\delta}\right)^{\frac{1}{5}}\left(\frac{H}{h}\right)^{\frac{4}{5}}\left(\frac{H}{\lambda}\right)^{\frac{-3}{20}} \qquad (5)$$

With δ the facing angle relative to the horizontal (18° ≤ δ ≤ 90) in Müller's experiments,

H is the maximum height of the wave, and λ is the wave length of which estimation is given by Stokes (Figure 6).

δ: The angle of the facing relative to the horizontal=45°.

H: The maximum wave height given by (3.40) =3.65 m to 10 m.

If we take the (λ / h) ratio (with λ the wavelength and h the height of water), then:

- (λ / h) ≤ 2, the deepwater wave or short wave;

- 2 < (λ / h) ≤ 20, the intermediate wave (the transition wave);

- (λ / h)>20, the shallow water waves or long waves;

- Short waves are usually studied using the Stokes theory, which consists of researching solutions in the form of a truncated series. In principle, the higher the order of development is, the better the accuracy will be, but it is necessary that the wavelength is relatively short for a rapid convergence to be ensured (Figure 7).

According to the STOKES theory (shortwave deep water theory) we get (Eq. (6)):

$$\frac{H\lambda^2}{h^3} = 26 \qquad (6)$$

λ: The wavelength $\sqrt{((26*(25)3)/(3.65))} = 10.45m$ (7)

h: Height of water=2.5 m for Moving about 250 m to the seabed.

Study of adequate protection solutions for the coast of Hammam Lif

Artificial Recharge and characterization of the Hammam Lif beach sands: Sampling by onsite coring: Artificial recharging requires a study of the nature of the existing materials in the study area and also a study of the materials from different possible borrow areas to ensure the right choice of sand that should have good size characteristics to be adequate for the Hammam Lif beach type [7].

Three coring holes SC1, SC2 and SC3 were made on the beach of Hammam Lif, each was 20 m deep (Figure 8). These core drilling are made using a hydraulic drill for geotechnical soil investigation and small water wells. (Teredo DC 123) showed in Figure 9.

Materials needed for grain size analysis

Metallic square-shaped ordinary dimensions sieves were fabricated in order to achieve reproduceable satisfactory results. An electric screening machine able to perform horizontal vibratory movement as well as vertical shakes along the sieves' column was used as well.

The procedure is defined by AFNOR standards (wet and dry) to establish a grading curve representing the respective proportions of the grains aggregate sizes classes.

A 200 g sample was split into several categories of decreasing grains diameters by means of a series of 21 square mesh sieves (AFNOR standard).

Determination of the shape parameters

The shape parameters provide information on the shape of the curve:

-The uniformity coefficient or Hazen coefficient (Cu) was determined as follows in Eq. (8):

Figure 8: Hydraulic drill for geotechnical soil investigation and small water wells (Teredo DC 123).

Figure 9: Location of coring holes SC$_1$, SC$_2$ and SC$_3$ drilled on the beach of Hammam Lif.

$$Cu = \frac{D_{60}}{D_{10}} \qquad (8)$$

-The curvature coefficient (CC) was determined as follows in Eq. (9):

$$Cc = \frac{D_{30}^2}{(D60*D10)} \qquad (9)$$

The geotextile Artificial submerged reef

Calculation of the transmission coefficient across the structure: The offshore underwater installed dykes in the sea grass deprived zone would favor not only the surge of strong waves, but also the dispersion of the masses breaking of water before reaching the dykes [8]. Let's introduce the state of the art theory on the design of a submerged breakwater. The theoretical transmission coefficient through the structure can be calculated by standard formulas of Angremond, Van der Meer and Jong (Eq. (10)), assuming that, the structure form of the design to be a single trapeze (Figure 10).

$$Kt = -0.4\left(\frac{F}{H_S}\right) + C_P\left(\frac{B^{-0.31}}{H_S}\right)*(1-\exp(-0.5\xi_B)) \qquad (10)$$

where:

Kt: is the transmission coefficient

F: is the water height between the water surface and the top of the structure

HS: is the incident significant height

B: is the width of the reef top

CP: is the coefficient of permeability (CP=0.64 in the case of a permeable structure, and

CP=0.80 in the case of a waterproof structure)

εB is the number of Iribarren unfurling as (Eq. (11)):

$$\xi_B = \frac{\tan(\alpha)}{\sqrt{\dfrac{H_S}{\lambda_0}}} \qquad (11)$$

With α the embankment slope and λ_0 the swell of the offshore wavelength (Eq. (12)):

$$\lambda_0 = gT^2 / 2\pi \qquad (12)$$

Results and Discussion

Rates interpretation of rain component

A close examination of the rain rates reveals a continuous increase over the years. In fact, during the 1931 storm; Q=2713l/s, as for the 1981 storm, Q=3824 l / s while that of 2003; Q=4000 l / s. The runoff flow has been on the increase after storms over the years, which has caused the flooding of the area.

The main causes of the flooding are essentially:

-The increase in soil sealing rates and in the urbanization rates through the years.

-The overwhelmed sewerage network of the area should generate a rate of 88900 m³/day of which 75% that is 66 675 m³/day is strictly waste water. Being filled with wastewater, the sanitation systems could drain a flow of 22,225 m³/day of rainwater which makes them unable to drain the whole quantities.

In 2012 the Q rain was 35 923.801 m³/day>22 225 m³/day. Therefore,

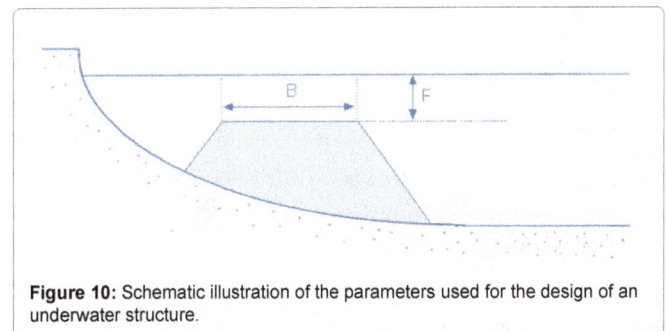

Figure 10: Schematic illustration of the parameters used for the design of an underwater structure.

the remainder of the water (13,698,801 m³/day) would run off and cause extensive flooding in the area HammamLif.

Unfortunately, as shown in Figure 11, the existing remedies, digging the casing of sewerage networks in the northern area of Hammam Lif to form a channel to discharge the polluted water into the sea, is harmful for the environment and the aesthetics of the beach.

The wave runs up component

Calculation of wave runs up from measurements of the sea foreshores and the corresponding tide: The following Table 1 and the Figure 12 shows the water mark observed and the values of the corresponding tides:

According to this diagram the wave run up is: R=2.50 m.

Study of adequate protection solutions for the coast of Hammam Lif

Artificial Recharge and characterization of the Hammam Lif beach sands: Sampling by coring on site: The core drilling performed along the Hammam lif Beach showed the following lithology (Figures 13-15):

According to the litho-stratigraphical drills achieved during this survey, the floor of the HammamLif coastline of is noticed to be

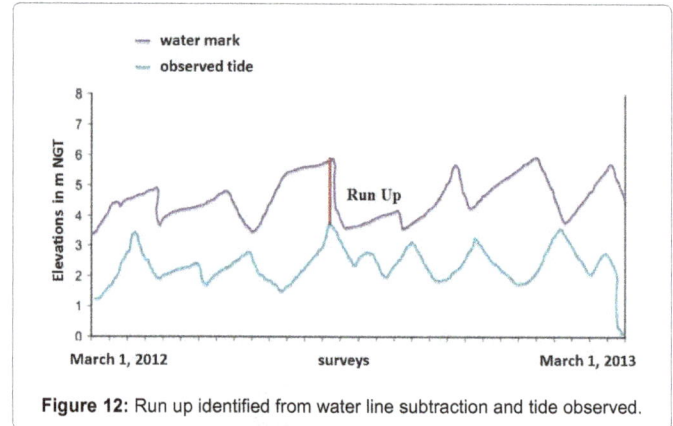

Figure 11: Hole in the drainage pipe forming a discharge channel of polluted water into the sea.

Days	The water mark (m)	Corresponding tide (m)
1 March 2012	3.3	1.2
5 April 2012	4.5	3.3
8 May 2012	3.8	1.7
16 June 2012	4	2
01 July 2012	3.5	2.5
02 August 2012	5	2.5
6 September 2012	5.5	3
28 October 2012	3.5	2.1
28 November 2012	3.5	2.5
31 December 2012	4.5	1.5
02 January 2013	5.5	2.5
06 February 2013	4.2	1.5
01March 2013	5	2

Table 1: The water mark and the values of the corresponding tides observed between 1st March 2012 and 1st March 2013 in Hammam Lif.

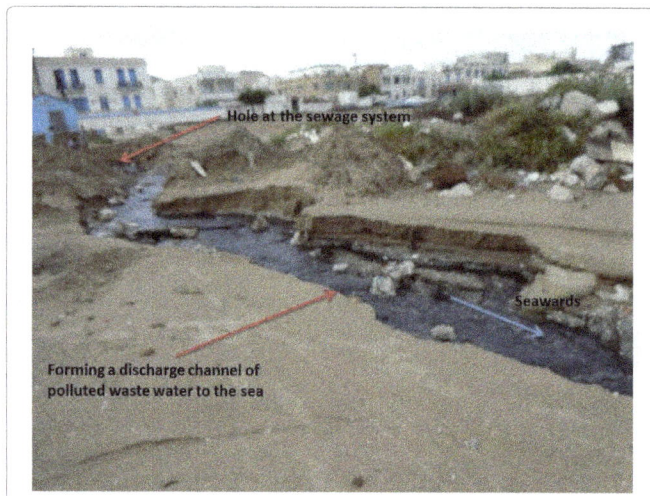

Figure 12: Run up identified from water line subtraction and tide observed.

homogenous for the different studied drills [9]. The lithological column of the field is characterized by a yellowish layer of fine shelly sand with a thickness varying between 0,70 m to 1.00 m followed by a layer of brown silty clay of 1.30 m thick, underneath which is a layer of silty sand that reaches 3.00 m depth. All of these layers lie on a layer of fine slimy sand of 3.00 m thick [10]. The lithological column ends with a layer of naturally grayish sand that starts from the depth of 6.00 m extending to the end of the exploration.

Results of grain size analysis performed on the samples

The Table 2 summarizes the particle size parameters of samples taken.

The Hammam Lif shoreline consists of bad quality sand which grain size is characterized by a high percentage of fine sand particles depriving the beach from any form of stability.

*2<Cu<5: the particle size is tightened.

*CC>2: This sand is characterized by a poor particle size with a dominance of fine fractions.

Artificial recharge is required using a coarser sand to promote the stability of the beach. The borrowing site should not be from the area of Hammam lif because of the bad size characteristics of this sand [11]. Hammam lif beach rather needs a size correction intake of better and mainly coarser sand from another area.

Research of borrow areas of possible suitable sand for beach nourishment of Hammam Lif

Charging the beach of Hammam Lif is necessary for its stability and its correction, Hammam Lif must return to its original appearance, bathing area with good sand.

The search for good sand requires the completion of sampling in different areas proposed in order to choose the right sand for beach nourishment.

Earthly origin sand borrow area (sand quarry DRIJET)

Characterization of the sands of the sand quarry DRIJET: surface sampling: The DRIJET sand quarry is located at a distance of 48.09 kilometers from the beach of Hammam Lif. She is old:

*M3: upper Miocene undifferentiated.

*O: Oligocene Limestone sandstone bioclastique, green clay and sandstone.

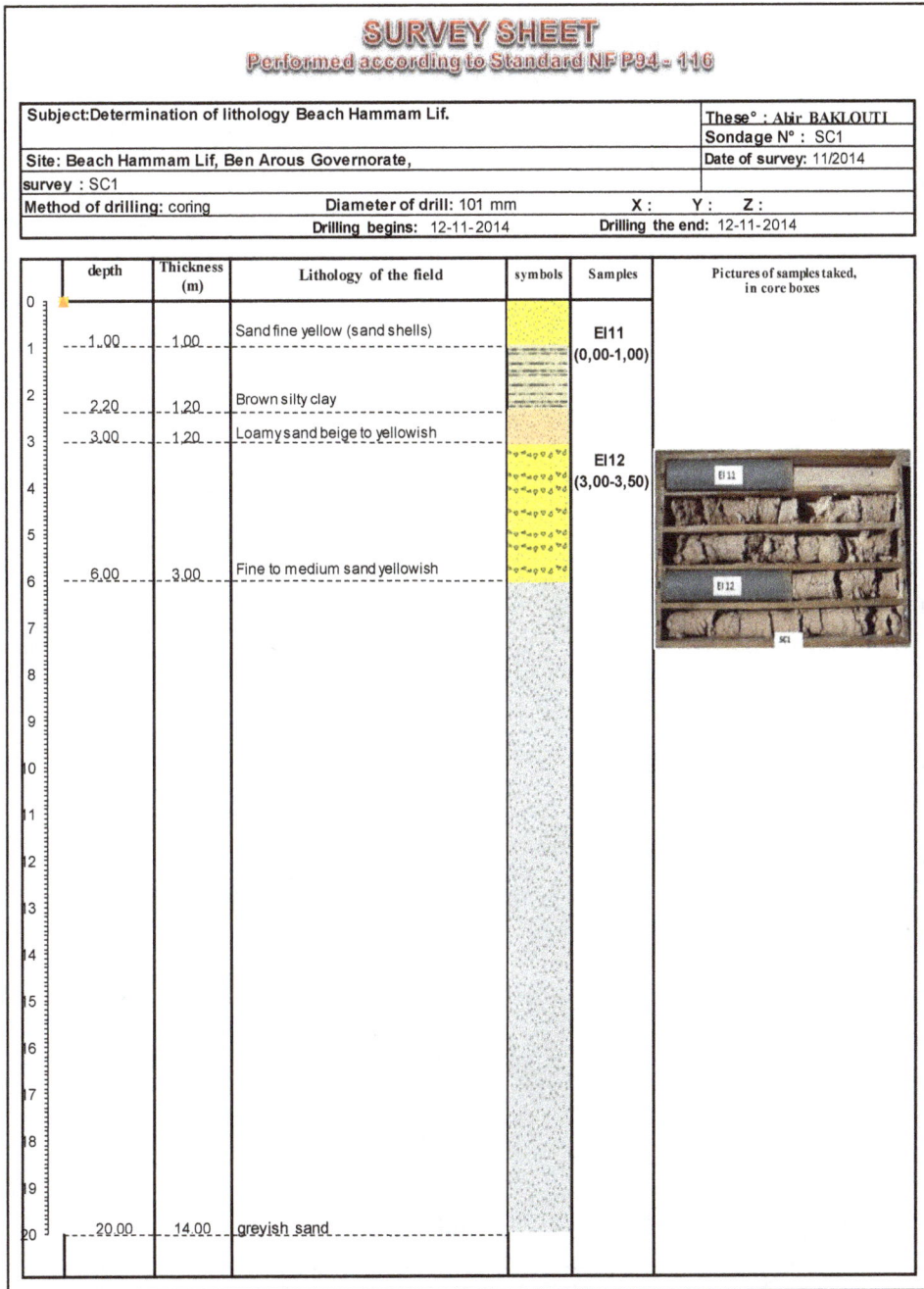

Figure 13: Lithology provided by the core sampling SC_1.

Three samples were taken from quarry sand DRIJET, redesigned Sample ER1d, ER2d and redesigned Sample revamped ER 3d, particle size analysis gave the results:

Interpretation: The sand of the quarry sand DRIJET is good sand with a particle size with a high percentage of coarse particles so this sand is a sign of stability of the beach.

*2<Cu<5: The particle size is tight.

*CC<2: was a dominance of the coarse fraction, the curve is asymmetrical.

The sand of DRIJET career is good sand for beach nourishment of Hammam Lif.

Earthly origin sand borrow area (sand quarry Borj hfaiedh)

Characterization sand quarry sand Borj hfaiedh: surface sampling: The BorjHfaieth sand quarry is located at a distance of 34.15 kilometers from the beach of Hammam Lif. She is old:

*M2S-1: Middle Miocene (Tortonian to Serravallian).

Three samples were taken from the quarry sand Borj Hfaieth, redesigned Sample ER1b, ER2b and redesigned Sample Sample revamped ER3b, particle size analysis gave the following results:

SURVEY SHEET
Performed according to Standard NF P94 - 116

Subject:Determination of lithology Beach Hammam Lif.	These° : Abir BAKLOUTI
	Sondage N° : SC2
Site: Beach Hammam Lif, Ben Arous Governorate .	Date of survey: 11/2014
survey : SC2	
Method of drilling: coring — Diameter of drill: 101 mm — X : — Y : — Z :	
Drilling begins: 12-11-2014 — Drilling the end: 12-11-2014	

depth	Thickness (m)	Lithology of the field	symbols	Samples	Pictures of sample taked in core boxes
0.70	0.70	Sand fine yellow (sand shells)			
2.00	1.30	Brown silty clay			
3.00	1.00	Loamy sand beige to yellowish		EI21 (2,50-3,00)	
6.00	3.00	Fine to medium sand yellowish			
20.00	14.00	greyish sand			

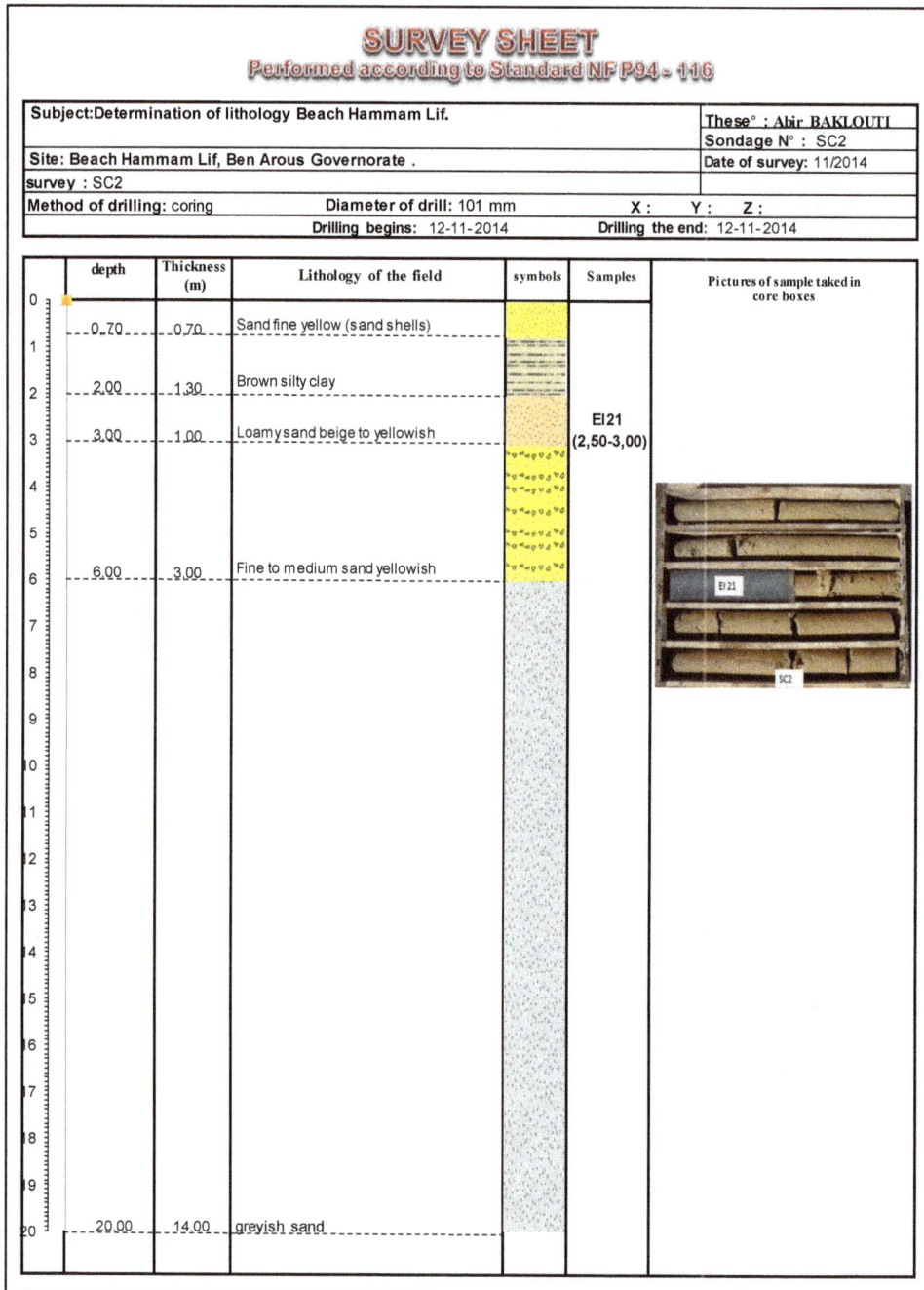

Figure 14: Lithology provided by the core sampling SC$_2$.

Interpretation: The sand of the quarry sand Borj hfaeith is good sand with a particle size with a high percentage of coarse particles so this sand is a sign of stability of the beach.

*2<Cu <5: The particle size is tight.

*CC<2: was a dominance of the coarse fraction, the curve is asymmetrical.

The sand of the quarry sand BORJ Hfaiedh is a good sandfor beach nourishment of Hammam Lif.

A geotextile underwater artificial reef

Calculation of the coefficient of transmission across the structure: We should have a good knowledge on the swell of the sea waves and their significant heights.

Indeed the positioning of the reef has to be adequately chosen. If it is chosen too close to the dyke, water would not have enough time to withdraw after the wave breaking and harmful "bagging" phenomenon would then be produced. Too far from the dike, the deshoaling effects would tend to amplify the wave transmitted to the back of the reef. Kt=0.170.

The method consists in considering a reef implanted at h=10 m deep, a hundred meters off the sea wall, with a varied wide side slope

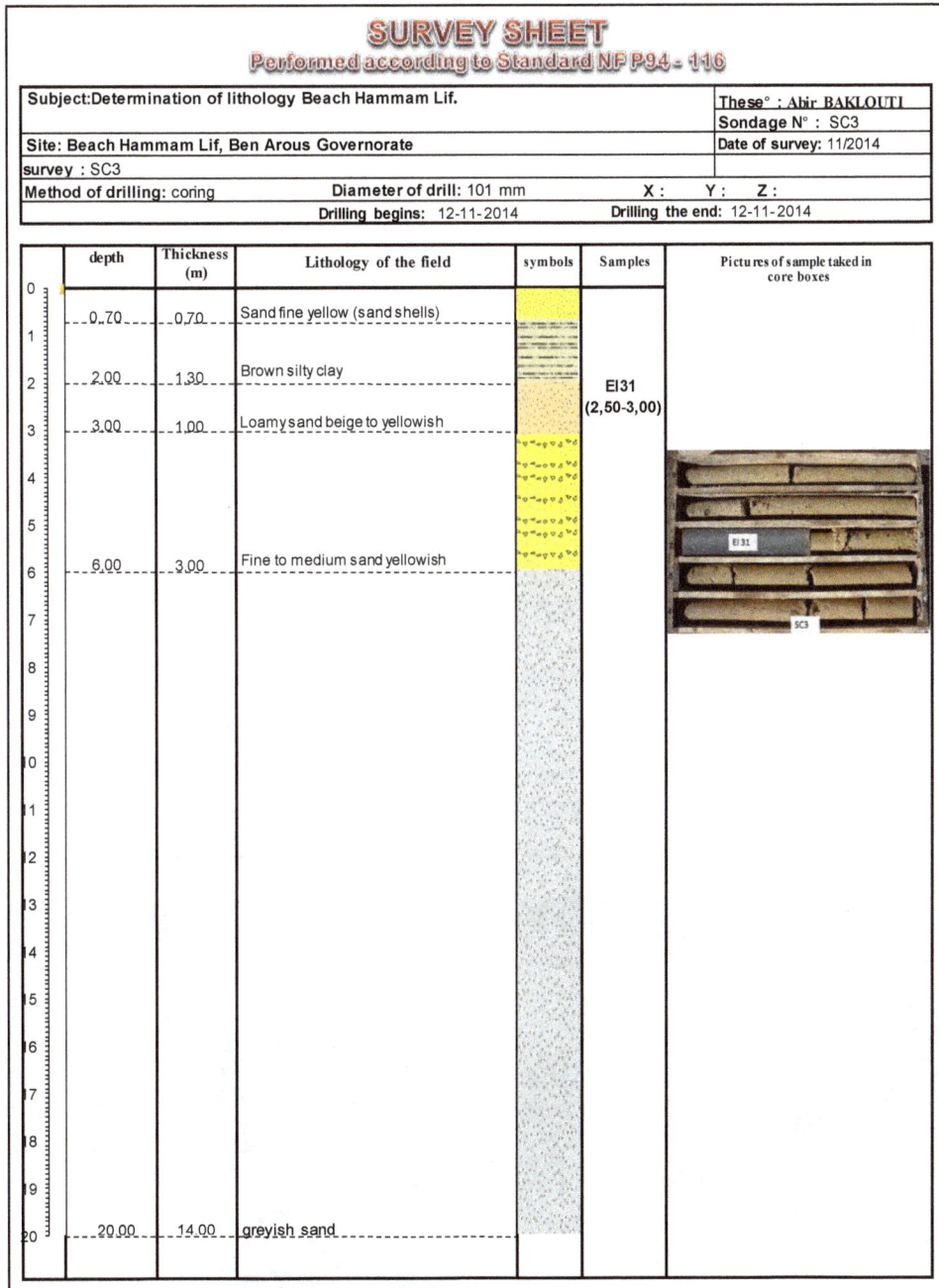

	depth	Thickness (m)	Lithology of the field	symbols	Samples	Pictures of sample taken in core boxes
	0.70	0.70	Sand fine yellow (sand shells)			
	2.00	1.30	Brown silty clay		EI31 (2,50-3,00)	
	3.00	1.00	Loamy sand beige to yellowish			
	6.00	3.00	Fine to medium sand yellowish			
	20.00	14.00	greyish sand			

Figure 15: Lithology provided by the core sampling SC$_3$.

Samples	Nature of samples	Depth (m)	D50	D60	D30	D10	Cu	Cc
EI11	Fine yellowish sand	0.00-1.00	0.17	0.20	0.10	0.075	2.66	3.33
EI12	Sandy yellowish means	3.00-3.50	0.18	0.22	0.13	0.085	2.58	3.47
EI21	Sandy loam	2.50-3.00	0.003	0.055	-	-	-	-

Table 2: Summary table of particle size parameters of samples taken.

(its slope rating side is set to 2/1), the draft (water withdrawal), and its berm length (Figure 16).

The characteristics of a successful artificial reef for this study in the Hammam Lif area are summarized in the Table 3.

The trapezoidal shape is not necessarily the most appropriate if the objective is to dissipate a maximum energy of incidental waves [12] (Figures 17 and 18) (Table 4). However, the benefit of such a shape is simplicity and ease to implement, and cost effective compared to a more complex solution, especially the goal here is to sufficiently reduce the

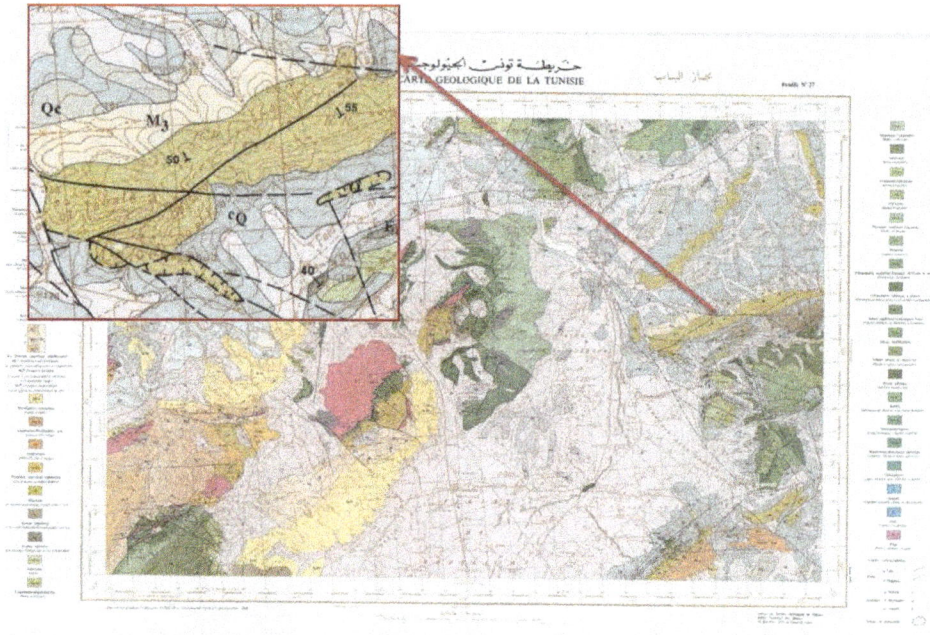

Figure 16: Geological map of Drijet sand quarry (drawn from national service TUNIS (OTC 1991).

Samples	Depth (m)	D50	D60	D30	D10	Cu	Cc
ER1d	0.00-1.00	0.29	0.3	0.19	0.09	3.48	1.05
ER2d	0.00-1.00	0.25	0.3	0.19	0.1	2.9	1.24
ER3d	0.00-1.00	0.3	0.4	0.23	0.14	3.08	0.85

Table 3: Size characteristics of the sand samples taken from the quarry sand DRIJET.

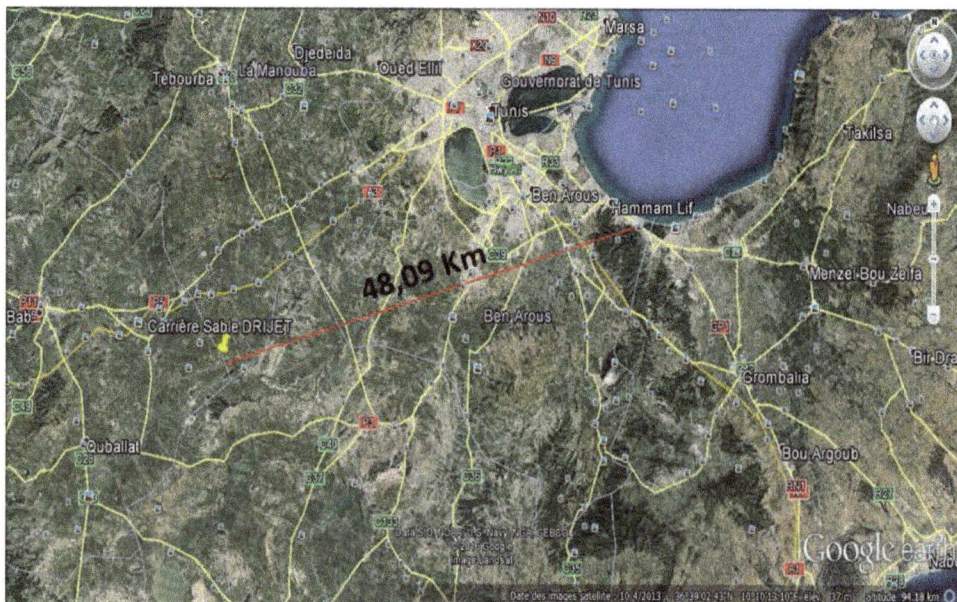

Figure 17 : Distance between sand quarry DRIJET and Hammam Lif.

significant height of the waves to prevent them from overtopping the dike [13] (Figures 19 and 20) (Table 5).

Conclusion

Faced with adverse characteristics, a highly urbanized coastal area of Hammam Lif with 90% urbanization coefficient, overwhelmed and clogged sewerage systems being a unit type draining both of rainwater and waste water, the beach needs immediate intervention. Added to that, the design and choice of protection structures to protect the beach are rather inadequate and badly conceived. This study confirmed

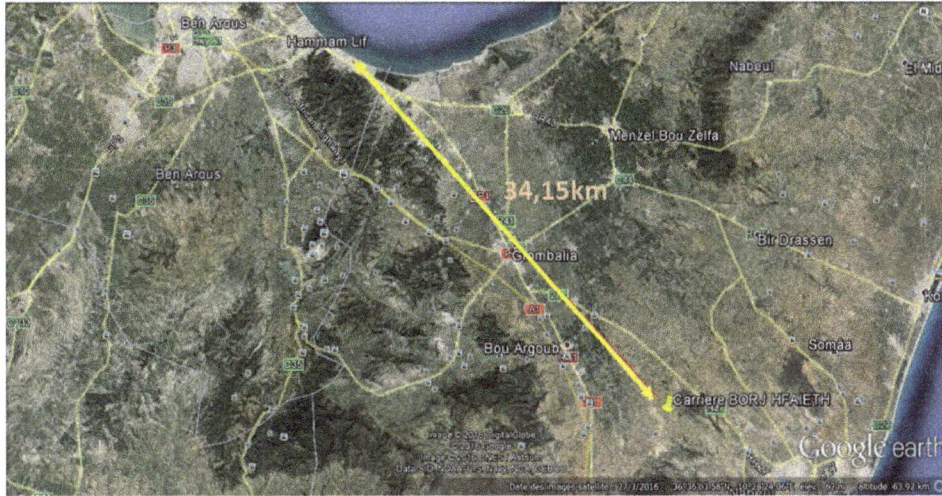

Figure 18: Distance between sand quarry BORJ HFAEITH and Hammam Lif.

Échantillons	Depth(m)	D50	D60	D30	D10	Cu	Cc
ER1b	0.00-0.50	0.23	0.27	0.18	0.1	2.83	1.04
ER2b	0.00-0.50	0.25	0.28	0.2	0.1	2.33	1.07
ER3b	0.00-0.50	0.29	0.32	0.22	0.15	2.29	0.98

Table 4: Size characteristics of the sand samples taken from the quarry sand BORJ HFAIETH.

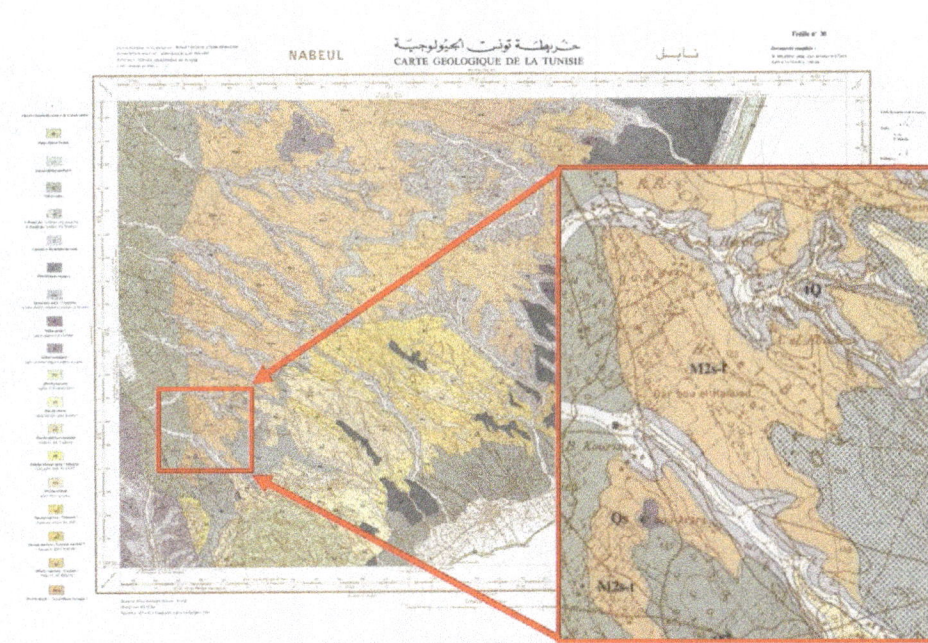

Figure 19: Geological map of the quarry sand BORJ Hfaiedh (drawn from national service TUNIS (OTC 1991).

Berm width	Slope	Reef width	Sectional area	Total volume for a 200m long reef	Transmission coefficient Kt
10 m	1-Feb	30 m	160 m²	32000 m³	0.170

Table 5: Dimensions selected for the estimated reef in the sea of Hammam Lif.

that the beach of Hammam lif is highly threatened by beach flooding and over time has presented major problems caused by flooding due primarily to the stormy rain, as well as the action of wave run up that crosses the existing breakwaters to populated areas.

The choice of a coastal protection solution has to be preceded by studies of the characteristics of the wave in the area of Hammamlif. Two solutions were remarked to be suitable: artificial recharge and artificial

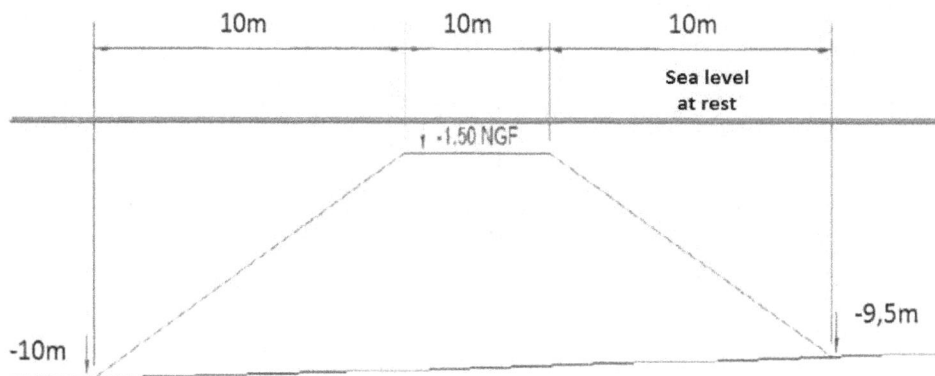

Figure 20: Sectional view of the proposed reef in the sea of Hammam Lif of a trapezoidal shape.

reef submerged geotextile. The application of one of these two solutions is enough to remedy against flooding due to the action of wave run up.

The artificial recharge is needed to Hammam Lif using coarse sand from one another than the sea to promote the stability of the beach. Generally the borrow site should not be of marine origin of the zone itself because the size characteristics are bad as the sand of Hammam Lif has rather it needs a size adjustment by supplying better and especially coarse sand of a other place.

The search of good sand borrows areas for Hammam lif is necessary.

Among the proposed sand areas, the area of OUTHNA but it is a closed area by the Ministry of Environment since 2014 because they became exhausted. The second area is that of DRIJET sand quarry. Samples taken from this area have good size characteristics (D50 is ranged between 0.25 mm and 0.30 mm). The third area is that of the BORJ HFAIETH sand quarry whose samples taken also exhibit good granulometric characteristics (D50 is varied between 0.23 mm and 0.29 mm).

So the tow borrow sites are adequate because their sands are coarser than the original sand of Hammam Lif beach so they promote its stability.

Acknowledgement

This work is undertaken as part of a research project funded by the THYNASONDAGE (Geotechnical Company) in cooperation with ENIS (National Engineering School of Sfax). Helpful contributions have been supplied by Chokri Yaich, Professor at the National School of Engineers of Sfax, Director of Sedimentary Dynamics Laboratory and Environment, Samir Medhioub, Professor of Civil Engineering at the Higher Institute of Technological Studies Of Sfax and Mohamed Bejaoui Manager geotechnical Company. Thynasondage and geotechnical Engineer. Their help is gratefully acknowledged.

References

1. AFNOR (Floor French standard: recognition and testing) (2000) NFP 94-041: Soil testing and recognition particle -Identification - sieving method wet, AFNOR 1.

2. Abdou K, Maria S, Saida N, Otmane R (2011) The Geographic Information System (GIS) as a tool for the assessment of coastal flood risk land due to climate change. If the coastline of Mohammedia. Adaptation Climate change and natural disasters cities, Phase 1: Risk Assessment current situation and 2030 for the coastal city of Tunis in North Africa, Egis BCEOM International / IAU-IDF / BRGM.

3. CEM (2003) Coastal Engineering Manual, U.S. Army Engineer Research and Development Centre, Coastal and Hydraulics Laboratory, Vicksburg, MS.

4. Hildebert I (1952) In: The seasonal distribution of rainfall in Tunisia. Annals of Geography 61: 357-362.

5. Ndébane IFB (1992) Dynamic Coastline on sandy coasts of Mauritania to Guinea-Bissau (West Africa): regional and local approaches by photo interpretation, image processing and analysis of old maps 4: 8.

6. Cariolet JM (2005) Inondation Low ratings and risks associated Brittany: towards redenition of hydrodynamic processes related to ocean weather conditions and sedimentary settings morpho.

7. Jentsje W, van der Meer (1998) Wave run-up and overtopping. Chapter 8.

8. Chini N (2003) Impact and effectiveness of artificial recharge beach by 2DV modeling. The case of Racou (Pyrénées Orientales, Mediterranean) HY.

9. Samir M, Fahmi N, Abir B, Chokri Y (2008) Impact of the dredging process on the granulometry of a shelly sand. Case study of TAPARURA project, Sfax, Tunisia.

10. Xuexue C, Bas H, Corrado A, Tomohiro S, Wim U (2014) Forces on a vertical wall on a dike crest due to overtopping flow. Coastal Engineering 95: 94-104.

11. Chen YY, Li MS (2014) Evolution of breaking waves on sloping beaches. Coastal Engineering Journal.

12. Bargaoui Z, Tramblay Y, Lawin EA, Servat E (2013) Seasonal precipitation variability in regional climate simulations over Northern basins of Tunisia. International Journal of Climatology 34: 235-248.

13. Van der Meer JW, Hardeman B, Steendam GJ, Schuttrumpf H, Verheij H (2010) Flow depths and velocities at crest and landward slope of a dike, in theory and with the wave overtopping simulator. Coast Eng Proc 1.

14

Analysis of Soil and Sub-Soil Properties around Veritas University, Obehie, Southeastern Nigeria

Youdeowei PO[1] and Nwankwoala HO[2]*

[1]Institute of Geosciences and Space Technology, Rivers State University of Science and Technology, Port Harcourt, Nigeria
[2]Department of Geology, University of Port Harcourt, Nigeria

Abstract

This study is aimed at assessing the soil and sub-soil characteristics around Veritas University, Obehie, Abia State, Nigeria. The study revealed that the soils are loose, coarse texture with 78 – 80% sandy fractions having single grain structure from unconsolidated materials of recent alluvial deposits, with soil pH values between 4.8 and 5.8 which could be attributed to leaching of the basic cations to lower depth. Organic carbon content ranged from moderately low 1.50% to high 2.38% due probably to high surface litter and vegetation cover. The low levels of total nitrogen could be attributed to the heavy losses through leaching and responsible for the very low levels of available phosphorus (0.96 - 2.81 µg/g). Nitrate levels are very low (4.33 - 7.80 µg/g) due to the combined effects of intensive cultivations, poor nitrification as well as high leaching processes which occurred in the soils as evidenced in the moderate/high carbon-nitrogen ratio (15 - 43) indicating slow mineralization and humidification process. The three boreholes reveal an overburden brown, medium grained, stiff consistency silty clay soil with thickness of 6-8 meters and underlain by 1 to 2 m depth of brown, medium grained clayey sand. Below this stratum is the aquiferous brown, coarse-grained, gravelly sand to the bored depth of 10 m. Results obtained from the grain size analysis of samples from the boreholes show that the fine to medium grained soils of silty clay and clayey sand have moderate to high fines passing sieve No. 200 (22 to 75%) while the coarse grained gravelly sands have much less silty/clay fractions (10 to 15%). The results show a low plasticity range (6 to 8.7%) for the cohesive soils, while the gravelly sands are non-plastic. The low values of plasticity indices of the plastic soils are an indication of their low water retaining capacity. The soils were classified under the unified soil classification system as SW, SC and CL implying well-graded gravelly sand, clayey sands and low plasticity clays. The permeability test results reveal low values of 1.65×10^{-8} to 1.60×10^{-5} cm/sec. for the clayey soils. The gravelly sand shows a high coefficient of permeability of 1.36×10^{-2} cm/sec.

Keywords: Sub-soil; Soil characterization; Aquifer; Permeability; Obehie; Abia state

Introduction

This study provides a detailed assessment of the suitability of the soils of the area, the sub-soil conditions and suggests relevant soil improvements where necessary as well as recommend appropriate foundation type and design parameters. Numerous studies have been carried out on geotechnical properties of the subsoils generally [1-4].

The study location is within the coast plan sand of the Niger Delta area. The area lies within a sub-horizontal geomorphologic terrain with a measure of undulations arising from uneven surface area erosion. Ground elevation ranges between 10-12 meters above mean sea level. The local geology is of the coastal plain sand, which is Miocene in age and form parts of the most strata of the outcropping Benin Formation. These consist of extensive thickness of brownish, coarse to medium sand with subordinate clay and silt. The area is associated with luxuriant freshwater vegetation typical of a tropical rainforest. Mean annual rainfall exceeds 2000 m.

The study area is endowed with the sedimentary rocks characteristic of the Niger Delta (Figure 1). The detailed geology of the area has been described by Allen [5], Reyment [6], Short and Stauble [7]. Litho-stratigraphically, the rocks are divided into the oldest Akata Formation (Paleoceone), the Agbada Formation (Eocene) and the youngest Benin Formation (Miocene to Recent). The wells and boreholes tap water from the overlaying Benin Formation (Coastal Plain Sands). This formation comprises of lacustrine and fluvial deposits whose thicknesses are variable. The Benin Formation has lithology consisting of sands, silts, gravel and clayey intercalations.

The hydrology of the study site is influenced by its high precipitation rate with a mean annual rainfall of over 2,500 mm, the over burden lithologic strata that over lie the aquifer, and the sometimes undulating topography. Surface waters are received from non-tidal seasonal fresh water flows. Recharge of the aquifer will be by rainwater that eventually moves through the over burden into the aquifer. Recharge depends on rainfall intensity and distribution and amount of surface runoff. Groundwater occurs under confined conditions at the site on account of the essentially clayey soil overlying the aquifer. The existence of this over burden-confining layer will determine whether or not groundwater contaminants introduced into the soil will reach the aquifer. During construction at the site, the protective soil vegetation is removed. Concentrated surface flow of rainwater rills the soil and changes the slope value, which may eventually result in sheet/gully erosion. The cohesive, stiff consistency of the over lying clayey soil will help to limit the degree of this environmental hazard.

Methods of Investigation

Field and laboratory methods

Soil samples were collected using the grid format and sampling location selected in such a manner as to adequately represent the ecological conditions of the study area [8]. At the grid intersection,

*Corresponding author: Nwankwoala HO, Department of Geology, University of Port Harcourt, Nigeria, E-mail: nwankwoala_ho@yahoo.com

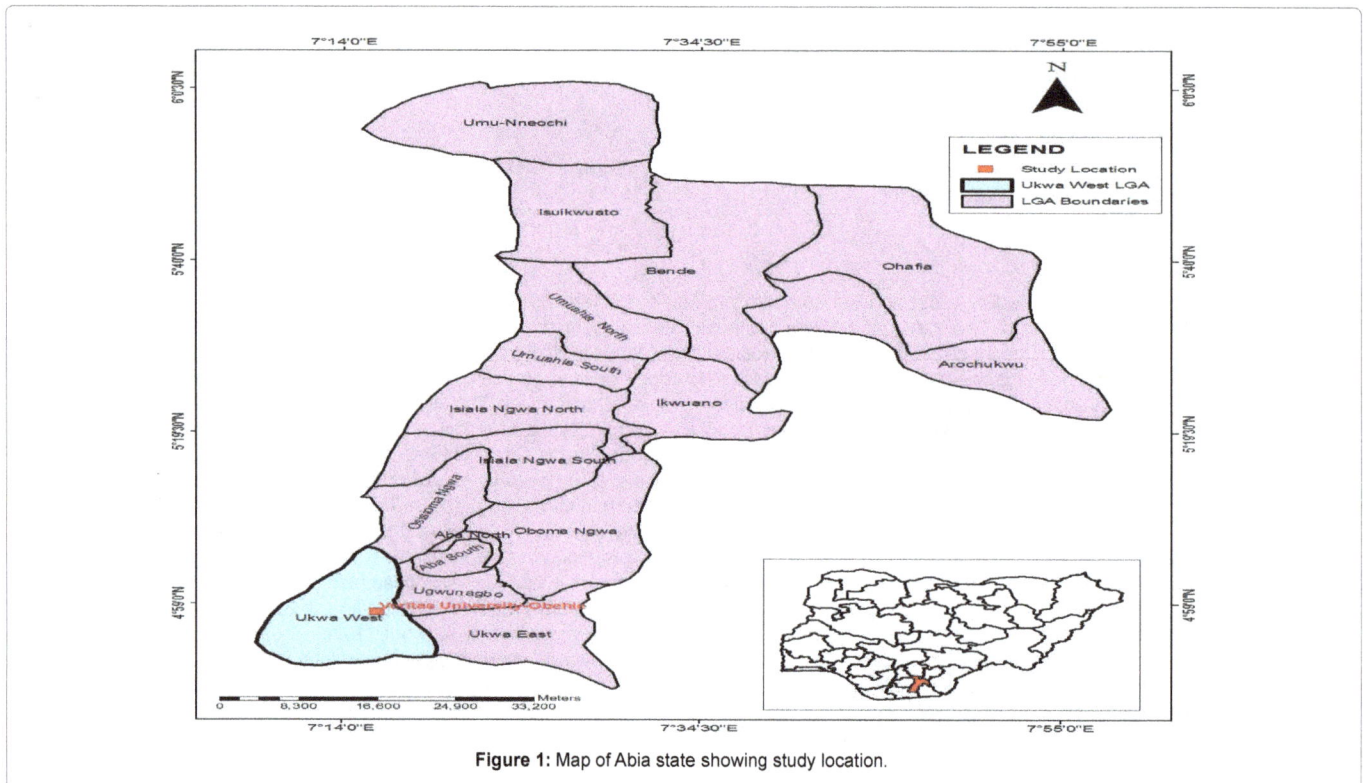

Figure 1: Map of Abia state showing study location.

soil samples were collected by taking about five auger borings at random around the sampling station to depths of 0 – 15 cm and compositing the soils from similar depth into well-labelled plastic bags. The quantity of composite samples collected was processed for analyses in the laboratory without sub-sampling in the field. This allowed for more accurate sub-samples that better represented the area and remove errors due to sample splitting and sub-sampling in the field. The analyses were performed on sub-samples of the air-dried soil samples using materials less than 2 mm diameter of the fine earth. Concentrations were expressed on a dry weight basis and the following physico-chemical parameters were determined: Soil pH, electrical conductivity, organic carbon, total and mineral nitrogen, exchangeable cations, available phosphorus, and particle size analyses. Analyses for oil content (THC) were measured using fresh soil samples.

Three water wells were bored at the site to obtain water quality data, water level monitoring and sub-soil analysis for infiltration characteristics. The boreholes were drilled with the use of light cable percussion rig to a depth of 10 m for each borehole. This drill type permits more accurate determination of groundwater levels and sampling of groundwater for quality analysis. The wells were logged on site with soil samples recovered at intervals where distinct changes in soil type occur, for laboratory analysis. Static water levels recorded in the boreholes were 9.40 m, 9.00 m and 9.00 m for boreholes 1, 2 and 3, respectively. The water level in the boreholes is subject to seasonal fluctuations. The values were observed during the rainy season. The physical properties of the soil samples recovered from the boreholes were examined to obtain parameters used as indices of the infiltration of the soils at the site.

Laboratory tests were carried out on representative sub-soil samples in accordance with British Standards 1377 [9], which are equivalent to the American Standards for Testing Materials (ASTM) [10]. The tests were conducted to enable the evaluation of the gradation, hydraulic

conductivity (coefficient of permeability) and consistency (water absorbing and adsorbing ability) properties of the soil samples, as well as their classification.

Results of the Study

Physico-chemical properties

The physico-chemical properties of soils of the study area are presented in Table 1. Typically the soils are loose, coarse texture with 78 – 80% sandy fractions having single grain structure from unconsolidated materials of recent alluvial deposits. The soils are strongly acid in reaction with pH values between 4.8 and 5.8 which could be attributed to leaching of the basic cations to lower depth. Organic carbon content ranged from moderately low 1.50% to high 2.38% due probably to high surface litter and vegetation cover. Total nitrogen ranged between 0.05 – 0.12%, the low levels could be attributed to the heavy losses through leaching which is responsible also for the very low levels of available phosphorus (0.96 – 2.81 µg/g). Nitrate levels are very low (4.33 – 7.80 µg/g) due to the combined effects of intensive cultivations, poor nitrification and high leaching processes occurring in the soils. This is evidenced in the moderate/high carbon-nitrogen ratio (15 – 43) indicating that mineralization and humification process will be slow.

The soils are moderate to high in exchangeable bases. Potassium, calcium and magnesium were dominant while sodium is low for Nigerian soils [11]. The moderate to high levels of exchangeable bases found in the soils could be attributed to the nature of the parent materials.

The cation exchange capacity is very low (2.67 – 6.28 Cmol/kg) indicating the type of clay minerals. The results revealed that fertility ratings of the soils are moderately low when compared with fertility indices of soils in Nigeria [12]. Total hydrocarbon (oil) concentration

S/N	Sample Station	Cm Depth	pH	µS/cm Electrical Cond.	mg/kg THC	% Org.	% Total N	C/N Ratio	µg/g AV.P	µg/g NO$_3$⁻	Cmol/kg K⁺	Na⁺	Ca⁺	Mg^{2+}	CEC	% Sand	% Silt	% Clay	Textural
1.	CW A 01	0-15	2.7	37	21.82	2.30	0.10	23	2.80	7.12	2.80	1.01	5.86	6.04	4.20	78	8	14	Sandy loam
2.		15-30	2.3	21	18.45	2.01	0.08	25	1.90	6.62	1.80	0.80	6.38	7.04	3.60	78	9	13	Sandy loam
3.	02	0-15	2.3	46	13.08	2.38	0.11	21	0.96	7.00	1.85	0.90	6.20	6.33	2.67	78	6	16	Sandy loam
4.		15-30	2.1	40	26.18	2.08	0.09	23	1.92	5.50	2.81	1.01	6.60	8.12	3.18	78	10	12	Sandy loam
5.	CW B 01	0-15	2.6	25	26.18	2.02	0.07	28	2.78	5.68	1.68	0.68	5.78	6.28	4.16	80	8	12	Sandy loam
6.		15-30	2.6	43	26.18	1.98	0.09	22	1.98	4.34	0.81	0.58	4.20	5.40	5.20	79	7	14	Sandy loam
7.	02	0-15	2.5	29	47.99	2.16	0.05	43	2.81	6.20	2.38	1.20	4.40	7.71	5.22	78	7	15	Sandy loam
8.		15-30	2.3	14	56.72	1.50	0.10	15	1.48	4.32	0.52	0.58	1.32	5.33	6.28	80	10	10	Sandy loam
9.	CW C 01	0-15	2.0	26	17.45	2.10	0.08	26	0.84	7.02	1.78	0.54	5.60	6.20	6.10	80	5	15	Sandy loam
10.		15-30	2.1	39	8.72	2.11	0.07	30	1.04	6.24	0.60	0.48	1.98	2.48	4.78	77	8	15	Sandy loam
11.	02	0-15	2.3	18	43.63	2.00	0.06	33	0.94	5.54	2.10	1.01	6.18	6.20	3.84	80	8	12	Sandy loam
12.		15-30	2.1	19	100.34	1.78	0.05	36	1.92	6.14	0.40	0.30	5.68	5.84	3.90	78	8	14	Sandy loam
13.	CW D 01	0-15	2.0	20	30.54	2.04	0.08	26	2.42	7.24	1.84	1.11	4.12	5.24	5.20	78	9	13	Sandy loam
14.		15-30	2.2	34	13.08	1.87	0.06	31	0.96	6.12	0.54	0.50	3.44	4.24	6.10	77	8	14	Sandy loam
15.	02	0-15	2.1	26	4.36	2.23	0.12	19	0.98	7.80	1.88	0.40	5.61	6.04	3.00	78	6.5	15.5	Sandy loam
16.		15-30	2.2	18	30.54	1.79	0.09	20	1.04	5.62	0.86	0.78	3.42	4.10	4.37	79	10	11	Sandy loam

THC=Total Hydrocarbon
Org. C=Organic Carbon
C/N ratio=Carbon-Nitrogen ratio
Av.P.=Available Phosphorus
CEC=Cation Exchangeable Capacity

Table 1: Physico- Chemical Properties of Soils around Catholic University, Obehie, Abia State.

Borehole Number	Sample Number & Depth	Soil Type	Grain Size Distribution (Percent Passing Sieves)				Atterberg Limits			Permeability (cm/sec.)	Classification
			NO. 4 (4.75 mm)	No. 10 (2.00 mm)	No. 40 (0.42 mm)	No. 200 (0.075 mm)	LL (%)	PL (%)	PL (%)		Unified Soil Classification System (U.S.C)
BH1	BH1 (0.5 m)	Silty Clay	92.0	84.0	80.0	80.0	75.0	16.6	7.9	1.65 X 10⁻⁸	CL
	BH1 (9 m)	Clay Sand	38.0	31.0	29.0	22.0	12.2	5.8	6.4	1.45 x 10⁻⁵	SC
	HH1 (10 m)	Gravelly Sand	26.0	22.0	18.0	15.0	-	-	Non Plastic	1-36 x 10⁻²	SW
BH2	BH2 (0.5 m)	Silty Clay	90.0	87.0	81.0	76.0	15.8	7.2	8.6	1.72 x 10⁻⁸	CL
	BH2 (7 m)	Clay Sand	35.0	33.0	31.0	30.0	12.4	6.0	6.4	1.60 x 10⁻⁵	SC
	BH2 (9 m)	Gravelly Sand	32.0	29.0	18.0	15.0	-	-	Non plastic	-	SW
BH3	BH3 (0.5 m)	Silty Clay	92.0	86.0	81.0	75.0	16.6	7.9	8.7	1.70 x 10⁻⁸	CL
	BH3 (7 m)	Clay Sand	32.0	31.0	30.0	28.0	12.2	6.2	6.0	1.50 x 10⁻⁵	SC
	BH3 (9 m)	Gravelly Sand	22.0	18.0	15.0	10.0	-	-	Non plastic	-	SW

Table 2: Summary of laboratory sub-soil test results (Veritas University Site, Obehie).

in the soil is low indicating very low hydrocarbon contamination. The amount detected could be attributed to biogenic sources.

Subsoil Results/Discussion

The study revealed from the logs a uniform correlation in the three boreholes of an overburden brown, medium grained, stiff consistency silty clay soil with thickness of 6-8 m. This is underlain by 1 to 2 m depth of brown, medium grained clayey sand. Below this stratum is the aquiferous brown, coarse-grained, gravelly sand to the bored depth of 10 m. Grain size analysis involved dry sieving on field obtained samples. Results obtained from the grain size analysis of samples from the boreholes show that the fine to medium grained soils of silty clay and clayey sand have moderate to high fines passing sieve No. 200 (22 to 75%) while the coarse grained gravelly sands have much less silty/clay fractions (10 to 15%). The fine grained nature of the clays mean that fluid flow through them will be slow, as the number of particles per unit area is relatively small and void spaces are fewer. The clayey

soils therefore have relatively low permeabilities on account of their fine grains.

Atterberg limits (also known as consistency limits) expresses the water absorbing and adsorbing ability of fine grained, cohesive soil, with the plasticity index indicating the range of water content, through which the soil remains plastic. The results show a low plasticity range (6 to 8.7%) for the cohesive soils, while the gravelly sands are non-plastic (Table 2). Atterberg limit tests are applicable only to fine grained, cohesive soils. The low values of plasticity indices of the plastic soils are an indication of their low water retaining capacity.

The soils were classified under the unified soil classification system as SW, SC and CL implying well-graded gravelly sand, clayey sands and low plasticity clays (Table 2). The permeability test results reveal low values of 1.65×10^{-8} to 1.60×10^{-5} cm/sec. for the clayey soils. The gravelly sand display a high coefficient of permeability of 1.36×10^{-2} cm/sec. Infiltration capacity of soil depends on the permeability, degree of saturation, vegetation and amount and duration of rainfall [13].

Conclusion

The study revealed that the soils are loose, coarse texture with 78 – 80% sandy fractions having single grain structure from unconsolidated materials of recent alluvial deposits. The soils are strongly acid in reaction with pH values between 4.8 and 5.8 which could be attributed to leaching of the basic cations to lower depth. Organic carbon content ranged from moderately low 1.50% to high 2.38% due probably to high surface litter and vegetation cover. The low levels of total nitrogen could be attributed to the heavy losses through leaching which is responsible also for the very low levels of available phosphorus (0.96 – 2.81 μg/g). Nitrate levels are very low (4.33 – 7.80 μg/g) due to the combined effects of intensive cultivations, poor nitrification and high leaching processes occurring in the soils, evidenced in the moderate/high carbon-nitrogen ratio (15 – 43) indicating slow mineralization and humification process.

The physical properties of the sub-soils were determined and used as indices of their infiltration capacity and classification. Boreholes logged to the maximum-drilled depth of 10 m reveal fine-grained, stiff consistency silty clay overlying medium grained clayey sand. Beneath this is the aquiferous coarse gravelly sand. The aquifer is confined and this condition may help to seal off the lower strata and aquifer from pollutants. However, environmental hazards that may occur at the site may include sheet erosion and flooding.

References

1. Nwankwoala HO, Amadi AN (2013) Geotechnical investigation of sub-soil and rock characteristics in parts of shiroro-muya-chanchaga area of niger state, nigeria. Int J Ear Sci Eng 6: 8-17.

2. Youdeowei PO, Nwankwoala HO (2013) Suitability of soils as bearing media at a freshwater swamp terrain in the Niger Delta. J Geol Min Res 5: 58-64.

3. Oke SA, Amadi AN (2008) An assessment of the geotechnical properties of the sub-soil of parts of Federal University of Technology, Minna, Gidan Kwano Campus, for foundation design and construction. J Sci Educ Tech 1: 87-102.

4. Oke SA, Okeke OE, Amadi AN, Onoduku US (2009) Geotechnical properties of the sub-soil for designing shallow foundation in some selected parts of Chanchaga area, Minna, Nigeria.

5. Allen JRL (1965) Late quaternary niger delta and adjacent areas - sedimentary environment and lithofacies. AAPG Bulletin 49: 547-600.

6. Reyment RA (1965) Aspects of Geology of Nigeria. University of Ibadan Press, Nigeria.

7. Short KC, Stauble AJ (1967) Outline of geology of the niger delta. AAPG Bulletin 51: 761-779.

8. Smith RT, Atkinson K (1975) Techniques in Pedology. A Hand Book for Environmental and Resource Studies. Elek Science, London.

9. http://infostore.saiglobal.com/store/details.aspx?ProductID=406410

10. ASTM (1979) Annual book of America society for testing and materials standards 1289, ASTM Tech Publication, Phildelphia.

11. Odu CTI, Esuruoso OF, Nwoboshi LC, Ogunwale JA (1985) Environmental study of the nigerian agip oil company operational areas. Soils and Freshwater Vegetation, Milian Italy.

12. Ayotade KA, Fagade SO (1986) Nigeria's program for wetland rice production and rice research, In: Juo ASR, Lowe JA The wetlands and rice in Sub-Saharan Africa, Ibadan Nigeria. IITA 201-260.

13. Todd DK (1980) Groundwater Hydrology (2ndedtn). John Wiley and Sons, New York.

15

Application of Electrical Resistivity and Ground Penetrating Radar Techniques in Subsurface Imaging around Ajibode, Ibadan, Southwestern Nigeria

Adelekan AO[1]*, Oladunjoye MA[1] and Igbasan AO[2]

[1]*Department of Geology, University of Ibadan, Nigeria*
[2]*Department of Applied Geophysics, Federal University of Technology, Akure, Nigeria*

Abstract

Integrated geophysical investigation involving Electrical Resistivity (ER) and Ground Penetrating Radar (GPR) techniques were carried out around a site underlined by Basement Complex rocks of southwestern Nigeria. The study was aimed at imaging the subsurface lithological units and delineating shallow geologic structures for the purpose of characterizing the area for construction suitability. A total of twenty five (25) Vertical Electrical Sounding (VES) data using Schlumberger array, ten (10) traverses of Electrical Resistivity Imaging (ERI) using Wenner array and Ground Penetrating Radar (GPR) surveys were carried out along the established traverse lines within the area. The VES data were quantitatively interpreted using partial curve matching technique and subsequently improve upon by inversion software using IPI2Win, to obtain the layer geoelectric parameters. The ERI data was inverted and interpreted using Res2Dinv and Res3Dinv inversion software's respectively to generate 2D and 3D resistivity image of the subsurface. The GPR data was processed into radar section using RadExplorer software. Vertical electrical sounding results delineates typically three to four geologic layers which are the topsoil/lateritic hardpan, weathered basement (consisting clay and sandy clay) and fractured/fresh basement with layer resistivity value ranges of 10-2684 Ωm, 12-242 Ωm and 229-3213 Ωm respectively and thickness value ranges of 0.5-2.1 m and 4.0-14.1 m respectively. 2-D inverted resistivity results also delineated three major geologic layers which are the topsoil, weathered basement and fresh basement and correlates well with the results obtained from the VES results. Layers 1 to 3 of 3D inverted resistivity slice results show high degree of variation in resistivity distribution at shallow depth, consisting of highly resistive material towards the eastern part with low resistivity material concentrating at the south-western part. Results of the GPR survey also delineated three to four geologic layers which include the topsoil/lateritic hardpan, weathered basement and fractured/fresh basement. The study area was categorized to have semi-competent to competent basement rock based on the resistivity value of the underlying material within the area. Bedrock depression delineated at some location could pose threat of differential settlement to construction works within the study area. Thus, it should be ensured that foundation is designed to sit comfortably on the competent bedrock or by employing suitable foundation work, such as piling to ensure foundation stability and prevent structural failure. Thus, electrical resistivity and ground penetrating radar techniques are versatile tools in site characterization.

Keywords: Subsurface; Topsoil; Bedrock; Seismic refraction; Amplitude

Introduction

The increasing rate in failure of structures such as road, buildings, dam and bridges has assumed an alarming dimension [1]. The need for pre-foundation studies has therefore become imperative before any structure is laid so as to prevent loss of lives and also goods and properties. Some earth materials such as sands and fresh rock provide firm support for solid foundation while other host material due to their nature cannot support solid and rigid structures [2]. Such improperly sited structures have failed due to lateral and vertical inhomogeneity in the subsurface earth materials [3-5]. Such fallout can however be prevented when necessary information aimed at better understanding the subsurface geometry and structural setting is available [6]. Geophysical investigation is one of the methods used in probing the subsurface for any construction activities. The deduced soil characteristics are used as preliminary information to determine the suitability of the site for a proposed structure. If this crucial step is omitted, concealed geologic features within the subsurface may precipitate excessive total or differential settlement leading to failure or collapse of structures [7]. Geophysical methods that have been found useful in pre-and post-construction geotechnical investigations include Electrical Resistivity, Seismic Refraction and Electromagnetic methods including the Ground Penetrating Radar (GPR) among others [1,6,8,9]. The main objective of this work was to investigate subsurface stratigraphic and structural setting of the study area and to define its competency for construction purposes using Electrical Resistivity and Ground Penetrating Radar methods.

Site description and geological setting

The study area, Ajibode is located in the North central part of Ibadan, southwestern Nigeria with Latitudes 7°28'32.6" and 7°28'36.6" North of the Equator and Longitudes 3°53'20.7" and 3°53'25.3" East of the Greenwich Meridian (Figure 1). It shares boundaries with the University of Ibadan and Orogun communities at the south and at the north with IITA, Ojoo and Shasha communities of Ibadan. The geology of the study area, Ibadan is the subset of the geology of the southwestern part of Nigeria. The dominant rock types in this region are granite, quartz schist of metasedimentary series, banded gneiss, granite gneiss, augen gneiss and migmatite complex. Quartzite outcrops as ridges

*Corresponding author: Adelekan AO, Department of Geology, University of Ibadan, Nigeria, E-mail: yemmylek@gmail.com

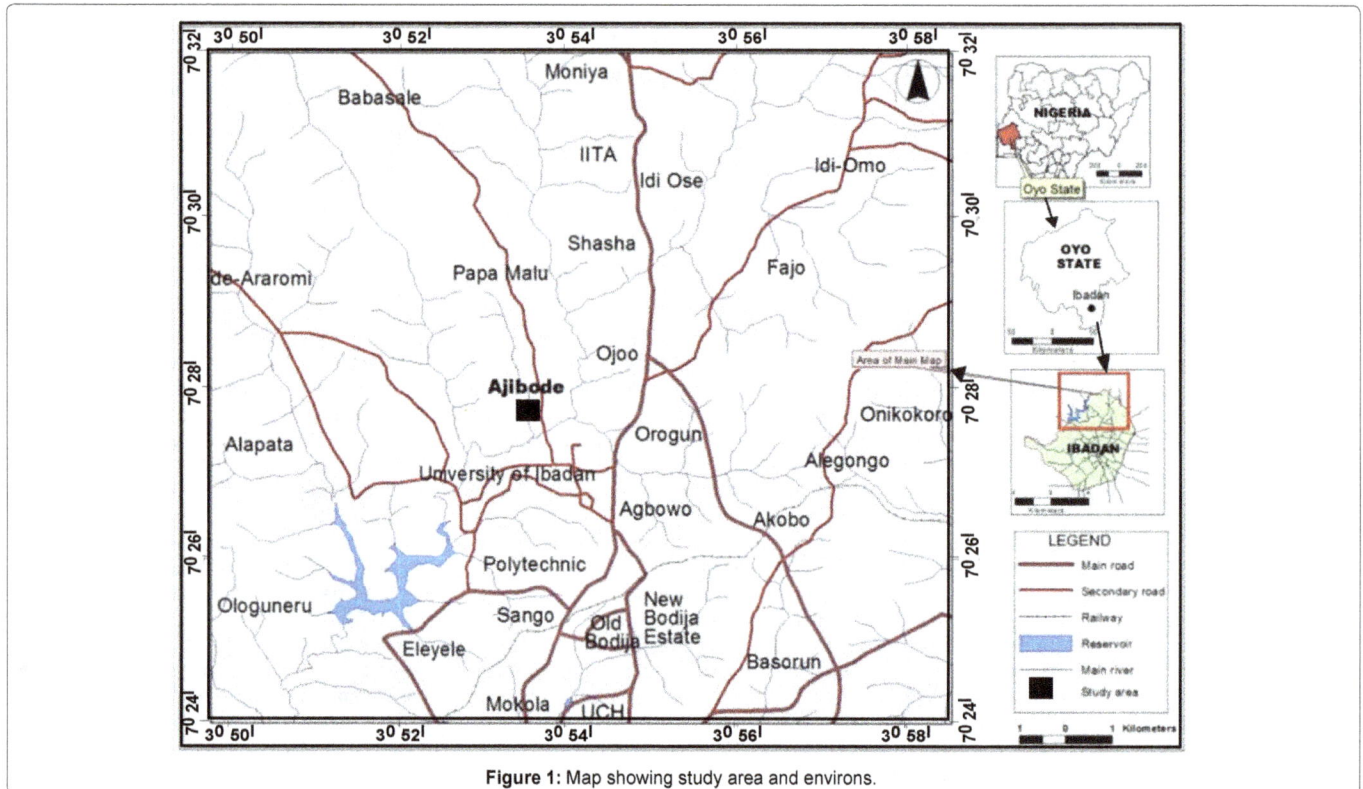

Figure 1: Map showing study area and environs.

with relatively high elevation with schistose structure. The strike line runs in the N-S direction between 340 and 350° consistently dipping eastwards with characteristic cross-cutting features [10], with dip angle of 47°E and characterized by joints and faults in some cases [11]. Field investigations of Ajibode area showed three main rock types which are banded gneiss, granite gneiss and augen gneiss [12] with the presence of fractures along the general strike of the rock (Figure 2). Banded gneiss occurs at the extreme east and west running south-east of the study area. The general strike of banded gneiss is NNW-SSE while the dips range from 42° to 53° East. The rock is medium grained, crystalline and contorted. The study area is well drained by river Ona and its tributaries. The drainage pattern is dendritic with irregular branching in all directions. The river flows in the southwestern direction and forms the boundary between the University of Ibadan campus and Ajibode village in the northern part.

Methodology

A total of ten (10) traverse lines were obtained with five (5) lines each in approximately NW-SE and NE-SW directions along which Electrical Resistivity and Ground Penetrating Radar data were acquired (Figure 3). Vertical Electrical Sounding (VES) data was acquired using Schlumberger electrode array. A total of twenty five (25) VES stations were occupied with each located at the grid points of the traverse line within the study area. The spread length of current electrode ranges from 1.0 up to 100 m in successive steps. Wenner array was employed for Electrical Resistivity Imaging (ERI) with spacing between adjacent electrodes represented by 'a', all the possible measurement made with Wenner array is of electrode spacing of "na" where n=1, 2....5 and 'a'=5 m. GPR data was acquired using "Mala RAMAC/GPR" geophysical survey equipment. The equipment consists of radar antennae housed in a box attached to the survey wheel which rolled along the traverse lines. The box was connected to the computer monitor unit which

automatically recorded and store the data as the survey wheel is being moved across the study area. The survey was carried out using two different antenna frequencies (250 MHz and 500 MHz) with antenna separation of 0.36 m and 0.18 m respectively. The start and end time ranges between 168-210 ns and 0-100 ns respectively for both 250 MHz and 500 MHz antenna frequencies and the mode of constant component is taken as "mean". The direct wave velocities were recorded as 30 cm/ns and the operation length as 52 ns for both frequencies. The GPR data positioning was calibrated using survey wheel.

Vertical Electrical Sounding data were interpreted by partial curve matching technique [13] with the result of the layer resistivity parameters derived from the partial curve matching technique being used as the initial interpretation of the field curves for the computer inversion using IPI2Win software. Forward inversion modeling technique using regularized least-square optimization method [14] of Res2Dinv and Res3Dinv inversion software's were used in the interpretation of the 2D and 3D data. GPR data was processed using "RadExplorer V1.4 [15] software" package which is commercially available. The processing parameters applied, as made available by the software and that which gives acceptable result on the radargram includes: DC removal, Time zero adjustment, Bandpass filtering and Amplitude correction.

Results and Discussion

Vertical electrical sounding results

Quantitative interpretation of the VES curves resulted in determination of geoelectric layer parameters (layer resistivity and thickness) for the subsurface characterization. Based on the approximate range of resistivity values of common rock types [16] and correlation with the pit information (having lithology ranging from topsoil/lateritic hardpan-weathered basement-basement rock) within

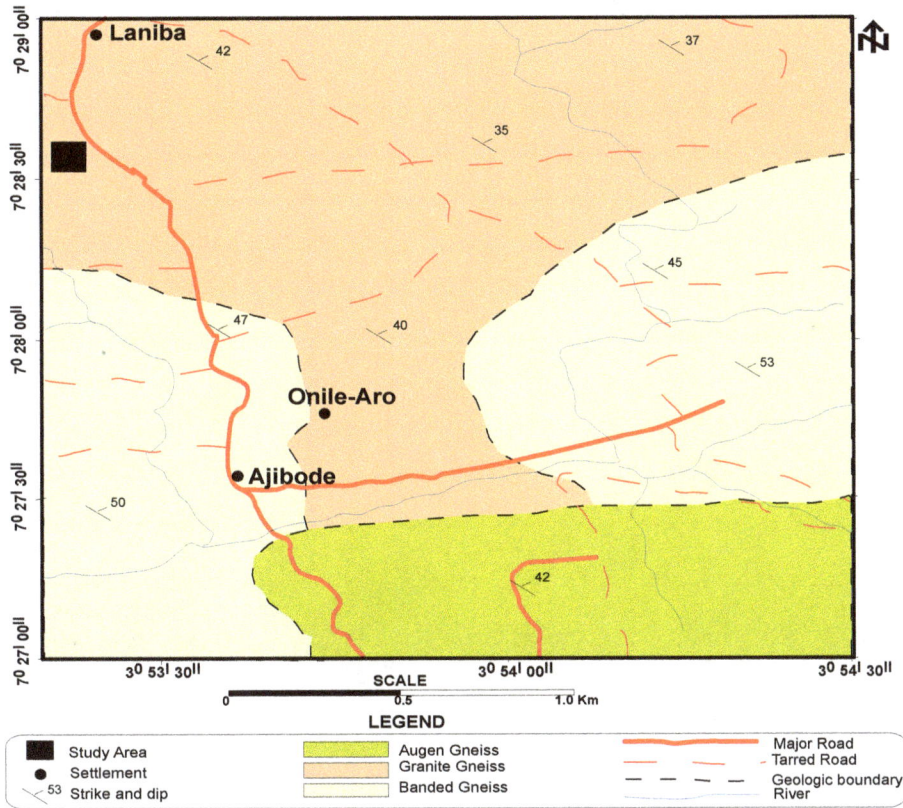

Figure 2: Geological map of Ajibode [12].

VES Station	Layer	Resistivity (Ωm)	Thickness (m)	Depth (m)	Lithology	Curve Type
1	1	761	1.4	1.4	Topsoil	H
	2	77	9.3	10.7	Weathered basement (Clayey)	
	3	1711	-	-	Fresh basement	
2	1	1013	1.1	1.1	Topsoil	QH
	2	133	3.4	4.5	Sand clay	
	3	20	10.8	15.2	Weathered basement (Clayey)	
	4	360	-	-	Partly weathered basement	
3	1	719	1.6	1.6	Topsoil	H
	2	45	4.9	6.5	Weathered basement (Clayey)	
	3	3213	-	-	Fresh basement	
4	1	565	1.1	1.1	Topsoil	H
	2	49	7.2	8.3	Weathered basement (Clayey)	
	3	1103	-	-	Fresh basement	
5	1	913	1.5	1.5	Topsoil	H
	2	60	7.3	8.8	Weathered basement (Clayey)	
	3	1026	-	-	Fresh basement	
6	1	1827	1.2	1.2	Topsoil (lateritic hardpan)	H
	2	94	13	14.2	Weathered basement (Clayey)	
	3	726	-	-	Fresh basement	
7	1	1772	1.3	1.3	Topsoil (lateritic hardpan)	QH
	2	233	2.7	4	Sandy clay	
	3	53	7.7	11.7	Weathered basement (Clayey)	
	4	555	-	-	Fresh basement	
8	1	2684	1.1	1.1	Topsoil (lateritic hardpan)	QH
	2	182	1.9	3	Sandy clay	
	3	94	4.5	7.4	Weathered basement (Clayey)	
	4	992	-	-	Fresh basement	

	1	1362	1.1	1.1	Topsoil (lateritic hardpan)	
9	2	281	2.6	3.7	Sandy clay	QH
	3	56	7.6	11.3	Weathered basement (Clayey)	
	4	1162	-	-	Fresh basement	
	1	1280	1.3	1.3	Topsoil (lateritic hardpan)	
10	2	76	7	8.3	Weathered basement (Clayey)	H
	3	1803	-	-	Fresh basement	
	1	2613	1.1	1.1	Topsoil (lateritic hardpan)	
11	2	81	6.4	7.4	Weathered basement (Clayey)	H
	3	1419	-	-	Fresh basement	
	1	1927	1.2	1.2	Topsoil (lateritic hardpan)	
12	2	200	1.6	2.8	Sandy clay	QH
	3	60	6.8	9.6	Weathered basement (Clayey)	
	4	558	-	-	Fresh basement	
	1	2541	1.3	1.3	Topsoil (lateritic hardpan)	
13	2	267	2	3.2	Sandy clay	QH
	3	69	7.8	11	Weathered basement (Clayey)	
	4	457	-	-	Fractured basement	
	1	2011	1.2	1.2	Topsoil (lateritic hardpan)	
14	2	133	2.1	3.3	Sandy clay	QH
	3	70	6.5	9.8	Weathered basement (Clayey)	
	4	607	-	-	Fresh basement	
	1	1294	0.8	0.8	Topsoil (lateritic hardpan)	
15	2	87	14	14.8	Weathered basement (Clayey)	H
	3	752	-	-	Fresh basement	
	1	2209	1	1	Topsoil (lateritic hardpan)	
16	2	165	1.6	2.6	Sand clay	QH
	3	20	7.8	10.3	Weathered basement (Clayey)	
	4	771	-	-	Fresh basement	
	1	269	0.7	0.7	Topsoil	
17	2	25	8.4	9.2	Weathered basement(Clayey)	H
	3	584	-	-	Fresh basement	
	1	436	0.8	0.8	Topsoil	
18	2	29	9.4	10.2	Weathered basement (Clayey)	H
	3	229	-	-	Fractured basement	
	1	1242	1.5	1.5	Topsoil (lateritic hardpan)	
19	2	91	8.8	10.3	Weathered basement (Clayey)	H
	3	397	-	-	Fractured basement	
	1	952	0.5	0.5	Topsoil	
20	2	51	1	1.6	Weathered basement (Clayey)	HA
	3	242	6.4	8	Partly weathered basement	
	4	777	-	-	Fresh basement	
	1	1860	1.1	1.1	Topsoil (lateritic hardpan)	
21	2	35	4	5.1	Weathered basement(Clayey)	H
	3	1934	-	-	Fresh basement	
	1	234	0.9	0.9	Topsoil	
22	2	25	6.2	7.1	Weathered basement (Clayey)	H
	3	1192	-	-	Fresh basement	
	1	10	2.1	2.1	Topsoil (Water saturated)	
23	2	54	7.6	9.8	Weathered basement (Clayey)	A
	3	561	-	-	Fresh basement	
	1	139	0.8	0.8	Topsoil	
24	2	12	2.1	2.9	Weathered basement (Clayey)	HA
	3	162	3.1	6.1	Partly weathered basement	
	4	1419	-	-	Fresh basement	
	1	119	0.7	0.7	Topsoil	
25	2	17	1.8	2.5	Weathered basement (Clayey)	HA
	3	158	3.1	5.6	Partly weathered basement	
	4	1474	-	-	Fresh basement	

Table 1: Summary of the VES interpretation from the study area.

the study area, three major geologic layers were delineated from the geoelectric layer parameters obtained from the study area (Figure 4) and these includes the topsoil/lateritic hardpan, weathered basement (clayey/sandy clay) and fractured/fresh basement. A summary of the VES interpretation is shown on Table 1.

Electrical resistivity imaging results

2D Wenner data results were presented as an inverted model resistivity section representing the image of the subsurface along each established traverse line within the study area. This shows both lateral and vertical variations in resistivity of the subsurface material. The 2D inverted resistivity sections generated along NW-SE and NE-SW (Figure 5) respectively within the study area delineated three major subsurface layers which are; the topsoil, weathered basement and basement rock. The inverted resistivity model section delineated

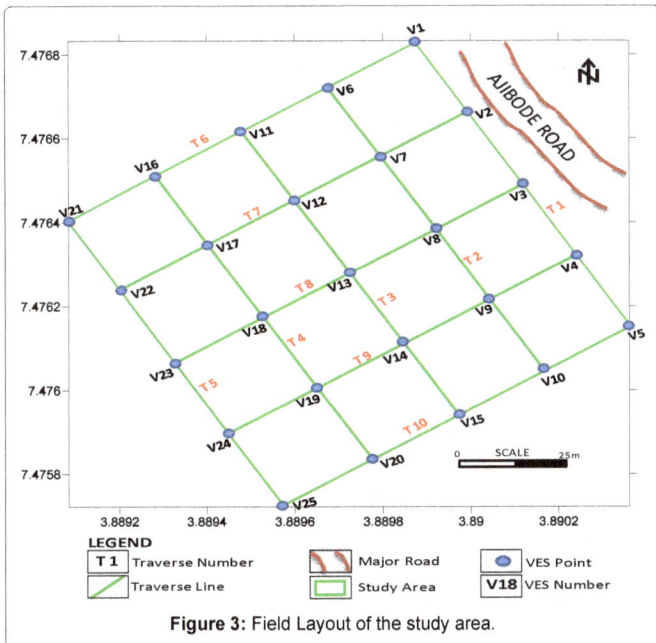

Figure 3: Field Layout of the study area.

topsoil layer with high degree of variations in resistivity value ranging between 10 and 1000 Ωm and layer thickness of approximately 3.5 m. Some areas of the topsoil have high resistivity value and are presumed to be dominated by lateritic hardpan material (T3, T7 and T8). The topsoil is underlined by a conductive material characterized by <100 Ωm resistivity which shows typically a clay-rich material and similar to the range established for similar geological provinces by Kearey et al. [16]. The thickness of the second layer ranges from 2-8 m, but could not be ascertained beneath some of the traverse lines as it forms depression into the basement. Such depressions were noticed beneath traverse line T3, T7, T8 and T10. The depressions are noticed at horizontal distance 35-45 m beneath traverse lines T3 and T7, between horizontal distance 35 and 60 m beneath T8 and at horizontal distance 40-55 m beneath T10. The weathered layer is underlined by basement rock with characteristic resistivity value of about 200 - 9036 Ωm. The depth to competent basement rock delineated by 2D inverted resistivity results within the area is approximately about 11 m, but could not be ascertained beneath regions where the weathered basement forms depression into the basement. Inverted resistivity section for traverse lines T3, T5, T7, T8 and T10 (Figure 5) generally show a continuous trend of depression into the basement which cut across the traverse lines.

The data acquired along all the ten gridded traverse lines with 2D Wenner profile were combined to produce the slice depth and a collective 3D view of the inverted resistivity models (Figure 6) for all the traverse lines (T1 to T10). X and Y coordinates represent the NW-SE and NE-SW directions respectively with slices been made along ZDirection to a maximum depth of 21.9 m. Layers 1 to 3 shows high degree of variations in resistivity distribution at shallow depth revealing formation consisting of highly resistive topsoil material (lateritic hardpan) at the eastern part with low resistivity (highly conductive) material concentrating at the south-western part of the study area. The high resistivity variations are attributable to the heterogeneity and unpredictable nature of the geology of a typical basement complex [17]. Material with characteristics blue to green color bands having resistivity value of <100 Ωm is presumed to be weathered unit and is

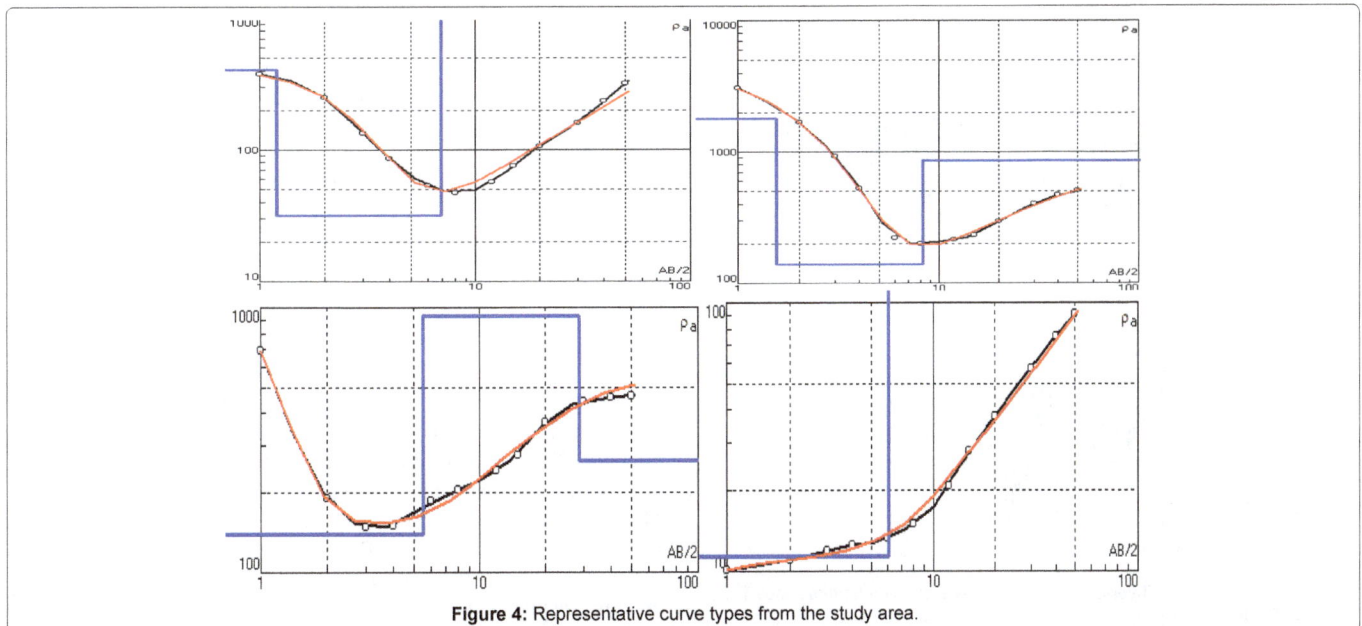

Figure 4: Representative curve types from the study area.

seen to be consistent across all the slices with an approximately NE-SW trend. Also, the underlain material within the area was presumed to be basement rock and was delineated by layers 4-6 with depth range of 8.68-21.9 m showing highly resistive material characterized by red to pink color bands towards the southern part of the area with resistivity values of >200 Ωm. The resistivity value within this region increases with layer depth.

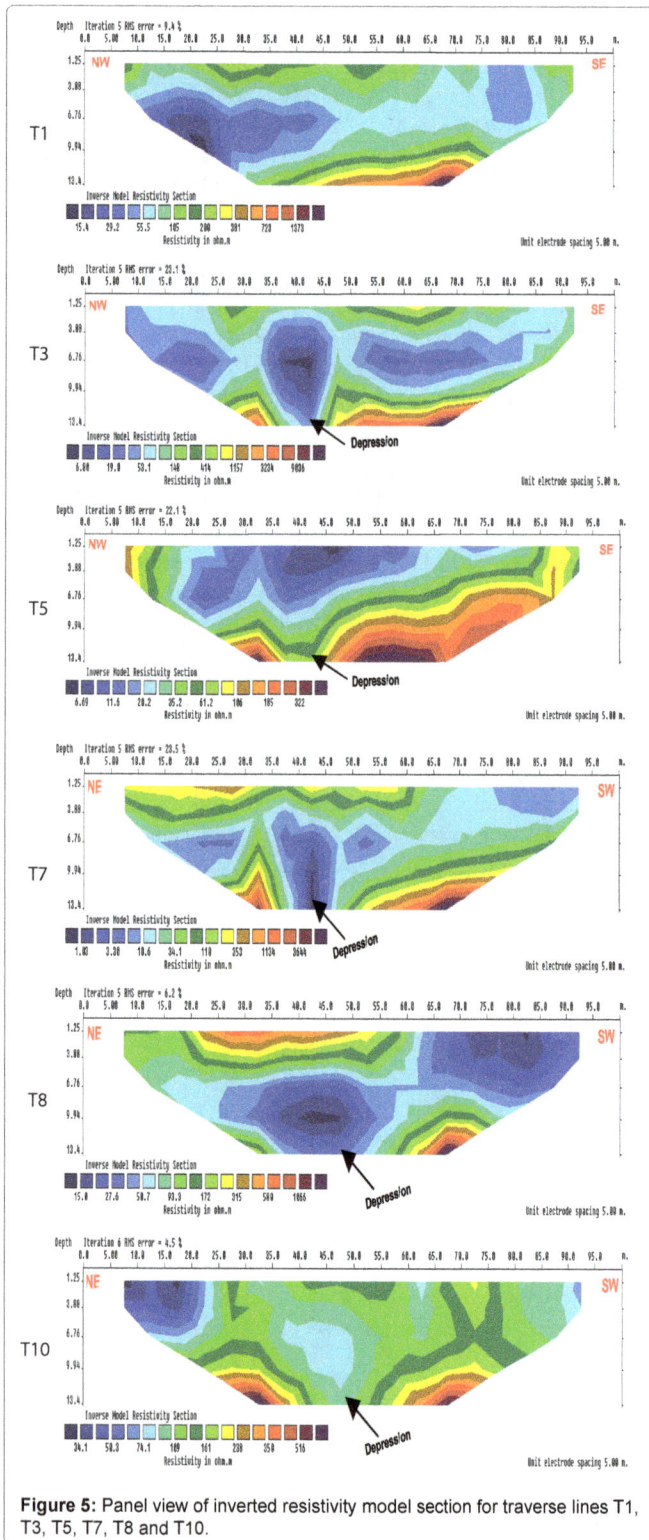

Figure 5: Panel view of inverted resistivity model section for traverse lines T1, T3, T5, T7, T8 and T10.

Ground penetrating radar results

Radar section results for 250 MHz antenna frequency as represented by traverse line 2 (Figure 7) respectively was able to image up to 10 m depth within the study area. Four radar facies were recognized and correlated to their lithological equivalence in geoelectric section and pit information within the study area. The upper layer (above the red dotted lines) shows high amplitude, parallel and horizontal reflections with good continuity and is interpreted as topsoil [18,19]. The thickness is approximately 0.7 m across the entire 250 MHz antenna frequency profiles. This is underlain by high amplitude, sub-parallel and chaotic with moderately continuous reflection characterized to be lateritic hardpan (between red and yellow dotted lines) across the 250 MHz antenna frequency profiles with thickness range between 0.3 m and 2 m except beneath traverse line 5 where the layer is absent. The third layer (between yellow and green dotted lines) depicts the weathered unit (clay/sandy clay) with low to moderate reflection amplitude, parallel and horizontal reflection with poor continuity and ranges in thickness between 4 m and about 8 m along the entire traverse lines. The weathered layer is underlain by fresh basement (from green dotted lines to infinity) consisting of low reflection amplitude with poor continuity and depth to fresh bedrock ranging from 6 m to 8.5 m across the traverse lines.

Figure 8, radar section for traverse line 8 shows a representative radar section result using 500 MHz antenna frequency. The antenna frequency generally delineated a maximum depth of 5 m. Three radar facies were recognized and correlated to the lithological equivalence from geoelectric section and pit information obtained within the study area. The high amplitude, parallel, horizontal and continuous reflections characterized by the upper layer (above red dotted lines) is interpreted as the topsoil with approximate depth of <0.5 m across the entire 500 MHz radar section traverse lines. The second layer (between red and yellow dotted lines) was interpreted as lateritic hardpan with high amplitude, chaotic reflectivity and moderate continuity with approximate thickness of 1.5 m except for traverse line 5 where lateritic hardpan was not delineated. The last layer (beneath yellow dotted lines) is characterized by low to moderate amplitude, parallel and horizontal reflection with moderate continuity and is interpreted within the area as weathered basement with depth at infinity.

Figures 9 and 10 shows representative correlation of the geophysical results obtained within the study area. After careful examination of all the results, the 2D resistivity models and ground penetrating radar section results interpreted beneath the traverse lines show some similarity. The 2D inverted resistivity sections, geoelectric section and radar section correlates well based on the delineated lithological sequences, approximate layer depth and structures underlying the study area.

Both techniques were able to delineate three to four geologic layers which include the topsoil/lateritic hardpan, weathered basement and fractured/fresh basement. The boundary between the weathered basement and fresh basement could not be effectively defined on the radar section, presumably due to the attenuation of radar signal by the conductive layer (clay-rich weathered basement) that overlies the fresh basement. The results of the 250 MHz antenna frequency was able to confirm the basement and it penetrates a depth of 10 m while the 500 MHz antenna frequency terminates within the weathered basement, as it penetrates a depth of 5 m. The 500 MHz antenna frequency gives better resolution of the subsurface information than the 250 MHz antenna frequency. Electrical resistivity results were able to map the

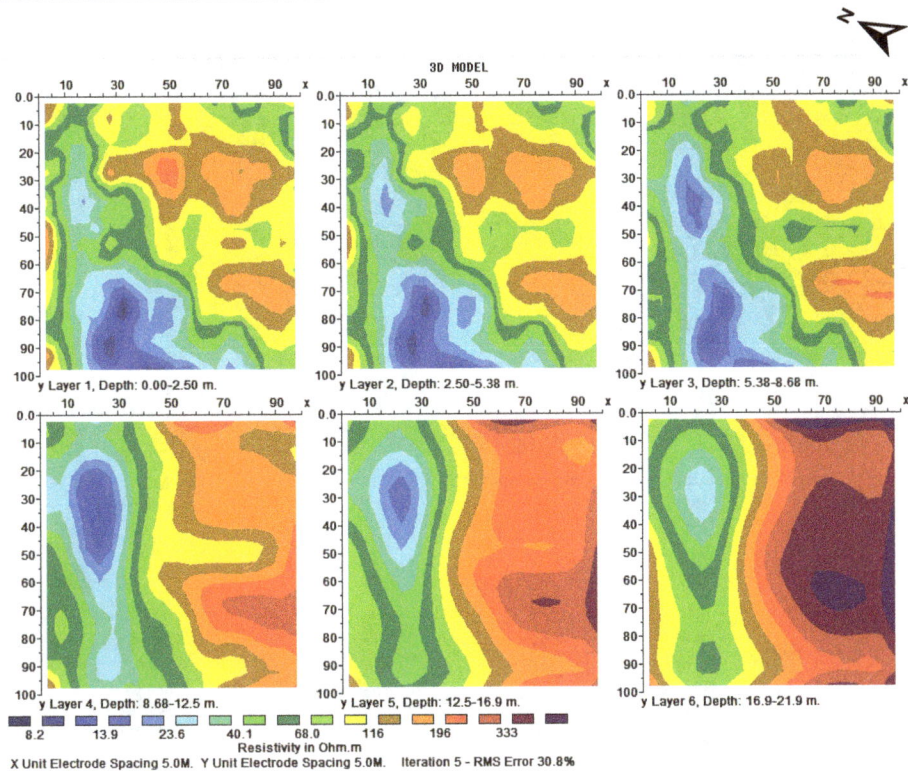

Figure 6: 3D inverted resistivity slices of the area.

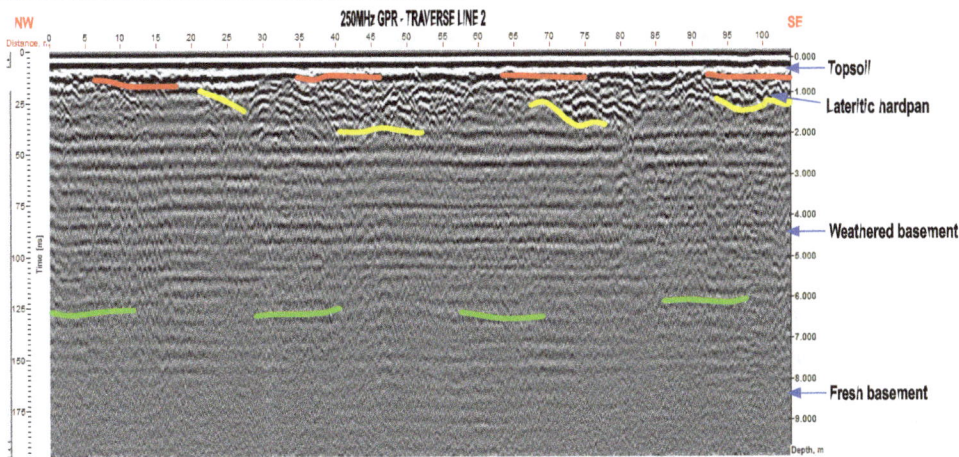

Figure 7: Radar section along traverse line 2 using antenna frequency of 250 MHz.

lithological sequences within the area, the presumed depression into bedrock and the direction along which the depression is trending. GPR results were able to delineate only the lithological sequences and could not ascertain the structural information due to low radar signal strength that gets to the bedrock.

Conclusion and Recommendation

Integrated geophysical techniques involving Vertical Electrical Sounding, Electrical Resistivity Imaging and Ground Penetrating Radar have been adopted in this study around Ajibode, southwestern Nigeria. The result of the Electrical Resistivity (ER) investigation has effectively delineated the subsurface geological information and gives detailed description of lithological setting within the area.

It was observed from the ER surveys that depth to competent basement varies across the study area with shallow depth towards the south-southeastern direction and as deep as 22 m towards the northwestern direction where presumed depression into the basement are delineated on resistivity data.

Thus, the area under study can be categorized to have semi-competent to competent basement rock considering the resistivity distribution of the underlying bedrock and the pit information from the area. The nature of the overlying material (clay-rich material) and depression into the bedrock at some location could pose threat of differential settlement to engineering construction works. Therefore, it should be ensured that subsurface regions where depressions were

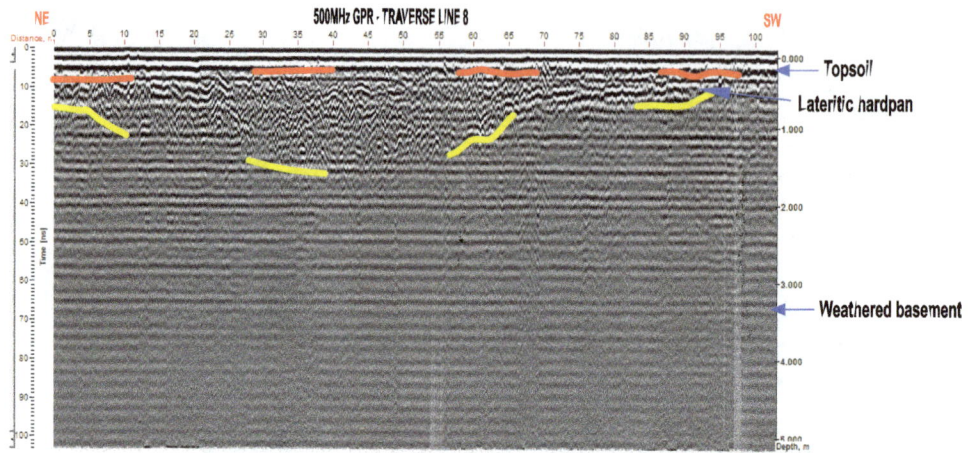

Figure 8: Radar section along traverse line 8 using antenna frequency of 500 MHz.

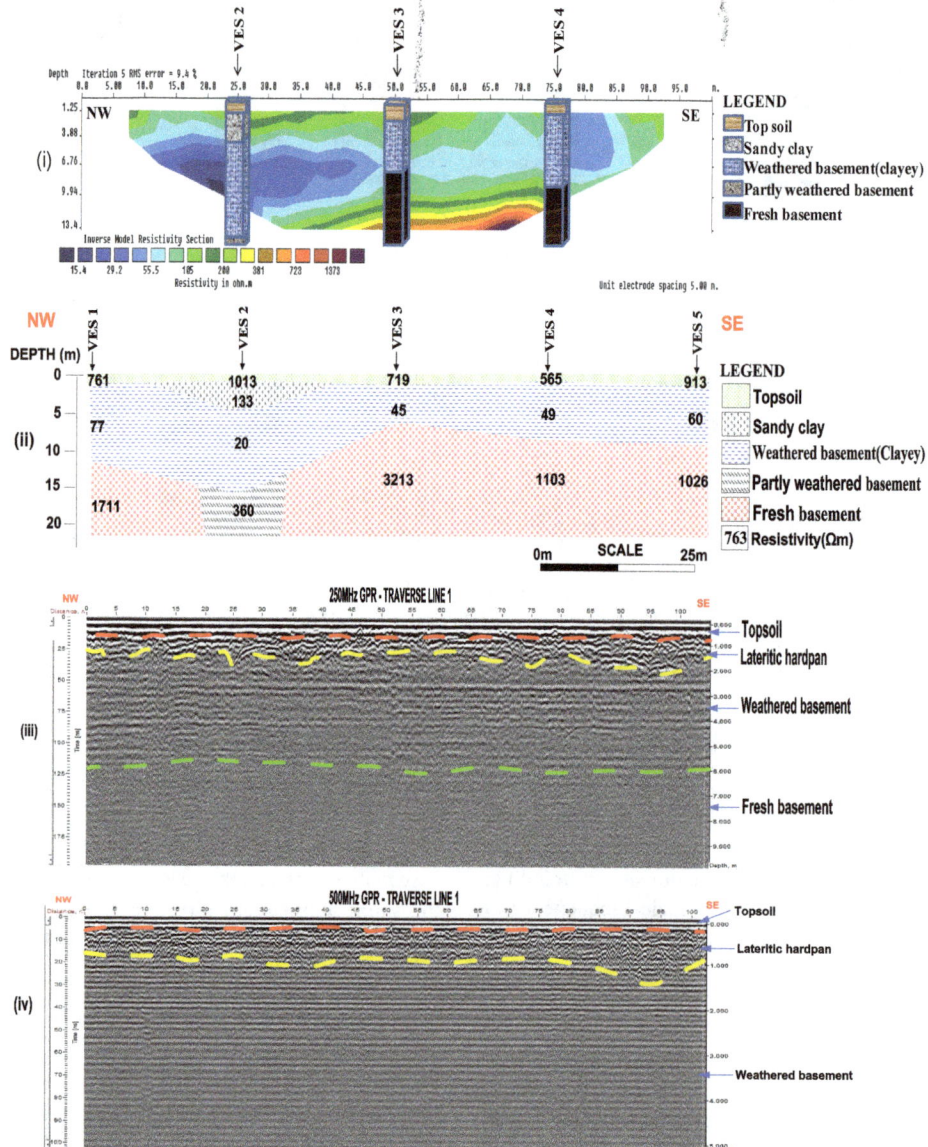

Figure 9: Correlation of (i) Inverted resistivity section (ii) Geoelectric section (iii) Radar section for 250 MHz antenna and (iv) Radar section for 500 MHz antenna beneath Traverse 1.

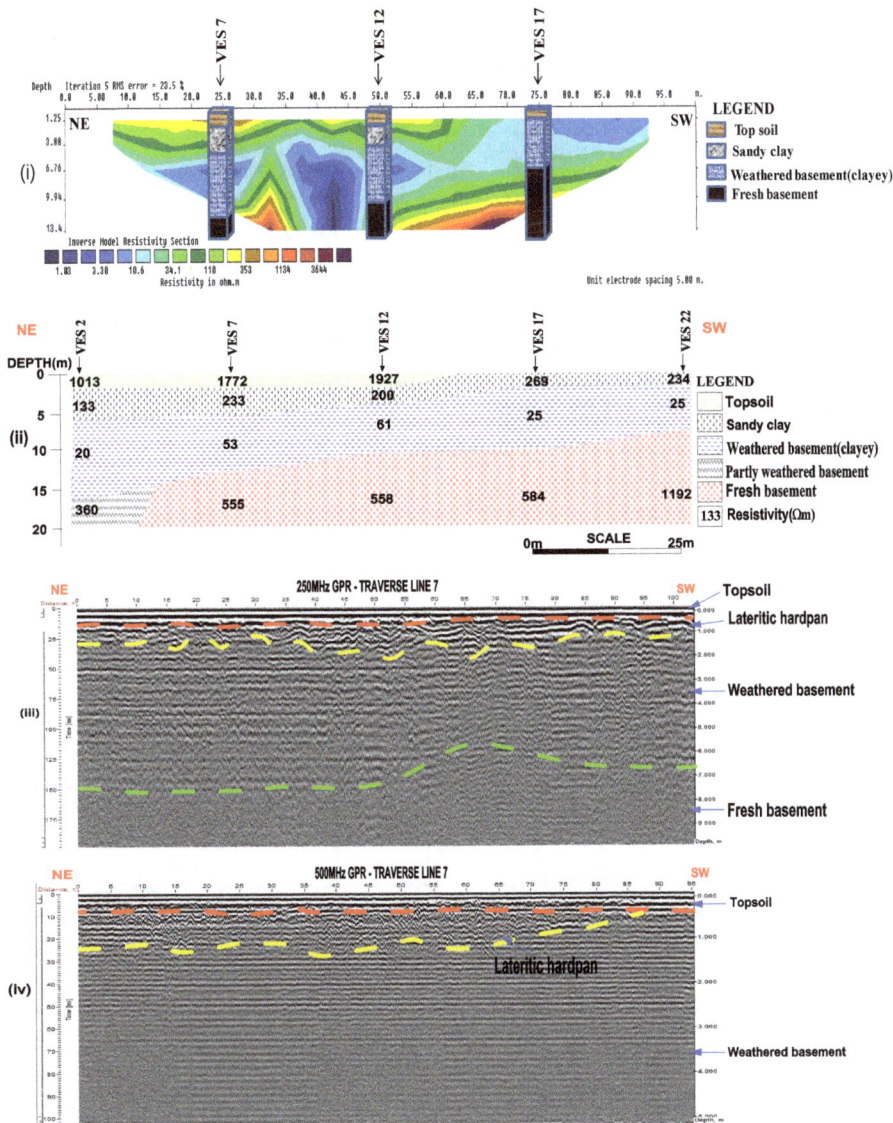

Figure 10: Correlation of (i) Inverted resistivity section (ii) Geoelectric section (iii) Radar section for 250 MHz antenna and (iv) Radar section for 500 MHz

noticed are avoided when sitting mega structures. Moreover for construction works within the area, foundation should be designed to sit comfortably on a competent bedrock or by employing suitable foundation method (such as piling) to ensure the stability of any proposed structure in the area in other to avoid foundation failure. Reliable information about the lithological characteristics of the study area has been deduced from the results of this research work. Thus, electrical resistivity methods and ground penetrating radar are versatile tools in shallow engineering site characterization and should therefore be incorporated in any pre-foundation or geotechnical investigation for better understanding of the subsurface geology.

Acknowledgement

We appreciate the CEO of Geoterrain Nigeria Limited and all the staff of the company for their contribution and assistance during this work by making the GPR equipment and processing software available for use and their assistance on data acquisition and processing. Our sincere thanks also goes to the management of Geo-sore Multi-transaction Limited, Nigeria for making the DC resistivity meter available for use in the course of this research work.

References

1. Elawadi E, El-Qady G, Nigm A, Shaaban F, Ushijima K (2006) Integrated Geophysical Survey for site investigation at a new dwelling area, Egypt. J Environ Eng Geophy 11: 249-259.

2. Coker JO, Makinde V, Mustapha AO, Adesokan JK (2013) Electrical resistivity imaging for foundation failure investigation at Remo Secondary School, Sagamu, Southwestern Nigeria. Int Sci Invest J 2: 40-50.

3. Carlsten S, Johansson S, Worman A (1995) Radar techniques for indicating internal erosion in embankment dams. J Appl Geophy 33: 143-156.

4. Adewumi I, Olorunfemi MO (2005) Using geoinformation in construction management. J Appl Sci 5: 761-767.

5. Oladapo MI, Olorunfemi MO, Ojo JS (2008) Geophysical investigation of road failures in the basement complex areas of southwestern Nigeria. Res J Appl Geophy 3: 103-112.

6. Fayemi O, Adepelumi AA (2012) Application of Ground Penetrating Radar and Electrical Resistivity Techniques for subsurface stratigraphic mapping in southwestern Nigeria. Extended abstract, 21st EM Induction workshop, Darwin, Australia pp: 25-31.

7. Fatoba JO, Salami BM, Adesida A (2013) Structural failure investigation

using electrical resistivity method: A case study of Amafor Ihuokpala, Enugu, Southeastern Nigeria. J Geo Min Res 5: 208-215.

8. Roth JS, Mackey JR, Mackey C, Nyquist JE (2002) A case study of the reliability of multielectrode earth resistivity testing for geotechnical investigations in Karst terrains. Eng Geo 65: 225-232.

9. Akintorinwa OJ, Ojo JS, Olorunfemi MO (2011) Appraisal of the causes of pavement failure along the Ilesa -Akure highway, Southwestern Nigeria using Remotely Sensed and Geotechnical data. Ife J Sci 13: 185-197.

10. Olayinka AI, Olayiwola MA (2001) An integrated use of geoelectric imaging and hydrogeochemical methods in delineating limits of polluted subsurface and groundwater at a landfill site in Ibadan area, Southwestern Nigeria. J Min Geo 37: 53-68.

11. Akuma N, Momin KN (1995) Recent discovery of prehistoric sites in the Ajibode area of Ibadan, Nigeria: A preliminary report. #3031, 1929 Plymouth Road, Ann Arbor Michigan p.38-42.

12. Oyediran IA, Adeyemi GO (2011) Geotechnical investigation of a site for landfill, Ajibode Southwestern Nigeria. Ozean J Appl Scie 4: 265-279.

13. Keller GV, Frischknecht FC (1966) Electrical Methods in Geophysical Prospecting, Oxford: Pergamon p.283.

14. Loke MH, Acworth I, Dahlin T (2003) A comparison of smooth and blooky inversion methods in 2D electrical imaging surveys. Explor Geophy 34: 182-187.

15. RadExplorer V1.4 (2005) The software for GPR data processing and Interpretation, User Manual.

16. Kearey P, Brooks M, Hil I (2002) An Introduction to Geophysical Exploration, 3rd Edition, Blackwell Science Limited.

17. Adelusi AO, Adiat KAN, Amigun JO (2009) Integration of surface electrical prospecting methods for fracture detection in precambrian basement rocks of Iwaraja area, southwestern Nigeria. Ozean J Appl Sci 2: 265-280.

18. Knight R (2001) Ground penetrating radar for environmental application. Ann Rev Ear Plan Sci 29: 229-255.

19. Win Z, Hamzah U, Ismail, MA, Samsudin AR (2011) Geophysical investigation using resistivity and GPR: a case study of an oil spill site at Seberang Prai, Penang. Bull Soc Mal 57: 19-25.

Origin of the Arima-type and Associated Spring Waters in the Kinki District, Southwest Japan

Hitomi Nakamura[1,2]*, Kotona Chiba[1], Qing Chang[1], Noritoshi Morikawa[3], Kohei Kazahaya[3] and Hikaru Iwamori[1,2]

[1]Japan Agency for Marine–Earth Science and Technology, 2-15 Natsushima-cho, Yokosuka-shi, Kanagawa 237-0061, Japan
[2]Department of Earth and Planetary Sciences, Tokyo Institute of Technology, 2-12-1 Ookayama, Meguro-ku, Tokyo 152-8551, Japan
[3]Geological Survey of Japan, AIST, 1-1-1 Higashi, Tsukuba, Ibaraki 305-8567, Japan

Abstract

Rare earth elements (REEs) of the spring waters upwelling in the non-volcanic fore-arc region of the Kinki district in southwest Japan were investigated to assess their upwelling processes and deep-seated origins. A principal component analysis of the REE data identified three principal components (PCs) that cover 89% of the entire sample variance: (1) PC-01, which corresponds to a dilution process by which fluids are introduced at low concentrations, previously represented by major solute binary trends, including $\delta^{18}O–\delta D$ systematics; (2) PC-02, which is a precipitation process of REEs from the brine; and (3) PC-03, which is an incorporation of REEs from country rock by carbonic acidity, although the types of country rocks may also have a significant impact on the spring water compositions. Based on these three PCs, together with the major solute concentrations and hydrogen, oxygen, and helium isotopic compositions determined in previous studies, five distinct types of spring waters in the Arima and Kii areas were identified: (i) "Tansansen", (ii) "Kinsen", (iii) "Ordinary Arima", (iv) "Ginsen", and (v) "Eastern Kii". These five types probably represent (ii) a deep brine, (iii) an evolved deep brine that precipitated REE-bearing minerals, (iv) a mixture of (iii) and meteoric water, (v) a meteoric water carbonated by deep gas derived from (ii), and (i) a spring water similar to (v) with a more significant influence of the country rock constituting the aquifer. A comparison of the spring waters in the Arima and Kii areas revealed systematic geographic distributions. The "Ordinary Arima"-type occurs along the Median Tectonic Line, and the "Eastern Kii"-type occurs in the eastern part of the Kii area. The latter seems to upwell in the restricted region where deep low-frequency tremors are observed. We suggest that the geographical distributions are linked to the tectonic setting and/or temporal evolution of fluid upwelling.

Keywords: Brine; Spring water; Subducting slab; Fluid; Arima; Kii

Introduction

Deep-seated geofluids play crucial roles in geological processes at subduction zones, including magmatism, hydrothermal activity, ore formation, and seismicity. For example, reductions in the melting temperatures of rocks by the introduction of slab-derived fluids are thought to be key to understanding arc magmatism with characteristic geochemical signatures, based on experimental and theoretical studies of igneous petrology [1]. However, direct observations and evidence of natural geofluids are rare because of their inaccessibility. The contribution of fluid to seismicity has been studied extensively because fluids may reduce the mechanical strength of rocks and plate boundaries [2]. In southwest and central Japan (Figure 1), deep low-frequency (DLF) tremors have been detected in an arc-parallel 600-km-wide region beneath the fore-arc area from the Shikoku district in the west to the Kinki and Chubu districts, including the Kii Peninsula, in the east, along the subducting Philippine Sea (PHS) slab, at a depth of approximately 30–40 km. This indicates that the tremors may have been caused by fluid generated by dehydration processes from the subducted PHS slab [3]. DLF-type earthquakes have also occurred locally in some locations (Figure 1), such as beneath Osaka Bay, Arima, and northern Kyoto in the Kinki district [4], and these are also thought, based on three-dimensional seismic tomography [5,6] to have been caused by slab-derived fluids. Despite these seismic observations, direct observations of such fluids have been rare, and more information on the physical nature and chemical compositions of these deep-seated geofluids is required.

The Arima-type brine is a candidate for being such a deep-seated geofluid. It may have originated from the subducted PHS slab [7,8] and could have some connection to DLF seismic events [9]. The Arima-type brine is defined as a spring water found in a non-volcanic fore-arc region [10] and is geochemically characterized by a high Cl content of approximately 40,000 ppm, specific $\delta^{18}O–\delta D$ isotopic ratios, and a

high $^3He/^4He$ ratio, comparable to the mantle value [7,11,12], unlike meteoric water or buried sea water [10,13,14]. In addition to these original definitions, the densest brine (the least diluted by meteoric water) in the Arima area shows essentially the same heavy isotopic compositions (Sr–Nd–Pb isotopic ratios) as the slab-derived fluid of the subducting PHS slab [8]. Nakamura et al. [8] have also found a similarity in the REE pattern between the Arima brine and a slab-derived fluid dehydrated at pressure and temperature conditions that correspond to those of the PHS slab beneath the Arima area, based on experimental and theoretical studies [7,15] that support its slab origin. Nakamura et al. [16] have analyzed both the dense and diluted brines of the Arima spring waters for REEs and have demonstrated that there are significant variations in the REE patterns that reflect several important processes during the ascent of deep brine. They identified at least two different patterns and processes that probably represent (a) significant mineral precipitation from the original deep brine and (b) mixing between the brine and the meteoric water, the former of which has not been identified solely from other elemental or isotopic studies.

Within this context, to investigate both the nature of the original deep brine and the geochemical processes that occur during its ascent, we extended our previous research to include more regional data on

*Corresponding author: Hitomi Nakamura, Japan Agency for Marine–Earth Science and Technology, 2-15 Natsushima-cho, Yokosuka-shi, Kanagawa 237-0061, Japan
E-mail: hitomi-nakamura@jamstec.go.jp

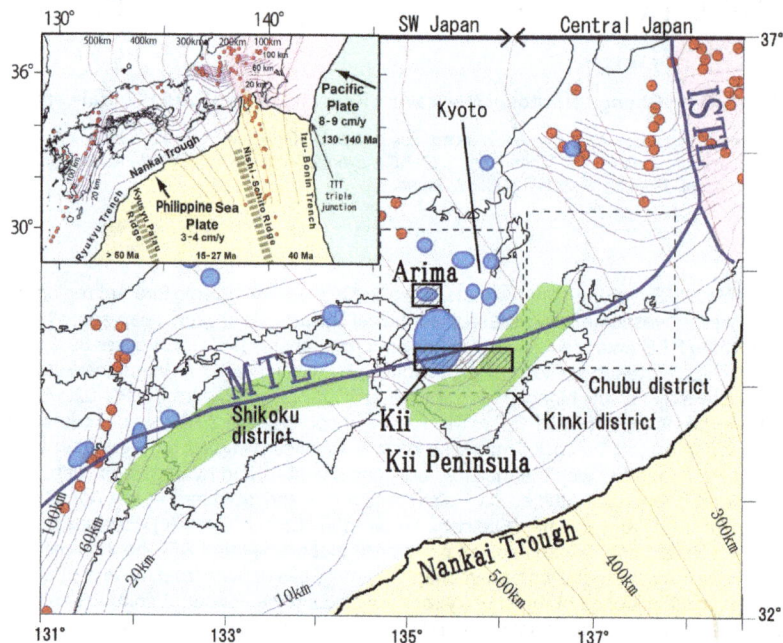

Figure 1: Tectonic setting involved in two continental plates, two oceanic plates, and two large tectonic lines (MTL and ISTL) around Japan as an inset. Enlargement is a map showing the location of the studied areas (Arima and Kii in the shaded square), DLF tremor (green belt) and DLF earthquakes (blue ellipse) based on several studies [3,4,9], the distribution of Quaternary volcanoes (red circles), and the geometry of the subducting Pacific and Philippine Sea slabs relative to the Itoigawa–Shizuoka Tectonic Line (ISTL) and the Median Tectonic Line (MTL). The pink contour lines indicate the depth of the upper surface of the Pacific slab (50 to 300 km depth at 50 km intervals), whereas the purple contour lines indicate that of the Philippine Sea slab (10 to 200 km depth at 10 km intervals). The aseismic parts are indicated by the dotted line.

the Arima-type brine, utilizing both the existing geochemical data from the Arima and Kii areas and new data from the Kii area. In this paper, we first describe the existing data set [7,8,16-18], and we then report new REE data from the Kii area (Figures 1 and 2). We then present the principal component analysis (PCA) results that identify the different sources and processes involved in the deep brine, in contrast to previous studies in which the sources and processes were identified visually. Finally, we discuss the geological and tectonic implications of the results, including the possible connection to basement rock types, fault structures, and seismicity.

Geological and tectonic setting of the studied area

In the Kii Peninsula, spring waters upwell along the Median Tectonic Line (MTL) and subsidiary faults formed in the Miocene (Figures 1 and 2) and exhibit variation in the compositions of the major solute elements and gas phases [7]. These springs appear to upwell through the MTL, dipping northward at approximately 30–40° to 24 km depth, according to seismic reflection profiling in Shikoku [19,20]. The tectonic setting in the southwestern Japan arc is associated with two oceanic plates, the Pacific plate and the PHS, which subduct beneath the studied area from the east at 9 cm/year and from the southeast at 4 cm/year, respectively. The slab surface depth of the Pacific slab is approximately 400 km beneath both the Arima and the Kii areas, whereas that of the PHS is 50–80 km (with considerable uncertainty associated with that estimated range) beneath the Arima area, 30 km beneath the eastern part of the Kii area, and 60 km beneath the western part of the western Kii area (Figure 1) [21,22]. Despite this active subduction, no Quaternary volcanoes have formed in these areas because the Pacific slab is too deep and the PHS is too shallow (Figure 1) to satisfy the physiochemical conditions for melting in the mantle wedge and/or slab [23].

As shown in the geological map of the Kii area (Figure 2), the basement is composed of metamorphic rocks and an accretionary complex associated with metamorphic belts, such as the Sambagawa, Chichibu, and Shimanto Belts, which are bounded by the Mikabu and Butsuzo Tectonic Lines, respectively [24]. The geological structure of the Arima area has been presented elsewhere [16]. The basement rocks are composed of late Cretaceous felsic volcanic rocks such as rhyolite of the Arima Group, granitic rocks including the Rokko granite, and late Eocene to early Oligocene non-marine sedimentary rocks with rhyolitic tuff layers of the Kobe Group [24].

Sample description and analytical method

Existing data set for the arima and kii spring waters

In this study, we utilized the existing data on the spring waters in the Arima area, which is the typical locality of Arima-type brine (Figure 1), and in the Kii area (Figures 1 and 2). We first describe the geochemical characteristics of the existing data from the Arima and Kii areas, and we then describe the sample localities and the analytical method used for the new data from the Kii area. For the Arima area, the data set includes data on the major solutes, $\delta^{18}O$–δD and $^3He/^4He$ isotopic ratios, Sr–Nd–Pb isotopic ratios, and REE abundances (Table 1) [7,8,16,18]. The Arima spring waters exhibit a wide range of $\delta^{18}O$–δD isotopic ratios and $^3He/^4He$ isotopic ratios, which are primarily explained by mixing of meteoric water and the deep brine [7,13,14]. The isotopic ratios also exhibit linear variations with major solute elements, such as Cl and Na (Figures 3A-3D), on the basis of which the composition of the deep brine end-member has been estimated [7,14]. The "Kinsen" hot spring is closest to the deep brine end-member, having the highest concentrations of Cl and other solute elements, and the "Tansansen" cold spring is closest to the composition of meteoric water as another end-member, with the lowest solute concentrations

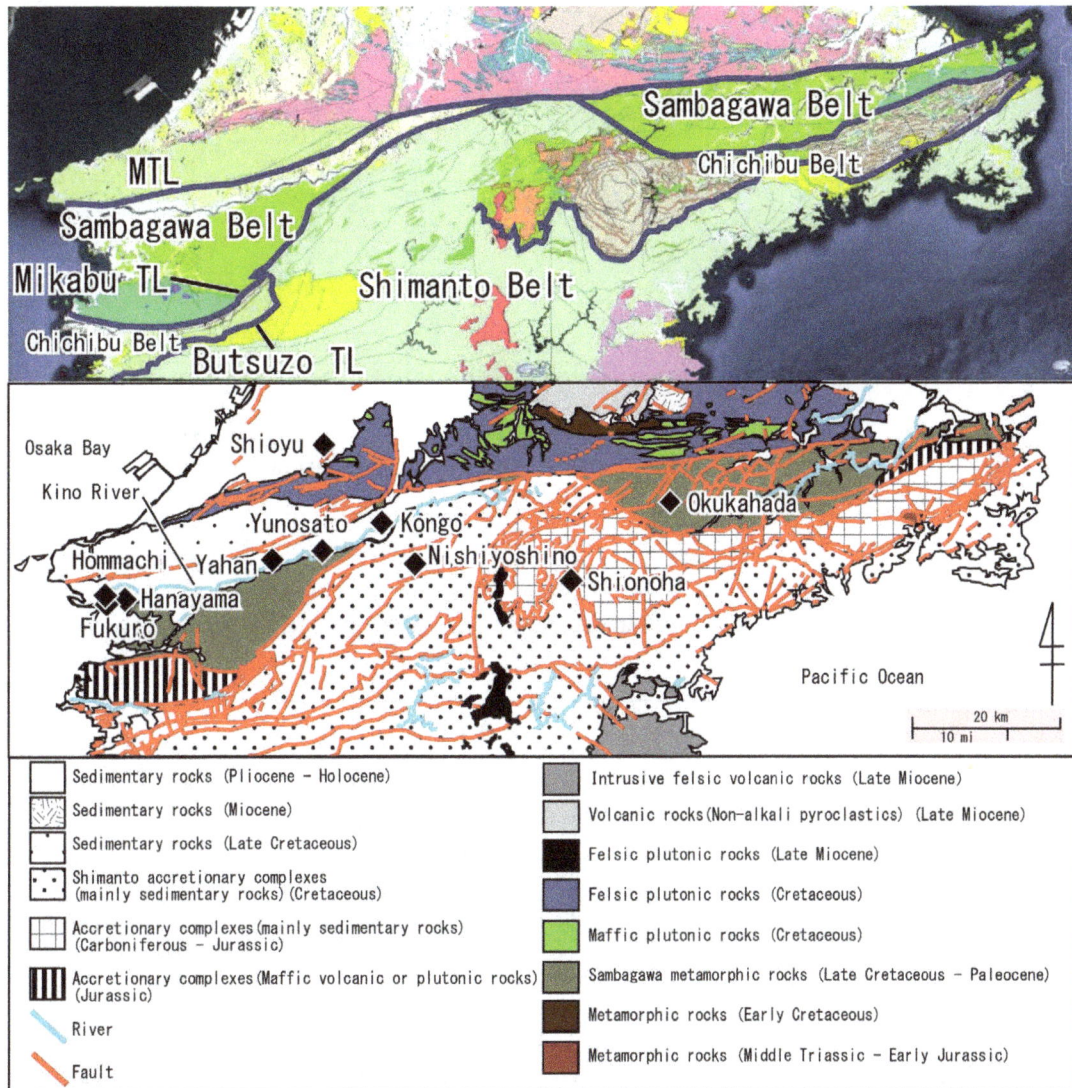

Figure 2: (A) Geologic map of the studied Kii area and three large metamorphic belts (Sambagawa, Chichibu, and Shimanto) exposed in the Kii Peninsula area, modified after Google Earth [35]. The Mikabu and Butsuzo tectonic lines (TL) are roughly aligned along the MTL. (B) Detailed geologic basements, faults, rivers, and locations of the studied spring waters in the Kii area, modified after the seamless digital geological map of Japan [24].

(Figures 3 and 4). Other spring waters with moderate Cl and cation concentrations plot between "Tansansen" and "Kinsen", as shown in Figures 3 and 4. It should be noted that the gas component, including He, behaves differently from the major solute elements and $\delta^{18}O$–δD isotopic systematics because the gas phase has been released from the deep brine to be regassed to "Tansansen", resulting in "Tansansen" having the highest $^3He/^4He$ ratio (close to the mantle value) (Figure 3D) [7].

Although simple mixing between the deep brine and the meteoric water is suggested by the above, recent REE studies have identified processes that cannot be observed from the major solute elements and $\delta^{18}O$–δD isotopic systematics [8,16]. Nakamura et al. [16] found four types in the REE pattern of the Arima spring waters, which may represent (a) mineral precipitation from the original deep brine in relatively deep parts and (b) mixing of the original brine or evolved brine that has undergone precipitation with the meteoric water. They inferred that carbonated spring water could have dissolved REEs from

the country rocks, which can be seen in the REE pattern of "Tansansen". These findings suggest that REEs in the spring water provide invaluable information about the processes that occur during the ascent of deep brine.

For the Kii area, a data set containing data on major solutes and $\delta^{18}O$–δD and $^3He/^4He$ isotopic ratios is available from previous studies (Table 1) [17,18]. However, the REE abundances have not yet been determined. In the Kii Peninsula, spring waters also exhibit a wide range of $\delta^{18}O$–δD and $^3He/^4He$ isotopic compositions, which can be explained by a mixing of meteoric water and upwelling of the deep-seated Arima-type brine, particularly along the MTL [9,14]. The $\delta^{18}O$–δD isotopes exhibit linear variations with the major solute elements, such as Cl and Na, although the slopes of the trends seem to be different from those of Arima (Figures 3B and 4A). For example, the slope of the $\delta^{18}O$–δD trend is steeper than that of the Arima spring waters (Figure 3A). The spring water "Honmachi", which has the highest $\delta^{18}O$–δD isotopic ratio, close to that of deep brine, has the highest contents of Cl

Region	Name of spring water	δ¹⁸O [SMOW]	δD [SMOW]	Cl [mg/L]	HCO₃ [mg/L]	³He/⁴He_Ra	Na [mg/L]	Ca [mg/L]	K [mg/L]	Mg [mg/L]
Arima	Ginsen	-7.27	-48.92	435	93	5.44	23	8	4	0
Arima	Gokuraku	2.07	-38.15	25088	26	2.33	11945	2192	2525	15
Arima	Gosho	-4.09	-45.92	8742	473	2.75	4700	627	1000	15
Arima	Gosya	-7.00	-51.05	3392	781	6.41	1743	491	208	35
Arima	Inari-kinsen	3.61	-35.11	36602	-	-	18379	2598	112	511
Arima	Kinsen	4.98	-34.12	40033	1293	4.78	20077	3006	1930	136
Arima	Tansansen	-7.88	-50.27	14	30	7.35	18	23	3	1
Arima	Tenjin	2.22	-37.71	25828	137	-	12189	2356	2593	17
Arima	Uwanari	0.77	-40.50	20786	203	2.38	10127	1845	2153	12
Kii	Fukuro	-	-	-	-	-	-	-	-	-
Kii	Hanayama	-5.75	-48.95	9186.24	3060.93	1.52	1177.06	2598.10	15.07	1953.60
Kii	Hommachi	1.04	-40.95	16089.99	4286.61	2.82	9438.73	763.37	101.32	986.46
Kii	Kongo	-	-	-	-	-	-	-	-	-
Kii	Nishiyoshino	-6.61	-54.12	5004.15	2004.51	1.03	3584.72	203.15	139.24	45.79
Kii	Okukahada	-7.98	-50.02	562.76	1635.34	4.91	551.78	196.67	19.14	77.26
Kii	Shionoha	-9.00	-58.25	775.22	1906.88	6.27	669.27	178.15	75.47	31.42
Kii	Shioyu	-2.73	-44.03	12612.91	3018.68	3.83	7549.10	80.17	87.04	868.94
Kii	Yahan	-2.26	-47.48	13378.30	4586.36	3.47	8487.99	582.46	115.12	599.28
Kii	Yunosato	-5.49	-46.75	6146.48	8097.35	1.79	6284.11	183.72	38.98	194.59

Table 1: Major solute elements and isotopic compositions discussed in this study. The data are from [7,17]. No data are available for "Fukuro" and "Kongo."

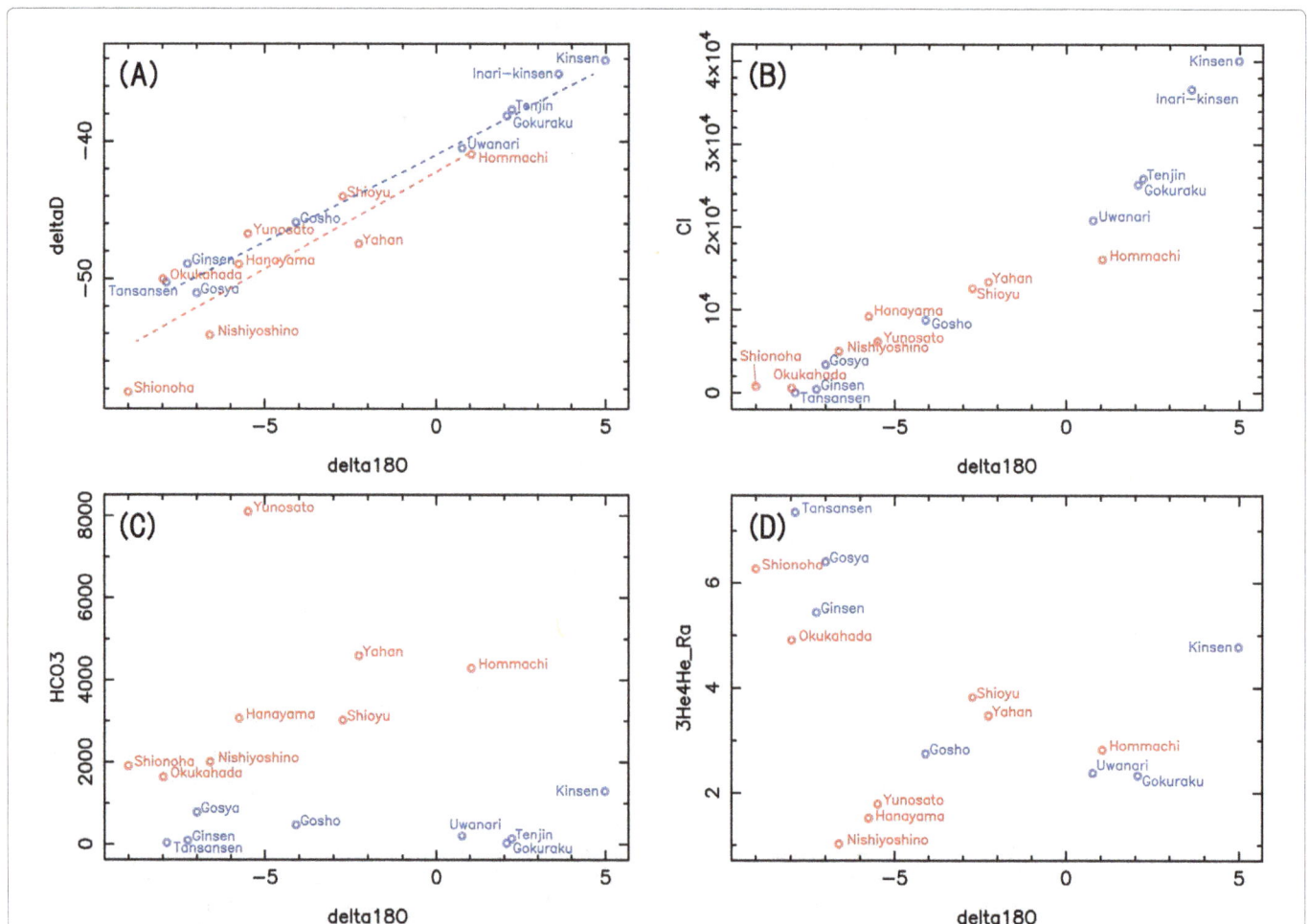

Figure 3: Correlations among δ¹⁸O isotopic ratio versus major solute elements (A) δD isotopic ratio, with a regression line for mixing of meteoric water and a slab-derived fluid as a function of the depth of the subducting slab's steeper slope of the Kii area that corresponds to shallower dehydration than Arima [7,16], (B) Cl abundances (ppm), (C) HCO₃ abundances (ppm), and (D) ³He/⁴He isotopic ratios of spring waters, with the names in blue and red for the Arima and Kii areas, respectively. The data were compiled from existing data and are listed in Table 1.

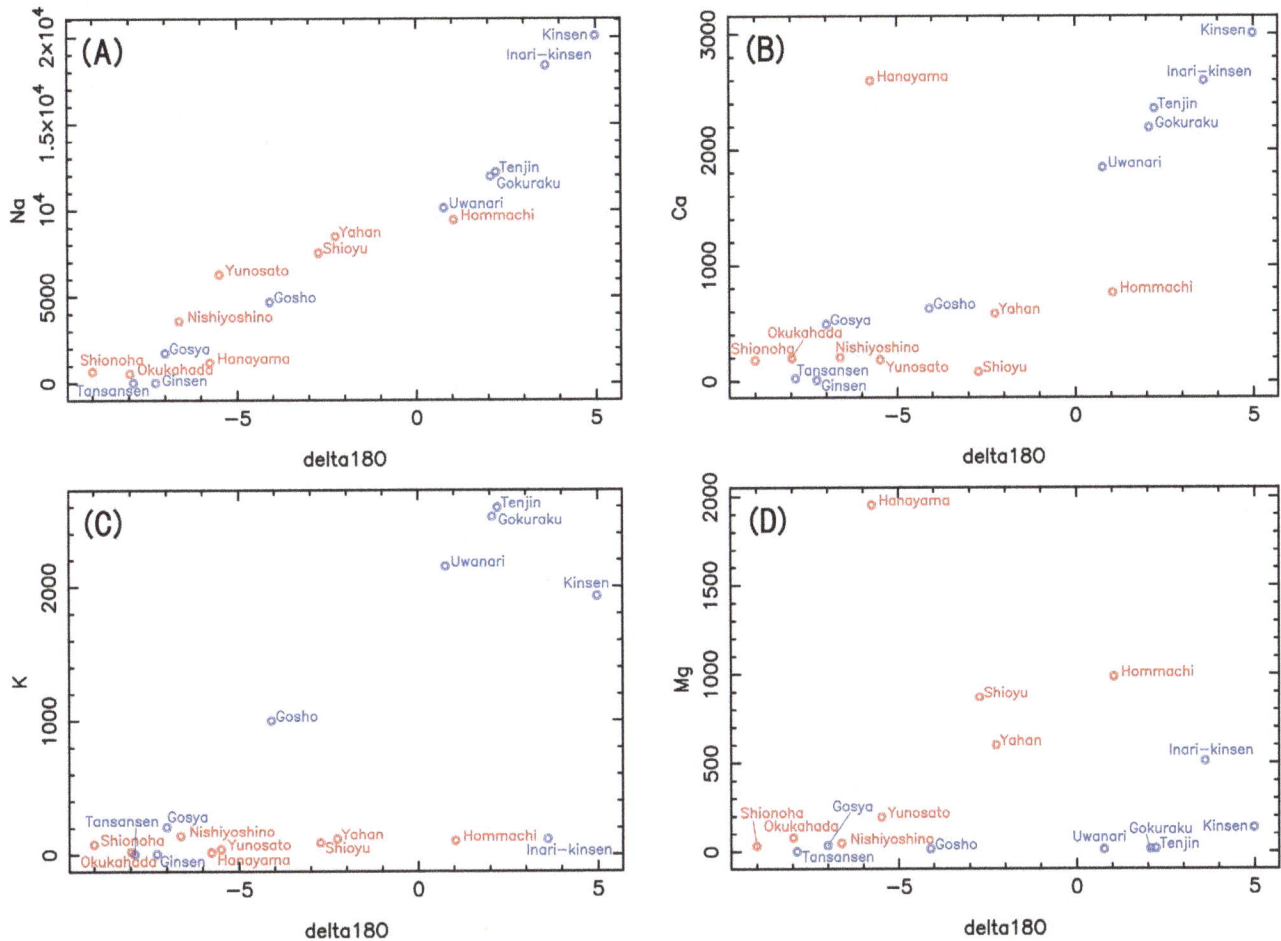

Figure 4: Correlation diagrams between δ18O isotopic ratio versus (A) Na, (B) Ca, (C) K, and (D) Mg in ppm. The color with the name of the type of spring water is the same as in Figure 3.

and other solutes, whereas the spring water "Shionoha", which has the lowest δ18O–δD isotopic ratio, has the lowest concentrations of Cl and lower cations. The concentrations of major solute elements for other spring waters plot between those of the dense "Honmachi" and diluted "Shionoha" waters, except for "Hanayama" (Figures 4B and 4D). Independent behavior of the gas phase, including He, is seen in the Kii area, which is similar to that seen in the Arima area. For example, two spring waters in the Kii area named "Shionoha" and "Okukahada" that have low δ18O–δD isotopic ratios exhibit higher $^3He/^4He$ ratios, equivalent to the mantle value [9]. Such variations are similar to those seen for "Tansansen" in the Arima area, which consists of meteoric water with the addition of gas and which may have originated from original deep brine.

Sampling and analytical method for REEs of the Kii spring waters

To supplement the existing data described above, we collected ten spring water samples from the Kii area (Figure 2) for REE analysis. Each sample of 1 to 2 L was collected in a bottle with as much air excluded as possible. No chemicals were added to the samples on site. The analytical method used for the REE analysis of the Kii spring waters followed the method described in Nakamura et al. [16] for the REE analysis of the Arima spring waters. The high salinity and high concentrations of solute elements in the brines interfered significantly with quantitative analysis of the REEs because of the matrix effect. In several cases in

which water samples contained visible particles, the particles were dissolved using nitric acid. Considering the matrix effect under high-salinity conditions, we applied the standard addition method for solutions with strong matrix effects [25] and analyzed the samples using inductively coupled plasma mass spectrometry (ICP-MS, iCAP-Qc, ThermoFisher Scientific) at the Japan Agency for Marine–Earth Science and Technology (JAMSTEC). The significant interference from Ba oxides against La, Eu, Nd, Sm, and Gd were quantitatively recalculated after being combined with ^{135}Ba for correction, following the procedure described in Nakamura et al. [16].

Results

The solute major elements and δ18O–δD–He isotopic ratios of the spring waters are summarized in Table 1, and the REE abundances are presented in Table 2 [7,16-18]. The REE compositions are plotted in Figure 5 in spidergram form, normalized with respect to the depleted MORB (mid-ocean ridge basalt) mantle (DMM) [26]. The REE abundances of the Kii and Arima spring waters are approximately three to five orders of magnitude lower than those of the DMM.

Principal component analysis of REE abundances

To examine the processes and/or sources contributing to the generation and upwelling of the spring waters, we performed a principal component analysis (PCA) of the REE abundances of the

spring water	La	Ce	Pr	Nd	Sm	Eu	Gd	Tb	Dy	Ho	Er	Tm	Yb	Lu	Dilution rate
Ginsen	0.2228	0.0093	0.0389	0.2089	0.0773	0.1467	0.0665	0.0161	0.1094	0.0411	0.1799	0.0429	0.3818	0.0769	10.3386
Gokuraku	0.1840	0.0132	0.0216	0.0229	0.1322	10.7723	0.3484	0.0576	0.0214	0.0051	0.0074	0.0474	0.7346	0.1454	1.0000
Gosho	0.0675	0.0013	0.0055	0.0113	0.0339	1.8843	0.0351	0.0061	0.0049	0.0014	0.0033	0.0043	0.0294	0.0099	1.0000
Gosya	0.0946	0.1068	0.0140	0.0557	0.0794	17.0435	0.1776	0.0358	0.1097	0.0700	0.1194	0.1505	0.1398	0.0490	9.9297
Inari-kinsen	0.2901	0.0062	0.0219	0.0115	0.2413	6.8933	0.3182	0.0301	0.0675	0.0184	0.0504	0.0422	0.2696	0.1118	9.6825
Kinsen	3.6382	5.2442	0.5676	2.5409	0.9609	27.8899	1.7651	0.3841	3.6653	0.8360	2.5692	0.3392	2.4161	0.2924	42.2693
Tansansen	4.1881	13.3532	1.9650	10.4090	2.7646	0.3982	4.2497	0.5709	4.1635	0.7909	2.6780	0.3121	2.2442	0.2057	49.9452
Tenjin	1.7895	0.0010	0.0300	0.0126	0.0244	30.1594	0.1479	0.0470	0.0628	0.0118	0.0107	0.0562	0.6806	0.3203	1.0000
Uwanari	0.1236	0.0705	0.0187	0.0599	0.0889	4.4356	0.0928	0.0162	0.0113	0.0038	0.0138	0.0162	0.0889	0.0421	1.0000
Fukuro	0.4664	0.2108	0.0197	0.1407	0.0880	0.9931	0.1371	0.0198	0.3601	0.0815	0.4588	0.1887	0.5590	0.0300	9.6049
Hanayama	0.2820	0.3230	0.0308	0.3936	0.2543	0.5177	0.2981	0.0208	0.3436	0.0640	0.2418	0.0506	0.2409	0.0085	9.5369
Hommachi	0.6209	0.1366	0.0276	0.0355	0.0998	2.4171	0.2168	0.0182	0.0410	0.0114	0.0669	0.0360	0.3804	0.0489	1.0000
Kongo	0.2645	0.0947	0.0208	0.1308	0.0802	0.5578	0.0612	0.0075	0.0310	0.0095	0.0377	0.0268	0.1256	0.0176	1.0000
Nishiyoshino	1.2530	0.0742	0.0041	0.0807	0.0687	6.6262	0.0807	0.1361	0.1887	0.0648	0.2674	0.0617	0.4931	0.0719	9.8453
Okukahada	0.0808	0.1108	0.0157	0.0828	0.0386	0.1289	0.0558	0.0099	0.0635	0.0171	0.0611	0.0111	0.0798	0.0157	9.4315
Shionoha	0.6237	0.4008	0.0578	0.3659	0.0976	1.5612	0.1675	0.0332	0.2384	0.0558	0.1632	0.0218	0.1214	0.0158	1.0000
Shioyu	0.6543	0.3879	0.0292	0.1780	0.1386	2.9333	0.2269	0.0250	0.0393	0.0109	0.0579	0.0407	0.2846	0.0812	9.9690
Yahan	7.1418	1.2716	0.1824	0.6851	0.6683	9.7843	0.3846	0.0968	0.4308	0.0956	0.3013	0.1654	0.3644	0.0528	48.5663
Yunosato	0.1664	0.2148	0.0244	0.1078	0.0258	0.0537	0.0454	0.0057	0.0630	0.0149	0.0542	0.0095	0.0446	0.0045	1.0000

Table 2: Rare earth element (REE) abundances in ppb of spring waters in the Arima and Kii areas.

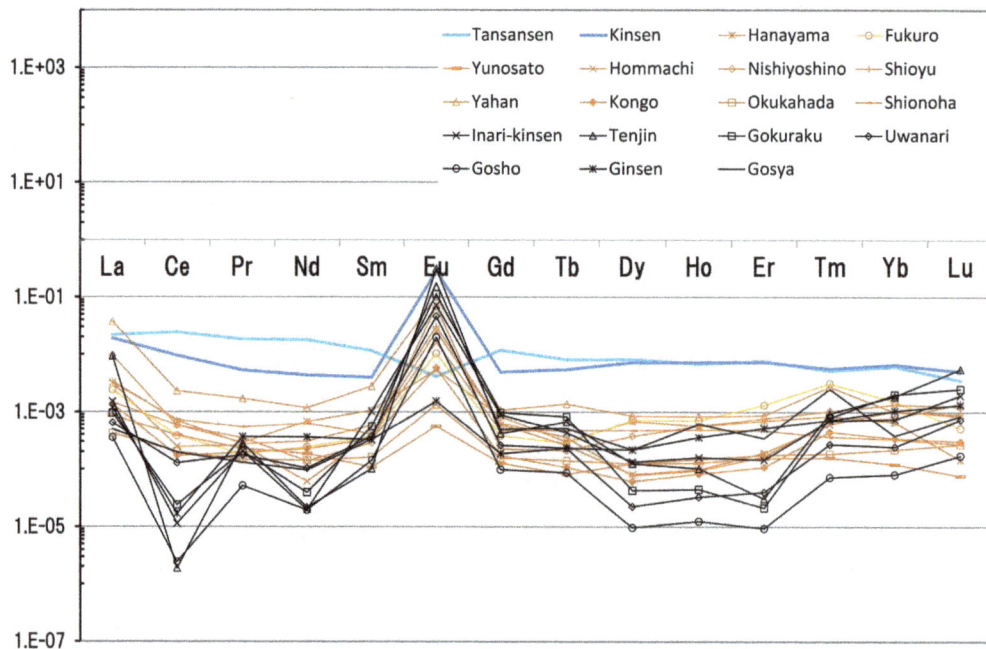

Figure 5: Depleted mid-ocean ridge basalt (MORB) mantle (DMM)-normalized rare earth element (REE) compositions for spring waters, from existing data and newly analyzed samples. The data are listed in Table 2.

spring waters in both the Arima and Kii areas (Table 2). The results show that the most powerful principal component (PC-01), which accounts for 66.8% of the entire sample variance, exhibits a flat REE pattern, as shown in the normalized eigenvector plot (i.e., eigen #1 in Figure 6A). This may indicate that the most dominant process controls the average concentration, moving the REE pattern upward and downward without significantly changing its shape, except for Eu: e.g., dilution by the introduction of fluids with low concentrations having flat REE patterns, such as meteoric water, or dissolution of elements (including REEs with a flat pattern) from source rocks. This

possibility was discussed by Nakamura et al. [16], based on visual REE analyses of the Arima spring waters. The second principal component (PC-02) accounts for 17.1% of the entire sample variance and exhibits a strongly convex-downward pattern with a strong positive Eu anomaly (i.e., eigen #2 in Figure 6B). This specific pattern implies differential behaviors among REEs, which provides information that is crucial to deciphering the processes and/or sources involved. Based on geochemical observations of spring waters in the Arima area, Nakamura et al. [16] argued that such a convex-downward pattern may represent a process of precipitation of minerals, such as coprecipitation

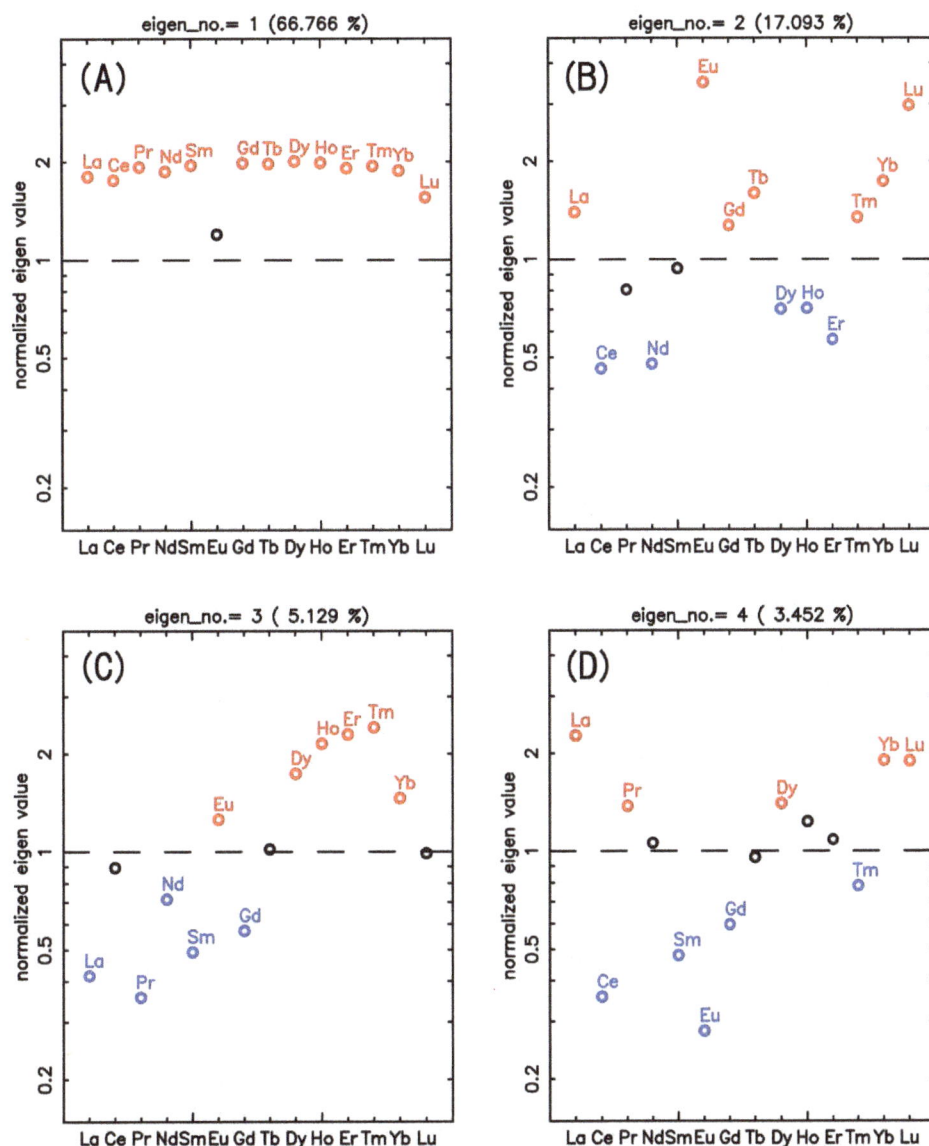

Figure 6: Normalized eigenvector plots. (A) The most powerful principal component (PC-01) accounts for 66.8% of the entire sample variance and exhibits a flat REE pattern (eigen #1). (B) The second principal component (PC-02) accounts for 17.1% exhibits a strongly convex-downward pattern with a strong positive Eu anomaly (eigen #2). (C) The third principal component (PC-03) accounts for 5.1% exhibits an undulation of the overall slope of the REE pattern (eigen #3). (D) The fourth principal component (PC-04) accounts for 3.5% and exhibits a depletion around MREEs, especially those associated with negative Ce anomalies (eigen #4).

of REEs with Fe oxyhydroxides. The third principal component (PC-03) accounts for 5.1% of the entire sample variance and corresponds to a contrasting behavior between light REEs (LREEs) and heavy REEs (HREEs)—in other words, an undulation of the overall slope of the REE pattern, shown as eigen #3 in Figure 6C. The apparent slope of this pattern is positive in Figure 6C, i.e., LREEs < HREEs in terms of eigenvalues. It should be noted that the eigenvectors of the PCs indicate a direction and magnitude of variance measured from the average value of the data and span the compositional space in the directions of both positive and negative signs of the eigenvectors. Therefore, in the case of PC-03, systematic variations in both LREEs < HREEs and LREEs > HREEs are described by PC-03 with opposite signs. The fourth principal component (PC-04) accounts for 3.5% of the entire sample variance and corresponds to a depletion around MREEs, especially those associated with negative Ce anomalies (eigen #4 in Figure 6D).

The principal components PC05 to PC13 are less significant in terms of the sample variance, accounting for 3.0, 1.3, 1.1, 0.81, 0.64, 0.35, 0.31, 0.11, 0.032, and 0.018%, respectively. Therefore, a significant portion of the data variation (89.0% of the sample variance) can be explained by the first three PCs (PC-01, 02, and 03). This indicates that several (but not many) processes and/or sources are potentially involved in determining the REE patterns of these spring waters during generation and upwelling in the Arima and Kii areas.

Classification of spring waters based on PCA

As shown in the score plots for individual samples (Figures 7A-7C), we can classify the spring waters based on the scores. Using the powerful first three PCs, we characterize the individual spring waters and the regional differences between the Arima and Kii areas. For the Arima spring waters, there is a wide range in PC-01 from "Gosho" (minimum) to "Tansansen" (maximum), whereas PC-02 exhibits a

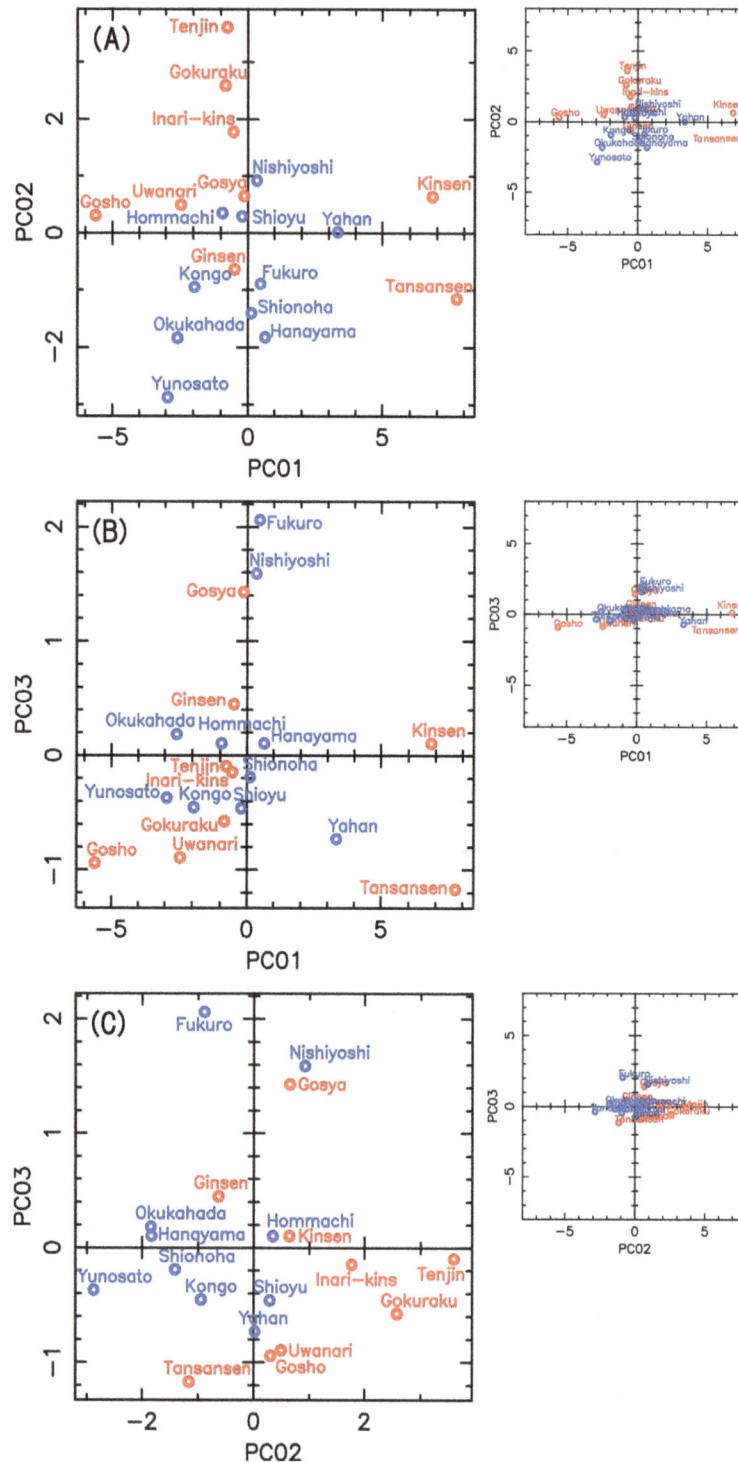

Figure 7: Score plots of individual samples (with names in the same colors as in Figure 3), with spring waters classified based on the scores. Combinations of the powerful three PCs are shown in (A) PC-01 vs. PC-02, (B) PC-01 vs. PC-03, and (C) PC-02 vs. PC-03.

smaller range from "Tenjin" (maximum) to "Ginsen"/"Tansansen" (minimum), as shown in Figure 7A. Two distinct spring waters, "Kinsen" and "Tansansen", between which other spring waters are broadly plotted in terms of their major solute concentrations and $\delta^{18}O$– δD diagrams (Figures 3A-3D and 4A-4D), have high REE abundances, as shown in the spidergram (Figure 5), and exhibit large positive PC-

01 scores. Because "Tansansen" plots very close to the meteoric water composition in terms of its major solute concentrations and $\delta^{18}O$– δD, PC-01 and the high REE abundances of "Tansansen" require specific processes and/or sources different from those identified by the conventional measures with the major components, including $\delta^{18}O$– δD, as will be discussed. The PC-03 score exhibits a relatively limited

range in the Arima area, from "Tansansen" (minimum) to "Gosya" (maximum).

For the Kii spring waters, there is a smaller range in PC-01 from "Yunosato" (minimum) to "Yahan" (maximum), as shown in Figure 7A. These limits correspond to the highest and lowest REE abundances, respectively (Figure 5). The overall range of PC-02 is smaller than that for the Arima spring waters, with four spring waters ("Hommachi", "Nishiyoshino", "Shioyu", and "Yahan") having positive scores and six spring waters ("Fukuro", "Hanayama", "Kongo", "Okukahada", "Shionoha", and "Yunosato") having negative scores. The large positive PC-02 score commonly observed in the Arima area is not seen in the Kii spring waters. The score variability of the Kii spring waters is smaller than that in the Arima area, which may indicate that the source and/or dilution process involved in the Kii spring waters is relatively uniform, even though the studied area in the Kii is wider than that in the Arima (Figure 1 and 2). Strong positive PC-03 scores were observed for two spring waters ("Nishiyoshino" and "Fukuro") in the Kii area, while a negative score was observed for "Yahan" (Figures 7B and 7C).

Based on these observations, coupled with information on the major solute elements, we can deduce the multi-dimensional compositional variations of the Arima-type brine. The correlations of major solute elements ($\delta^{18}O$, Cl, HCO_3, and $^3He/^4He$ in Ra units, indicating a multiple of the present-day atmospheric value 1.38×10^{-6}) with the individual PC scores (PC-01, PC-02, and PC-03) are shown in Figures 8A-8L, respectively. The plot of PC-01 vs. $\delta^{18}O$ clearly discriminates "Kinsen" from "Tansansen", although both have a high PC-01 score, and five spring waters ("Gokuraku", "Gosho", "Inari-kinsen", "Tenjin", and "Uwanari") are plotted in the area of relatively high $\delta^{18}O$ and negative PC-01 (Figure 8A). The remaining two spring waters of Arima ("Ginsen" and "Gosya") and almost all of the spring waters of the Kii area are plotted in the area of lower $\delta^{18}O$ with a smaller PC-01 score. Therefore, at least four types of spring water can be recognized, which is consistent with the score plot of PC-01 vs. PC-02 (Figure 7A). This classification is also observed in the plots of Cl and HCO_3 (Figures 8B and 8C). However, there is no clear correlation indicated by the plot of PC-01 vs. $^3He/^4He$, indicating that the gas phase behaves independently (Figure 8D). The $\delta^{18}O$ variation exhibits a moderately positive correlation with PC-02 (Figure 8E), with a higher value for the Arima spring waters and a lower one for the Kii spring waters, in contrast to that shown in the PC-01 plot (Figure 8A). This correlation is also observed in the Cl plot (Figure 8F), which is divided into a positive group for the Arima area and a negative group for the Kii area with respect to PC-02 for a similar Cl content. The involvement of PC-02 in a precipitation process at a meteoric aquifer seems to be related to a source for or process of deep brine. The plot of PC-02 vs. HCO_3 (Figure 8G) obviously shows a higher HCO_3 content in the Kii spring waters than in those in the Arima area, which also suggests that the PC-02 score is not linked to the HCO_3 behavior. The $^3He/^4He$ content shows no correlation with PC-02 in either area (Figure 8H). The plot of PC-03 vs. $\delta^{18}O$ illustrates the variation in the spring waters in the Kii area, while a positive correlation from "Tansansen" to "Kinsen" for the five spring waters (with PC-03 < 0) is observed in the Arima area (Figure 8I). The remaining two Arima spring waters, "Ginsen" and "Gosya", are plotted separately, along with the Kii spring waters. A similar variation (in terms of scatter for the Kii spring waters and the positive trend in the Arima area) is also confirmed in the plot of Cl (Figure 8J). Although the solute HCO_3 content exhibits no clear correlation with PC-03, there is a moderate correlation with the $^3He/^4He$ isotopic ratios in the two areas, with two exceptions ("Tansansen" and "Nishiyoshino"). Based on these observations, we can classify the spring waters into several types,

possibly induced by three process or origin, that differ from a simple mixing process between the deep brine and meteoric water previously suggested by the major solute and $\delta^{18}O–\delta D$ isotopic systematics [7].

Discussion

Origin and upwelling process of spring waters in Kii

One of the differences between the Arima and Kii areas is that the slope of the Kii spring waters is distinct from that of the Arima spring waters in the $\delta^{18}O–\delta D$ plot (Figure 3A). This difference may be attributed to a difference in the depth of dehydration at the subducting slab: the steeper slope in the Kii area corresponds to shallower dehydration, compared to the Arima area, which has a gentler slope [7,18]. This difference can also be observed in other plots, such as the $\delta^{18}O–Cl$ plot in Figure 3B: the slope of the trend is gentler in the Kii area than in the Arima area. These observations suggest that the dense end-member in the Kii area has lower salinity that in the Arima area, possibly reflecting a different chemistry of the slab-derived fluids. On the other hand, the variable trends of major solute cations, such as K and Mg (Figures 4C and 4D), may represent local mechanisms within individual areas of Arima and Kii. The K content is almost constant for the Kii area, whereas it exhibits a large range in the Arima area, and the Mg content increases toward the denser spring water in the Kii area, whereas that in the Arima spring waters is almost constant and at a low level (Figure 4D). These observations could reflect differences in the geology and basement rock types in the two areas, especially those surrounding the meteoric aquifers through which spring waters upwell [16]. In the Arima area, a positive Eu anomaly correlated with the K content is observed, which may be the result of an interaction between hot water and wall rock containing feldspar [27].

We now discuss the origin and upwelling process of spring waters in both areas, especially the Kii spring waters, for which REE data are reported for the first time in this paper. The PCA results show that the PC-01 score is involved in the dilution process. In this regard, PC-01 is expected to have a high correlation with $\delta^{18}O–\delta D$ isotopic ratios, which are robust and commonly used indices for specifying a mixing ratio of meteoric water. As Figure 8A shows, most of the Arima spring waters, including the most primitive deep brine "Kinsen" and five hot spring waters, exhibit a clear positive correlation. However, the cold spring waters "Tansansen", "Ginsen", and "Gosya", as well as most of the Kii spring waters, plot off the Arima trend and exhibit significant scatter. "Tansansen" has the highest PC-01 and lowest $\delta^{18}O$, which is interpreted as being due to the introduction of REEs from the aquifer country rocks into the highly carbonated waters such as "Tansansen" [16]. The same is true for "Ginsen" and "Gosya", but to a lesser extent. The Kii spring waters, most of which are highly carbonated, should have undergone such an effect, which may explain the relatively high REE abundances, represented by moderate PC-01 scores.

The common Arima-type brine, with moderately high $\delta^{18}O–\delta D$ isotopic ratios and moderately high Cl concentrations, exhibit positive PC-02 scores. The spring waters "Honmachi", "Nishiyoshino", "Shioyu", and "Yahan", located along the Kino River, which is associated with the MTL (Figure 2) and five spring waters in the Arima area, can be classified as belonging to this "Ordinary Arima"-type brine. These spring waters have a convex-downward REE pattern detected as eigen #2 (Figure 6B), with a moderately high $\delta^{18}O–\delta D$ signature, indicating that (i) the dilution effect of the deep brine component by meteoric water is limited, which is consistent with the moderate values of PC-01 (Figure 6A), and (ii) the precipitation of REEs, possibly by oxyhydroxides [16], does not affect the $\delta^{18}O–\delta D$ systematics. This

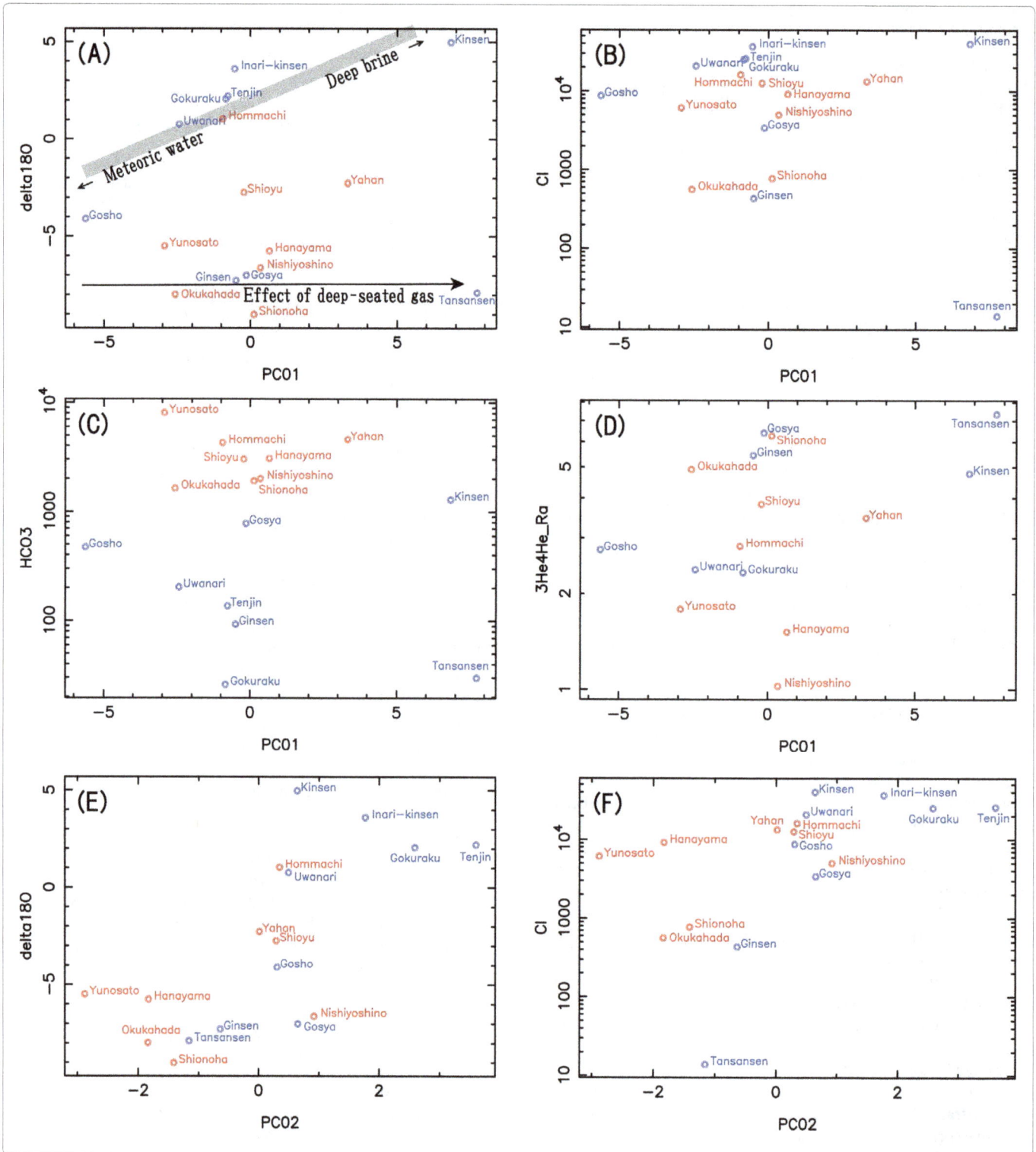

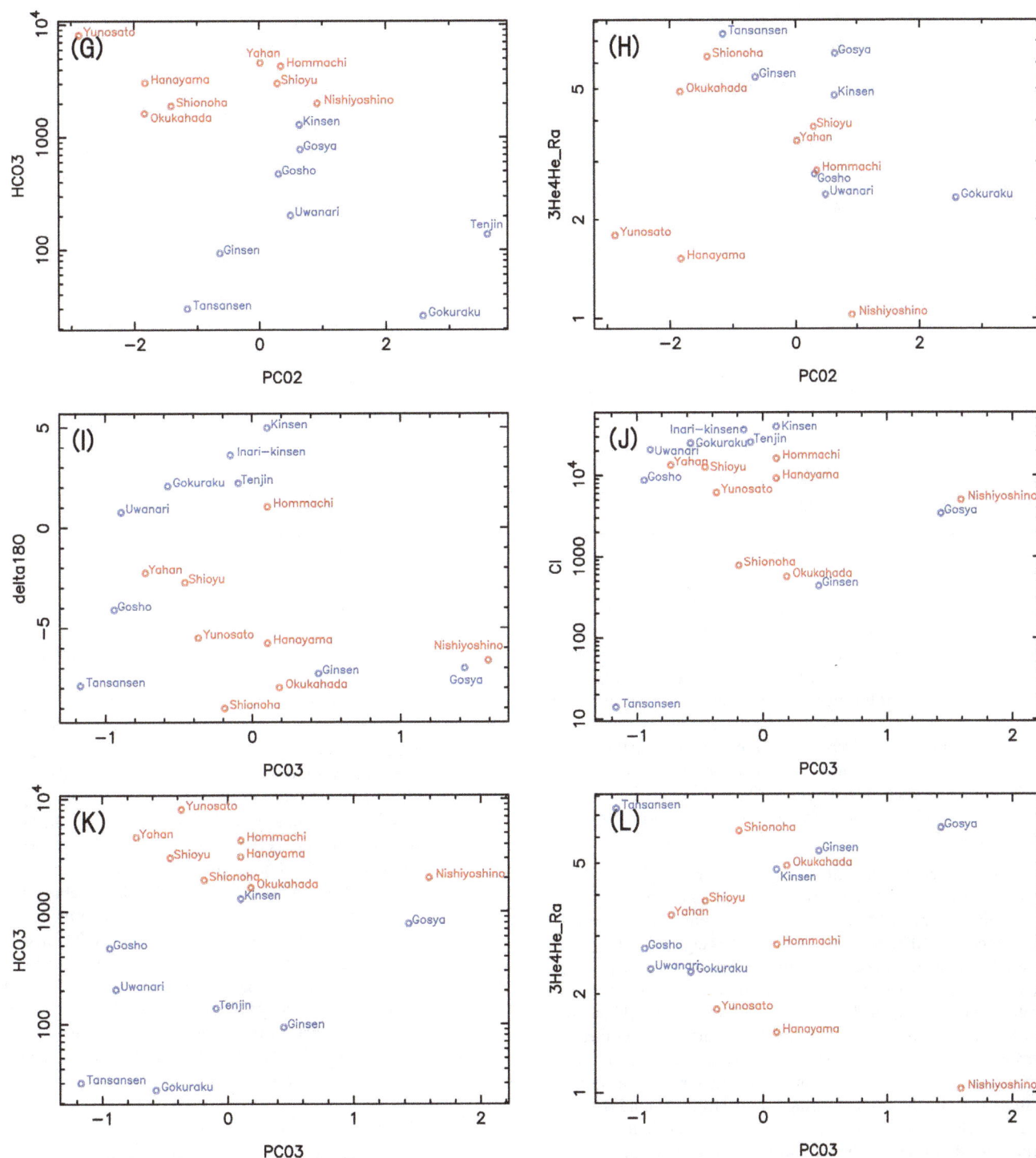

Figure 8: Correlation diagrams for PC-01 vs. (A) ^{18}O, (B) Cl abundances (ppm), (C) HCO$_3$ abundances (ppm), and (D) ^{3}He/^{4}He in Ra. (E) ^{18}O, (F) Cl abundances (ppm), (G) HCO$_3$ abundances (ppm), and (H) ^{3}He/^{4}He in Ra are plotted against PC-02 on the x axis. (I) ^{18}O, (J) Cl abundances (ppm), (K) HCO$_3$ abundances (ppm), and (L) ^{3}He/^{4}He in Ra are plotted against PC-03 on the x axis. The names of the types of spring water are shown in the same colors as in Figure 3.

"Ordinary Arima"-type brine corresponds to the NaCl-type spring water classified by Masuda et al. [14] and Morikawa et al. [17]. The positive correlation between PC-02 and the Cl concentration (Figure 8F) may indicate that the precipitation is enhanced for a denser brine with higher REE contents, (high contribution of slab-derived fluid), such as the Arima hot spring waters. On the other hand, most of the

Kii spring waters originate from less dense brines and could have undergone less precipitation.

It is worth noting that regional variation is clearly observed in the PC-02 vs. HCO$_3$ plot (Figure 8G). This is partly due to the different degrees of variation in PC-02 for the Arima and Kii areas, as mentioned

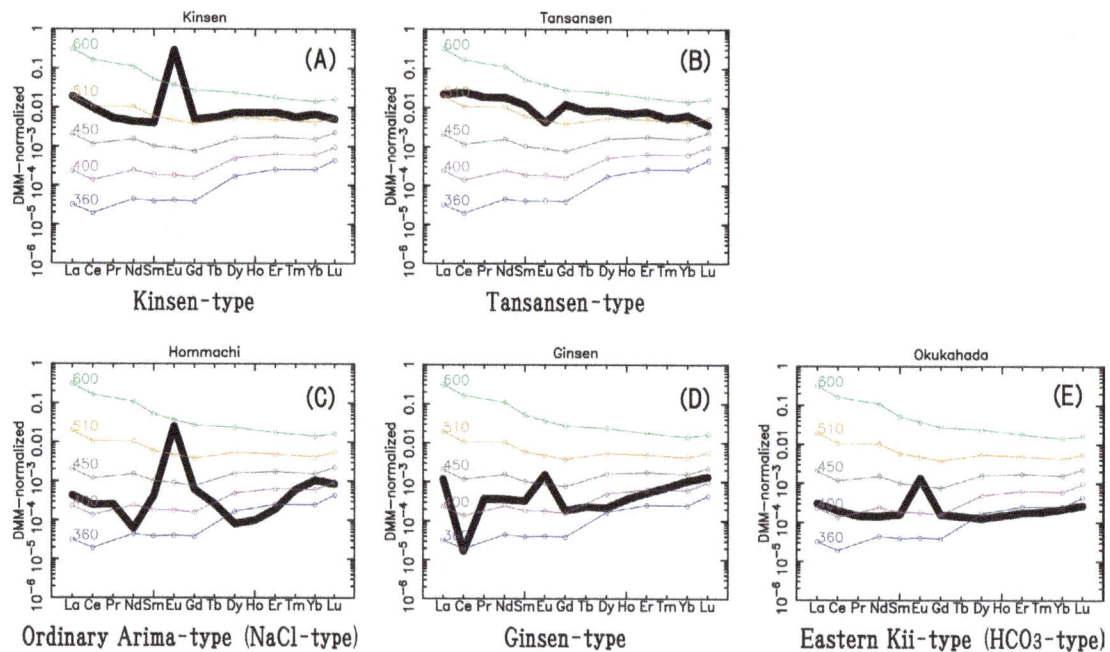

Figure 9: Typical REE patterns shown in spidergram form for (A) "Tansansen"-type, (B) "Kinsen"-type, (C) "Ordinary Arima"-type, (D) "Ginsen"-type, and (E) "Eastern Kii"-type spring waters. The color contours show the composition of fluid derived from altered oceanic crust at 360 to 600° [26-28].

above, but is mostly due to the distinct range in HCO_3. The mechanism that differentiates the areas exhibiting such distinct HCO_3 behavior is unknown at this stage but could be related to the tectonic and geological structure that defines the characteristics of fluid ascent and reactions. For example, fluid ascent through rocks with lower permeability may limit the upwelling velocity which would allow complete separation of the gas component from the deep brine and results in carbonation of the overlying aquifer. The basement rock types in the Kii area are different from those in the Arima area, as will be discussed in section 5.2, and may affect such processes and the amount of REE incorporation that occurs at the aquifer level.

PC-03 is not correlated with either PC-01 or PC-02 (Figures 7B and 7C), which suggests the existence of an independent process involving the origin, partitioning, and source, resulting in an overall contrast between LREEs and HREEs detected as eigen #3 (Figure 6C). This variation may be induced by the presence of gas phases, especially carbonated gas, which enhances the partitioning among host rocks and saline water [16,28,29]. The plot of PC-03 vs. $^3He/^4He$ (Figure 8A) shows a broad positive correlation, which supports the hypothesis of the involvement of gas phases from deep brine, as well as incorporation of significant amounts of REEs from the country rock as a result of the carbonic acid nature of the spring waters [16]. It should be noted that "Tansansen" has a negative PC-03 value but that significant REE incorporation has occurred because of the high carbonic acidity and high $^3He/^4He$ ratio of this water [16]. Because the overall slope of the LREEs/HREEs ratio represented by PC-03 is affected not only by partitioning behavior upon incorporation of REEs but also the REE pattern of the source country rock, in the case of "Tansansen", the involvement of source rock with a high LREEs/HREEs ratio (i.e., a negative PC-03) is suggested, which is consistent with the suggestion that the basement rocks consist of felsic igneous rocks with high LREEs/HREEs ratios [16]. In the Kii area, "Nishiyoshino" has a positive PC-03 score and the lowest He isotopic ratio, which suggests possible contamination of crustal He.

Although the HCO_3 ion content in a water sample may not be directly related to a flux of carbonaceous gas, the HCO_3 abundance in the Kii spring waters exhibits a moderately strong correlation with the $\delta^{18}O$ content that increasing toward the dense end-member (Figure 3C). In the eastern part of the studied area of the Kii Peninsula, two spring waters, "Shionoha" and "Okukahada", are classified as negative PC-02 and moderately positive PC-03, with distinctly high He isotopic ratios (Figures 8H and 8L) and low $\delta^{18}O$–δD isotopic ratios (Figure 3A). These two spring waters referred to as being of the "Eastern Kii" type are similar to "Gosya" in the Arima area, which corresponds to the HCO_3-type classified by Masuda et al. [14] and Morikawa et al. [17]. Morikawa et al. [17] presented the distributions of the Cl concentrations of the spring waters in the Kii Peninsula area and classified them as being Na-Cl-, Na-HCO_3-, Ca-Cl-(or SO_4-), and Ca-HCO_3-dominant types. The distribution of HCO_3-dominant types, including the "Okukahada" and "Shionoha" waters considered in this study, exhibits a moderately positive PC-03 and high He gas, which seems to correspond to the DLF tremor belt detected by seismicity tests [3], whereas the distribution of NaCl-types (that is, "Ordinary Arima"-type brine), distinguished by a positive PC-02, including the "Honmachi", "Nishiyoshino", "Shioyu", and "Yahan" waters considered in this study, are located along the MTL where DLF earthquake activity has been observed (Figure 1). The possible link is discussed in the next section.

Scenarios concerning origin and upwelling of the deep brines and gases

Based on the arguments presented previously, we discuss two scenarios in this section. When the deep brine derived from the PHS slab at a depth of 35–40 km ascends and arrives at a shallow crustal level (a depth of approximately 2000 m), the gas phase formed by decompressional degassing (as suggested by Morikawa et al. [17], based on the gas depth profile) can be separated from the deep brine and ascend to dissolve into meteoric water in the near-surface aquifer system. Such deep gas-bearing water may dissolve REEs from the aquifer country rocks, resulting in a positive PC-03 and a high He gas isotopic ratio, and would be classified as an "Eastern Kii"-type (HCO$_3$-type) of spring water (Figure 9).

In this case, there are two possible scenarios for the subsequent evolution of the original deep brine that ascended to a depth of approximately 2000 m. One scenario assumes a transient state, i.e., that the original deep brine is still ascending after degassing but has not yet reached the surface. This scenario is consistent with absence of an "Ordinary Arima"-type (NaCl-type) spring in the region where "Eastern Kii"-type (HCO$_3$-type) springs are dominant (Figures 2 and 9E). It is worth noting that in this "Eastern Kii"-type spring region, DLF tremors and their regional migration over a time scale of days to weeks have been observed (Figure 1), suggesting that deep-seated fluids (possibly deep brine from the slab) could have been repeatedly supplied beneath the region. Such repeated supplies may cause transient upwelling of deep brine that follows the gas phase, which ascends faster than the brine, as a precursor phase of surface effusion of deep brine, although there is no direct evidence of brine ("Ordinary Arima"-type spring) effusion having occurred in the region in the past. If this is the case, an "Ordinary Arima"-type (NaCl-type) spring may occur in the (near) future and coexist with the "Eastern Kii"-type (HCO$_3$-type) spring, as in the present-day Arima area.

Another steady-state scenario is also possible. In this scenario, the deep brine remains at a depth of approximately 2000 m, where it is degassed and loses some of its buoyancy. In this case, the deep brine ("Ordinary Arima"-type, NaCl-type spring) will never appear on the surface and could disperse. Unlike the "Ordinary Arima"-type (NaCl-type) spring region along the MTL, where many spray faults are present, the fault system is relatively undeveloped in the "Eastern Kii"-type (HCO$_3$-type) spring region (Figure 1). Accordingly, the fluid pathways along which the deep brine is guided to ascend to the surface are poorly developed, which could be one reason for the absence of "Ordinary Arima"-type (NaCl-type) springs in the eastern Kii area along with the HCO$_3$-type spring. Slab-derived fluids will exhibit variable REE abundances depending on the temperature of dehydration. With slab dehydration occurring at lower temperatures beneath the fore-arc Kii region (Figure 1), REEs in the slab-derived fluids will exhibit significantly lower abundances because of the strong temperature dependence of the partition coefficient between fluid and residual solids (Figures 9A-9E for slab fluids derived from altered oceanic crust at 360 to 600 degrees) [30,31]. Although the rare occurrence of the original deep brine (i.e., a brine that has not undergone REE-bearing deposition of minerals) prevents us from estimating the physical conditions for slab dehydration accurately, the overall REE level, especially the Lu and La contents, of the "Ordinary Arima"-type (NaCl-type) spring waters in the Kii area is thought to originate from slab-derived fluids generated by dehydration of the slab at approximately 450°C (Figure 9C).

In any case, the original deep brine precipitates REE-bearing minerals to produce spring water variations represented by PC-02 (Figure 7A). As shown in previous studies of the Arima area [16], when the deep brine enters a meteoric aquifer, the precipitation process is likely to be triggered by a change in temperature and oxygen fugacity conditions, simultaneously generating a gas phase that will ascend separately and be added to the shallower aquifer. Within the shallow aquifer, HCO$_3$-type spring water is produced, enhancing reactions with the basement rocks because of its carbonic acidity. In the Arima area, the basement granite is continuously metasomatized by the intrusion of deeply originated hot water, which makes granite turn to decomposed granite called "Masado" and also enriches the river water with REE [32]. Positive Eu anomalies are commonly observed

for both the "Ordinary Arima"-type (NaCl-type) and "Ginsen"-type (HCO$_3$-type) spring waters in the Arima area, except Tansansen, although the basement granite exhibits a clear negative Eu anomaly [33-35]. When the deep brine encounters near-surface water under oxidizing conditions in the aquifer, the overlapping effects associated with elution of Eu^{2+} from plagioclase of the granite and deposition of other REEs with oxyhydroxides will enhance the strong positive Eu anomaly with a low abundance of Ln^{3+}, as observed in the spidergrams (Figures 5 and 9). This suggests that the basement rock composition and oxidation conditions may have strong influences on the spring water composition. The variability in terms of major solute elements and REEs in the Kii area, illustrated in Figures 3 and 4, could be partly attributed to such an effect, although quantitative modeling to confirm this remains to be performed.

Conclusions

To characterize the Arima-type brine and associated spring waters in the Kii area, we performed REE analyses of the spring waters in a wide area of the Kii Peninsula. In addition, we compiled data on the major solute elements, REE abundances, and δ^{18}O–δD, He, and Sr–Nd–Pb isotopic systematics from previous studies encompassing a broad region of the Arima and Kii areas in southwestern Japan. Based on this data set, we classified the spring waters into several types described below.

The results of a principal component analysis (PCA) of the existing and new REE data from the Kii and Arima areas showed that three principal components (PCs) explain 89% of the entire sample variance. Comparisons between the major solute components (including δ^{18}O–δD isotopic ratios) and the REE systematics represented by these three PCs allow us to categorize the spring waters into the following five types:

(i) "Tansansen" -type: A flat pattern with a negative Eu anomaly, large positive PC-01, large negative PC-02 (Figure 7A), and negative PC-03, observed uniquely as "Tansansen" (Figure 9A) in the Arima area. This spring water exhibits affinities with the HCO$_3$-type described below but is distinct because of the significant impact of basement granitic rocks.

(ii) "Kinsen" -type: An overall slightly convex-downward REE-pattern with a moderately positive Eu anomaly and large positive PC-01 and PC-02 scores (Figure 7A) and a small positive PC-03, observed uniquely as "Kinsen" (Figure 9B) in the Arima area. This spring water exhibits geochemical features closest to those of the original deep brine from the subducted Philippine Sea slab.

(iii) "Ordinary Arima" -type: A strongly convex-ward pattern with a strong positive Eu anomaly, moderately negative PC-01, positive PC-02, and negative PC-03 (Figures 7A and 8I) and exhibiting the PC-01–δ^{18}O trend shown in Figure 8A. This type of spring water is derived from type (ii) original brine that has undergone REE-precipitation and is commonly observed in the Arima area, as well as in the Kii area (Figure 9C), although in the Kii area, this type of spring water seem to have some influence from (iv) below that obscures the PC-01–δ^{18}O trend. This spring water has been classified as being of the NaCl-type (Figure 9C).

(iv) "Ginsen" -type: A slightly convex-downward pattern with a moderately positive Eu-anomaly, moderately negative PC-01, moderately negative PC-02, and positive PC-03 (Figures 7A and 8I), departing from the PC-01–δ^{18}O trend of type (iii) toward a positive PC-01 due to the gas phase effect (Figures 8A and 8D). This spring water has affinities with both the NaCl-type (deep brine signature) and HCO_3-type (gas-added meteoric water signature), represented as "Ginsen" (Figure 9D), although it is rather diluted.

(v) "Eastern Kii" -type: An almost flat pattern with a moderately positive Eu anomaly, both positive and negative PC-01, negatives PC-02, a moderately negative PC-03, and a high $^3He/^4He$ isotopic ratio (Figures 7A and 8L). These features are similar to those of type (iv) (the "Ginsen" type), except that type (v) and all other Kii spring waters exhibit high HCO_3 contents (Figures 8C, 8G, and 8K) unlike type (iv) and all other Arima spring waters. These types of spring waters, which occur only in the eastern part of the studied Kii area, are represented by "Okukahada" and "Shionoha" and have been conventionally classified as belonging to the HCO_3-type (Figures 9E).

It should be noted that the five types (i) to (v) described above are accounted for by a combination of a smaller number of processes and sources: (1) mixing between the slab-derived deep brine and meteoric water (represented by major solute binary trends, including $\delta^{18}O$–δD systematics, as well as a part of PC-01), (2) precipitation of REEs from the brine (represented by PC-02), and (3) incorporation of REEs from the country rock by carbonic acidity (represented by PC-03), although the type of country rocks (i.e., granitic rocks, metamorphic rocks) may also have a significant impact on the spring water composition. In addition, the compositional variability of slab-derived fluid with temperature and depth of slab dehydration probably influences the regional differences observed between the Arima and Kii areas. In comparing the spring waters in the Arima and Kii areas, we detected systematic geographic distributions of the NaCl-type and HCO_3-type waters in the Kii area. The former upwells along the MTL, whereas the latter upwells in the eastern part of the studied area, where DLF tremors have been observed. This suggests that the geographic distributions are linked to the tectonic setting and/or temporal evolution of upwelling of the slab-derived fluid.

Acknowledgments

This work was supported by JSPS KAKENHI Grant Number 25400524 and the Earthquake Research Institute Cooperative Research Program of the University of Tokyo.

References

1. Tatsumi Y, Eggins S (1995) Subduction zone magmatism. Blackwell, Oxford.

2. Nakajima J, Tsuji Y, Hasegawa A (2009) Seismic evidence for thermally-controlled dehydration reaction in subducting oceanic crust. Geophys Res Lett 36, L03303, doi:10.1029/2008GL036865.

3. Obara K (2002) Nonvolcanic deep tremor associated with subduction in southwest Japan. Science 296: 1679-1681.

4. Takahashi H, Miyamura J (2009) Deep low-frequency earthquakes occurring in Japanese islands. Geophysical Bulletin of Hokkaido University 72: 177-190.

5. Zhao D, Kanamori H, Negishi H, Wiens D (1996) Tomography of the source area of the 1995 Kobe earthquake: Evidence for fluids at the hypocenter? Science 274: 1891-1894.

6. Zhao D, Tani H, Mishra OP (2004) Crustal heterogeneity in the 200 Tottori earthquake regions: Effect of fluids from slab dehydration. Phys Earth Planet Inter 145: 161-177.

7. Kusuda C, Iwamori H, Nakamura H, Kazahaya K, Morikawa N (2014) Arima hot spring waters as a deep-seated brine from subducting slab. Earth Planets Space 66(1):119. doi: 10.1186/1880-5981-66-119.

8. Nakamura H, Fujita Y, Nakai S, Yokoyama T, Iwamori H (2014) Rare earth elements and Sr–Nd–Pb isotopic analyses of the Arima hot spring waters, Southwest Japan: implications for origin of the Arima-type brine. J Geol Geosci 3:161. doi: 10.4172/2329-6755.1000161.

9. Kazahaya K, Takahashi M, Yasuhara M, Nishio Y, Inamura A, et al. (2014)

10. Matsubaya O, Sakai H, Kusachi I, Satake H (1973) Hydrogen and oxygen isotopic ratios and major element chemistry of Japanese thermal water systems. Geochem J 7: 123-151.

11. Nagao K, Takaoka N, Matsubaya O (1981) Rare gas isotopic compositions in natural gases in Japan. Earth Planet Sci Lett 53: 175-188.

12. Sano Y, Wakita H (1985) Geographical distribution of 3He/4He ratios in Japan – implications for arc tectonics and incipient magmatism. J Geophys Res 90: 8729-8741.

13. Tanaka K, Koizumi M, Seki R, Ikeda N (1984) Geochemical study of Arima hot-spring waters, Hyogo, 408 Japan, by means of tritium and deuterium. Geochem J 18: 173-180.

14. Masuda H, Sakai H, Chiba H, Tasurumaki M (1985) Geochemical characteristics of Na-Ca-Cl-HCO3 type waters in Arima and its vicinity in the western Kinki district, Japan. Geochem J 19: 149-162.

15. Ayers J (1998) Trace element modeling of aqueous fluid–peridotite interaction in the mantle wedge of subduction zones. Contrib Mineral Petrol 132: 390-404.

16. Nakamura H, Chiba K, Chang Q, Nakai S, Kazahaya K, et al. (2015) Rare earth elements of the Arima spring waters, Southwest Japan: implications for fluid–crust interaction during ascent of deep brine. J Geol Geosci 4:217. doi:10.4172/jgg.1000217.

17. Morikawa N, Kazahaya K, Takahashi M, Inamura A, Yasuhara M, et al. (accepted) Widespread distribution of ascending fluids transporting mantle helium in the fore-arc region and their upwelling processes: Noble gas and major element composition of deep groundwaters in the Kii Peninsula, southwest Japan.

18. Takahashi M, Kazahaya K, Yasuhara M, Tsukamoto H, Sato T, et al. (2011) A database of deep groundwater chemistry. Geol Surv Japan, AIST 532.

19. Ito T, Kojima Y, Kodaira S, Sato H, Kaneda Y, et al. (2009) Crustal structure of southwest Japan, revealed by the integrated seismic experiment Southwest Japan 2002. Tectonophys 472: 124-134.

20. Kawamura T, Onishi M, Kurashimo E, Ikawa T, Ito T (2003) Deep seismic reflection experiment using a dense receive and sparse shot technique for the deep structure of the Median Tectonic Line (MTL) in east Shikoku, Japan. Earth Planets Space 55: 549-557.

21. Nakajima J, Hasegawa A (2007) Subduction of the Philippine Sea plate beneath southwestern Japan: Slab geometry and its relationship to arc magmatism. J Geophys Res 112, B08306, doi:10.1029/2006JB004770.

22. Hirose F, Nakajima J, Hasegawa A (2008) Three-dimensional seismic velocity structure and configuration of the Philippine Sea slab in southwestern Japan estimated by double-difference tomography. J Geophys Res 113, B09315, doi:10.1029/2007JB005274.

23. Iwamori H (2007) Transportation of H_2O beneath the Japan arcs and its implications for global water circulation. Chem Geol 239: 182-198.

24. Geological Survey of Japan, AIST (2015) Seamless digital geological map of Japan 1: 200,000. Geological Survey of Japan, National Institute of Advanced Industrial Science and Technology.

25. Uemoto M (2008) Principles and practices of ICP emission and ICP mass spectrometry, Ohmsha, ISBN 978-4-274-20539-2, Tokyo (in Japanese).

26. Workman RK, Hart SR (2005) Major and trace element composition of the depleted MORB mantle (DMM). Earth Planet Sci Lett 231: 53-72.

27. Teranishi K, Isomura K, Yano M, Chayama K, Fujiwara S, et al. (2003) Measurement of distribution of rare earth elements in Arima-type springs using preconcentration with chelating resin/ICP-MS. Bunseki Kagaku 52: 289-296.

28. Ohta A, Kawabe I (2000) Rare earth element partitioning between Fe oxyhydroxide precipitates and aqueous NaCl solutions doped with NaHCO: Determinations of rare earth element complexation constants with carbonate ions. Geochem J 34: 439-454.

29. Ohta A, Kawabe I (2000) Theoretical study of tetrad effects observed in REE distribution coefficients between marine Fe-Mn deposit and deep seawater, and in REE(III)-carbonate complexation constants. Geochem J 34: 455-473.

30. Spandler C, Hermann A, Arculus R, Mavrogenes J (2003) Redistribution of trace elements during prograde metamorphism from lawsonite blueschist to

Spatial distribution and features of slab-related deep-seated fluid in SW Japan. Journal of Japanese Association of Hydrological Science 44: 3-16.

eclogite facies; implications for deep subduction-zone processes. Contrib Mineral Petrol 146: 205-222.

31. Kessel R, Schmidt MW, Ulmer P, Pettke T (2005) Trace element signature of subduction-zone fluids, melts and supercritical liquids at 120–180 km depth. Nature 437: 724-727.

32. Nakajima T, Terakado Y (2003) Rare earth elements in stream waters from the Rokko granite area, Japan: Effect of weathering degree of watershed rocks. Geochem J 37: 181-198.

33. Ishihara S, Chappell BW (2007) Chemical compositions of the late Cretaceous Ryoke granitoids of the Chubu District, central Japan-Revisited. Bull Geol Surv Japan 58: 323-350.

34. Terakado T, Fujitani T (1995) Significance of iron and cobalt partitioning between plagioclase and biotite for problems concerning the Eu^{2+}/Eu^{3+} ratio, europium anomaly, and magnetite-/ilmenite-series designation for granitic rocks from the inner zone of southwestern Japan. Geochim Cosmochim Acta 59: 2689-2699.

35. Google Earth (2013) accessed April 10, 2013.

17

Recognition of Lithostratigraphic Breaks in Undifferentiated Rock Units Using Well Logs: A Flow Chart

Shaaban FF* and Al-Rashed AR

College of Basic Education, PAAET, Kuwait

Abstract

Location of subsurface lithostratigraphic boundaries (breaks) in a drilled sequence requires high precision in identification of the faunal and lithofacies of the sediments or rocks above and below them. Occasionally, difficulties face well site geologists and stratigraphers in locating these boundaries due to the absence of guide faunal assemblages and/or major lithological changes. A new approach is established to delineate breaks between undifferentiated subsurface rock units and to verify the most likely cause for their occurrence using Natural Gamma Ray Spectrometry and High Resolution Dipmeter logs. This depends on the physical response of these tools to the overall characteristics of the rocks penetrated below and above the concerned break. Abrupt changes in the SGR, CGR, Th/K and Th/U of the NGS logs indicate variations in the contents of the radioactive minerals, and consequently changes in geological conditions of deposition of the rocks penetrated i.e. lithostratigraphic breaks. Changes in the quality, density, and regularity of the Dipmeter log patterns of strata are also detected below and above break. A flow chart, illustrating the steps followed for the recognition of lithostratigraphic breaks, is suggested. Ambiguities that may arise in differentiation between unconformities and faults have been discussed.

Keywords: Lithostratigraphic breaks; Undifferentiated rock units; Well logs

Introduction

Modern stratigraphy pays more attention to the breaks in the sedimentary and stratigraphy records. Measured formations can be subdivided into parastratigraphic units, which can be dated directly if characteristic flora and fauna can be detected in core samples or cuttings [1]. These units are delimited by marker levels above and below. Of substantial extent, these markers are approximately parallel, suggesting a continuous sequence of deposition. Depending on the refinement of the division into stratigraphic units, it may be possible to identify units which correspond to very specific periods in the geological history of a basin, thus providing stratigraphic markers (breaks) of considerable significance. The stratigraphic breaks can vary in their duration in time and in their geographic extent and can be due to non-deposition or erosion or both [2]. In general, these breaks may be surfaces of normal bed boundary between conformable lithostratigraphic sequence, unconformities or faults (Figure 1a). Identification of these breaks in a drilled sequence is of great importance for both academic and practical reasons, especially in reservoir characterization and management.

In many cases, well-site geologists as well as stratigraphers face difficulties in locating subsurface lithostratigraphic break between undifferentiated rock units having no faunal (marker) assemblages and/or without major lithological changes. Location of these breaks becomes possible, utilizing Natural Gamma Ray Spectrometry (NGS) and Dipmeter (HDT or SHDT) logs, based on the physical response of these tools to the overall characteristics of the rock units below and above the break. The use of these logs to identify this phenomenon involves analyzing each discontinuity on a curve in order to establish the most likely cause for its occurrence.

Major stratigraphic breaks are usually recognizable on dipmeter logs by their dip changes and on NGS log by a change in the thorium/potassium (Th/K), thorium/uranium (Th/U) ratios or by an anomalous peaks of CGR and SGR (Figures 1b,c and 2a). Nevertheless, many important breaks are far more subtle [3]. Other types of well logging tools such as resistivity, self potential, neutron, density and sonic can be used for detecting breaks but with limited conditions (Figure 2b andc). This is because the types of fluids in the borehole and the logged formations influence these tools. For example, a self-potential or resistivity log discontinuity may arise if there is a change in the salinity and types of drilling mud. Neutron, density or sonic log discontinuities may also arise if there is a gas-oil or oil-water contact.

Materials and Methodology

A number Natural Gamma Ray Spectrometry (NGS) and High Resolution Dipmeter (HDT or SHDT) logs of several deep exploratory wells have been tested for applying the proposed flow chart in identification the subsurface breaks. These logs have been examined and interpreted to construct a system of sequential steps for tracing subsurface lithostratigraphic breaks between undifferentiated rock units and to establish the most likely cause for break occurrence. This depends on the physical response of these tools to the overall characteristics of the rocks penetrated below and above the concerned break. Two wells of them (FG88-10 and AS418-1X) located at the Geisum and Asharafi oil fields SE the Gulf of Suez and NS21-1 at the off-shore area of northern Sinai, clarify the idea of the research. Locations of these wells are shown in Figure 3.

A curve discontinuity or break is any significant response change occurring over a depth interval not exceeding the vertical resolution of the log tool. The break will appear sharper when the depth scale is more compressed and the resolution of the tool is good [4]. Thus it is easier to identify on logs of depth scale 1/1000 than on a 1/500 or 1/200, and if a micro device rather than a macro device records the log data.

---

*Corresponding author: Shaaban FF, College of Basic Education, PAAET, Kuwait
E-mail: fouadshaaban60@gmail.com

Abrupt changes in the SGR, CGR, Th/K and Th/U of the NGS logs indicate variations in the contents of the radioactive minerals, and consequently changes in geological conditions of deposition of the rocks penetrated i.e. lithostratigraphic breaks. In addition, changes in the quality, density, and regularity of the Dipmeter log patterns of strata are also expected below and above break.

Any curve break is an indication of a major change in, at least, one of the factors affecting the response of the tool. This is why a break is significant and why we have to try to determine the reason for it. In addition, any major change in one of the geological parameters will provoke a response change, and thus a discontinuity, but only on those logs which measure parameters susceptible to such changes [4].

Results and Discussions

Lithostratigraphic breaks on wireline logs fall into two major categories as follows:

I- Breaks corresponding to a major change in lithology

A change in lithology represents a major change in sedimentation conditions and may or may not be a part of a sequential pattern. In the first case, the lithological change only represents the passage from one element of the sequence to the next, e.g. from dolomite to anhydrite. In the second case, several reasons may explain such a change. The choice between them depends on detailed analysis of various log types and making use complementary information on the general geology of the region. These reasons are; *unconformity, transgression, erosion, tectonic accident or diagenesis together with mineralogical change* [4].

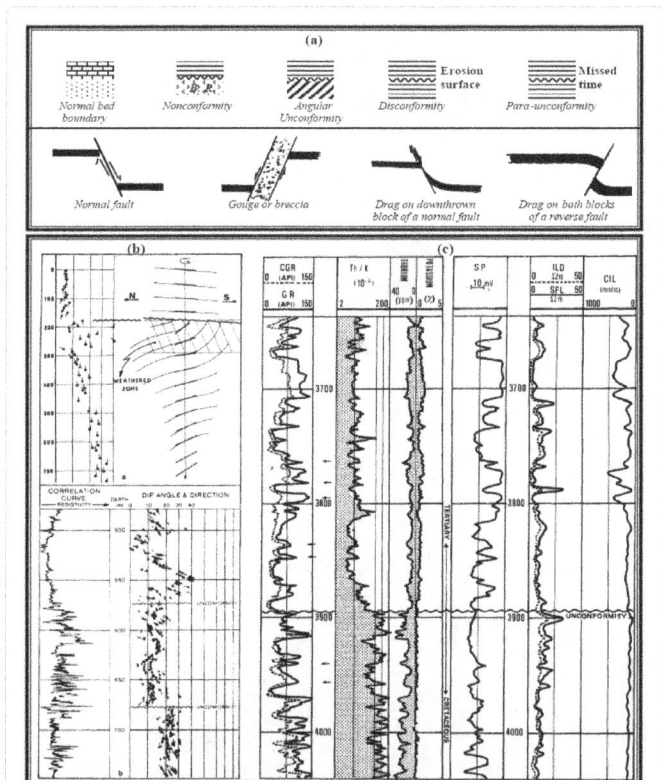

Figure 1: A) Anomalous gamma ray peak at an unconformity, radioactivity is probably due to uranium concentrated in phosphate nodules or organic matter [3]. B) Examples of unconformities identified by shale SP baseline shift [13]. C) Detection of unconformities from compaction profiles [11-13].

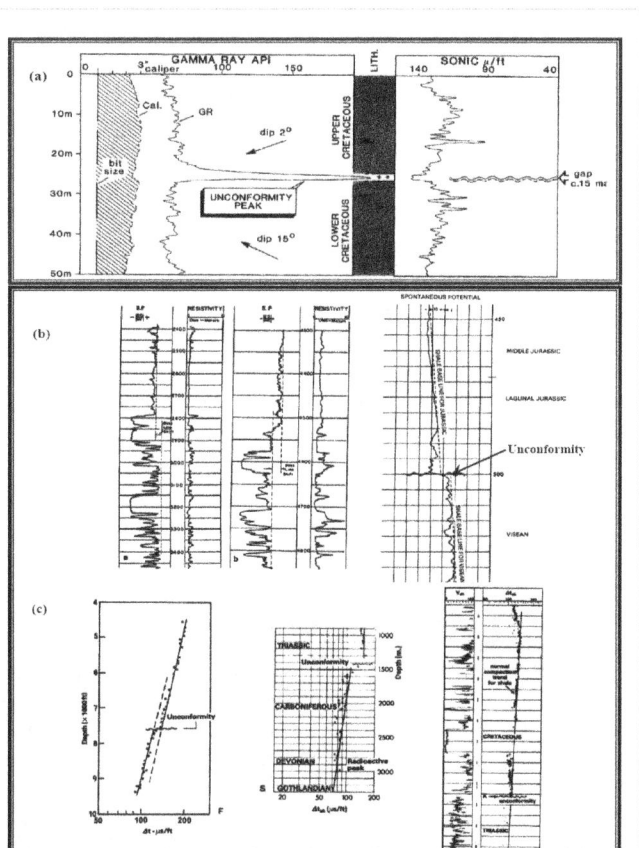

Figure 2: A) Anomalous gamma ray peak at an unconformity, radioactivity is probably due to uranium concentrated in phosphate nodules or organic matter [3]. B) Examples of unconformities identified by shale SP baseline shift [13]. C) Detection of unconformities from compaction profiles [11-13].

II- Breaks without major lithological changes

The possible causes for these breaks are the following:

- *Change in the type of fluid*, which appears if there is a gas-oil or oil-water contact and seen by using resistivity, self-potential, neutron, density and sonic logs.

- *Textural change*, a change in sorting or in cement percentage will affect the porosity and consequently all measurements that depend on it, such as density, hydrogen index, sonic travel time and resistivity.

- *Diagenesis*, this phenomenon occurs frequently in carbonate sequence. Studying the behaviour of the apparent matrix porosity and transit times from neutron will show up the changes and sonic logs, respectively.

- *Erosion* may be indicated by a sudden change in the textural parameters. But it may also be the result of an abrupt change with opposing trend on the resistivity curves.

- *Tectonic accidents*, which may bring into contact two identical lithologies with different petrophysical properties.

- *Unconformity*, a change in shale baseline, possibly associated with a change in radioactivity may indicate an unconformity. Dipmeter analysis should confirm this.

III- Recognition of lithostratigraphic breaks from NGS logs

The geological significance of radioactivity lies in the distribution of three elemental sources; the radioactive elements of the uranium-radium family, thorium family and the radioactive isotope of potassium ^{40}K [5]. The natural gamma ray spectrometry logs give the amount of each individual radioactive elements in a formation. The abundance of these radioactive elements is controlled by their geochemical behaviour, as follows:

Uranium behaves as an independent constituent, so it has a very heterogeneous, original, sedimentary distribution. Typically, on the logs, uranium is shown by irregular, high peaks corresponding to its uneven distribution. Due to the unusual requirements of its original conditions of deposition, these peaks are associated with unusual environments such as that found in condensed sequence or at unconformity.

Thorium, like uranium, has its origin in the acid and intermediate igneous rocks. It is extremely stable and, unlike uranium, will not generally pass into solution. For this reason it is found in bauxite's (residual paleo-soils). Thorium and its minerals usually find their way into sediments principally as detrital grains of heavy minerals. Moreover, because of its detrital nature and consequent transport by currents, thorium shows an affinity for terrestrial minerals and, amongst the clay minerals. *Potassium* is both chemically active and volumetrically common in naturally occurring rocks. It is generally chemically combined in clay silicate structure and in evaporites as salt, and in rock-forming minerals such as feldspars.

The principal use of the natural gamma ray spectrometry logs (SGR, CGR, Th, U, K and their ratios) is the identification of the depositional environments of shales. The affinity of uranium for shales of marine origin has been documented in contrast to the affinity of thorium for terrestrial sediments [6]. Consequently, it has been proposed that the content of uranium in shales compared with that of thorium (Th/U) gives an index of the amount of marine influence in the environment of deposition [5]. Marine shales should have a low Th/U ratio (<2) whereas the reverse is true (>6) for the continental shales [3]. Application of these rules helped in identification the depositional environments of subsurface Cretaceous section in north Sinai [7].

By contrast, the Th/K ratio is a function of the mineralogical composition of the shale. During the weathering processes, thorium and potassium have a different history according to the stability of their host minerals. Thorium bearing minerals are generally more stable than the potassium bearing minerals due to the fact that thorium is insoluble and potassium soluble. Therefore, the stronger the weathering the more the potassium, present in the sediments, will be eliminated and the higher will be the Th/K ratio. Consequently, the Th/K ratio can be used as a compaction indicator [8].

The lithostratigraphical break may represents an interface between different depositional environments. Since thorium, potassium and uranium are environmental indicators, the Th/U and Th/K ratios of the NGS logs can be used for environmental identification. An abrupt changes in the mean Th/K ratio are generally indicative of important variations in the proportion of radioactive minerals that occur when there are changes in geological conditions of deposition (Figure 1c). These correspond to unconformities [3].

Usually high gamma ray values often occur as narrow, isolated peaks (Figure 3a). Considering the geochemistry of the radioactive minerals, these peaks are generally associated with uranium concentrations. As stated above, on the geochemical behaviour of uranium, its concentrations indicate extreme conditions of deposition. Experience has shown that these conditions frequently occur around unconformity where a long passage of time is represented by little deposition. The minerals associated may be uranium-enriched organic matter or phosphate nodules [3].

An example for locating lithostratigraphic breaks, using NGS tools, between subsurface rock units in the AS-418-1X well, Gulf of Suez, is illustrated in Figure 4. This figure shows abrupt changes in shape and magnitude of the NGS logs (SGR, CGR, Th, K, U and their ratios) at three depth intervals in the examined section, separated by the depths 1901 m and 1957-1966.5 m (Figure 4c and d).

At the depth 1901, a distinct decrease in the magnitude of the NGS logs corresponding to variation in the lithology from marl and clay (80->100 GAPI) of the Kareem Formation to salts (<5 GAPI) of the South Gharib Formation with no characteristic gamma ray peak (unconformity) for the break itself at this depth (Figure 4c and d). This unconformity was identified palaeontologically at the same depth (Figure 3b). On the contrary, comparison of the NGS logs of the Rudies and Kareem Formations shows, in general, developed CGR, SGR, Th, K, U logs and their ratios with distinct decrease in the uranium content and the Th/K ratio in the Kareem Formation than those of the underlying Rudies Formation (Figure 4c and d). This NGS pattern suggest different depositional environments of the Rudies/Kareem Formations and confirms the presence of a break (unconformity) between them at depth 1966.5 m, which was identified palaeontologically (Figure 4b). But the NGS logs added that this unconformity has a vertical (upward) extension to a depth of 1957 m (i.e. 9.5 m thick) as deduced from the anomalous CGR and SGR peaks (Figure 4c). This phenomena is matched with the presence of about 9.5 m thick of coarse sandstones separating the examined formations (Figure 4a).

IV- Recognition of lithostratigraphic breaks from HDT or SHDT logs

As dipmeter logs have specific patterns of each sedimetological, stratigraphical or structural phenomena, they are used for locating discontinuities or breaks between undifferentiated lithostratigraphic

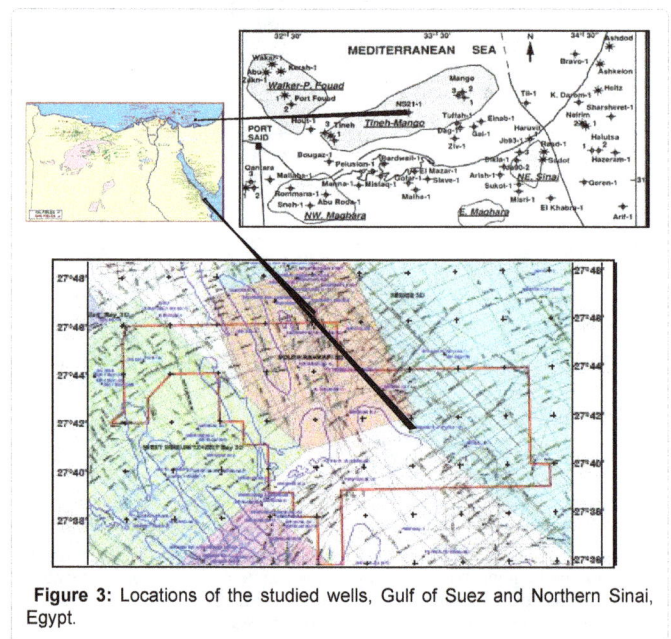

Figure 3: Locations of the studied wells, Gulf of Suez and Northern Sinai, Egypt.

units. Unlike the NGS log, dipmeter goes to establish the most likely causes for occurrence of these breaks. The following items describe the dipmeter patterns of the common lithostratigraphic breaks.

1- **Normal bed boundary:** Since this is an interface between normal lithological sequence, it has a parallel dipmeter pattern (without discontinuity) of constant magnitude, unless there is an obvious difference in dip magnitude and/or direction of the strata (due to differential compaction) below and above it (Figure 5a). Varying conditions of deposition may produce secondary dips not parallel to the enclosing strata. This "cross stratification" will produce a confused DM picture where the parallel dips of the main strata will be mixed with dips of the depositional features (Figure 5b and c). Examination of the NGS log in this case is of utmost importance for solving these ambiguities and locating bed boundary.

2- **Unconformity:** An unconformity is a hiatus in the normal geological sequence caused by a break in the process of deposition, by erosion, or by structural deformation. It results in a missing amount of sediments corresponding to a missing "geological time" as compared to the normal sequence [9]. It is made of two different series of strata separated by a surface (or zone) "surface of unconformity". The main types of unconformities that shall be concerned in this work are: *angular unconformity* (in which the strata above and below are not parallel), *disconformity* (in which strata are parallel on both sides, but

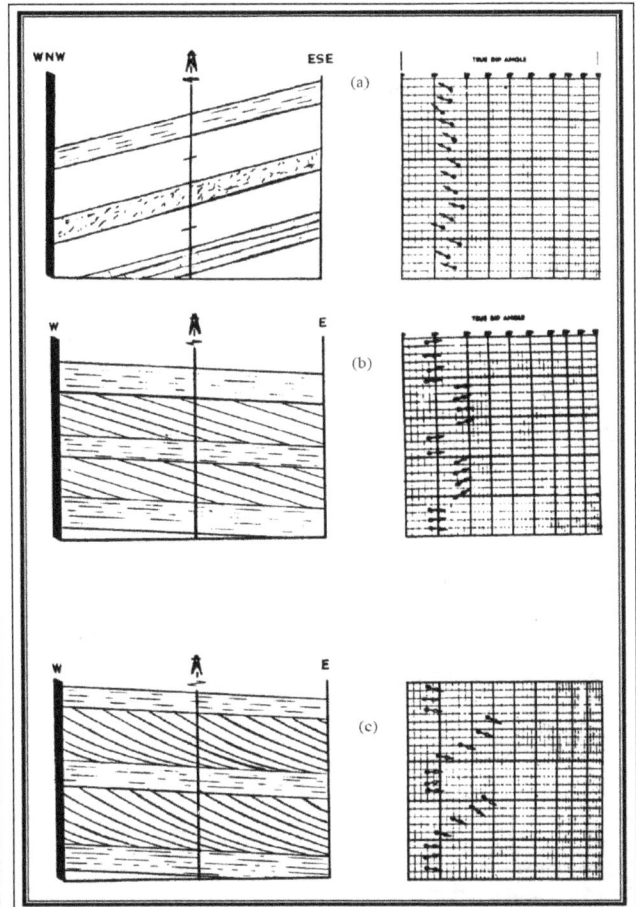

Figure 5: Dipmeter patterns of homoclines and cross stratifications. A) Regular dips reflect parallelism. B) Regular cross bedding. C) Green and blue dipmeter patterns of foreset beds [14].

there is an erosion surface), and *para-unconformity* (in which strata are also parallel on both sides, but some time is missed).

a) Detection of angular unconformity: Where the dip of bedding plane above an angular unconformity differs from that of the bedding plane below. Like faults, angular unconformities are characterized by a change of dip trend and magnitude (Figure 6a). In many instances, it is not easy to distinguish a fault from an angular unconformity on a dipmeter log, when there is an increasing or decreasing dip pattern, below the disconformity surface, resulting from erosion of the pre-existing structure. Presence of drag, with blue and/or red DM pattern on one or both blocks of the fault, may be the only way for differentiating the fault from the angular unconformity.

b) Detection of disconformity: Where there is no change in dip trend between the upper and lower strata of the disconformity, it may go unnoticed on the dipmeter log, especially when the unconformity zone is thin (Figure 6b). Nevertheless, a disconformity may be detected by one of the following features:

- Change in the quality, density, or regularity of the dips of strata above and below the break.

- Weathering zone, occurring immediately above or below the break surface which indicated by an interval of incoherent (Random) dipmeter tadpoles (Figure 6c).

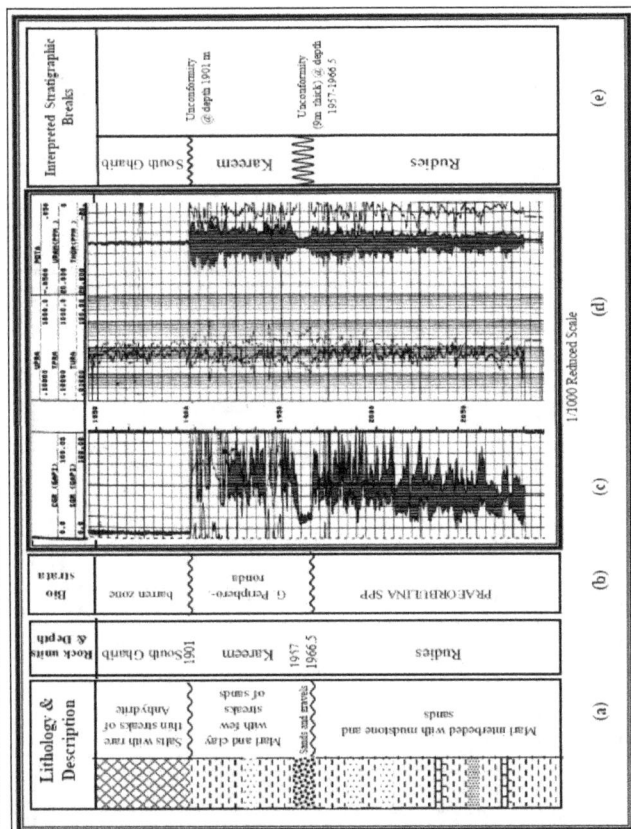

Figure 4: An example showing the applicability of the NGS log for detecting lithostratigraphic breaks between differentiated rock units in the AS-418-1X well, Gulf of Suez. A) The examined sequence (Rudies/Kareem/South Gharib). B) Biostratigraphical identification of the sequence. C) The SGR and CGR logs (1/1000 depth scale) of the examined sequence. D) The curves of Th, K, U and their ratios of the examined sequence. e) The examined sequence after locating the lithostratigraphic breaks.

- Local erosion, which may result in a local high or local low at the disconformity surface (Figure 6d).

In all cases it must be referred to the NGS log to confirm this interpretation by the presence of an anomalous peak(s) at the suggested break level.

c) Para-unconformity on dipmeter logs: A para-unconformity separates parallel strata on both sides, but some time is missed. It appears, in this case as a normal lithological sequence that consists of a series of strata exhibiting the same attitude with parallel, regular dipmeter pattern (Figure 5a). Therefore, break of para-unconformity is almost unnoticed on the dipmeter log, but it may be predictable on the NGS logs if there is a contrast in the depositional environment of the strata below and above it. An anomalous NGS peak may or may not developed at the break level itself.

3- Fault plane (or zone) on dipmeter logs: One of the most important tools for the recognition of faults is the dipmeter. Under favorable conditions it can help to identify the fault, and even provide a means of estimating the strike and dip of the fault plane. Generally,

the more drag there is along the fault, the higher the amount of rotation of the two blocks, and the presence of brecciated zone the easier a fault can be seen on dipmeter results [10]. Figure 7a shows how the dipmeter may help in fault identification under various conditions. Figure 7b shows an example of fault prediction using the dipmeter logs in the Gulf of Suez.

Sometimes, the fault does not show corrlateable anomalies on the dipmeter log (if there is no dip contrast of the faulted blocks) rather than a single large dip arrow (tadpole) when the fracture produces a single clean cut fault plane (Figure 8a). A change of dip from one block to another may be the only distinct evidence of a fault (Figure 8b and c). Generally, a zone of progressive distortion is associated with the fault (Figure 8d).

An example for locating lithostratigraphic breaks in the AS-418-1X well, based on dipmeter tools, and is shown in Figure 9. This figure shows a break (unconformity) between the Kareem and South Gharib Formations represented by incoherent dip pattern at the depth interval from 1901-1904 m (Figure 9c and d). This pattern indicates the

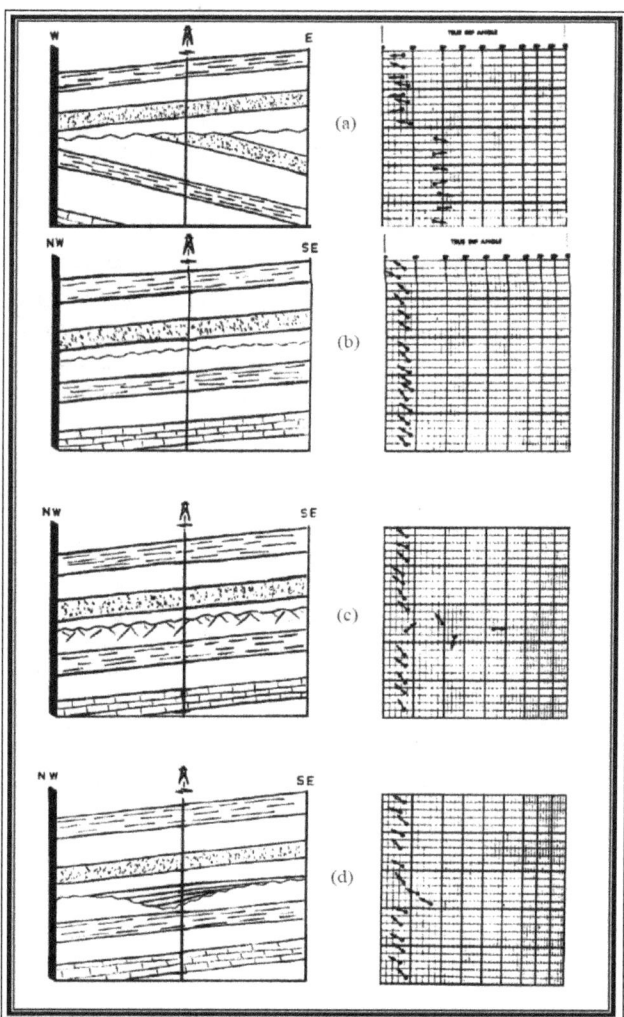

Figure 6: Dipmeter patterns of unconformities. A) Angular unconformity. B) Unnoticed thin disconformity due to absence of dip contrast of the upper and the lower strata. C) Incoherent (Random) dip pattern at disconformity surface due to the presence of weathering zone. D) Increasing dip above a surface of disconformity due to local erosion [14].

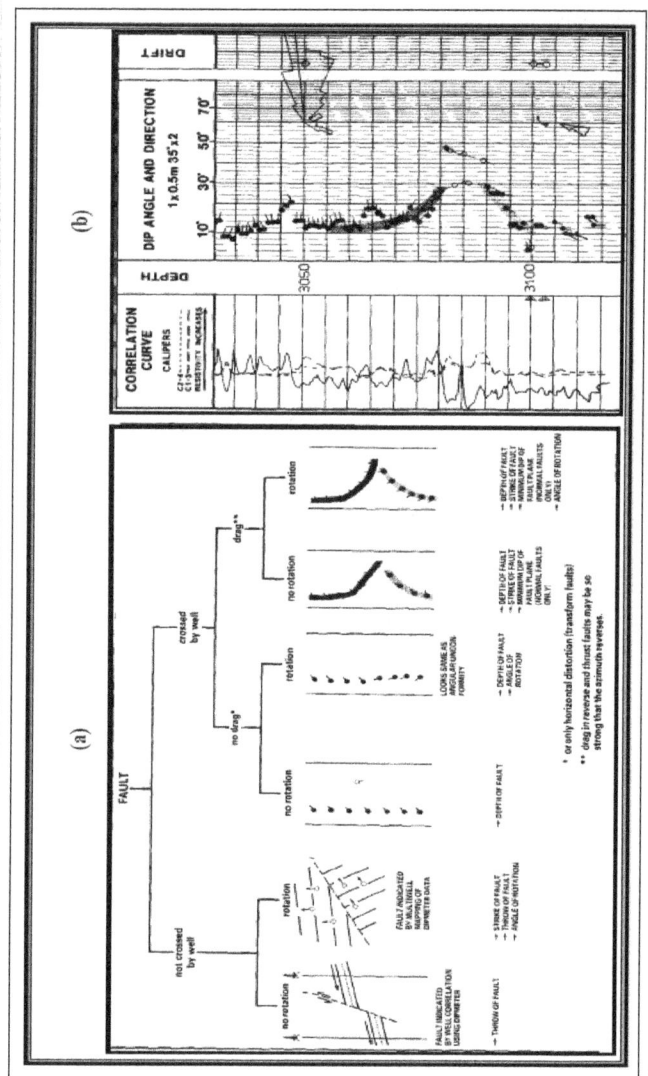

Figure 7: A) Idealized sketches showing how the dipmeter can contribute to the recognition of faults. B) SHDT survey across a fault from the Gulf of Suez [10].

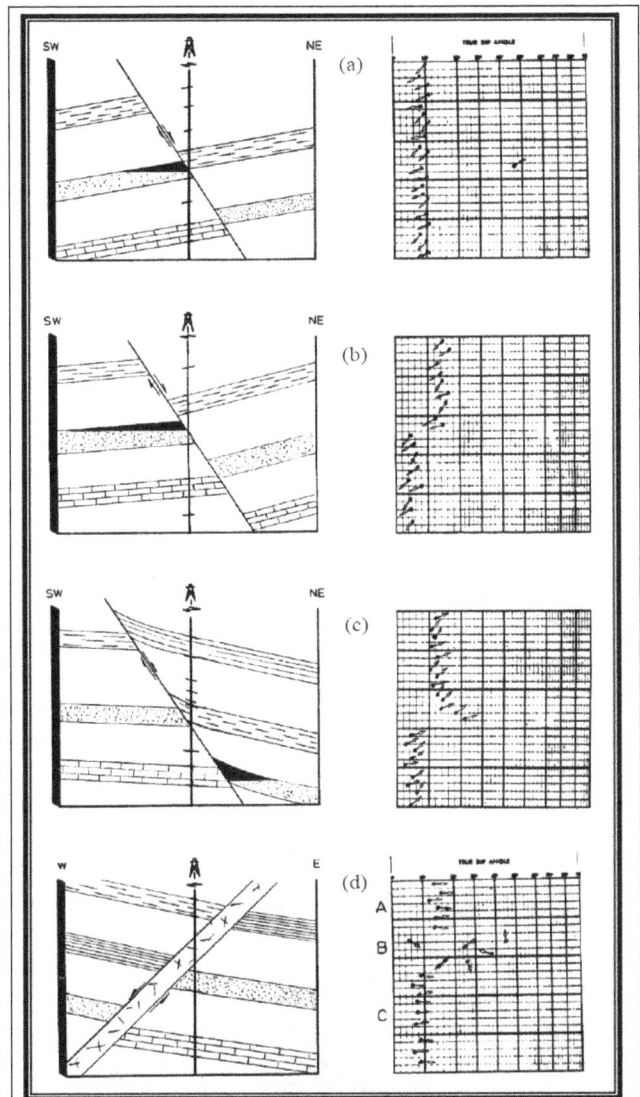

Figure 8: Dipmeter patterns of faults. A) Single large dip arrow resulting from clean-cut fault plane. B) A change of dip from one block to another may be evidence of fault. C) A zone of progressive distortion (draging) is associated with the fault. D) Breccia or gouge zone (E) around the fault [14].

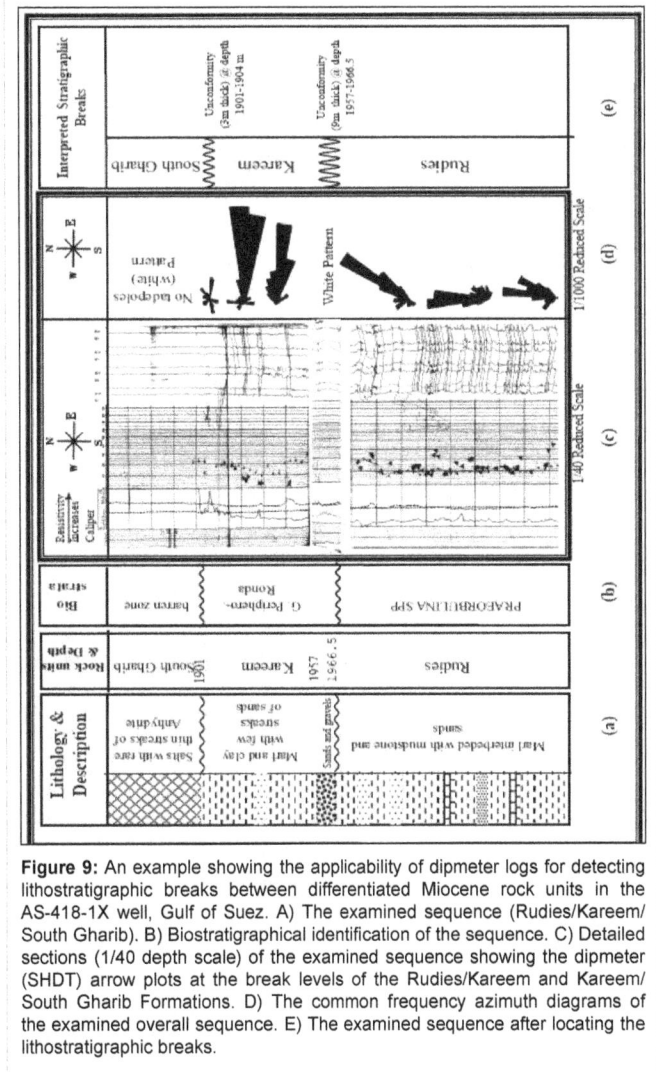

Figure 9: An example showing the applicability of dipmeter logs for detecting lithostratigraphic breaks between differentiated Miocene rock units in the AS-418-1X well, Gulf of Suez. A) The examined sequence (Rudies/Kareem/ South Gharib). B) Biostratigraphical identification of the sequence. C) Detailed sections (1/40 depth scale) of the examined sequence showing the dipmeter (SHDT) arrow plots at the break levels of the Rudies/Kareem and Kareem/ South Gharib Formations. D) The common frequency azimuth diagrams of the examined overall sequence. E) The examined sequence after locating the lithostratigraphic breaks.

presence of weathered zone of about 3 meters between the examined formations. Above this break, a white DM pattern characterizes the salt section of the South Gharib Fm due to the absence of stratification (Figure 9c). Below this break, the Kareem Formation shows a common frequency azimuth of major dip direction eastward (Figure 9c and d). The dipmeter pattern of Kareem Formation differs from that of the underlying Rudies Formation which dips generally northward and NE ward at its top (Figure 8c and d). These formations are separated by a break of white DM pattern at depth ranges from 1957-1966.5 (9.5 m thick) which indicates presence of weathered zone [11-13]. The distinct difference in the dip azimuth that characterizes both the Kareem and Rudies Formations, with the presence of an anomalous NGS peak indicates that this break is an angular unconformity (Figures 4c and d and 9c and d).

Recognition of the lithostratigraphic breaks, in the examined sequence of the AS-418-1X well, using the dipmeter logs shows a great

similarity in depth levels with those detected from the NGS logs. In addition, the use of the azimuth frequency diagrams and the tadpole plots of the dipmeter go to establish the thickness of these breaks and the most likely causes for their occurrence [14].

A flow chart has been constructed to illustrate, sequentially, the recommended steps for recognizing breaks between undifferentiated formations that previously identified to time-stratigraphic units (Figure 10). Table 1 summarized the NGS and dipmeter patterns of the common breaks. Examination of these patterns carefully is helpful for solving confusions in interpretation that may arise between unconformities and faults [15].

These rules, illustrated in Figure 10 and Table 1, are followed for detecting the stratigraphic breaks in selected examples of undifferentiated rock units in boreholes of the Gulf of Suez and north Sinai (Figures 11 and 12).

Conclusions

The present study introduces a new approach for delineating subsurface breaks in undifferentiated rock units and to establish the most likely cause for their occurrence using natural gamma ray

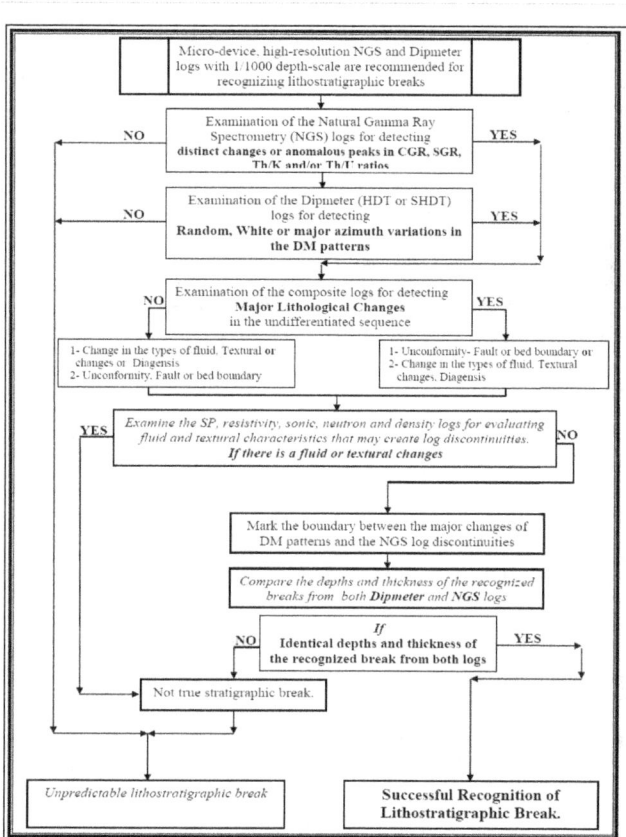

Figure 10: Flow chart illustrating the steps followed for recognizing lithostratigraphic breaks.

Break		NGS	HDT or SHDT
Normal Bed Boundary		No Characteristic anomalous peak for the break itself. A change of radioactivity of the strata above and below it may be the only distinct evidence for locating the break.	No random or white patterns characterize the break itself. A dip change of the strata above and below it may be the only distinct evidence for locating the break.
Unconfirmity	Angular Unconfirmity	Distinct anomalous peak or zone with change in Th/K and Th/U ratios above and below the break.	Zone of few, scattered tadpoles of different azimuth forming random or occasionally white pattern separating DM pattern of higher magnitude of the underlying strata than the overlying one.
	Disconfirmity	Distinct anomalous peak or zone with change in Th/K and Th/U ratios above and below the break.	Zone of incoherent dips or occasionally white pattern separating two identical DM patterns of the strata above and below it.
	Para-unconfirmity	No anomalous peak with or without change in Th/K and Th/U ratios of the strata above and below it depending on their radioactivity and depositional environments.	No distinct pattern for the break itself with identical DM patterns of the strata above and below it. Therefore, it is almost unnoticed on the dipmeter logs.

		NGS	Dipmeter
Faults	Brecciated Fault	Zone of distinct anomalous SGR and CGR with or without change in Th/K and Th/U ratios from one block to another.	Zone of no tadpoles (white) pattern due to the absence of stratification. The random pattern is rare.
	Clean-cut fault	No distinct anomalous peak with or without change in Th/K and Th/U ratios from one block to another.	Single large dip arrow due to the presence of clean cut fault plane.
	Normal Fault	Different CGR, SGR, Th/K and Th/U patterns from one block to another below and above the break.	Two identical DM patterns above and below the tracked breack (or an increasing, i.e. Red DM patterns below the break due to dragging on downthrwon block of a normal fault.)
	Reverse Fault	Distinct repetition of CGR, SGR, Th/K and Th/U patterns on both sides of the break.	An increasing (Red) and decreasing (blue) DM patterns above and below the traced break, respectively due to dragging that usually affects both blocks of reverse fault.

Table 1: Natural Gamma Ray Spectrometry and Dipmeter log responses to various lithostratigraphic breaks.

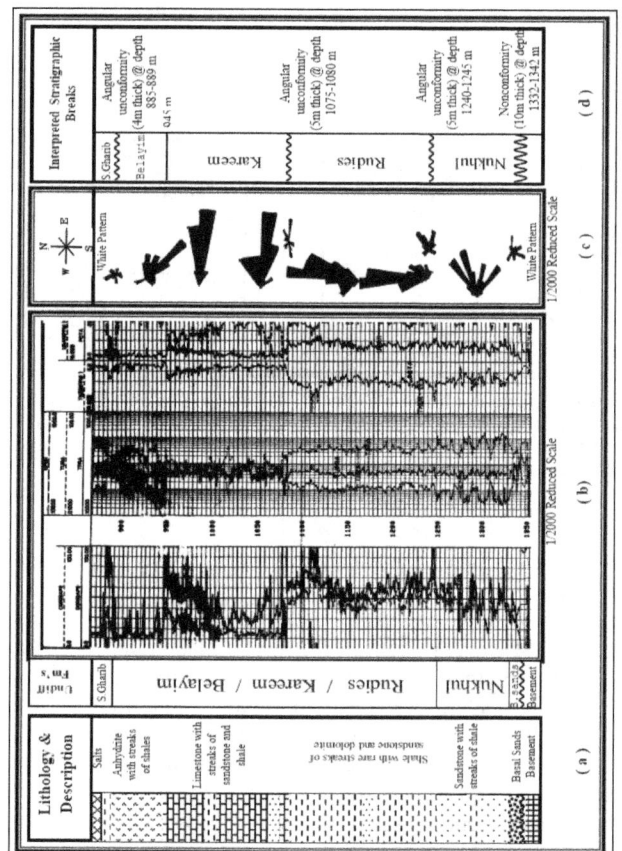

Figure 11: Detection of lithostratigraphic breaks between undifferentiated Miocene rock units of the Fg 88-10 well, Gulf of Suez. A) The examined sequence (Nukhul/Rudies/Kareem/Belayim) before differentiation. B) The SGR, CGR records, Th, K, U and their ratios of the NGS Log. C) The common frequency azimuth diagrams of the examined sequence. D) The examined sequence after locating the lithostratigraphic breaks.

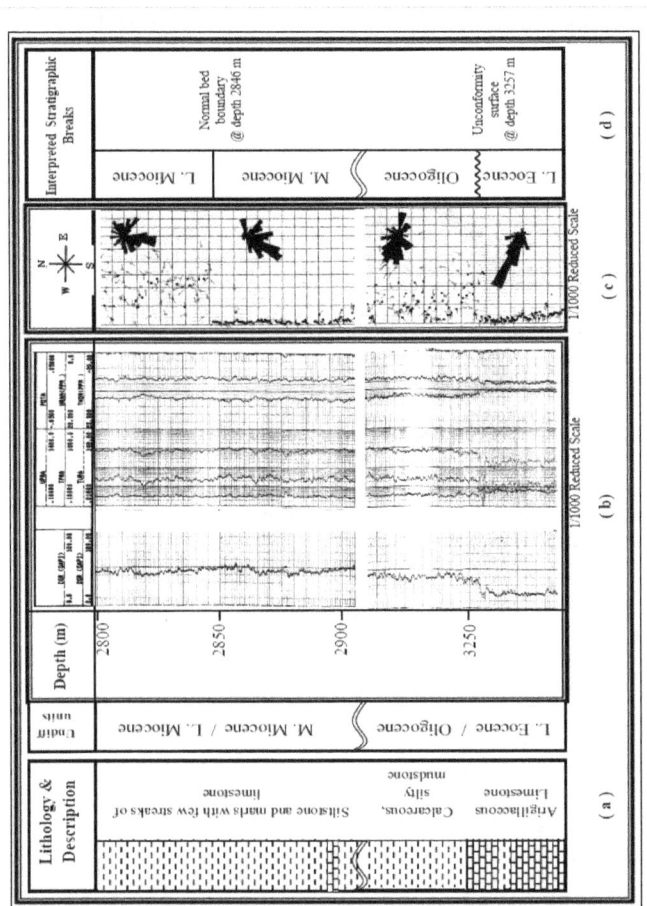

Figure 12: Detection of lithostratigraphic breaks between undifferentiated Miocene, Eocene and Oligocene rock units in the NS 21-1 well, north Sinai. A) The examined sequence (Late Eocene/Oligocene and Middle-Late Miocene) before differentiation. B) The SGR, CGR records, Th, K, U and their ratios of the NGS Log. C) The arrow plots at the breaks of the Late Eocene / Oligocene and the Middle / Late Miocene with their common frequency azimuth diagrams. D) The examined sequence after locating the lithostratigraphic breaks.

spectrometry and dipmeter logs. This approach deals essentially with already identified time-stratigraphic units, but undifferentiated by boundaries, where there is a difficulty in locating the breaks due to the absence of guide faunal assemblages and/or major lithological changes.

Identification of breaks begins with examination of the given lithology of the undifferentiated rock units, followed by examination of the dipmeter patterns and NGS log curves through a sequential steps of specific targets. A flow chart is constructed, summarizes the steps followed in the break recognition. This is depending on the physical response of both gamma ray and dipmeter tools to the overall characteristics of the rock units below and above the concerned break. Solved examples are tested and proved that this approach is powerful for tracing breaks. Undifferentiated subsurface rock units in boreholes from the Gulf of Suez and north Sinai are examined carefully and lithostratigraphic breaks are successfully recognized.

Acknowledgements

The authors are greatly indebted to the authorities of the Egyptian General Petroleum Corporation (EGPC) for supplying the wireline logs on which the present work has been based.

Nomenclature

SGR: Total gamma-ray reading

CGR: Gamma-ray without uranium

DM: Dipmeter

NGS: Natural gamma-ray spectrometry

HDT: High Resolution Dipmeter Tool

SHDT: Stratigraphic High Resolution Dipmeter Tool

UPRA: Uranium/Potassium ratio

TPRA: Thorium/Potassium ratio

TURA: Thorium/Uranium ratio

SP: Self potential

GAPI: Unit of gamma ray log reading (American Petroleum Institute)

References

1. Busson G (1972) Principles, methods et resultats d'une etude stratigraphique du Mesozoique saharien. Mem Museum Nat, Hist Natur, nouvelle serie C.

2. Sloss LL (1984) Comparative anatomy of cratonic unconformities. In: International Unconformities and Hydrocarbon Accumulation. Am Assoc Pet Geol Mem 36: 1-36.

3. Rider MH (1986) The geological interpretation of well logs. John Wiley and Sons, New York.

4. Serra O (1986) Stratigraphy, tectonics and multi-well studies using wireline logs. Schlumberger Copyright, France.

5. Adams JA, Weaver CE (1958) Thorium-uranium ratios as indicators of sedimentary process: example of concept of geochemical facies. Bull Am Assoc Petrol Geol 42: 387-430.

6. Hassan M, Housein A, Ombaz A (1976) Fundamentals of the differential gamma ray log interpretation technique. Society of Professional Well Log Analysts, 17th Ann Log Symp, Trans Pap H.

7. Shaaban FF, El-Shahat A, El-Belqasi M (2000) Depositional environments, facies and structural characteristics of the subsurface Cretaceous sequence, northern Sinai, using wireline logs. Proceeding of the 2nd International Conference on Basic Sciences and Advanced Technology, (BSAT-II) Assiut University.

8. Schlumberger (1982) Natural gamma ray spectrometry. Essentials of N.G.S. interpretation. Schlumberger Limited, New York, USA.

9. Bennison GM (1985) An introduction to geological structures and maps. (3rd edition), Edward Arnold Publ Ltd.

10. Schlumberger (1984) Well Evaluation Conference Egypt. Schlumberger, Middle East SA.

11. Fertel WH (1976) Abnormal formation pressures. Development in petroleum sciences, Elsevier Amsterdam.

12. Ghiarelli A, Serra O, Gras C, Mass P, Tison J (1973) Etude automatique de la sous-compaction des argiles par diagraphies differees. Rev Inst Franc Petrole 28: 19-36.

13. Serra O (1972) Diagraphies et stratigraphie. In: Mem BRGM 77: 775-832.

14. Schlumberger (1971) Fundamentals of dipmeter interpretation. Schlumberger Limited, New York, USA.

15. Serra O (1985) Sedimentary environments from wireline logs. Schlumberger Copyright, France.

18

Geotechnical Evaluation of Foundation Conditions in Igbogene, Bayelsa State, Nigeria

Nwankwoala HO[1]* and Adiela UP[2]

[1]Department of Geology, University of Port Harcourt, Nigeria
[2]Geosciences Department, Nigerian Agip Oil Company, P.O Box 923, Port Harcourt, Nigeria

Abstract

Geotechnical studies were carried out to investigate the foundation conditions in Igbogene, Etelebou in Bayelsa State, Nigeria. The evaluation was carried out by means of three (3) number boreholes to a maximum depth of 30 m below the existing ground level using the cable percussive rig. Field and laboratory investigations reveal a near surface stratigraphy of clay to an average depth of 6 m underlain by loose silty sand to a depth of 10 m below the existing ground level. Underlying this clay layer, the formation presents a stratum of sand which extends to the maximum depth of investigation. Field and laboratory analysis carried out on relatively undisturbed soil samples of the silty clay showed the undrained shear strength of this near surface soil to lie between 40 and 56 kPa with a mean value of 47 kPa. However, the 1.0 m thick peat embedded between 3.0 m and 4.0 m will great increase the compressibility of this clay. Pile foundation is recommended, considering the anticipated load and the very high compressibility of peat under imposed load. Piles should be straight-shaft, closed-ended steel pipe piles and driven into the medium dense sand. Pile load test should be carried out on all piles to confirm working load and estimated settlements.

Keywords: Borehole; Engineering geology; Foundation; Subsoil; Stratigraphy; Bayelsa state

Introduction

Site investigations in one form or the other is always required for long tern stability of structures [1]. The knowledge of the geotechnical characteristics is very desirable for design and construction of foundation of civil engineering structures in order to minimize adverse effects and prevention of post construction problems. Civil engineering projects are dedicated to the realization of efficient and economical works in a short time which requires an acceptable risk increasingly low. Geotechnical studies are highly important in such projects. Thus, a good estimate of the risk associated with geotechnical parameters has become a major issue since most of the new structures are located on sites with difficult conditions [2]. Some studies have been carried out on geotechnical properties of the subsoil's generally [3-5].

The study area (Figure 1) lies in the coastal Niger Delta sedimentary basin. The area is endowed with the sedimentary rocks characteristic of the Niger Delta. The detailed geology of the area has been described by Reyment, Short and Stauble [6,7]. Litho-stratigraphically, the rocks are divided into the oldest Akata Formation (Paleoceone), the Agbada Formation (Eocene) and the youngest Benin Formation (Miocene to Recent). The wells and boreholes tap water from the overlaying Benin Formation (Coastal Plain Sands) (Figure 2). This formation comprises of lacustrine and fluvial deposits whose thicknesses are variable but basically exceeds 1970 m. The Benin Formation has lithologies consisting of sands, silts, gravel and clayey intercalations.

The area is within the coastal zone. The coastal zone which comprises the beach ridges and mangrove swamps is underlain by an alternating sequence of sand and clay with a high frequency of occurrence of clay within 10 m below the ground surface (Figure 3). Because of the nearness of these compressible clays to the surface, the influence of imposed loads results to consolidation settlement. The impact of the imposed load is exacerbated by the thickness and consistency of the compressible layer. This, in addition to other intrinsic factors contributes to the failure of civil engineering structures [8,9]. For the purpose of generating relevant geotechnical data inputs for the design and construction of foundations for structures, it is imperative that the area be geo-technically characterized through sub-soil investigation. The interest of this study therefore is to identify, quantify and take into account the physical and mechanical characteristics of soils for a better estimate of the geotechnical risk (settlement and stability), in the area.

Methods of Investigation

Three (3) geotechnical boreholes with soil sampling and measurement of water table were investigated. The boreholes were advanced using a cable percussion boring rig. All three (3) boreholes were terminated at a depth of 30 m below the existing ground. The boreholes were drilled by the shell and auger cable percussive drilling method, using a hand rig. Detailed laboratory investigations were carried out on representative undisturbed and disturbed samples obtained from the boreholes for the classification tests and other tests (Figure 4). All tests were carried out in accordance with BS1377 [10] – Methods of test for soil for civil engineering purposes. Representative samples were taken at regular intervals of 0.1 m depth, and also when a change in soil type was observed. The samples were used for a detailed and systematic description of the soil in each stratum in terms of its visual and haptic properties and for laboratory analysis. In the cohesive soils, a large number of undisturbed samples were taken for examination and laboratory analysis [11]. Standard Penetration Tests (SPT) was carried out at regular intervals of depth in the granular sediments in order to assess their *in situ* densities. In this test, the number of blows required to drive the standard sampling spoon 300 m penetration after the initial sitting drive was recorded as the SPT (N) value.

*Corresponding author: Nwankwoala HO, Department of Geology, University of Port Harcourt, Nigeria, E-mail: nwankwoala_ho@yahoo.com

Figure 1: Map of Bayelsa state showing Igbogene.

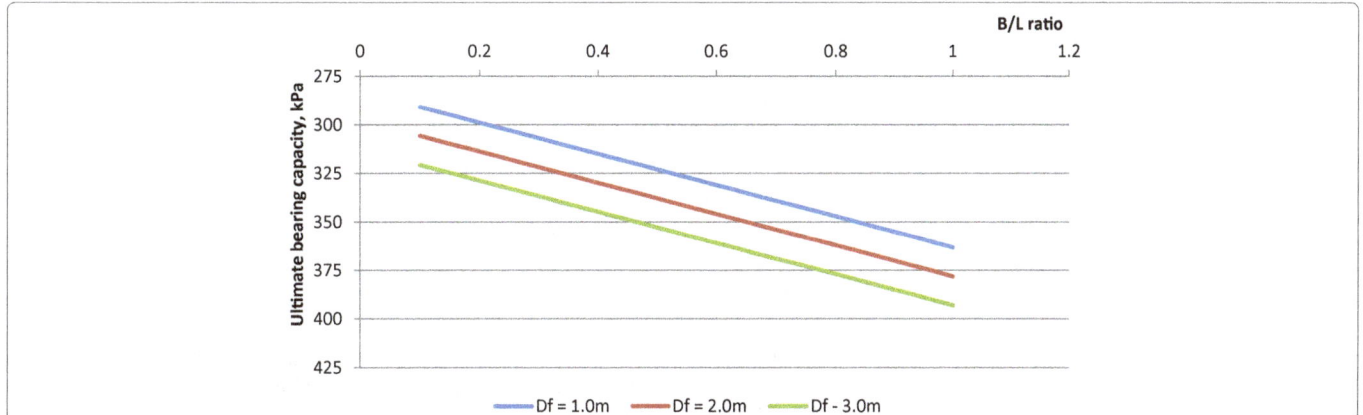

Figure 2: Chart of ultimate bearing capacity.

Atterberg consistency limit tests were carried out on the cohesive samples (Figure 5). The particle size distributions of a number of representative samples of the cohesion less soils were determined by sieve analysis [12]. Unconsolidated Undrained triaxial (UU) tests were performed on relatively undisturbed samples obtained from the boreholes. Laboratory consolidated tests were also carried out on relatively undisturbed samples with the aim of determining the compressibility properties of the soils [13].

Results and Discussion

Soil properties and stratigraphy

The lithology reveals a near surface clay layer about 6.0 m thick. The silty clay layer overlay a stratum of sand that extends to the final depth of investigation. The loose sand with occasional gravel increases in density to becoming medium dense sand as the borehole advances to the final depth. The results revealed that the samples are low to medium

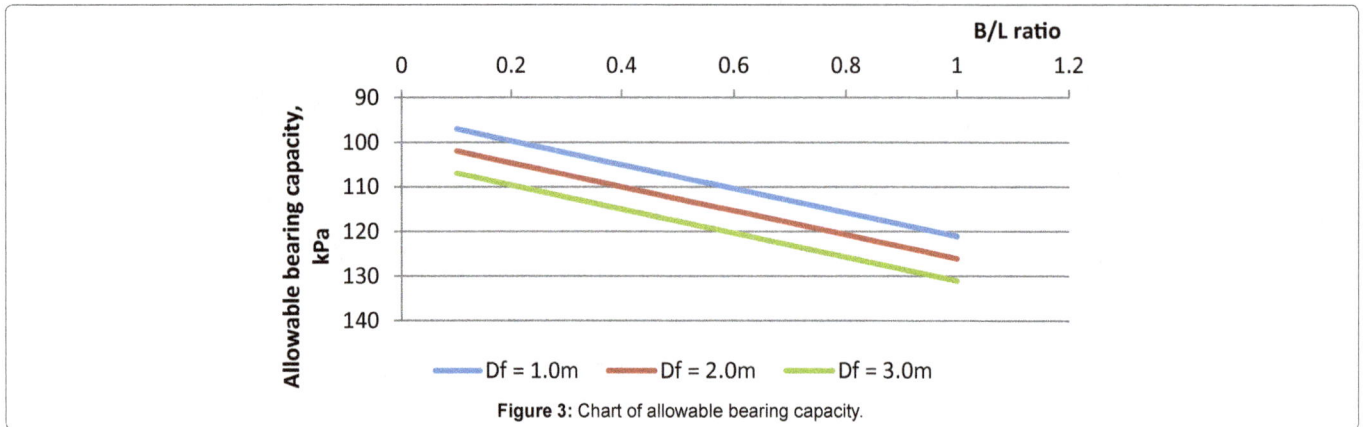

Figure 3: Chart of allowable bearing capacity.

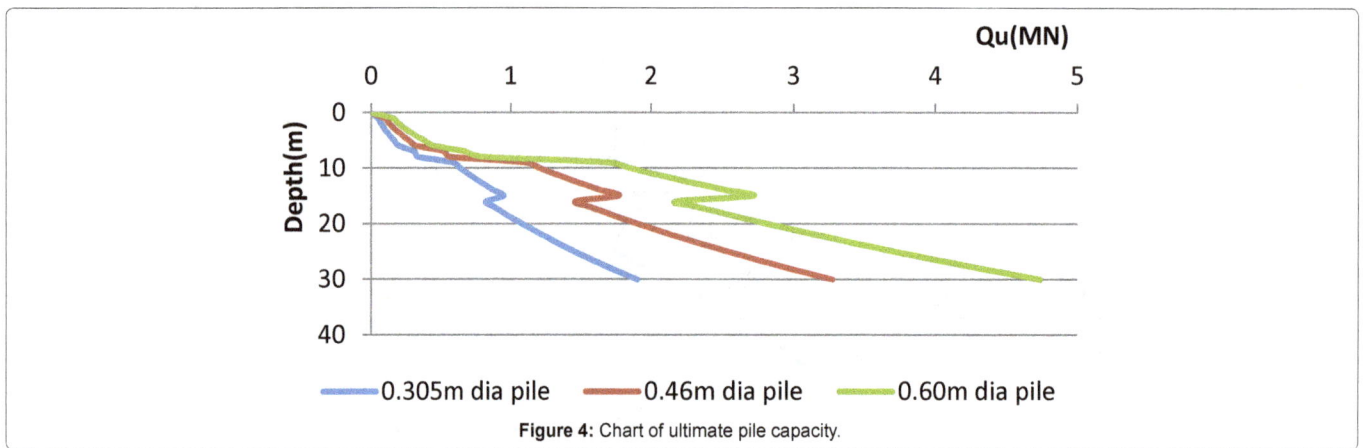

Figure 4: Chart of ultimate pile capacity.

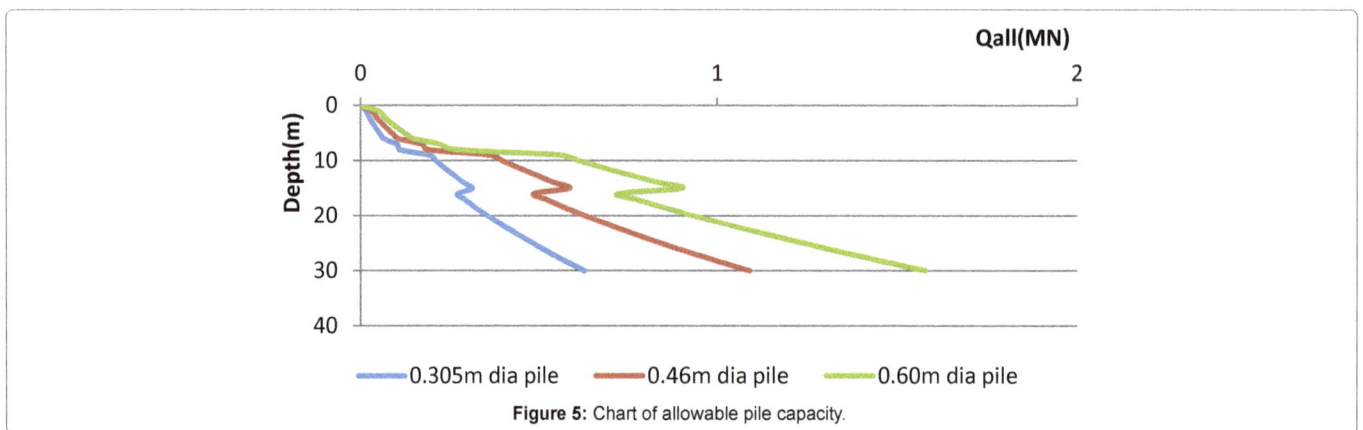

Figure 5: Chart of allowable pile capacity.

plasticity silty clay [14]. The results also disclosed that the samples are predominantly, fine to medium and medium sands. The plot of void ratio (e) against effective pressure (p) for the samples tested, are shown in Table 1, together with calculated values of the coefficients of consolidation (Cv) and of the coefficients of compressibility (Mv). Test results show that the samples are of moderately high compressibility and predominantly exhibiting negligible swelling potentials.

The near surface firm clay

The near surface soil encountered during the investigation is firm clay extending from the ground level to an average depth of 6.0 m below the ground surface [15]. This firm clay has embedded in it a thin layer of peat at a depth between 4.0 m and 5.0 m. The clay peaty

formation is characterized by high compressibility, moisture content and low undrained strength. The range of variations in the index and engineering parameters of this near surface soil are shown in Table 1.

Loose and medium dense sand

The sand encountered beneath the near surface silty clay soil increases in density with depth. Immediately below the clay, the strata present a loose sand formation. This loose sand extends to a depth of 8.0 m beneath the existing ground level. From this depth to the maximum depth of investigation, the strata presents a formation of dense sand loosening to becoming medium dense and continues in this density to the final depth of investigation. The ranges of variations of the geotechnical parameters are shown below.

Implications of foundation conditions

The investigation was carried out with the aim of determining the shear strength of the area for the design of the cellar slab for the rig positioning at the Gas Gathering Station at Igbogene in Bayelsa State. The scope for the investigation requires the determination of the relevant soil parameters for the design of the foundation. The near surface soil encountered during the investigation is firm clay extending from the ground level to an average depth of 6.0 m below the ground surface (Table 2). This clay has embedded within it a 1.0 m thick layer of peat between 4.0 m and 5.0 m. The clay and peaty formation is characterized by high compressibility, moisture content and low undrained strength. From consideration of the nature of the intended structure, the anticipated load and the very high compressibility of the peat layer embedded in the upper clay stratum, considerable settlement should be expected from the cellar slab as the rig comes on it. To avoid the unpredictable settlement characteristics of peat under imposed loads, raft foundation should be avoided on this location and pile foundation should be used. Using a safety factor of 3 on the ultimate bearing capacity, the chart for the allowable bearing capacity is as presented in Figure 2.

Safe bearing capacity, Qs

Table 3 shows the allowable, Q_{all} (kPa) and safe Q_s (kPa) bearing capacities for various foundation width, B (m) of raft footing at different foundation depth, D_f, m. The safe bearing capacity for the raft

Parameters	Min	Max	Mean
Natural moisture content (%)	23	40	34
Liquid limit (%)	43	52	48
Plastic limit (%)	16	33	26
Plasticity index (%)	15	27	21
Liquidity index	0.23	0.53	0.40
Consistency index	0.77	0.47	0.60
Bulk unit weight (kN/m³)	18.23	19.43	18.65
Dry unit weight (kN/m³)	14.65	15.44	14.94
Final void ratio	0.58	0.79	0.68
Final porosity (%)	46	48	48
Undrained strength (kPa)	40	56	47
Coefficient of consolidation m²/yr	2.76	7.46	4.5
Coeff. of compressibility, mv, m²/MN	0.20	0.65	0.42

Table 1: Geotechnical properties.

Parameters	Min	Max	Mean
Effective particle size, d_{10} (mm)	0.08	0.25	0.16
Mean particle size, d_{30} (mm)	0.15	0.45	0.29
Particle size, d_{60} (mm)	0.26	9.8	0.95
Coefficient of uniformity, $Cu=d_{60}/d_{10}$	2.34	41.52	5.56
Coefficient of curvature, $Cc=d_{30}^2/d_{10}.d_{60}$	0.04	1.53	1.07

Table 2: Subsoil properties.

Foundation depth, D_f (m)	Allowable, Q_{all} (kPa) and Safe Bearing Capacity, Q_s (kPa), for Various Width, B(m) of Raft Foundation					
	2 m		5 m		10 m	
	Q_{all}	Q_s	Q_{all}	Q_s	Q_{all}	Q_s
1	97	85	100	50	107	45
2	102	90	105	60	112	55
3	107	105	110	75	117	70

Table 3: Allowable bearing capacities for various foundations.

Soil type	δ	Skin friction, f (kPa)	N_q	Unit End Bearing, q(kPa)
Very loose sand	10	48	8	1900
Loose sand	15	67	12	2900
Medium dense sand	20	81	20	4800
Dense sand	25	96	40	9600
Very dense sand	30	115	50	12000

Table 4: Design parameters for cohesion less soil, according to American Petroleum Institute (API).

Pile depth (m)	Safe Pile Capacity, kPa, for Various Pile diameter		
	305 mm (12") Pile	460 mm (18") Pile	610 mm (24") Pile
10	213	401	618
15	317	591	904
20	361	636	937
25	487	847	1238
30	635	1091	1580

Table 5: Allowable bearing capacity and specific pile diameter for specific depth.

foundation is limited by a maximum settlement value of 50 mm (Tables 4 and 5).

Conclusion

The investigation was carried out by means of three (3) number boreholes to a maximum depth of 30 m. The boreholes were carried out using the cable percussive rig. Field and laboratory investigations reveal a near surface stratigraphy of clay to an average depth of 6 m underlain by loose silty sand to a depth of 10 m below the existing ground level. The laboratory analysis carried out on relatively undisturbed soil samples of the clay gave undrained shear strength between 40 and 56 kPa with a mean value of 47 kPa. However, the 1.0 m thick peat embedded in the clay between 3.0 and 4.0 m depth will greatly increase the compressibility of the clay. Considering the nature of the intended structure, the anticipated load and the very compressibility of the peat, pile foundation is recommended to take the imposed load from the cellar to the underlying sand stratum. All piles are engaged should be driven and pile load test carried out on them to confirm the working load and the estimated settlement. All piles employed should be driven piles. However, pile load test should be carried out on all driven piles to confirm working load and the estimated settlement.

References

1. Adebisi NO, Oloruntola MO (2006) Geophysical and geotechnical evaluation of foundation conditions of a site in Ago-Iwoye area, Southeastern Nigeria. J Min Geol 42: 79-84.

2. Haddou MB, Essahlaoui A, Boujlal M, Elouali A, Hmaidi A (2013) Study of the geotechnical parameters of the different soils by correlation analysis and statistics in the Kenitra Region of Morocco. J Ear Sci Geotech Eng 3: 51-60.

3. Nwankwoala HO, Amadi AN (2013) Geotechnical Investigation of Sub-soil and Rock Characteristics in parts of Shiroro-Muya-Chanchaga Area of Niger State, Nigeria. Int J Ear Sci Eng 6: 8-17.

4. Oke SA, Amadi AN (2008) An assessment of the geotechnical properties of the sub-soil of parts of Federal University of Technology, Minna, Gidan Kwano Campus, for foundation design and construction. J Sci Edu Tech 1: 87-102.

5. Oke SA, Okeke OE, Amadi AN, Onoduku US (2009) Geotechnical properties of the sub-soil for designing shallow foundation in some selected parts of Chanchaga area, Minna, Nigeria. J Sci Edu Tech 1: 103-110.

6. Reyment RA (1965) Aspects of Geology of Nigeria. University of Ibadan Press, Nigeria. p133.

7. Short KC, Stauble AJ (1967) Outline of Geology of the Niger Delta. AAPG Bull 51: 761-779.

8. Youdeowei PO, Nwankwoala HO (2013) Suitability of soils as bearing media at a freshwater swamp terrain in the Niger Delta. J Geol Min Res 5: 58-64.

9. Amadi AN, Eze CJ, Igwe CO, Okunlola IA, Okoye NO (2012) Architect's and geologist's view on the causes of building failures in Nigeria. Mod Appl Sci 6: 31-38.

10. BS 1377-2 (1990) British standard methods of test for soils for civil engineering purposes. Published by the British Standards Institution p.8-200.

11. Etu-Efeotor JO, Akpokodje EG (1990) Aquifer systems of the Niger Delta. J Min Geol 26: 279-284.

12. Meyerhof GG (1951) The Ultimate Bearing Capacity of Foundations. Geotech 2: 532-539.

13. Ngah SA, Nwankwoala HO (2013) Evaluation of Geotechnical Properties of the Sub-soil for Shallow Foundation Design in Onne, Rivers State, Nigeria. J Eng Sci 2: 08-16.

14. Nwankwoala HO, Warmate T (2014) Geotechnical Assessment of Foundation Conditions of a Site in Ubima, Ikwerre Local Government Area, Rivers State, Nigeria. Int J Eng Res Develop 9: 50-63.

15. Oghenero AE, Akpokodje EG, Tse AC (2014) Geotechnical Properties of Subsurface Soils in Warri, Western Niger Delta, Nigeria. J Ear Sci Geotech Eng 4: 89-102.

19

An Overview of CBM Resources in Lower Indus Basin, Sindh, Pakistan

Nazeer A[1*], Habib Shah S[1], Abbasi SA[2], Solangi SH[2] and Ahmad N[3]

[1]*Directorate of Asset Operations, Pakistan Petroleum Limited (PPL), Islamabad, Pakistan*
[2]*Center of Pure and Applied Geology, University of Sindh, Jamshoro, Sindh, Pakistan*
[3]*Ex GM Exploration, Pakistan Petroleum Limited (PPL), Islamabad, Pakistan*

Abstract

Pakistan is energy deficient and underdeveloped country but it contains wide resources of low quality coal. The contemporaneous models for Coal Bed Methane (CBM) in low-rank coals have changed dramatically in recent years due to the growth of commercial CBM activity in the Powder River Basin (PRB). The CBM models are still evolving because the CBM evaluation concepts are on steep learning curve based on proven and tested commercial activities. Coal is an unusual lithology in that it is both an excellent source and reservoir rock. CBM resource has also been found in commercial quantities in the Cambay Basin of India. The CBM resource of Cambay Basin and Powder River Basin (PRB) are similar in age and rank to most of Pakistan's coal. The success in the above mentioned basins provoked geoscientists in Pakistan to re-look into Sindh's CBM resource.

Thar coal is considered as the largest reserves of low ranking coal in Pakistan. Preliminary geological investigation was carried out; results show that low ranked coal seams of class Lignite B to High Volatile B bituminous coal exists in Sindh. The rank specified above is better in quality from Powder River Coal Deposits, so it warrants further evaluation to firm up further exploration and subsequent exploitation. Interactive wireline correlations between several wells have been carried out explicitly. Results show that isolated coal seams of Bara Member (Paleocene) and Sonari Member of Laki Formation (Eocene) exists a few kilometers in sub-surface. The thickness of coal seams is thickest in Thar area with better prospect for gas adsorption capacity. The dedicated CBM studies also reveal that the bituminous coal exists in Badin, Sonda, Thatta and Jherruck areas. Depositional Model of Thar coal deposit has been prepared using plate reconstruction.

Keywords: Coalbed methane; Lower Indus basin; Sindh; Pakistan

Introduction

CBM is found as trapped and adsorbed on the surface of coal seams found in the subsurface. It is used as fuel gas and is mainly composed of methane (CH_4) as its name suggests. Coalbed methane (CBM) is also known as coal seam gas (CSG) in Australia suggesting that there is no unequivocal terminology for gas extracted from coals [1]. Porosity and permeability in coals exists due to the presence of matrix porosity (micropores) and natural fractures known as cleats [2]. According to Ahmad et al. [3], 10% of the world's coal deposits are present in Sindh, Pakistan, it may be considered promising for CBM potential of the country.

Coal is located in the southern, mid-western and northern parts of Pakistan (Figure 1). Based on geological mapping, coal resources in the south (Sindh) are approximately 185 billion tones and are mostly lignitic to sub-bituminous, while the estimated resources in the north (Potwar Plateau and Salt Range) are 235 MM tones and are Lignitic to Bituminous. Analogy for the coals in the south is Powder River basin (Wyoming and Montana) where extensive CBM operations and production has been taking place.

This study has taken a solid step forward by initiating by evaluating CBM Resources on the basis of available Coal properties. This study shows that coals in the North are much more favorable for CBM due to their maturity and sub-bituminous to bituminous nature.

Previous work

Pakistan is endowed with large reserves of coal deposits ranging from lignite to high volatile bituminous. The 7th largest lignitic coal reserves (Figure 2); need to be exploited for provincial as well as national development. Coal assessment was first initiated by US Geological Survey with the financial help of of US Aid. The largest part of measured deposits is associated with Southern Sindh Monocline [4]. Figures 1 and

2 are explicitly depicting coal deposits of Sindh monocline. Ahmad et al. [3] presented Badin and Thar Coal as potential candidates for CBM and compared coal composition, seam thickness and confined aquifer within and below coal zone with its analog on Indian side. Ahmad et al. [3] proposed strong cooperation among industry and research and development (R & D) organizations/universities in sharing technical information and experience and recommend government support. Siddiqui et al. [2] and Siddiqui et al. [5] presented the results of detailed Scanning Electron Microscopic (SEM) study carried out on coal samples from various locations of Thar and Lakhra coalfields of Sindh to evaluate the porosity and permeability patterns. The study suggested that the Thar and Lakhra coals may have best permeability for the storage of CBM provided other geological factors required for the CBM generation are favorable. Coal deposits in Sindh are present nearby gas pipeline infrastructure (Figure 3). Biggest challenge is to appraise and produce gas at economic rates.

Coal Reserves of Pakistan

Malkani [4] presented the details of coal reserves of Pakistan along with coal properties based on various projects carried out by Geological

*Corresponding author: Nazeer A, Directorate of Asset Operations, Pakistan Petroleum Limited (PPL), Islamabad, Pakistan
E-mail: a_ nazeer@ppl.com.pk; adeelnazeer@gmail.com

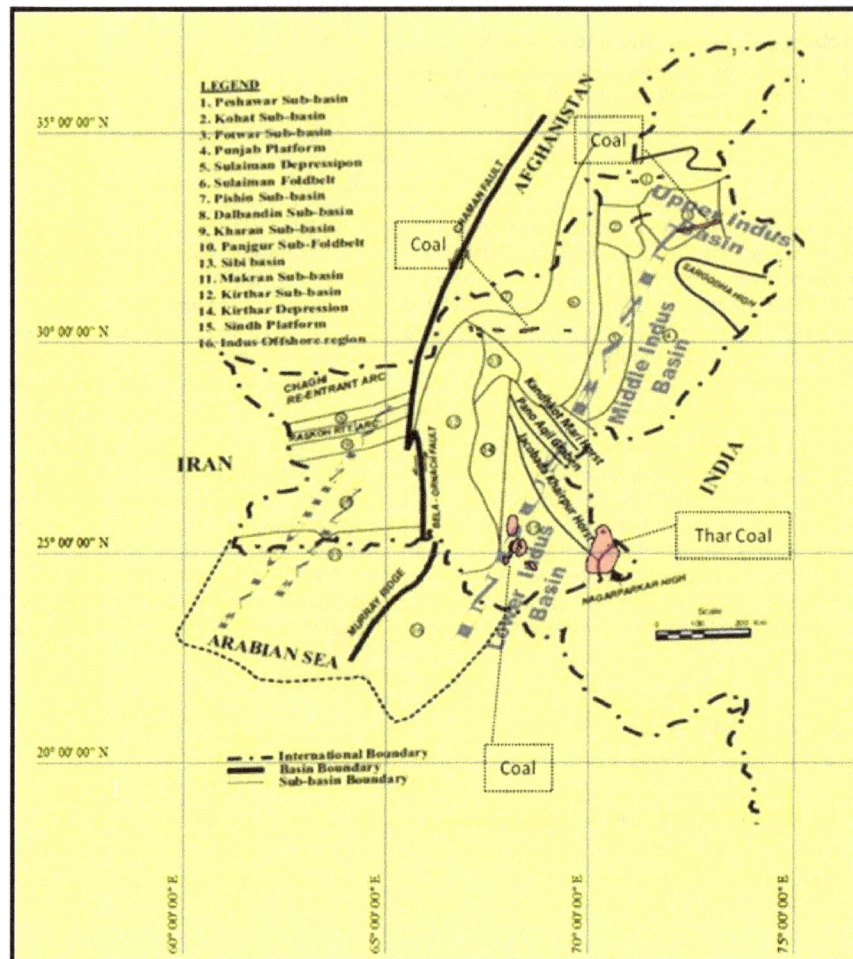

Figure 1: Figure is showing coal deposits in different basin of Pakistan [14].

Survey of Pakistan (GSP). CBM potential is evaluated on the basis of data used by Malkani [4]. Operational coal mines aren't depicting the true coal potential of the country (Figure 4). However, coal reserves measured on the basis of geological mapping adds valuable addition in total reserves. Therefore total reserves of coal are 186 billion tones in Pakistan, out of which most coal is associated with Sindh Monocline. This paper also emphasizes on assessment of CBM resource based on various properties of coal. The coal resources of the major administrative units of Pakistan [4] are as under,

- Tertiary coal is being exploited from Early Paleocene, Hangu Formation (in Makerwal and Surghar areas) and Late Paleocene, Patala Formation (Central and Eastern Salt Range) in Punjab province. Total reserves of Punjab Coal are about 235 million tons [4].

- Working coal mines in Sindh are Lakhra and Meting-Jhimpir coalfields whilst non-developed coalfields are Sonda-Thatta, Jherruck, Ongar, Indus East, Badin and Thar coalfields with total reserves of about 185,457 million tones [4]. Coal seams are associated with Paleogene rocks of Sindh Monocline.

- Working coal mines in KPK are Hangu/Orakzai, Cherat, Dara Adamkhel and Gulakhel coalfields. Non-developed coal field in the same region is the Shirani coalfield with total reserves of about 122.99 million tones [4].

- The coal deposits of Baluchistan are situated between, Quetta-Duki, it is comprised of commercial deposits over a large area, roughly coincident with the eastern flank of the Sanjawi Arch, and a narrow elongate area between Quetta and Johan. Coal bearing host rock in Baluchistan province is Toi Formation (Eocene). The Eocene coal fields are, Khost-Shahrig- Harnai, Johan,Ghazoe, Ghar, Dewan, Narwel-Dab and Kingri. In the Sor Range near Quetta, the Toi Formation consists of calcareous sandstone, conglomerate, calcareous claystone and carbonaceous shale with commercially exploitable coal seams. Cretaceous or Cretaceous-Tertiary (K-T) coal is present in Maastrichtian age. Total reserves of Baluchistan Coal are about 458.5 million tones.

- Total reserves of Azad Kashmir and Kotli coalfields are about 8.72 million tones (Table 1).

Cleats in coal deposits of Lower Indus basin (Lib)

Coal contains dual porosity; it contains micropores (matrix) and network of natural fractures, also known as cleats [6].

Analogy of coal fields in lower Indus basin and Powder River basin

The Powder River Basin (PRB) is located in northeastern Wyoming

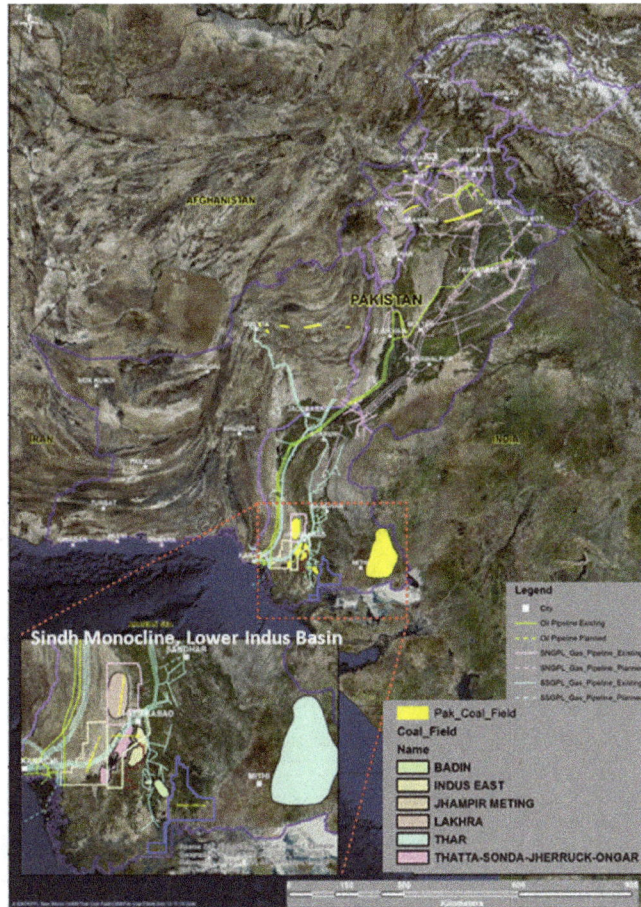

Figure 2: Figure is showing coal deposits in Sindh Monocline, Lower Indus basin and route of national gas pipeline infrastructure.

and southern Montana. It covers an area of approximately 66,822.69 square miles with majority of the potentially productive coal zones ranging from about 137 m to over 1981 m below ground surface. Gas Reserves range from 7 to 12 TCF (PPL in-house evaluation). In 2002, wells in the Powder River Basin produced about 823 million cubic feet (Mcf) per day of coalbed methane [7]. Most of the coal in the Powder River Basin is subbituminous in rank, which is indicative of a low level of maturity. Some lignite, lower in rank, has also been identified. The thermal content of the coals found in the Powder River Basin is typically 8,300 British thermal units per pound. Coal is found in the Paleocene, Fort Union Formation and Eocene, Wasatch Formation of Powder River Basin. Most of the Coalbeds in the Wasatch Formation are continuous and thin (six feet or less) although, some localized thicker deposits have also been found.

In Sindh, coal was first discovered from Lakhra in 1853. Most of Sindh's coal fields are associated with Bara Member of Ranikot Group (Paleocene). However, in some parts, Sohnari Member of Laki Group (Eocene) hosts coal. The thickest coal beds are present in Thar coal field, where maximum thickness is 22.8 m. The average thickness of coal seams range from 0.3 to 2.5 m.

Thar coal field is considered as the 7th largest lignite deposits of world. Thar coal isn't exclusively Lignite rather it ranks from Lignite to Sub-bituminous-A with heating values (BTU) ranging from 6,244-11,045 BTU/lb, which is higher than Powder River basin.

The previous Geological and Geophysical (G and G) work is very limited and the true thickness of the seam can't be concluded from these studies. Preliminary evaluation shows that coal seams distribution is in the form of isolated bodies with chances of thick coal seams in area other than Thar Coal Field. A 6 m thick coal seam is reported in Jheruck coal field and heating value of the seam is 8,800-12,846 BTU/lb which is also higher than Powder River Basin coal (PRB). A brief comparison between coal fields (characteristic) of Lower Indus Basin (LIB), Sindh, Pakistan and Powder River Basin, USA is given in Table 2 and discussed rank wise in Figure 5.

CBM resources in Barmer Basin (India)

The Western Rajasthan Shelf which forms an integral part of Indus Basin, and is composed of three main sub basins which are separated from each other by basement ridges/faults. These Sub-basins are:

1. Jaisalmer Sub-basin

2. Bikaner- Nagaur Sub-basin

3. Barmer-Sanchor Sub-basin

Barmer Basin [8] was considered as having extension of coal of Sindh and CBM has been exploited CBM in North Gujrat, nearby Pakistan Border (Figure 6). CBM Exploration in the acreage falling in Barmer-Sanchor Basin, covering an area of about 790 sq. Km in Banaskantha district of North Gujarat was taken up by ONGC as operator [9]. In this area, coal seams are confined to the Middle Tharad

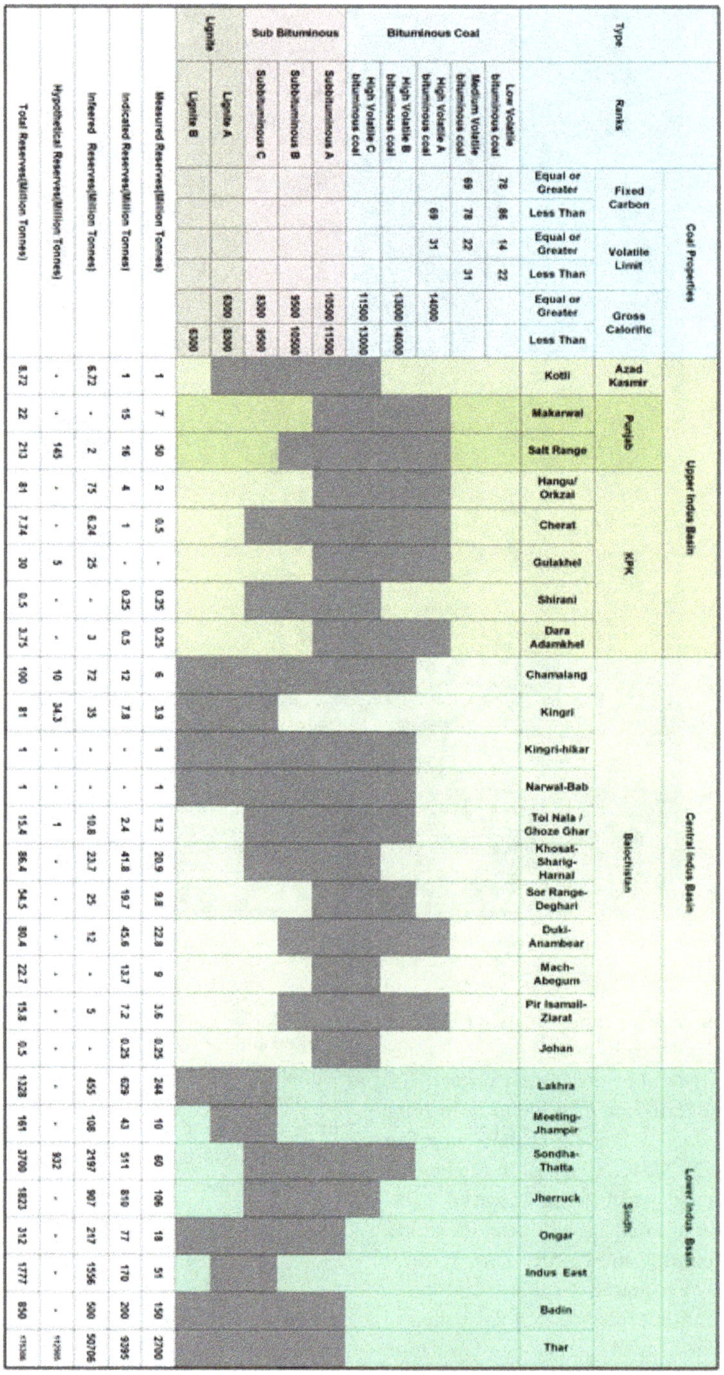

Figure 3: Figure is showing coal characteristic of Pakistan [4].

Formation of Middle Eocene age (Figure 7). The coal ranks as Lignite and is similar to lignite of Sindh. The Thard Formaton (Lower to Middle Eocene) is correlatble with Laki Group on the basis of overall lithology and its lower part may considered as extension of Sohnari Member with coal bearing facies in Pakistan but thorough research is needed to firm it up. The initial syn-rift basin fill sediment is of Paleocene age and is equivalent to Syn–rift basin fill sediment of Paleogene in Sindh Monocline.

According to Rao et al. [10], lab analysis data showed that Barmer Basin coals are high in moisture (9-23%), very low ash (2-9%), content with high volatile matter (30-55%) with low calorific value and were assigned to Lignite rank. Low Vitrinite reflectance (0.28 to 0.4) values were observed, indicating very low thermal maturity. The test wells in the study area did not produce coalbed gas, even after dewatering for one year, showing little accumulation of gas in the drilled wells. This is due to the low rank and low thermal maturity, lack of cleat and fracture, high degree of under saturation, great difference between reservoir pressures and critical desorption pressures. Whereas, Sindh coal is older than Barmer coal, which suggests that it got more time for maturation from Lignite to Bituminous, at places. Interestingly, cleats are present in Sindh coal contrary to Barmer coal, which was the main reason of deterioration of reservoir quality in Barmer coal.

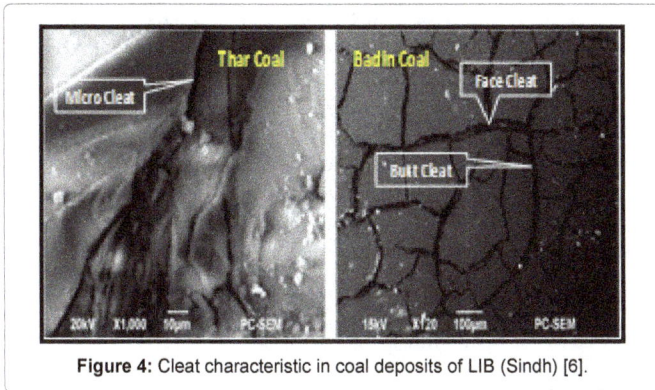

Figure 4: Cleat characteristic in coal deposits of LIB (Sindh) [6].

Coal bearing facies in paleogene of lower Indus Basin (Sindh) Pakistan			
Early Eocene	Laki Group	Laki limestone	Dominantly limestone
		Meeting shale	Claystone, sand with bed of limestone and sand
		Meeting limestone	Dominantly limestone
		Sohnari formation	Varicolored lateritic clay and shale with locally coal seams
Late Paleocene	Ranikot Group	Lakra formation	Limestone, sandstone, shale interbeds
Middle Paleocene		Bara formation	Claystone, shale, Early Eocene siltstone, sandstone, coal, carbonaceous
Early Paleocene		Khadro formation	Sandstone, shale

Table 1: Coal bearing facies in lower Indus Basin (Sindh).

Composition	Powder River Basin		Lower Indus Basin (Sindh)		
	Illinois Bituminous	Wyoming Sub-bituminous	Lignite	Sub-bituminous	Bituminous
Heating value (Btu/lb)	12,770	8,683	5,219	11	13,555
Ash (%, dry)	17.4	6.6	0.4-4.9		
Moisture (%)	3.1	23.8	9-55.5		
Sulfur (%, dry)	4.2	0.4	0.4-15		
CaO (%)	2.7	28.3	NA		
MgO (%)	1	4.5	NA		

Table 2: Analogy between lower Indus Basin, Sindh, Pakistan and Powder River Basin (PRB), USA (Internal reports of PPL).

Coal bearing facies in wells drilled in Sindh Monocline

Subsurface data of multiple wells shows that the coal bearing facies are mostly associated with Paleocene (Bara Formation) and Eocene (Sohnari Member of Laki Formation) in the surrounding of the Meeting-Jhimpir and Sonda-Thatta- Jherruck-Ongar areas. Figure 8 is showing thickness of Paleocene coal seams drilled in the petroleum wells situated in the surrounding of Paleocene coal fields of Sindh. Similarly Figure 9 is showing thickness of Eocene coal seams drilled in the wells present in the surrounding of Eocene coal fields (Sindh). Average thickness of Eocene coal is 1-3 m.

Quality of Sindh coal

Qualitative analysis of coal bearing facies in Sindh province has been done on the basis of heating values of coal. The data used to arrive on the conclusions was used from the published report of Malkani

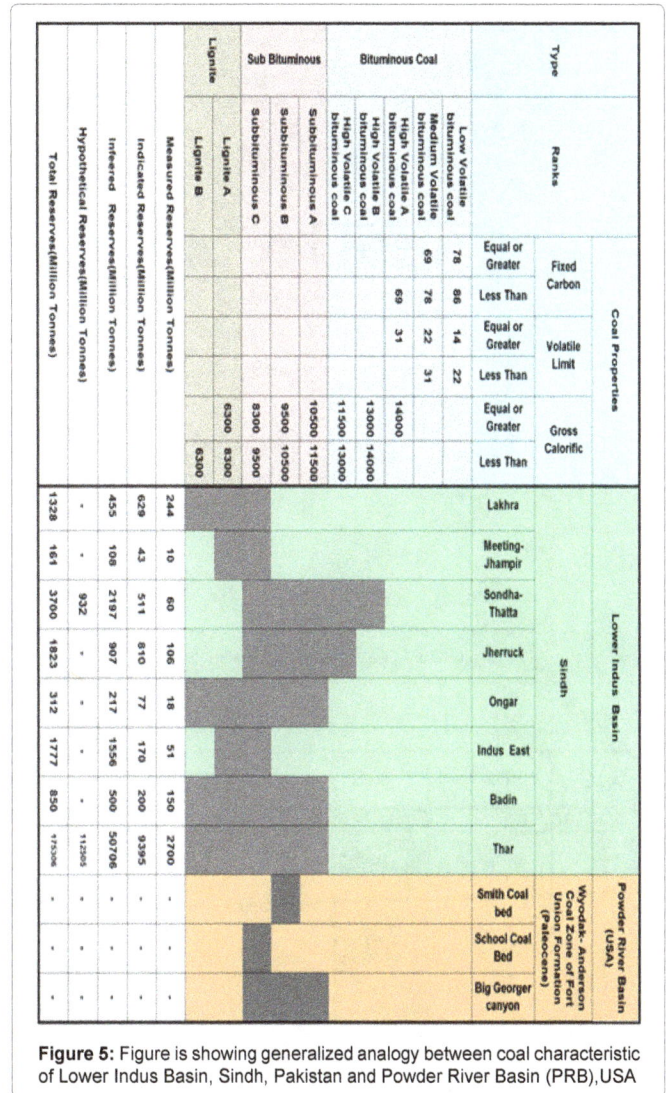

Figure 5: Figure is showing generalized analogy between coal characteristic of Lower Indus Basin, Sindh, Pakistan and Powder River Basin (PRB), USA

[4] of Geological Survey of Pakistan (GSP). Lakhra, Meeting-Jhampir and Sonda East are low quality coals, because their rank is restricted to Lignite to subbituminous C only. Best quality of coal is situated in Sonda-Thatta-Jherruck-Ongar areas. Eocene Coal is also present in the surrounding of Sonda-Thatta- Jherruck –Ongar areas. Figure 10 is showing quality of coal seams of Paleocene/Eocene coal fields of Sindh, on the basis of heating values. This heating value is directly proportional to Methane in coal.

Lakhra and Sodha East contain low ranks of coal, of varying quality. The coal is not lignite throughout, but also matures at places up to Sub Bituminous C, which increases the tendency for adsorption of gas. Similarly, Thar Coal is not only limited to lignite rank throughout, it is also mature up to Bituminous-A rank. Results show that the Thar Coal is the best candidate for the exploration of CBM project as pilot project because (i) Heating value is higher than Powder River Basin and Barmer Basin Coal. (ii) Thickest Coal Seam is present in Thar. (iii) Thar Coal should be extended to deep in basin as deepest is considered, as the best part for exploitation of CBM gas. (iv) Micropores or intrinsic pore spaces within the coal matrix stores the methane gas in Thar Coal [6].

Ash quality of coal of various regions of Pakistan is also analyzed in this paper. Sonda-Thatta-Jherruck – Ongar coals produce highest

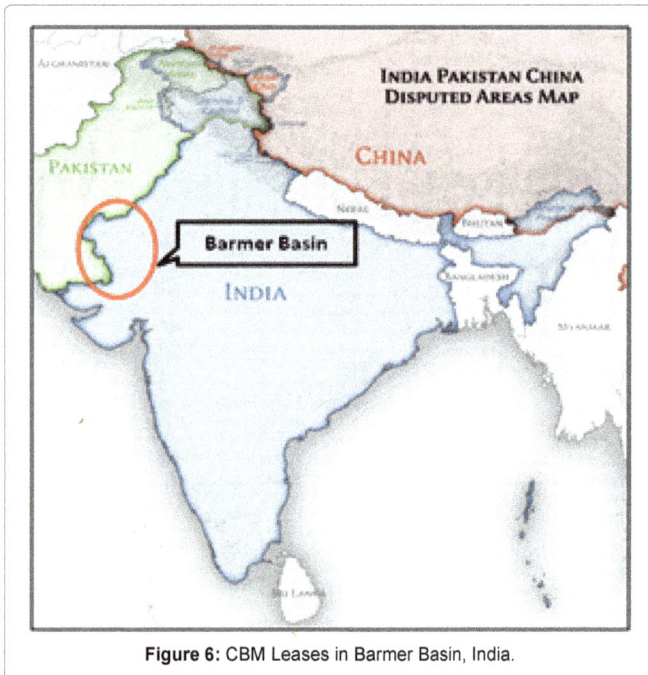

Figure 6: CBM Leases in Barmer Basin, India.

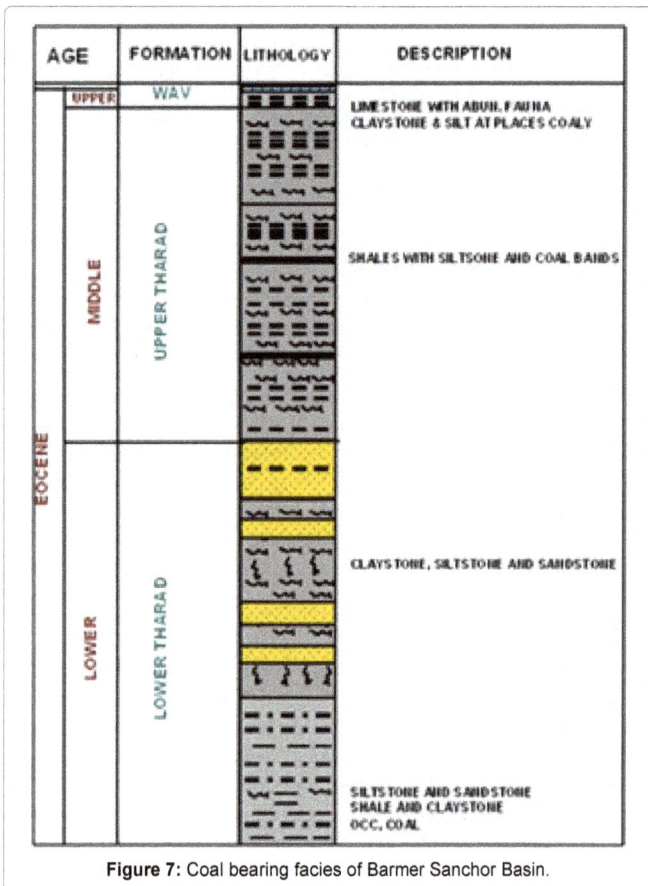

Figure 7: Coal bearing facies of Barmer Sanchor Basin.

proportion of ash as compared to other region's coals. Figure 11 is showing distribution of ash in coal seams of Paleocene/Eocene coal fields (Sindh). Additionally, distribution of sulfur and moisture was also considered and the Figures 12 and 13 are showing distribution of Sulfur

and moisture in coal seams of Paleocene/Eocene coal fields (Sindh) respectively. Results show that moisture content is comparatively less in Sonda.

Facies study on wireline logs

The depositional model shows that Bara Formation consists of sediments of deltaic environment. The predominant lithology is sand with intercalations of clay and shale (Figure 14). The formation can be divided into two parts,

• Interbeds of sand and shale

• Mostly shaly part with sand at the base

Figure 15 is showing coal bearing facies in the mudlog as well as wireline of Meeting-01 well. The average thickness of coal is 1 m; it can be clearly seen on the mud log and wireline log. The Coal seams as interpreted on wireline log shows that the coals are well isolated seams. Whereas Figure 14 is showing stratigraphic section of coal bearing facies of Laki Group (Sohnari Member).

Eocene coal is present in the SSW of Meeting-Sonda-Thatta-Jherruck Fields. Meeting X-1 is only well in this project which contains both Paleocene and Eocene Coals. Figure 16 is showing wireline correlation of Coal facies in Sonhari Formation (Eocene), Lower Indus Basin, Sindh, Pakistan. Correlations show that isolated coal seams are present and probably deposited along the fringes of river dominant deltaic deposits. Figure 17 is showing Wireline log correlation of Coal Facies in Bara Formation (Paleocene), Lower Indus Basin, Sindh, Pakistan. Three types of facies are identified in wireline correlation.

Depositional model

According to Malkani [4], Thar coalfield rests on Pre-Cambrian shield rocks and is covered by sand dunes. The coal thickness varies from 0.20-22.81 m. There are maximum 20 coal seams in the area. The most common depth for coal seams is 150-203 m. The thickness of overburden varies from 114-245 m above the top coal seam. Malkani [4] considered Reed as origin of Thar Coal which is commonly used for grass like plant in wetland. The thick coal seam with low ash and sulfur indicate stable upper delta environments [7]. Kumar [11] presented origin of organic matter as mixture of pollens, spores, algae, cuticles, fungi and wood. Kumar's [11] work suggests that palynological assemblages recommend a warm and exceptionally humid climate.

According to Hakkro [12], Bara Formation (Middle Paleocene) consist of fine-grained sandstone and subordinate amount of shale, carbonaceous shale, siltstone and coal beds of variable thicknesses in boreholes of Thar coal field and investigation of major elements (Si, Ti, Al, Na, Ca, K, Mn, Mg, Fe+3,). The major elemental ratios and their correlation coefficient (r2) show that the origin of its constituting silica content is detrital, which is further confirmed by differences in the source of silica and alumina [13]. Therefore, the studied sediments have potentially deposited along the fringe of basin of deposition in deltaic to near shore depositional environment. They assumed that sedimentation took place under humid climatic conditions, with relatively fast rate of sedimentation, showing better conditions for the growth, accumulation and preservation of organic source material of its coal. They proposed that sediments are potentially derived from igneous and metamorphic rock.

In this paper, Plate Reconstruction was carried out using GPlate software, results show that the greater Indian peninsula was approaching towards line of tropical humid environment (Figure 18) at the time of Paleogene. At that time, Indian Shield element was probably exposed

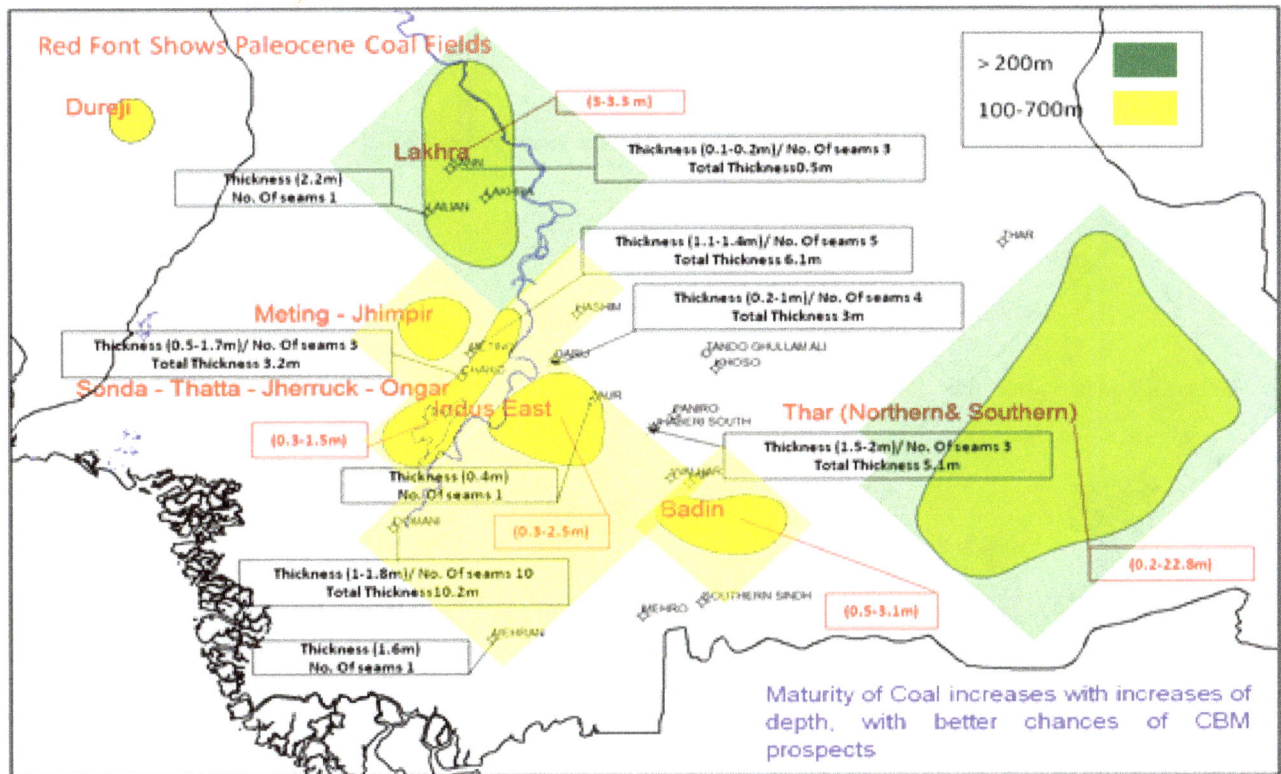

Figure 8: Thickness of Paleocene coal seams drilled in the wells present in the surrounding of Paleocene coal fields (Sindh).

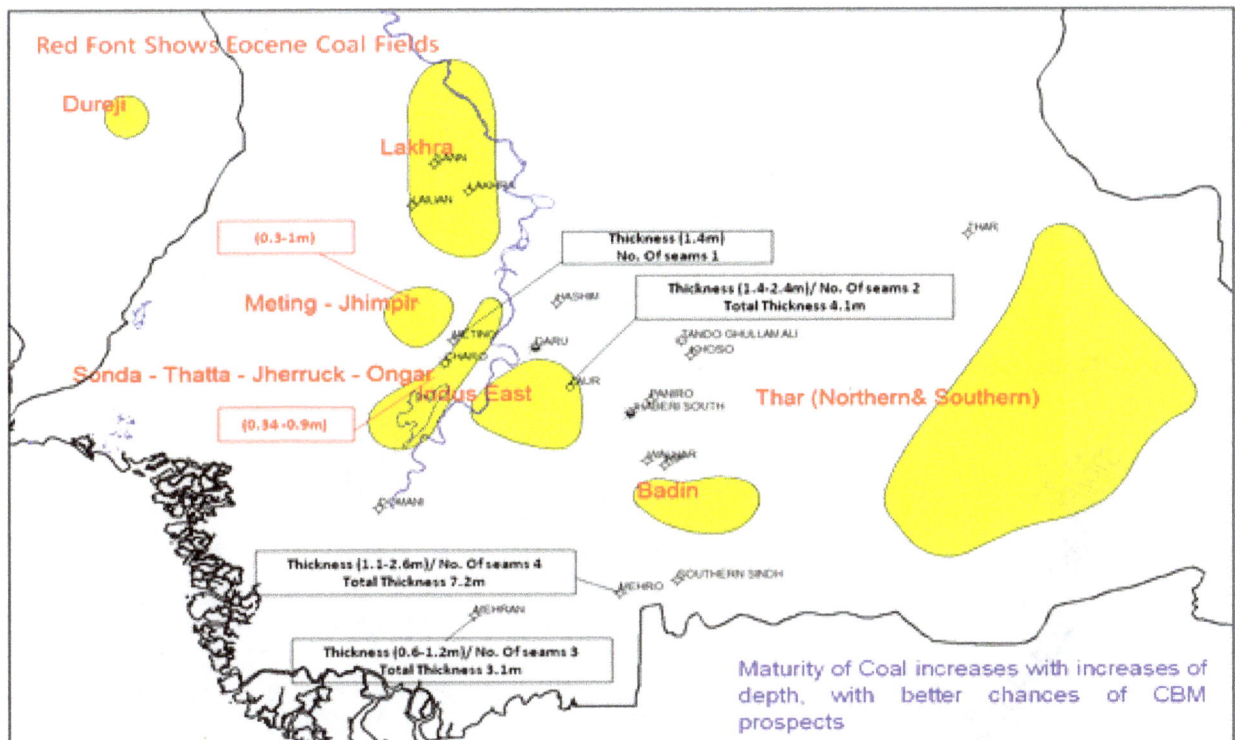

Figure 9: Thickness of Eocene coal seams drilled in the wells present in the surrounding of Eocene coal fields (Sindh).

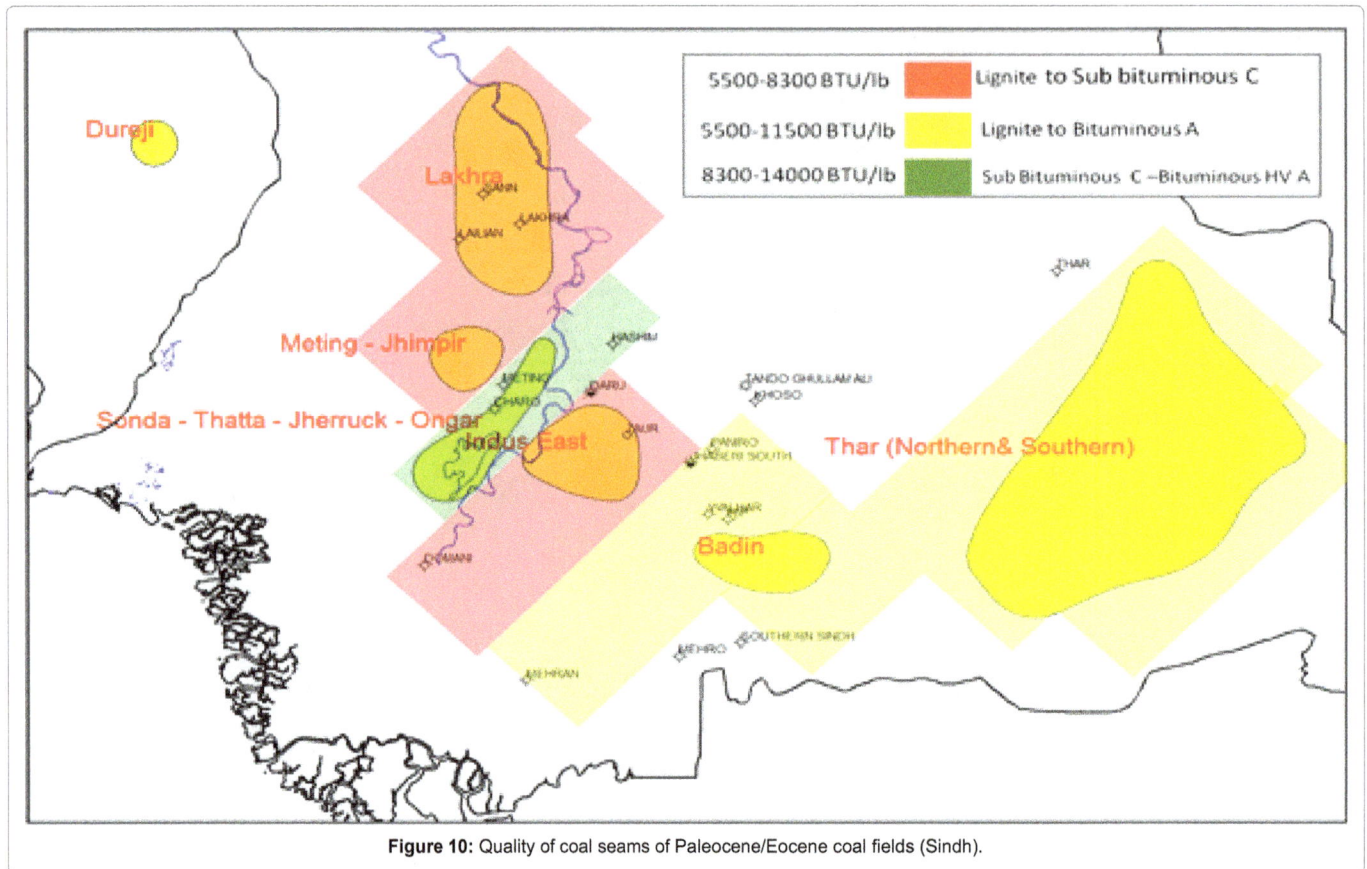

Figure 10: Quality of coal seams of Paleocene/Eocene coal fields (Sindh).

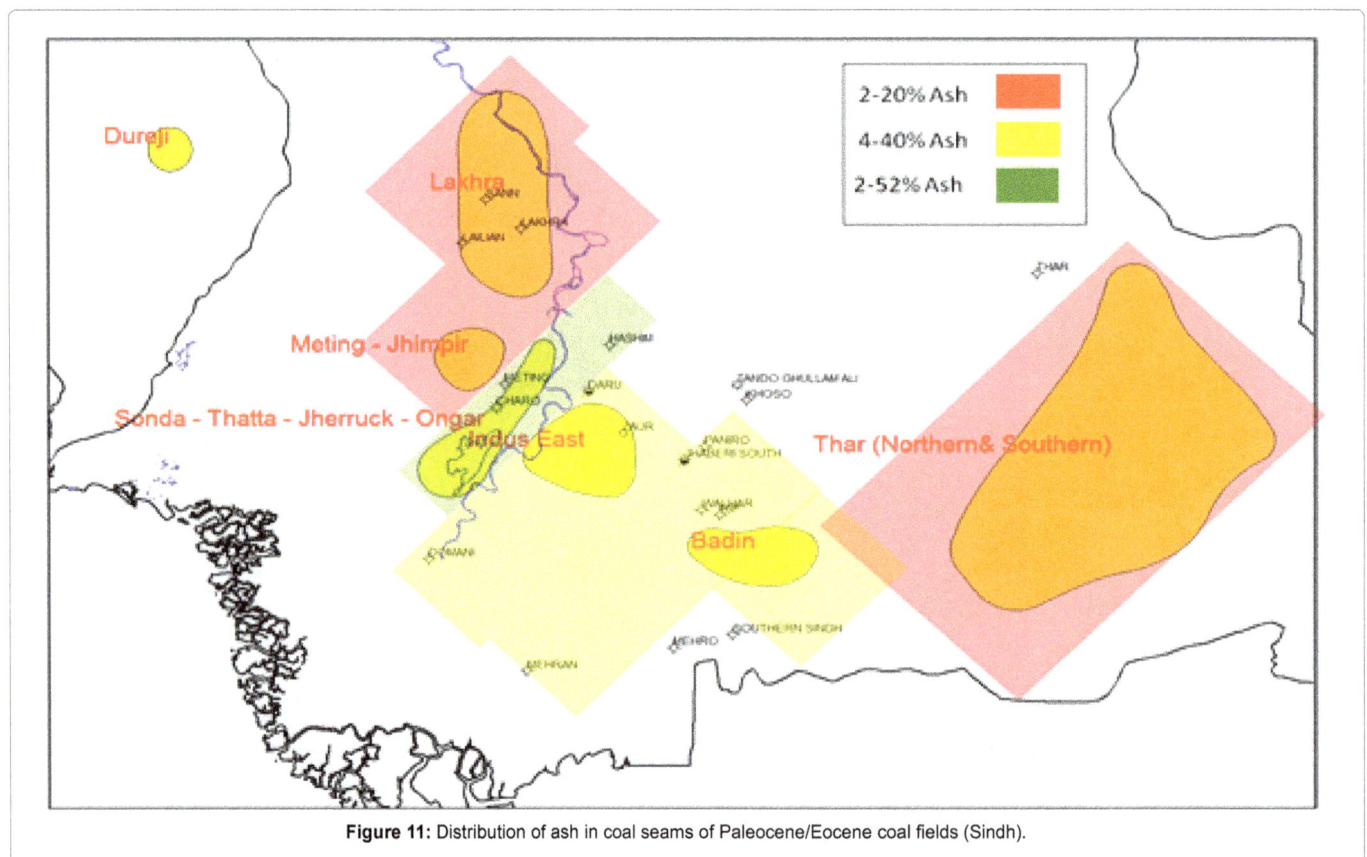

Figure 11: Distribution of ash in coal seams of Paleocene/Eocene coal fields (Sindh).

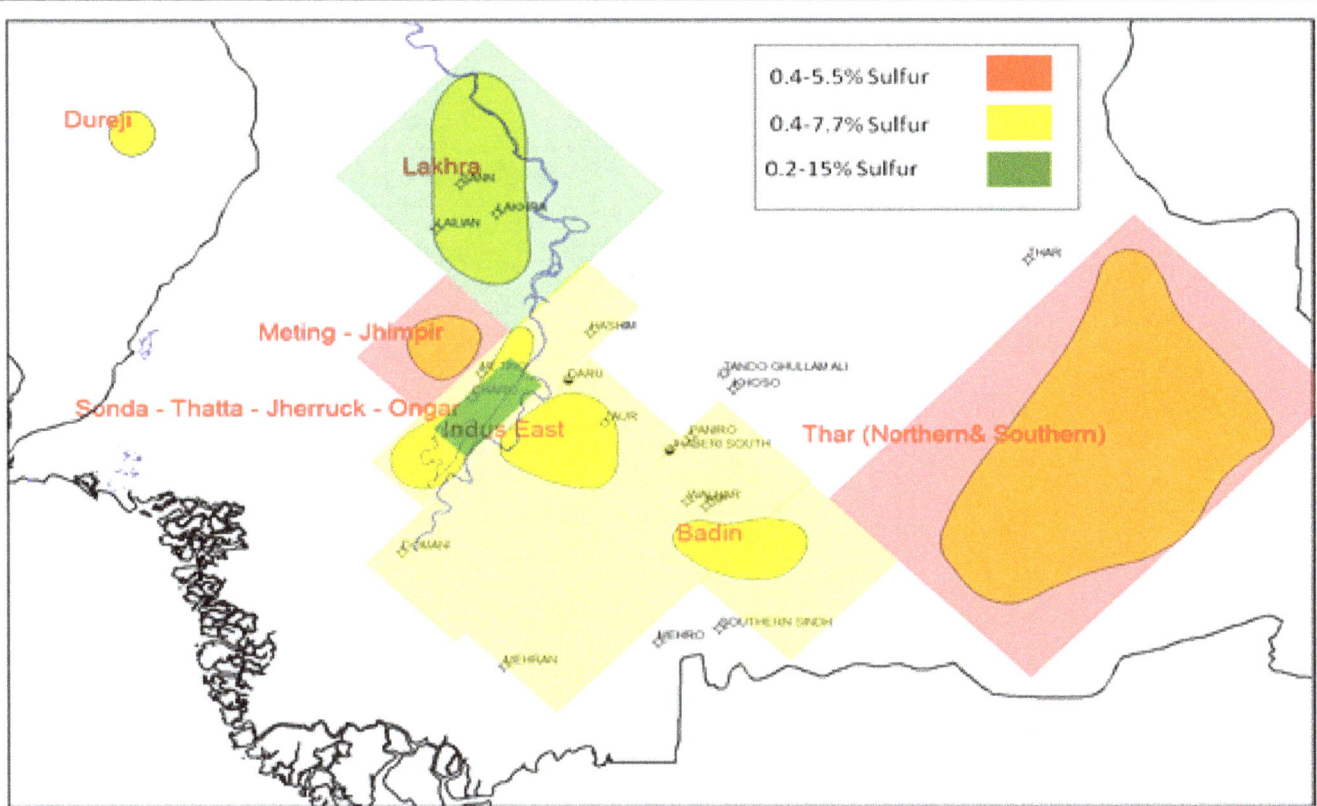

Figure 12: Distribution of sulfur in coal seams of Paleocene/Eocene coal fields (Sindh).

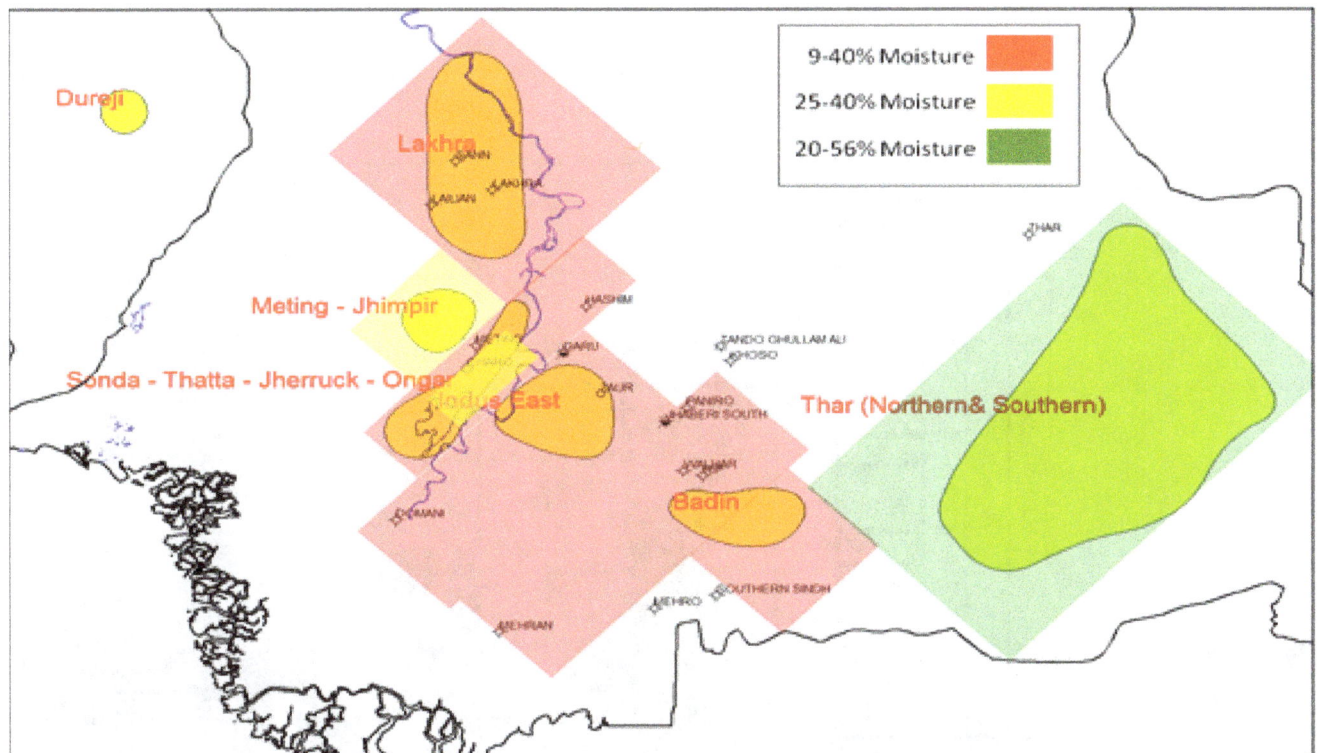

Figure 13: Distribution of moisture in coal seams of Paleocene/Eocene coal fields (Sindh).

MEHRAN-01					
No	Top(MD)	Base (MD)	Top (SS)	Base (SS)	Thickness
1	171.6	172.8	158.8	160	1.2
2	190.7	192	177.9	179.2	1.3
3	249.4	250	236.6	237.2	0.6
Total					3.1

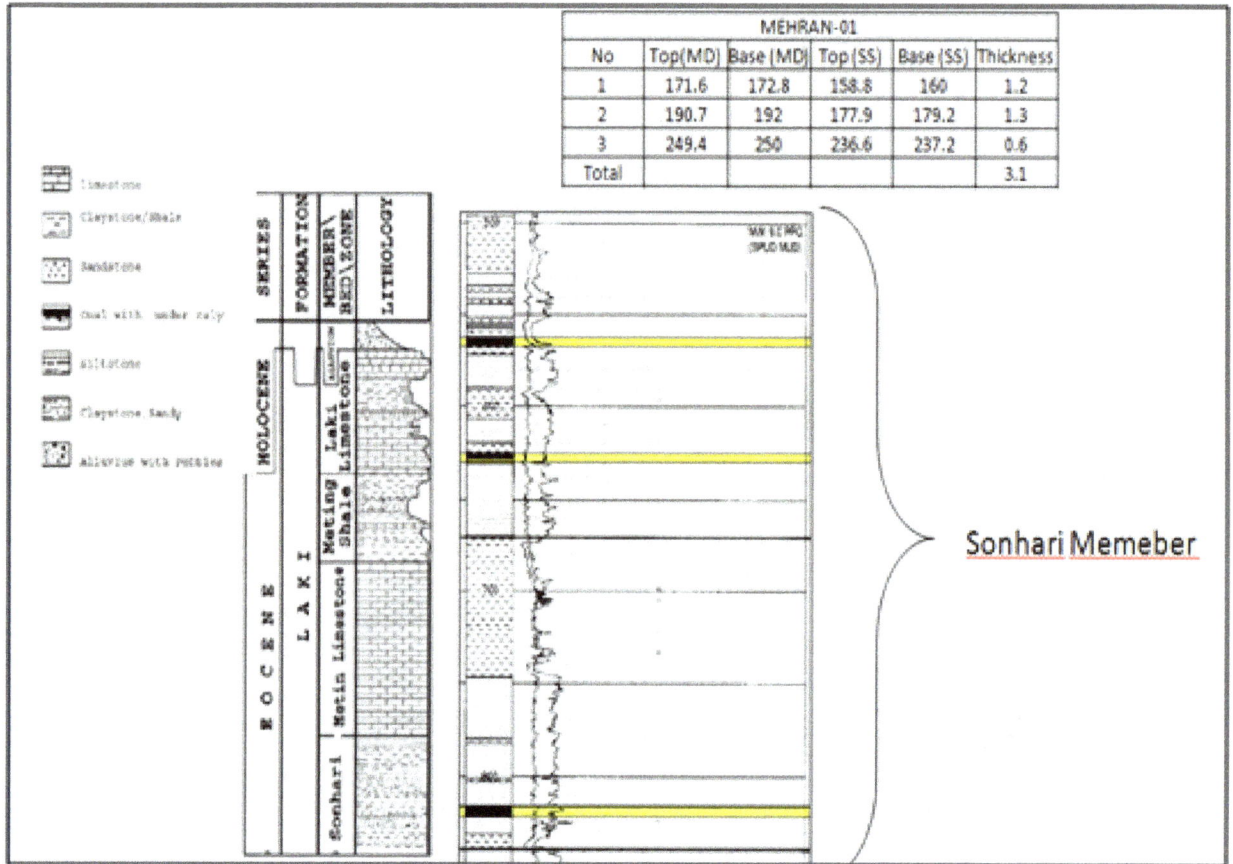

Figure 14: The fig is showing coal bearing facies encountered in Sonhari Formation (Eocene) in Mehran-1.

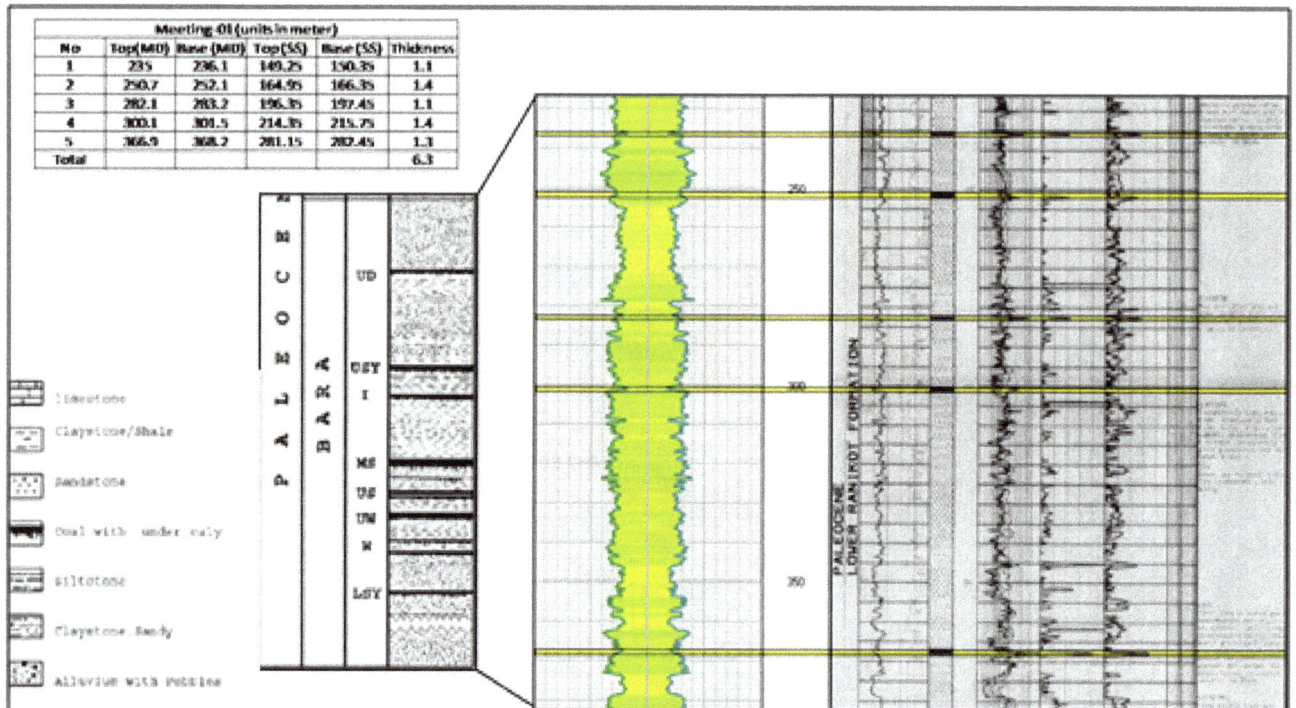

Meeting-01 (units in meter)					
No	Top(MD)	Base (MD)	Top (SS)	Base (SS)	Thickness
1	235	236.1	149.25	150.35	1.1
2	250.7	252.1	164.95	166.35	1.4
3	282.1	283.2	196.35	197.45	1.1
4	300.1	301.5	214.35	215.75	1.4
5	366.9	368.2	281.15	282.45	1.3
Total					6.3

Figure 15: The fig is showing coal bearing facies encountered in Bara Formation (Paleoocene) in Meeting-1.

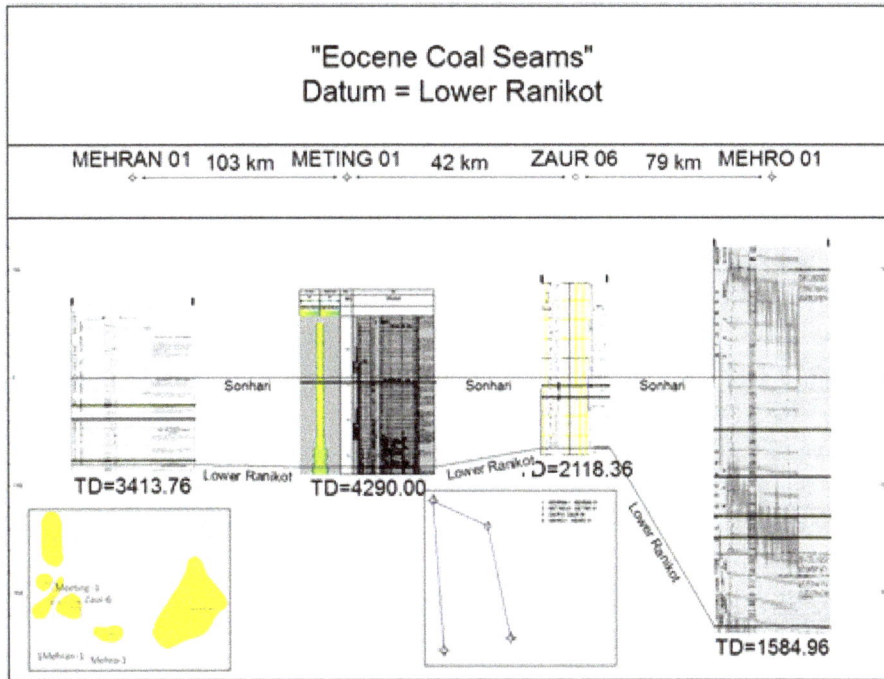

Figure 16: Wireline correlation of coal facies in Sonhari Formation (Eocene), Lower Indus Basin, Sindh, Pakistan.

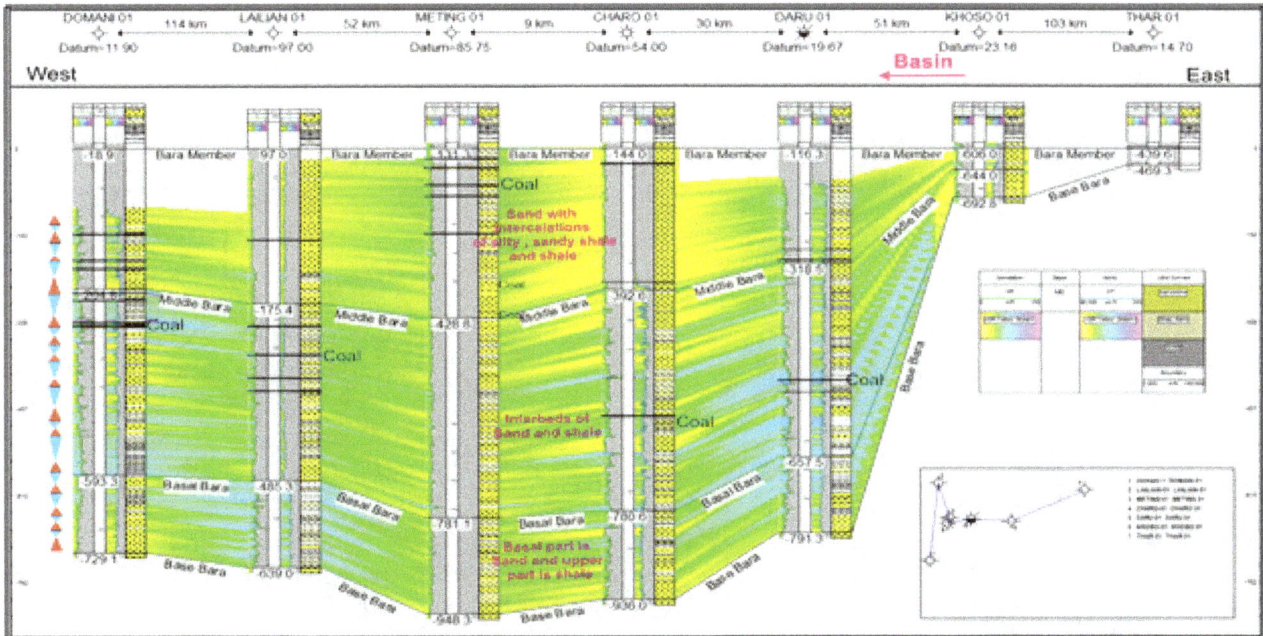

Figure 17: Wireline correlation of coal facies in Bara Formation (Paleocene), Lower Indus Basin, Sindh, Pakistan

and eroded due to rise and fall of sea and existing humid climatic conditions, as suggested by previous workers. As a result sediments were deposited along Indian Shield Element. As plate moved toward north, mangroves as discussed by Kumar [11] and thick vegetation had been grown in area of Thar Coal Field. Presence of carbonaceous shales show the presence of marshy area, which existed for some time. However, whole area was under influence of humid environment and tidal waves. As a result, thick vegetation grew in Thar Coal field area.

Probably, Thar Coal area is relatively high as compare to its surrounding and shows possible reasoning for origin of thick vegetation in limited aerial extension. Figure 19 is showing gross depositional model based on wireline modeling and prevailing literature. Author believes that coal seams are not continuous in wire line logs showing isolated seams in subsurface along monoclonal slope of Sindh Monocline. In Lakhra, Sonda-Thatta, Ongar, Jherruck, and Sondha East, deposition of coal took place along margins of fluvial deltaic tributaries. Vegetation is not thick as compare to Thar in these areas.

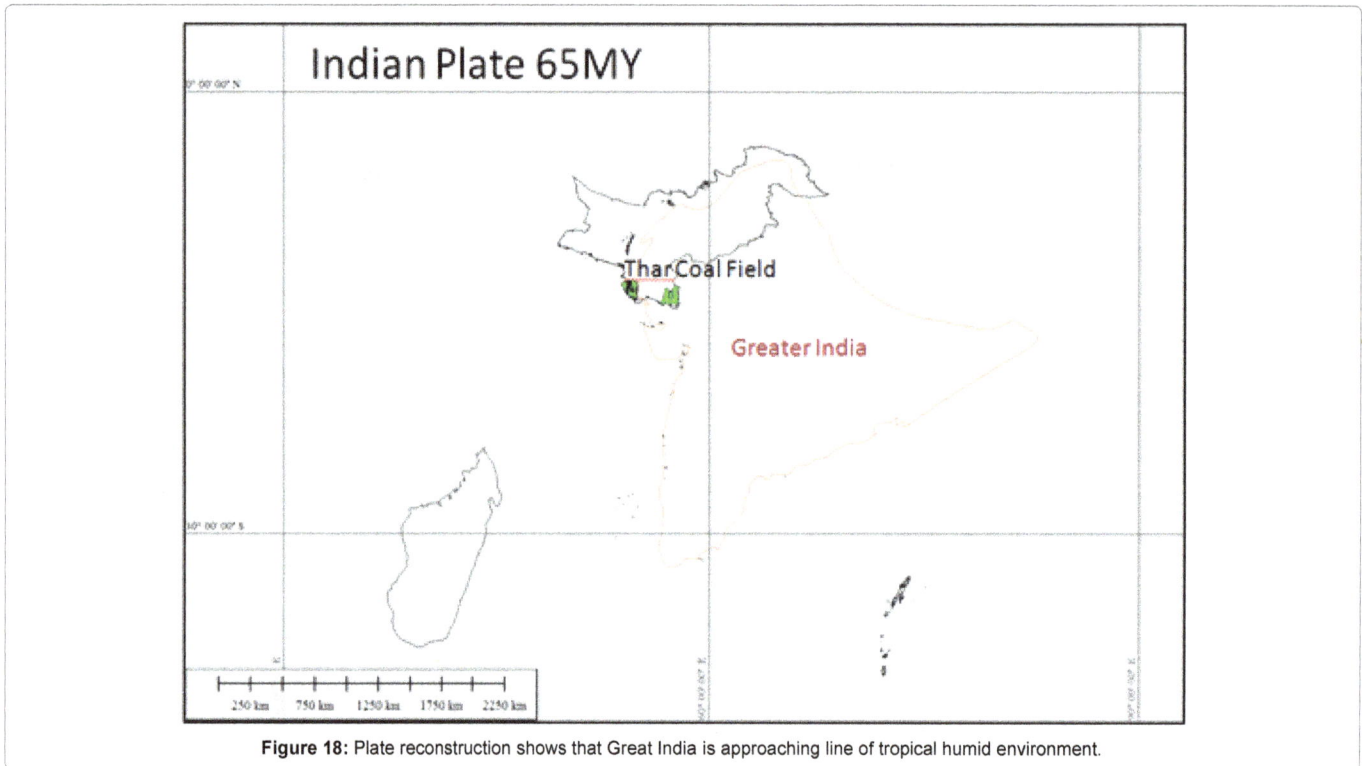

Figure 18: Plate reconstruction shows that Great India is approaching line of tropical humid environment.

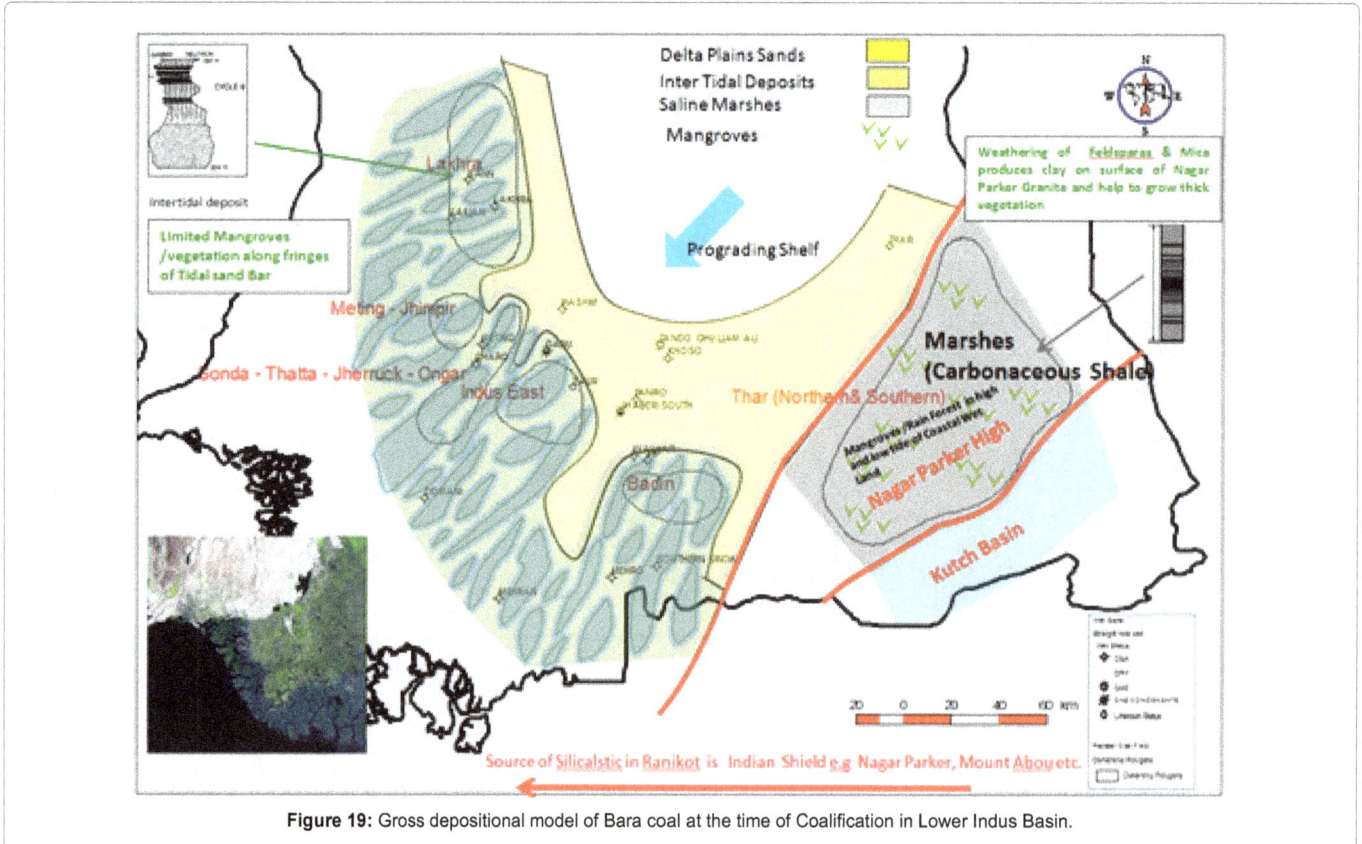

Figure 19: Gross depositional model of Bara coal at the time of Coalification in Lower Indus Basin.

Pore diameter of Thar coal

Professor Dr. Imdad Siddiqui of Sindh University recently studied the characteristics of pores and its effect on probable occurrence of coalbed methane in Thar coal field Pakistan and found that pore; diameter study shows that Thar coalfield has dual porosity. Some coal seams have nearly micropores or meso-pores, while other seams possess meso pores. On the basis of pore's micropore, the pore volume is also determined and it shows that the pore volume vary between 0.06 and 2.36 cc/g for seams of Thar coalfield. This study shows that the Thar Coal contains pores to harbor free as well as adsorbed gas.

Conclusion

1. Coal is associated with Bara Formation (Middle Paleocene) and Sohnari Member (Early Eocene).

2. Quality of coal has been determined on the basis of heating values, published in literature.

3. Best quality of coal (Sub-bituminous-C, bituminous HV A) is associated with Sonda-Thatta- Jherruck.

4. Ongar, Badin and Thatta coal quality belongs to Lignite to bituminous A.

5. Lakhra-Meeting –Jhampir coal rank is Lignite to Sub-bituminous.

6. Mixed quality of coal is present in all over Lower Indus Basin, Sindh, Pakistan.

7. Coal seams are present in number of wells and could not be correlated to each other; showing isolated seams of coal.

8. Quality of coal will increase with depth along slope of Sindh Monocline. The reason for enhanced coal quality is attributed to overburden and maturation.

9. Thar Coal is considered as the biggest lignite deposits of Pakistan and overall study shows that Thar coal also extends into subsurface along monoclonal slope. Further investigations are also recommended.

10. Presence of cleats in Badin and Thar increases prospectively of coal.

11. Powder River Basin (USA) is considered as analogy of Sindh Coal Fields.

Recommendation

Authors strongly recommend Government of Pakistan, Sindh Coal Authority, National and International Donor Agencies and E&P companies to carry out Pilot project to exploit CBM resources in Lower Indus Basin, Sindh. Gas Pipe Line is already available nearby by prospect area. Seismic Mapping is helpful to mark prospect area for experimental drilling in subsurface. Gas adsorption properties will be established for CBM reserves on the basis of core studies taken during pilot project. We also recommend for development of legal frame work for the exploitation of CBM resources of Pakistan, incentive for exploitation through pilot projects and gas sales agreement as national policy.

Acknowledgement

We are grateful to the management of Pakistan Petroleum Limited to allow us to publish this paper. We also acknowledge Mr. Salim Muhammad Jandula (GM) and Atta Mohammad Khakwani because present publication is highly appreciated and supported by them. We also acknowledge Mr. Anjum Shah, Senior Geologist, Saif Energy Ltd, Islamabad for providing valuable input during proof read of this article.

References

1. Levine JR (1993) Coalification: the evolution of coal as a source rock and reservoir rock for oil and gas. AAPG 38: 39-77.

2. Siddiqui I, Solangi SH, Mahmud SA (2010) Cleat fractures and matrix porosity in coals of Sindh. PAPG/SPE Annual Technical Conference 2010 November 10-11, Islamabad.

3. Ahmad A, Ahmad H, Solangi SH (2010) Badin and Thar Coals: Potential candidates for CBM. PAPG/SPE Annual Technical Conference 2010 November 10-11, Islamabad.

4. Malkani SA (2012) A review of coal and water resources of Pakistan. Sci Tech Dev 31: 202-218.

5. Siddiqui I, Solangi SH, Samoon MK, Agheem MH (2011) Preliminary studies of Cleat Fractures and Matrix Porosity in Lakhra and Thar coals, Sindh, Pakistan. J Himalayan Ear Sci 44: 25-32.

6. Siddiqui I (2012) Characteristics of pores and its effect on probably occurrence of coalbed methane in Thar Coalfield Pakistan. Internal Report, Center for Pure and applied Geology, University of Sindh, Jamshoro-Pakistan.

7. Ghaznavi MI (2002) An overview of coal resources of Pakistan. GSP, Pre Publication Issue of Record. 114: 167.

8. Ayaz SA, Haider BA, Ismail K, Smith PM (2012) Unconventional Hydrocarbon Resource Plays in Pakistan: An Overview Awakening a South East Asian Sleeping Giant-Technological Solutions to Unlock the Vast Unconventional Reserves of Pakistan. Search and Discovery Article #80216.

9. Chakraborty A, Tiwari PK, Singh AK (2011) Coal bed Methane Exploration in A Tertiary Lignite Basin, North Gujarat, India. The 2nd South Asian Geoscience Conference, Greater Noida, New Delhi, India.

10. Rao PLS, Rasheed MA, Hassan SZ, Rao PH, Harinarayana T (2014) Role of Geochemistry in Coal bed Methane-A Review. Geosci 4: 29-32.

11. Kumar P (2012) Palynological investigation of coal bearing deposits of the Thar Coal Feld Sindh, Pakistan. Department of Geology, Lund University, UK.

12. Daahar Hakro AAA, Khokhar QD, Solangi SH, Siddiqui, I, Agheem MH, et al. (2015) Geochemical study of Bara Formation from boreholes SB-14-TC and ST-24-TC, Thar Coalfield, Sindh, Pakistan. Sindh Univ Res J 47: 275-282.

13. SanFilipo JR (2000) USGS open-File Report 00-293, "A primer on the occurrence of coalbed Methane in low-rank coals, with Special reference to its potential Occurrence in Pakistan".

14. Nazeer A, Solangi SH, Brohi IA, Usmani P, Napar LD, et al. (2012-2013) Hydrocarbon Potential of Zinda Pir Anticline, Eastern Sulaiman Foldbelt, Middle Indus Basin, Pakistan. Pakistan J Hydrocar Res 22 and 23: 73-84.

15. Flores RM (2004) Coal bed Methane in the Powder River Basin, Wyoming and Montana: An Assessment of the Tertiary-Upper Cretaceous Coal bed Methane Total Petroleum System, Chapter 2 of "Total Petroleum System and Assessment of Coal bed Gas in the Powder River Basin Province, Wyoming and Montana", By USGS Powder River Basin Province Assessment Team.

20

Characterization of Low Grade Natural Emerald Gemstone

Reshma B, Sakthivel R and Mohanty JK*

CSIR-Institute of Minerals and Materials Technology, Bhubaneswar, Odisha, India

Abstract

The low quality gemstones need suitable treatment for its quality enhancement before being marketable. Hence, a detail characterization is essential to assess the quality of a particular gemstone. In this context, low grade emerald from Baripada, Odisha is characterized for its physical and chemical properties by various characterization techniques to provide strategy to improve its quality. Optical microscopic and XRD studies reveal the presence of mineral (silicate and oxide) impurities in emerald. FTIR, micro-Raman, micro CT, UV-Visible DRS and Photoluminescence techniques are used to characterize the emerald. ICP-OES is used to quantify the impurities such as Fe, Cr, V, Co, Ni etc in the sample.

Keywords: Low grade natural emerald; Raman spectra; Photoluminescence

Introduction

Beryl is a group consists of five gemstones of different color. Out of five, emerald is the most important one and is the third most valuable gemstone after diamond and ruby. Though it occurs in many countries of world, Columbia supplies most of the world's emerald. The green color of emerald is due to trace amounts of chromium and/or vanadium replacing aluminum at the y site. Beryl is relatively rare because of very little Be in the upper continental crust [1]. Unusual geological and geochemical conditions are required for Be and Cr and/or V to be present. It is a general belief that Be bearing pegmatite interacts with Cr-bearing mafic/ultramafic rocks. However, looking at the emerald deposits world over, it is observed that a combination of mechanisms (magmatic, hydrothermal and metamorphic) are required to bring Be into contact with the chromophores.

Gemstones are mainly used in jewelry for ornamental and astrological purposes. Therefore, demand for gemstones is increasing day by day. Gemstones have limited occurrences in India and other parts of the world. Gemstones from various part of world vary in their quality due to differences in geological milieu in which they are formed. Availability of good quality gemstones is scarce in India in general and Odisha in particular. The low quality gemstones need quality enhancement by different techniques before being marketable. Hence, a detail characterization is essential to assess the quality of a particular gemstone.

In India, emerald was first discovered in Rajasthan in 1943. Emerald bearing veins occur at the contact between pegmatite and talc schist. In 1995, emerald was discovered from Tamil Nadu. The emerald is Cr-dominant and occurs in a mica schist belt [2]. Emerald has been reported for the first time from a narrow zone of talc-tremolite schist in Mayurbanja district, Odisha.

The emerald mineralization area is confined to 22° 15'00″ to 22° 20'00″ N: 86° 30'00″ to 86° 39'25″E and belongs to Toposheet No 73 J/11 of Survey of India. The mineralized area is, 40 km from Baripada, and around 260 kms from the state capital, Bhubaneswar. The area is located to the south of Singhbhum shear zone and forms a part of Singhbhum-North Orissa Craton of Eastern India shield. The rocks of the area belonging to Iron Ore Super group of Precambrian age overly the older metamorphics. This paper gives an account of the optical and spectroscopic characterization of low grade emerald that will help to adopt suitable techniques for value addition.

Materials and Methods

Emerald samples were collected from Bangiriposi area around 40 km from Baripada, Mayurbhanj district, Odisha. Pocket deposits spread over around 20 km². Pits are dug at Ghasabani, Jharguda, Kairakocha, Saraskona areas in and around Bangiriposi. The emerald grains are embedded in host rocks such as Talc-Tremolite schist, Talc schist and metamorphosed ultramafic rocks. Emerald bearing rock samples were collected from the pits. The emerald grains are small in size and impure. They were separated from the host rock by size reduction, followed by washing and heavy liquid separation. These separated grains were characterized by using a combination of optical microscopy (Leica), XRD and spectroscopic methods such as FT-IR, Raman, UV-DRS and PL. To indentify various inorganic phases in the sample X-ray power diffractometer (XRD) of X PERT' PRO PANalytical, Netherland was used. It is equipped with CuK_{a1} radiation having 0.154056 nm wavelength and diffractometer was operated at 40 KV and 30 mA. Diffractogram was recorded in the 2θ range from 10° to 80°. For the identification of functional groups present in the emerald sample, Fourier Transform Infrared (FTIR) spectrometer was used to record the spectra from 400 to 4000 cm⁻¹ by using Perkin Elmer Spectrophotometer (model: Spectrum Gx). KBr was used as reference. Raman Spectra of the sample were recorded from 200 to 3000 cm⁻¹ by using dispersive type micro-Raman spectrometer (renishaw inVia spectrometer UK). The sample was ground to have the average particle size of ~10 μm. Absorbance spectrum was made with UV-visible spectrophotometer (Perkin Elmer Lambda 9) between 300 and 800 nm wavelengths. The photoluminescence spectra of sample were obtained with help of Fluoromax 4P Spectrofluorometer. Trace elements such as Fe, V, Cr, Ni etc. are analyzed by ICP-OES techniques (Perkin Elmer OPTIMA 2100DV) after triacid digestion.

***Corresponding author:** Mohanty JK, CSIR-Institute of Minerals and Materials Technology, Bhubaneswar-751013, Odisha, India
E-mail: jkmohanty@immt.res.in

Results and Discussion

Megascopic characterization

Representative rock samples covering all the lithounits of the area were collected. It is observed that talc-tremolite schist is the most important host rock where appreciable amount of tiny emerald grains are present in a haphazard manner (Figure 1a). Megascopic picture of emerald stones separated out from the rock is shown in Figure 1b. It shows emerald stones have different shape and size, majority of them have size below 1 cm. Shape varies from anhedral to subhedral. However, few well developed stones are also present. Its colour varies from dull green to dark green. The refractive index and specific gravity of the emerald stones are found to be 1.57 (ε) to 1.58 (o) and 2.75 respectively.

Microscopic characterization

The thin section study of the emerald stones indicates that stones have low birefringence and first order light green to light yellow interference color. Polished sections are examined under monochromatic light to decipher the mineralogical attributes of emerald. Figure 2 is a talc-tremolite schistose rock having well developed rhombohedral and hexagonal emerald crystals. Figure 3 shows presence of magnetite (bright) and silicate (Grey) in the emerald sample as impurities. A few chromite grains are also found to be present. The chromite grains are mostly altered at grain boundaries to ferritchromite (Figure 4). Chemical analysis by FE-SEM indicates that the chromite grains are in fact chromian magnetite. The gemological microscopic pictures of emerald stones are depicted in Figure 5. It shows the distinct green

Figure 1a: Talc tremolite schist showing presence of small emerald grains.

Figure 1b: Megascopic picture of emerald stones separated out from the rock.

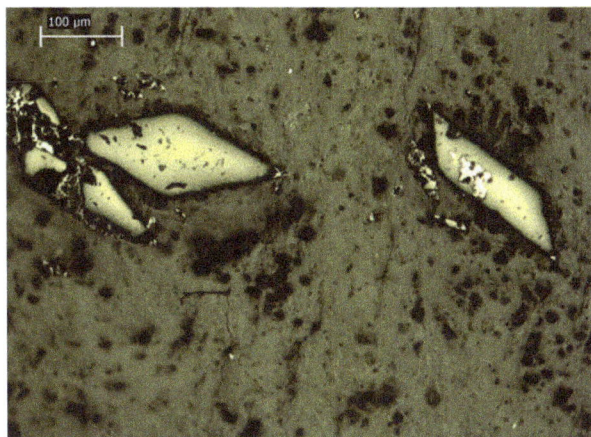

Figure 2: Euhedral emerald crystals in talc-tremolite schist.

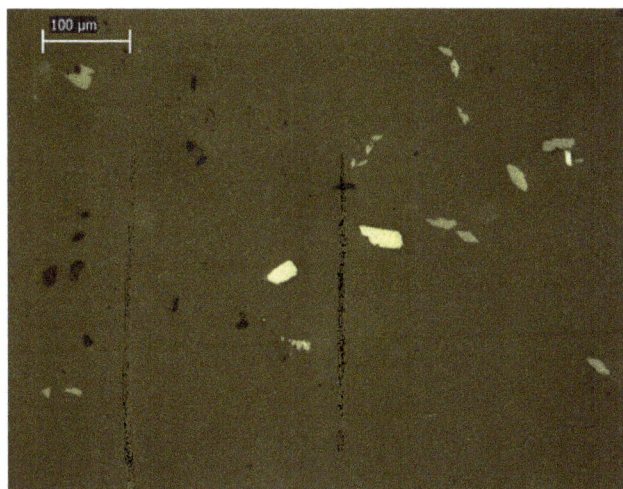

Figure 3: Irregular grains of magnetite (white) and silicate (Grey) in emerald.

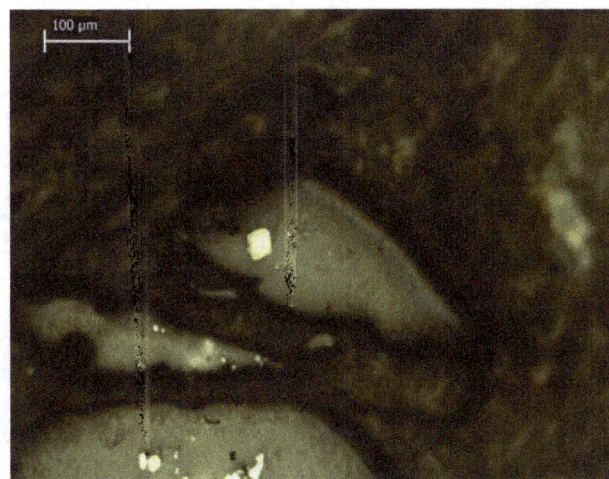

Figure 4: Small altered chromite grain in emerald grain.

colour and co-existence of some impurities (various shades of colour other than green) in the emerald stones. Micro CT examination was carried out to find out the shape and size of the impurities and their spatial distribution in emerald sample. From the Figure 6, it is observed that the impurities are of irregular shape and size and show a definite pattern of distribution in the sample.

Phase characterization

X-ray powder diffraction pattern of low grade emerald sample is shown in Figure 7. It shows very prominent diffraction peaks indicating for good crystallinity. The diffraction peaks were observed at $2\theta°$ (11.003), (19.291), (22.260), (27.288), (29.3), (30.9), (35.39), (39.15), (41.6), (41.8), (45.4), (52.4), (60.7), (63.9) and (73.6). These peaks very well match with standard emerald (JCPDS-01-083-1023). In addition to emerald phase, minor phases such as feldspar and magnetite/chromian magnetite are also identified.

Emerald has hexagonal structure and its lattice parameters are calculated from the diffraction pattern using following equation.

$$d^2_{hkl} = \frac{3a^2}{4(h^2 + hk + k^2)}$$

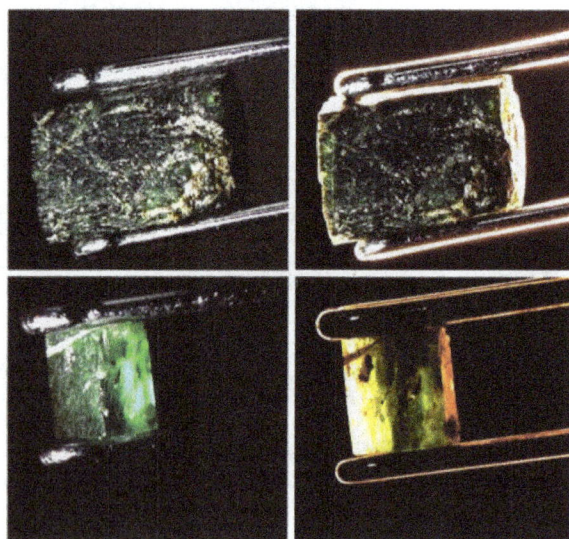

Figure 5: Gemological microscopic pictures of emerald stone.

Figure 6: Micro CT of low grade natural emerald sample.

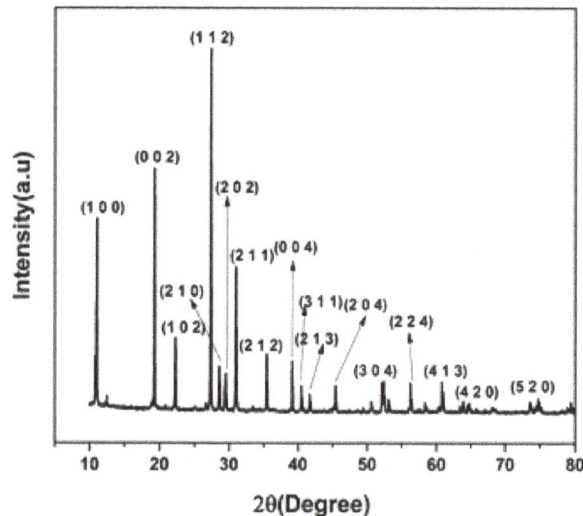

Figure 7: X-ray diffraction pattern of low grade natural emerald gemstone.

The obtained lattice parameter values are a=9.258 Å and c=9.152 Å and these values are in good agreement with the literature [3]. The c/a ratio (0.98) very well matches with the ICSD pattern value (0.99). It indicates that emerald sample belongs to the tetrahedral series [4]. For further confirmation of phase, Raman spectrum is obtained and it is depicted in Figure 8. It shows various Raman bands which are given in Table 1 along with their spectral assignments. The characteristic vibrational modes for Be-O and Si-O, Al-O, Al-O-Si are observed which support the emerald mineral phase. Bands due to moisture and hydroxyl groups are also noticed and this observation is well supported by the following FTIR results.

FTIR characterization

In order to know the different functional groups present in the low grade emerald sample, FTIR spectra are recorded from 400 to 4000 cm^{-1} and shown in Figures 9a and 9b. It shows many absorption bands at different wavenumbers. The spectral assignments of all the bands are given in Table 2. Among all bands, the bands observed for Si-O-Al, Be-O and Si-O vibrations are providing evidence for emerald phase. Further, it also provides the information about physically adsorbed water molecule and also hydroxyl groups associated with lattice.

UV-Visible (diffuse reflectance) spectroscopic study

To know the optical absorption properties of low grade emerald sample, UV-visible spectrum is recorded in reflectance mode and the obtained spectrum is shown in Figure 10. It shows very strong absorption band around 310 -330 nm wavelength and weak absorption bands at 430 and 600 nm wavelengths. These absorption bands are attributed to the various impurities present in the sample. It has been reported that absorption band positions observed in the case of natural emerald sample vary compared to the synthetic emerald sample. However, those bands are attributed to the Cr^{3+} ion which is substituted at Al site. In this study, low grade natural emerald shows absorption bands at 430 and 600 nm wavelengths which are corresponding to the $4A_2 \rightarrow 4T_1$ and $4A_2 \rightarrow 4T_2$ transitions associated with Cr^{3+} ion respectively [18]. Another prominent absorption band observed around 310-330 nm wavelength is attributed to charge transfer transitions from O^{2-} to Fe^{3+} in the tetrahedral Be^{2+} site [14].

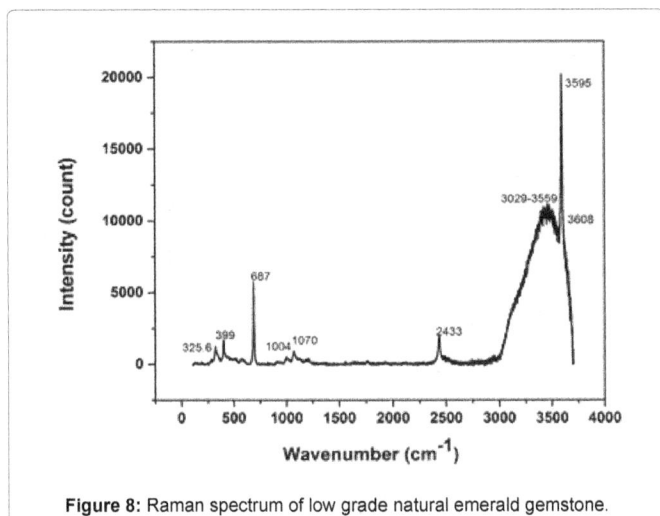

Figure 8: Raman spectrum of low grade natural emerald gemstone.

Raman Peak position (cm^{-1})	Spectral assignments	References
321-326	Ring vibration of emerald	[4]
325 and 395	The deformation mode of the tetrahedral ring	[4-6]
687	Signature of ring stretching of Be-O bond	[4,5,7,8]
1007	(1002.5-1007cm^{-1}) correspond to Si-O stretching	[4]
1070	Attributed to Be-O stretching and Si-O stretching of vibration beryl crystal	[4,5,8,9]
3595	Water type-II (water molecules with alkali ions nearby)	[10]
3608	Water type-I (those water molecules without any presence of alkali ion nearby)	[17]

Table 1: Raman spectral observation of low grade natural emerald sample.

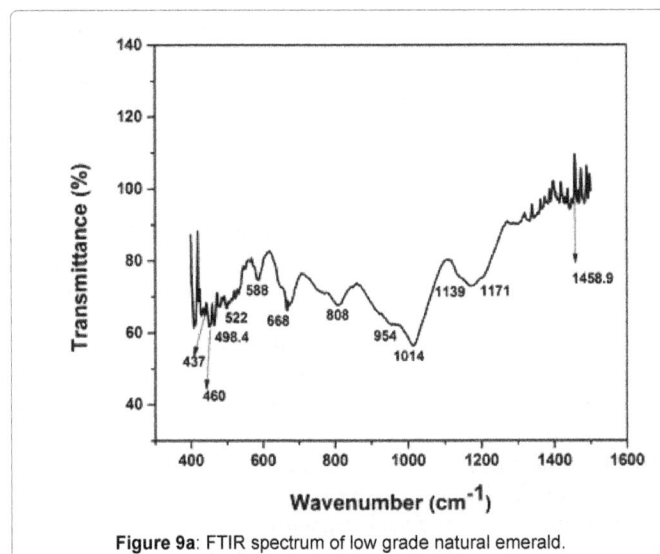

Figure 9a: FTIR spectrum of low grade natural emerald.

Photoluminescence property

Excitation and emission spectra of low grade natural emerald are shown in Figures 11 and 12 respectively. Figure 11 shows weak

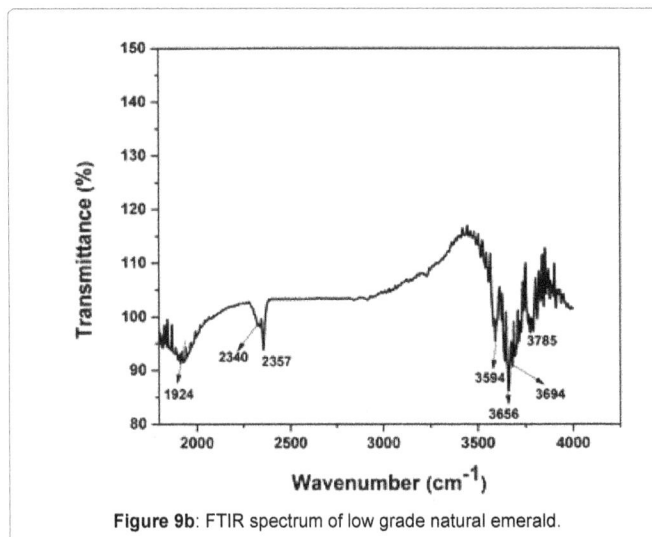

Figure 9b: FTIR spectrum of low grade natural emerald.

FTIR Peak position (cm^{-1})	Spectral assignments	References
437	Metal oxygen bond vibration	
460	Si-O-Si bending vibration	[11,13]
492-522	Si-O-Al bending vibration	[14]
808	Si-O and Be-O bond vibration	[11,13]
954	Si-O vibration of beryl	[11,13]
1014	Internal Si-O stretching vibration	[11,13]
1171-1203	Be-O vibration and Si-O stretching vibration	[12,14]
2340 and 2357	Band of CO_2 molecules	[15]
3592-3655	Water molecules type-II (water molecules which are bound to alkali ions)	[17]
3595	Stretching vibration involving H_2O-Na-OH_2	[11]
3655	One water molecules or OH group bonded to alkali ions	[11]
3694	Water type-I (physically adsorbed water molecules)	[6,14,16,17]

Table 2: FTIR band positions and their spectral assignments of low grade natural emerald sample.

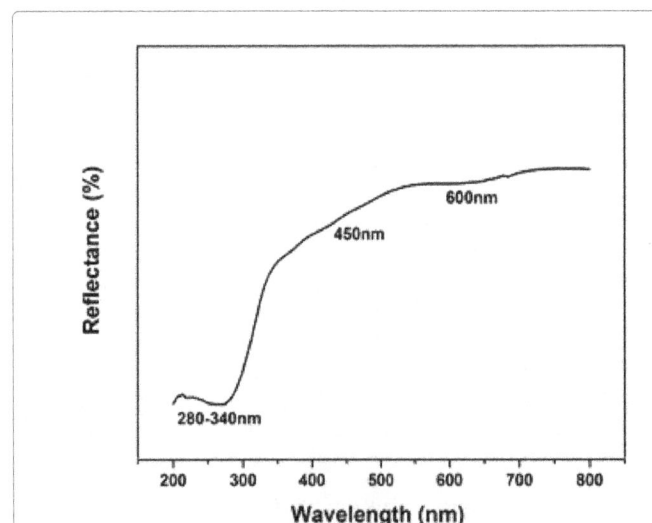

Figure 10: UV-Visible (diffuse) reflectance spectrum of low grade natural emerald.

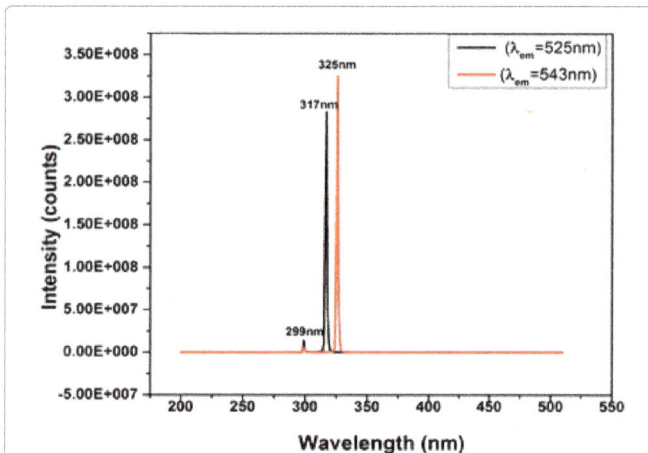

Figure 11: Photoluminescence excitation spectra of low grade natural emerald obtained with two different emission wavelengths.

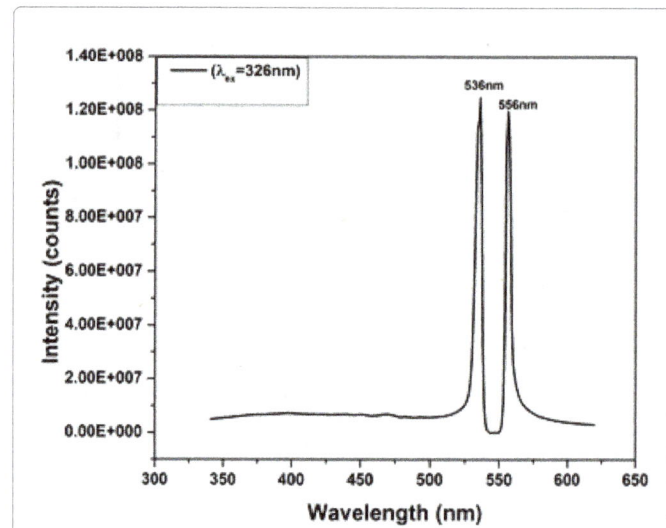

Figure 12b: Photoluminescence emission spectra of low grade natural emerald obtained with 326 nm excitation wavelength.

Elements	Weight Percentage (%)
Fe	3.057
Cr	0.315
Mn	0.037
Ti	0.23
Ni	0.028
V	0.016
Zn	0.018

Table 3: Trace elements present in the low grade natural emerald sample.

Table 3 indicates the weight percentage of different trace elements (Fe, Cr, Mn, Ti, Ni and Zn) present in the emerald sample.

Conclusion

From the above characterization results, it is concluded that emerald sample collected from the field is a low quality one with a lot of impurities. Optical microscopic study reveals mineral (silicate and oxide) impurities. X-ray powder diffraction study reveals it contains emerald as the major phase with minor amounts of feldspar and magnetite/chromian magnetite phase. FTIR spectrum shows characteristic vibration modes for Si-O-Si, Al-O-Si, stretching and bending mode for hydroxyl groups present in the sample. In addition, spectrum provides the presence of physically adsorbed water molecules in the sample. Laser Raman spectroscopic study support the findings of XRD and FTIR characterization. UV-Visible diffuse reflectance spectrum shows prominent absorption band at 500 and 320 nm wavelength. Photoluminescence study shows doublet emission bands between 510 and 550 nm wavelengths. This emission band position shifts with different excitation wavelengths. Emission behavior of low grade emerald is attributed to the impurities present in the sample. Distribution of mineral impurities in the emerald grain is observed by micro CT. The above studies give an idea about mineralogical as well as chemical characteristics of the low grade emerald which provides the scope to enhance its quality by adopting suitable purification strategy.

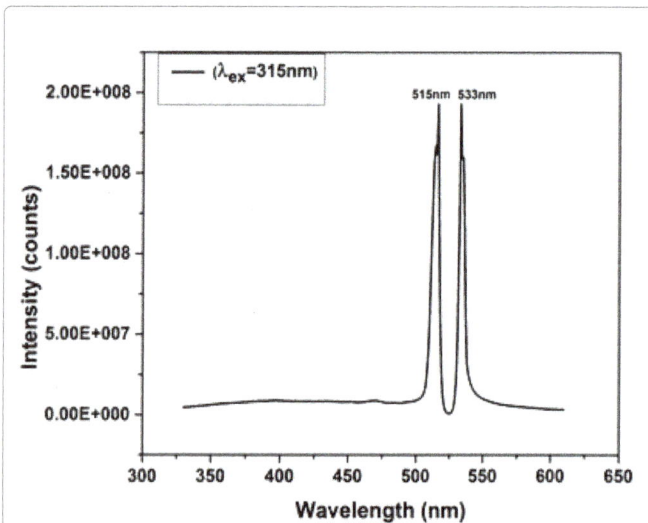

Figure 12a: Photoluminescence emission spectra of low grade natural emerald gemstone obtained with 315 nm excitation wavelength.

excitation bands at 299 nm and very prominent excitation bands at 317 and 325 nm wavelengths [19]. These bands are possibly due to the charge transfer from O^{2-} to Fe^{3+} in the tetrahedral Be^{2+} site. This observation is very similar and matching well with the absorption bands noticed in the UV-Visible spectrum. In order to study the emission characteristics of low grade emerald, emission spectra were recorded from 330 to 650 nm wavelengths by exciting the sample at 315 and 326 nm wavelengths. Figure 12a shows the emission spectrum obtained at 315 nm wavelength. It shows very strong doublet emission bands at 515 and 533 nm wavelengths. When it is excited at 326 nm wavelength, the emission bands are shifted towards longer wavelengths, 536 and 556 nm, (Figure 12b) indicating the red shift. These emission bands are attributed to Cr^{3+} ion substituted at Al site, electronic transitions involved for Cr^{3+} ion corresponding to $4A_{2g}(F) \rightarrow 4T_{2g}(F)$ according to the literature [20].

Trace elements characterization by ICP-OES

Trace-element analysis of low grade emerald sample was carried out inductively coupled plasma-optical emission spectrometry (ICP-OES).

Acknowledgements

The authors express their sincere thanks to Director, CSIR-IMMT, Bhubaneswar for his keen interest in the work and permission to publish the paper. Sincere thanks are due to colleagues who have helped in various instrumental analyses for this study. We also acknowledge the Council of Scientific and Industrial Research, New Delhi for financial support.

References

1. Taylor SR, Mclennan SM (1995) The geochemical evolution of continental crust. Rev Geophy 33: 241-265.

2. Panjikar J, Ramchandran KT, Balu K (1997) New emerald deposits from southern India. Aus Gemmologist 19: 427-432.

3. Gilberto A, Romano R, Kenny S, Pier Francesco Z (1993) Structure refinements of beryl by single-crystal neutron and X-ray diffraction. Am Mineral 78: 762-768.

4. Adams DM, Gardner IR (1974) Single-crystal vibrational spectra of beryl and dioptase. J Chem Soc Dalton Trans 14: 1502-1505.

5. Moroz I, Roth M, Boudeulle M, Panczer G (2000) Raman microspectroscopy and fluorescence of emeralds from various deposits. J Raman Spectroscopy 31: 485-490.

6. Charoy B, Donato P, Barres O, Pinto-Coelho C (1996) Channel occupancy in an alkali-poor beryl from Serra Branca (Goias, Brazil): Spectroscopic Characterization. Am Mineral 81: 395-403.

7. Hagemann H, Lucken A, Bill H, Gysler-San J, Stalder HA (1990) Polarized Raman spectra of beryl & bazita. Phys Chem Miner 17: 395-401.

8. Kim CC, Bel MI, McKeown DA (1995) Vibrational analysis of beryl ($Be_3Al_2Si_6O_{18}$) and its constituent ring (Si_6O_{18}). B Condens Matt 205: 193-208.

9. Huong LTT (2008) Microscopic, chemical and spectroscopic investigations on emeralds of various origins. Ph. D. Thesis - University of Mainz Press.

10. Schmetzer K, Kiefert L (1900) Water in beryl-a contribution to the separability of natural and synthetic emeralds by infrared spectroscopy. J Gem mol 22: 215-223.

11. Gervais F, Piriou B, Cabannes (1972) Anharmonicity of infrared vibration mode in beryl. Phys Status Sol 51: 701-712.

12. Manier-Glavinaz V, Couty R, Lagache M (1989) The removal of alkalis from beryl: Structural adjustments. Can Mineral 27: 663-671.

13. Plyusnina II, Surzhanskaya EA (1967) IR spectrum of beryl. J Appl Spectrosco 7: 611-616.

14. Aurisicchio C, Grubessi O, Zecchini P (1994) Infrared spectroscopy and crystal chemistry of beryl group. Can Mineral 32: 55-68.

15. Stockton CM (1987) The separation of natural from synthetic emeralds by Infrared spectroscopy. Gems Gemol 23: 96-99.

16. Wood DL, Nassau K (1968) The characterization of beryl and emerald by visible and infrared absorption spectroscopy. Am Mineral 53: 777-800.

17. Wickersheim KA, Buchanan RA (1959) The near infrared spectrum of bery. Am Mineral 44: 440-444.

18. Sánchez-Alejo MA, Hernández-Alcantara JM, Flores Jimenez C, Calderon T, Murrieta SH (2011) 22nd Congress of the International Commission for Optics: Light for the Development of the World.

19. Edgar A, Vance ER (1977) Electron paramagnetic resonance, optical absorption, and magnetic circular dichroism studies of the CO_3^- molecular-ion in irradiated natural beryl. Chem Mineral 1: 165-178.

20. Skvortsova V, Mironova-Ulmane N, Trinkler L, Merkulov V (2015) Optical Properties of Natural and Synthetic Beryl Crystals. IOP Conf Series Mat Sci Eng 77.

21

Land Magnetic Investigation on the West Qarun Oil Field, Western Desert-Egypt

Khashaba A*, Mekkawi M, Ghamry E and Abdel Aal E

National Research Institute of Astronomy and Geophysics (NRIAG), Helwan, Cairo-Egypt

Abstract

In this study we delineate the subsurface structures within an area of about (35 × 25) km² in the Western part of Qarun Concession, using magnetic method. The main goal is to understand the role of subsurface structures and tectonics in the petroleum processes. Land magnetic survey has been carried out along profiles covering the area under study. The magnetic data set is processed using horizontal Gradient and tilt derivative. Also, 3D Euler deconvolution and 2D power spectrum have been used as fast techniques for depth estimation. The results indicate that the most predominant tectonic trends are generally aligned in NE-SW for the major structures, while the minor structures are aligned in NW-SE.

The depth to the regional basement estimated range around 3900 m, and the shallower structures range around 750 m. The obtained results are in good agreement with the data from drilling wells in the area under study. The RTP magnetic anomalies range between -116 nT and 145 nT. The high values strongly suggest that the near structures (ferromagnetic minerals) accompany the basement along the West Qarun concession. There is good correlation between the structures deduced from the magnetic analysis and the known geological information. Most of these oil accumulations are restricted to the major tectonic trends with a NW-SE and NW-SE directions. We conclude that oil accumulation is structurally controlled by faulting, probably as a result of tectonic regimes during Cretaceous and Jurassic periods.

Keywords: Land magnetic survey; West Qarun oil field; Subsurface structures

The Geologic and Tectonic Setting

The Western Desert of Egypt covers an area more than 600,000 km² and comprises almost two thirds of the whole area of Egypt. The North Western Desert represents an important part of the unstable shelf of Northern Africa. It has been subjected to different tectonic regimes since the Paleozoic time, which were able to form the construction of many basins, sub-basins, ridges, troughs and platforms. The West Qarun concession is located to in the north part of the Western Desert (Figure 1). It is close to the Nile Delta region and the convergence boundary of three plates (Euroasian, Africa and Arabian). Also, it is affected by the opening of the Red Sea and its branches (Gulf of Suez and Gulf of Aqaba). Thus the seismic activity there is due to the interaction between the three plates. These structural features encourage foreign companies to establish condensed exploration works in this region. The Western

Desert still has a significant hydrocarbon potentiality as some oil and gas discoveries indicate [1,2]. The geologic information of the Qarun oil field and its reservoir are summarized in Table 1. El-Nady [3] reported that the oils from Khatatba and Alam El Bueib formations are mature, derived from source rocks containing marine and terrestrial organic matter, respectively. The source environments and maturity of the oil from Khatatba reservoir are similar to the Khatatba source rock. The Khatatba formation is ranging in age from Upper Cretaceous to Middle Jurassic. Figure 2 represents the potential source rocks of the studied oilfield. In the present work, land magnetic survey is carried out for the 1st time, covering an area of about 800 km² around West Qarun oil field and lies between latitudes 29.70°-30.00° and longitudes 30.45°-30.85°. The area is not leveled and the ground elevation ranges between 116 and 175m above the sea level. There are major faults and minor folds; the land rises to the complex geological history [4]. The complicated

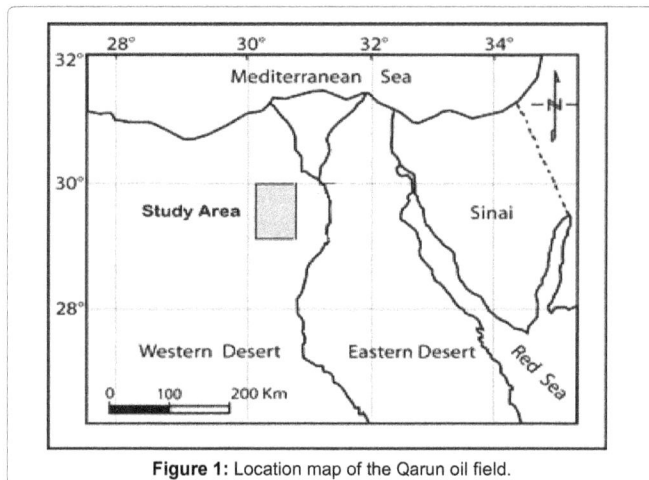

Figure 1: Location map of the Qarun oil field.

Depth (ft)	17786
Reservoirs	KhatatbaFm.
Age	M. Jurassic
Lithology	Sandstone

Table 1: Geologic information of the Qarun oilfield in the North Western Desert, Egypt [15].

*Corresponding author: Khashaba A, National Research Institute of Astronomy and Geophysics (NRIAG), 11721 Helwan, Cairo-Egypt
E-mail: khashaba80@yahoo.com

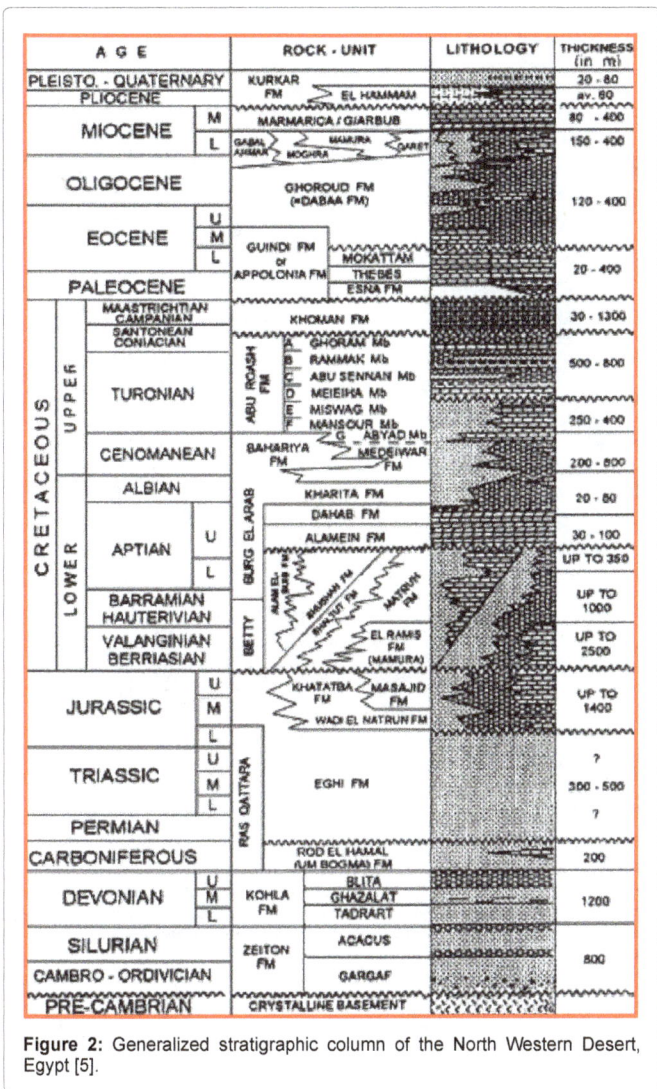

Figure 2: Generalized stratigraphic column of the North Western Desert, Egypt [5].

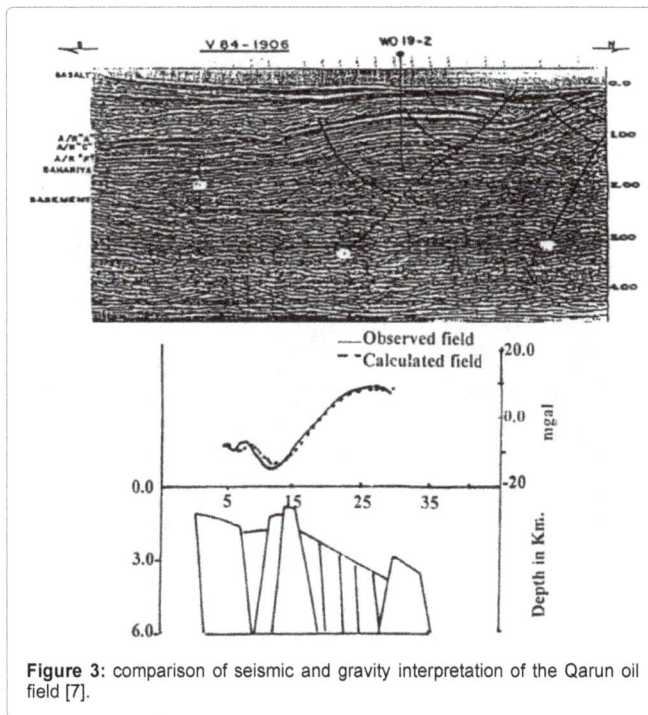

Figure 3: comparison of seismic and gravity interpretation of the Qarun oil field [7].

structure and the great sedimentary deposition (2) are similar to the Abu Ghardig basin [5]. In the Qarun concession both shallow and deep structures are present along faulted anticlines that generally aligned in NE-SW direction [6].

Qarun Oil Field

The area is characterized with different tectonic events, affecting the north western desert and occurred during the Late Cretaceous to Quaternary. It is related to the movement of the North Africa plate towards Europe. This compression movement caused the elevation, major faulting and minor folding in the northern western desert. Historically, the west Qarun oil field was discovered in 1994 by Phoenix and its partner Apach companies. The first significant hydrocarbon zones were found in kharita and Baharia formations as primary reservoirs at depths between 2900-3200 m. They lie between northwest flank of the Ginidi basin and Kattaniya uplifts. Previous seismic and gravity studies suggested that these thrusts faults are nearly vertical (Figure 3) and rooted in the basement rocks [7]. The reverse faults and the associated echelon fold are bisected by two NW-SE trending normal faults, downthrown to the SW [6].

Magnetic Data

This work is done on behave of the geological understanding of the site settings that associated with hydrocarbon and subsurface structures. Generally, this work represents a part of the geological and geophysical studies conducted in West Qarun concession. The magnetic method is shearing in the work to resolve the subsurface structure pattern and the ambiguity between the available geological and geophysical data. The land magnetic survey has two objectives: the first is to obtain the subsurface structures of the fracture basement, while the second is to estimate depth of hydrocarbon reservoir zone.

The magnetic method has an advantage that it is a fast and effective technique for mapping the fracture basement [8]. The field strength is measured with a magnetometer placed a few meters above the ground level. Measurements can be taken either along profiles or on a grid (mesh of points). High sensitivity devices are utilized Proton Precision Magnetometers for the measurement of the total magnetic field of the earth. In homogeneities in the magnetic properties of soils will cause variations in the measured field which can easily be detected.

Data acquisition

Before the survey, a base station, for diurnal correction is placed at a homogenous place lying near the investigated area. The magnetic survey is conducted using two proton magnetometers (Figure 4). One of them is used to survey the area (mobilized), while the second magnetometer is considered a base station to record the magnetic field diurnal variations. The magnetic profiles are taken along the available Paths around and crossing the Qarun oil field. Figure 5 shows the profiles and stations positions using GPS system navigation and the direction of traverses of each profile.

The measured magnetic data have been subjected to two kinds of corrections: removing the diurnal variation and subtracting 3.8 nT/km toward north direction (latitude correction). However, the survey area is not leveled but has gentle slope as the topography has minor variations

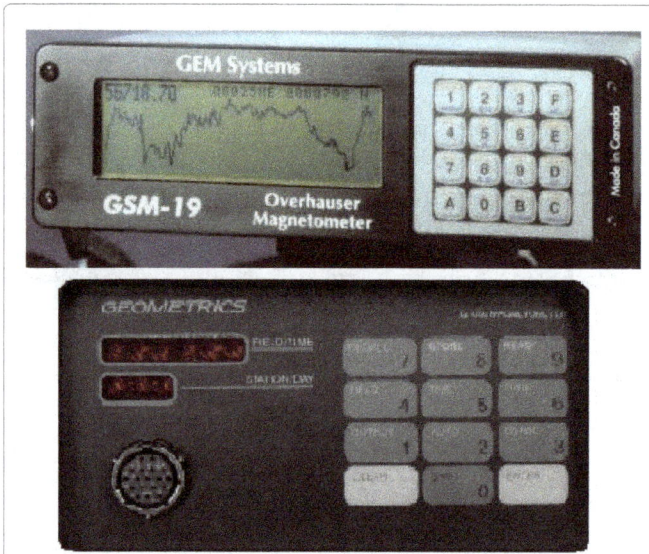

Figure 4: Proton-Precession Magnetometer (G856-AX, [16]) and Overhauser Magnetometer [17].

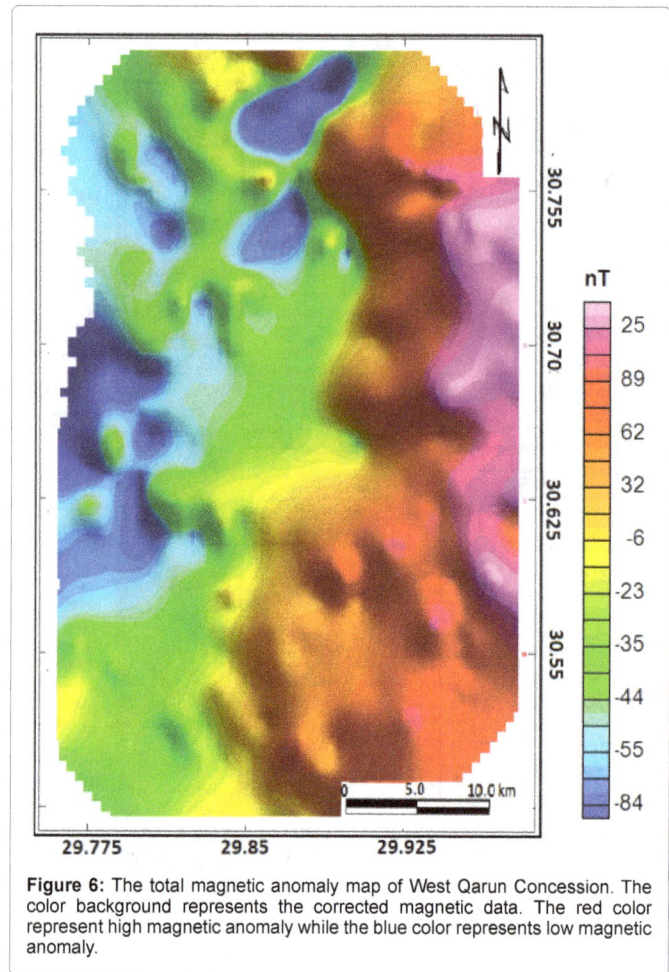

Figure 6: The total magnetic anomaly map of West Qarun Concession. The color background represents the corrected magnetic data. The red color represent high magnetic anomaly while the blue color represents low magnetic anomaly.

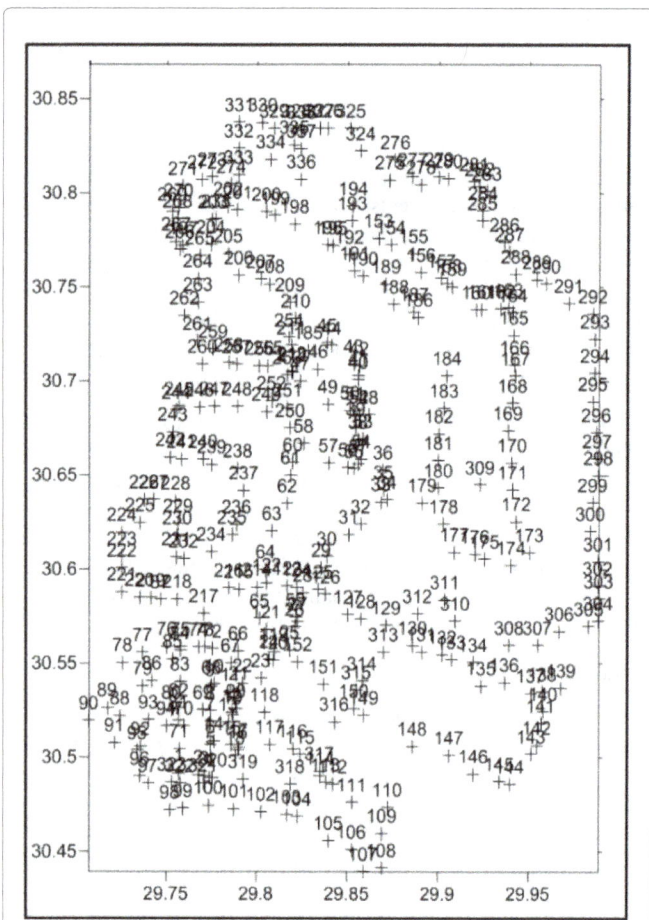

Figure 5: Showing the magnetic profiles and stations location.

profiles and has corrected from diurnal and latitude corrections, we are able to produce the variation in magnetic field which directly is related to the magnetic materials within the area (Figure 6).

Magnetic data analysis and interpretation

The present study is based on qualitative and quantitative analysis of the magnetic data to delineate both shallow and deep structures. In that regards, reduce to the magnetic pole (RTP) is applied to the magnetic data using [9]. The RTP magnetic anomaly map (Figure 7) displays a major high magnetic anomaly about 145 nT (red color) in the eastern side and low magnetic anomaly about-116 nT (blue color) in the western side. The high values strongly suggest that the near structures (contact/fault) contain ferromagnetic materials that accompany the basement along the West Qarun concession. It should be stated that magnetic structures do not occur randomly, but are generally aligned along definite and preferred axes forming trends that can be used to define fracture basement. These anomalies could be related to significant susceptibility contrast between the highly magnetic basement rocks and the non-magnetic sediments.

In order to trace the high magnetic anomaly from the RTP map, the Horizontal Gradient (HG) method has been used intensively to locate the contacts of magnetic susceptibility contrast. It is also robust method to delineate shallow and deep sources in comparison with the vertical gradient, which is useful only for the shallower structures. The amplitude of the horizontal gradient [10] is expressed a

from 144 to 175 m above sea level. The profiles are choosing along tracks and main roads (Figure 5). Once that data has been carried out along 13

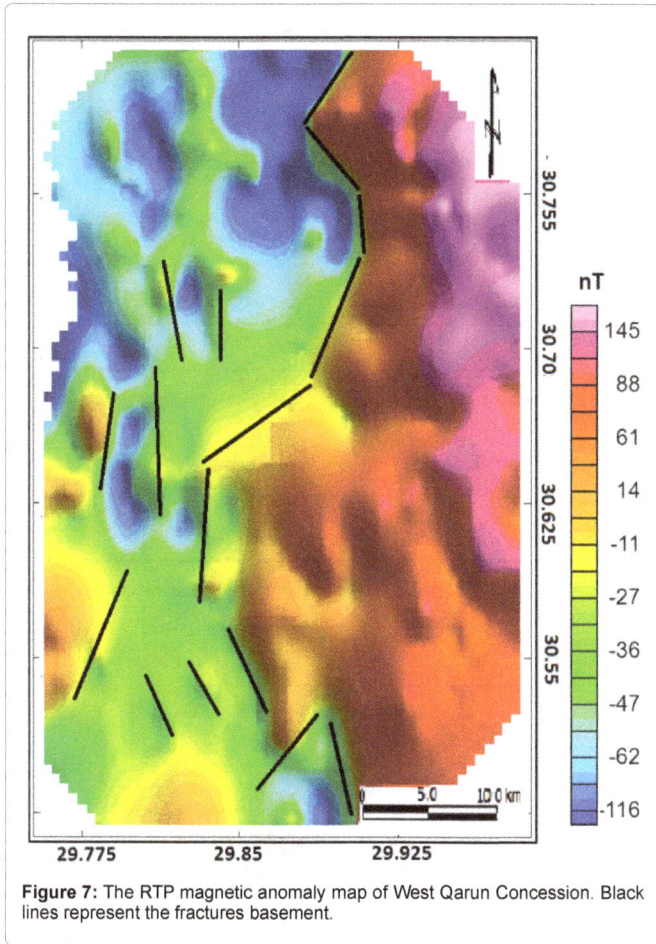

Figure 7: The RTP magnetic anomaly map of West Qarun Concession. Black lines represent the fractures basement.

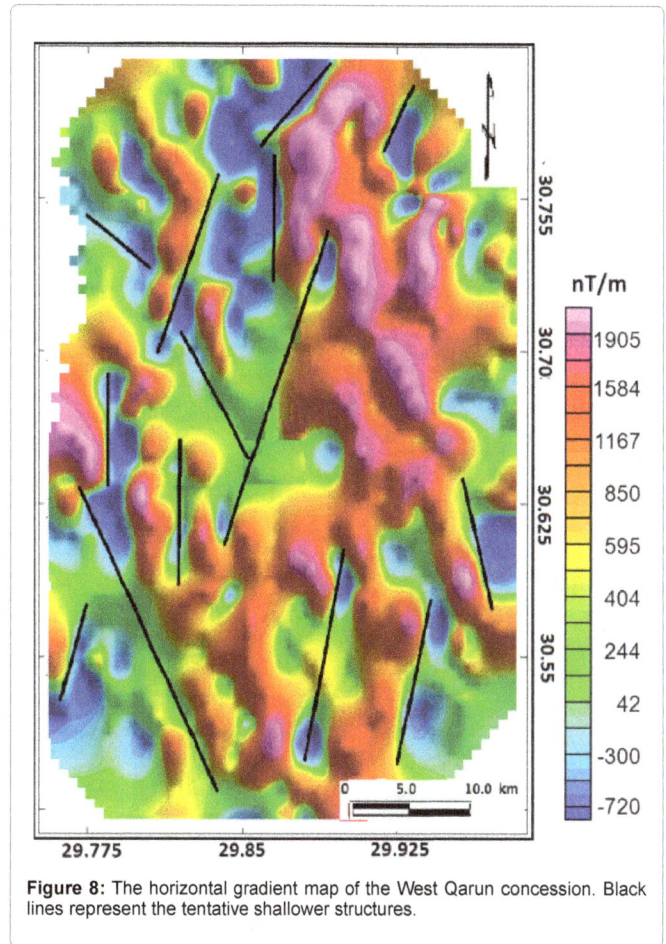

Figure 8: The horizontal gradient map of the West Qarun concession. Black lines represent the tentative shallower structures.

$$HG = \sqrt{\left(\frac{\partial T}{\partial X}\right)^2 + \left(\frac{\partial T}{\partial Y}\right)^2} \qquad (1)$$

Where $\frac{\partial T}{\partial X}, \frac{\partial T}{\partial Y}$ the horizontal derivatives of the magnetic are field in the x and y directions. The amplitude of the horizontal gradient of the magnetic data of the West Qarun field area is calculated in (Figure 8). It shows a tentative qualitative interpretation of the horizontal gradient data. Generally, the area under study may be dissected by major faults striking in the NE-SW and NW-SE directions. The most interesting result is that the location of the reservoir field is well correlated with the horizontal gradient anomalies. This indicates that hydrocarbons in West Qarun concession area are structurally controlled, especially for the shallower basement.

The tilt derivative filter (TDR) is estimated by dividing the vertical derivative by the total horizontal derivative Verduzco [11] as

$$TDR = \tan^{-1}\left\{\frac{VD}{THD}\right\} \qquad (2)$$

Where $\left(VD = \frac{\partial T}{\partial Z}\right)$ and $\left(THD = \sqrt{\left(\frac{\partial T}{\partial X}\right)^2 + \left(\frac{\partial T}{\partial Y}\right)^2}\right)$ are first vertical and total horizontal derivatives, respectively.

The most advantage of the TDR is that its zero contour line is on or close to the fault/contact location. Figures 9 and 10 shows the TDR map and the faults interpretation of the magnetic data. The most prominent trend is NE-SW and NW-SE directions (Figure 10).

Depth of the fracture basement zone

The depth to the source of magnetic anomalies is valuable and important information in any geophysical interpretation of subsurface structures. In order to estimate the depth of fractures basement zone, 3D Eulardeconvolution and 2D Power Spectrum methods are applied to the magnetic data. The 2D Power Spectrum has been applied to land magnetic survey data [12]. It is used to determine the depths of basement complex. The Fast Fourier Transform (FFT) is applied on the magnetic data to calculate a two-dimensional power spectrum curve (Figure 11) shows two main average levels (interfaces) at depth 750 m (shallow intrusions) and 3900 m (deep basement layer).

The 3D form of Euler's equation can be defined [13] as

$$x\frac{\partial T}{\partial x} + y\frac{\partial T}{\partial y} + z\frac{\partial T}{\partial z} + \eta T = x_\circ\frac{\partial T}{\partial x} + y_\circ\frac{\partial T}{\partial y} + z_\circ\frac{\partial T}{\partial z} + \eta b \qquad (3)$$

Where $\frac{\partial T}{\partial x}, y\frac{\partial T}{\partial y}, z\frac{\partial T}{\partial z}$ are the derivatives of the magnetic field in the x, y, and z directions, η is the structural index value that needs to be chosen according to a prior knowledge of the source geometry. The results of the 3D Eular solutions are shown in Figure 12 using structure index 1.0 for dikes and faults. The depths are between 500 3750 m. The Eular's depths are calculated from analytical signal magnetic map using Roest et al. [14] formula.

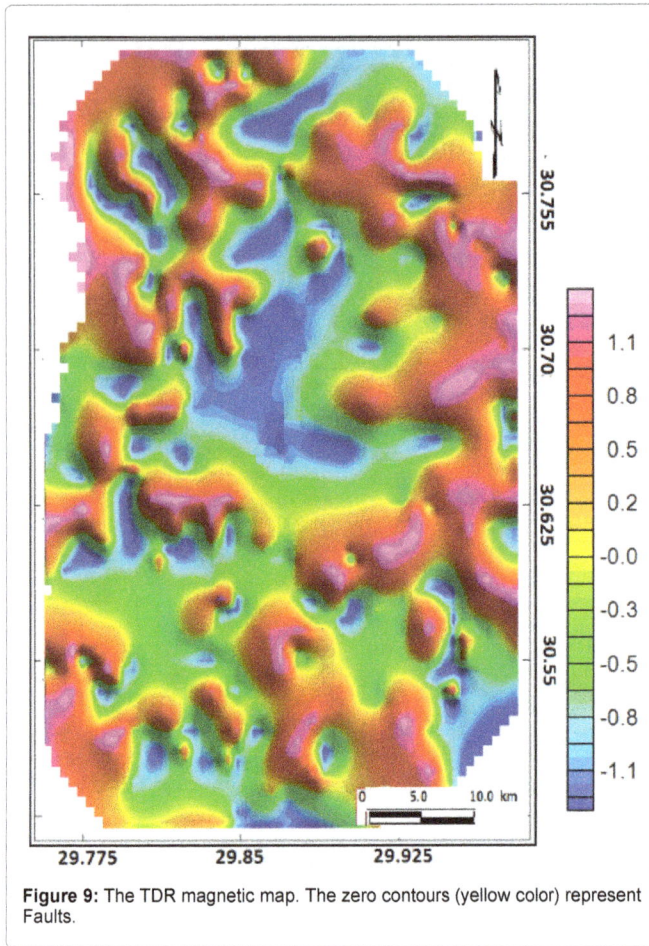

Figure 9: The TDR magnetic map. The zero contours (yellow color) represent Faults.

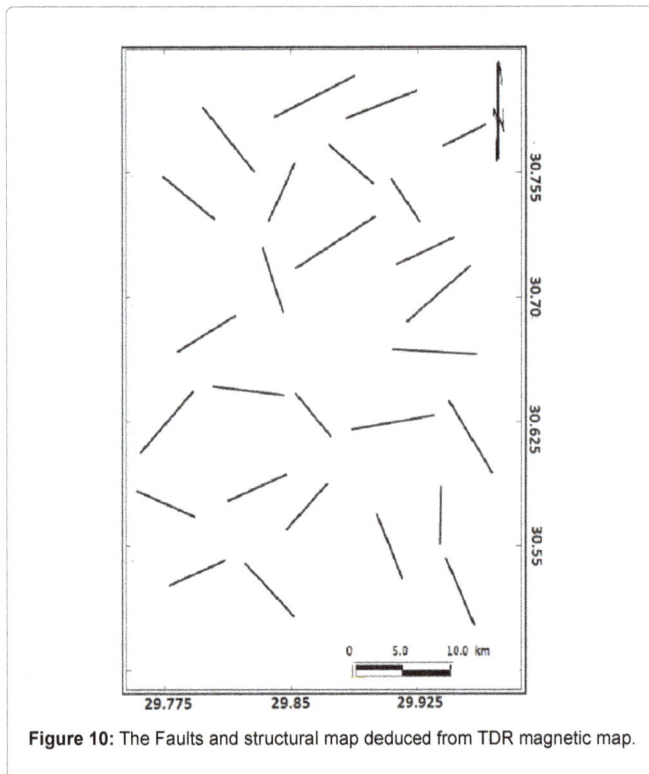

Figure 10: The Faults and structural map deduced from TDR magnetic map.

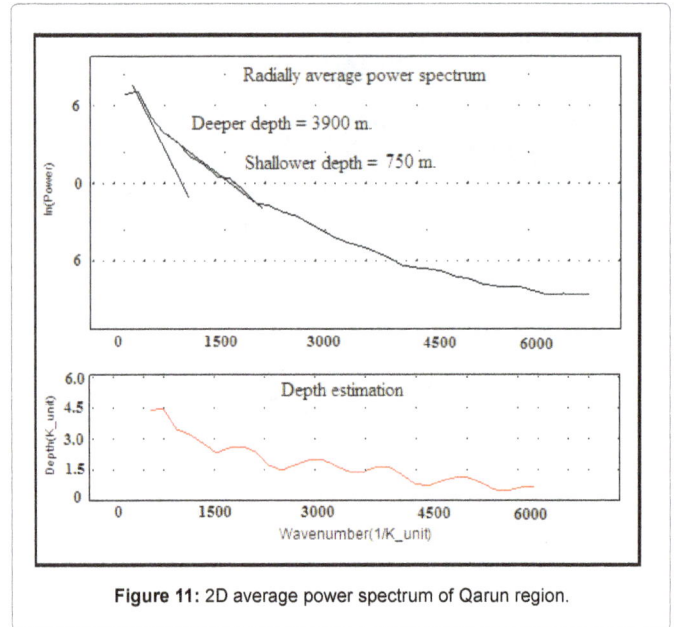

Figure 11: 2D average power spectrum of Qarun region.

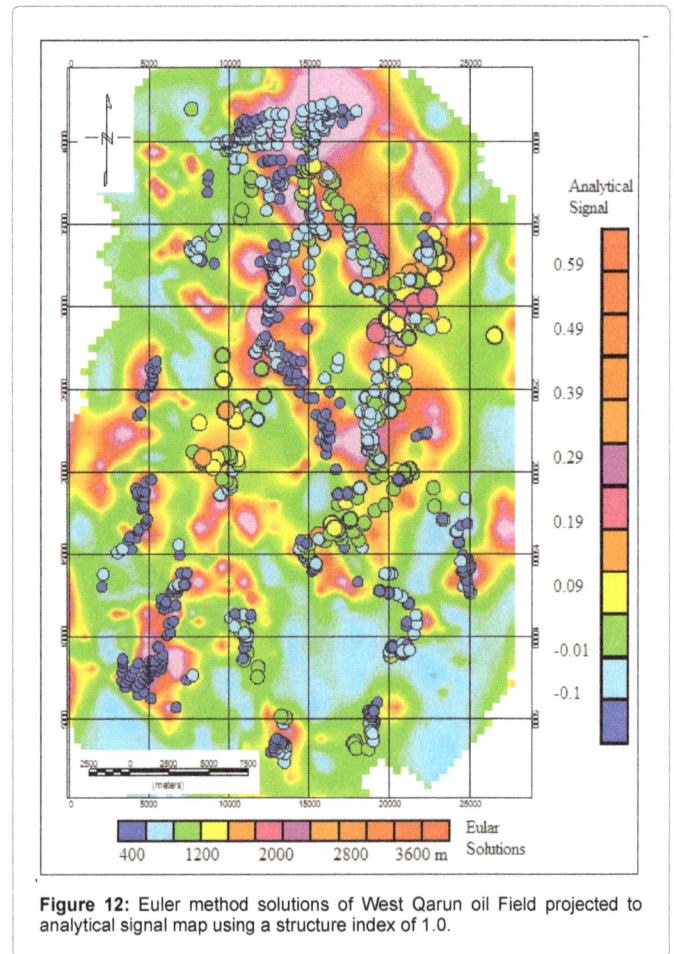

Figure 12: Euler method solutions of West Qarun oil Field projected to analytical signal map using a structure index of 1.0.

Results and Conclusion

The previous magnetic maps should be converted in terms of local structures in the area (Figures 7-10). The solid black lines are expecting extensions of fractures in the basement rocks with the major trend NE-SW and NW-SE. The study is based on application of Horizontal gradient, tilt derivative, 3D Euler and 2D Power Spectrum methods. Results of these methods help to define the main geological trends and depths of subsurface structures. Also add a new insight on the subsurface structural settings of West Qarun concession from the land magnetic data.

References

1. Dolson JC, Shann MV, Matbouly S, Harwood C, Rashed R, Hammouda H (2001a) In: Downey MW, Threet JC, Morgan WA (eds.) The petroleum potential of Egypt: Petroleum provinces of the twenty-first century. AAPG Memoir 74: 453-482.

2. Dolson JC, Shann MV, Matbouly SI, Hammouda H, Rashed RM (2001b) Egypt in the twenty-first century: petroleum potential in offshore trends. GeoArabia 6: 211-230.

3. El-Nady MM (2009) Biomarkers assessment of crude oils and extracts from Jurassic-Cretaceous rocks, North Qattara Depression, North Western Desert, Egypt. Petrol Sci Tech 26: 1063-1080.

4. Said R (1990) The Geology of Egypt. Elsevier, London, New York, 337.

5. Schlumberger company (1995) Well evaluation conference, Egypt. Schlumberger Technical Editing Services, Chester. p.87.

6. Nemec M (1996) Qarun oil field, Western Desert, Egypt. EGPC 14th Petrol. Explor Prod Conference, Cairo, Egypt p.140-164.

7. Abdel Aziz W, Zahra H (2003) Delineation of the subsurface structures and tectonics of Qarun area using potential field data. Egyptian Geophys Soc EGS 1: 17-29.

8. Ahmed F, El-bohoty M, Abdel Aal E (1998) Magnetic acquisition and geophysical interpretation for the area around Qarun oil field. NRIAG Bull B.

9. Geosoft program (7.01) (2007) Processing and analysis geophysical data, Canada.

10. Cordell L, Grauch VJS (1985) In: Hinze WJ (Ed.) Mapping basement magnetization zones from aeromagnetic data in the San Juan Basin, New Mexico. The utility of regional gravity and magnetic anomaly maps. Soc Explor Geophys p.181-197.

11. Verduzco B, Fairhead JD, Green CM, MacKenzie C (2004) New insights into magnetic derivatives for structural mapping. Lead Edge 23: 116-119.

12. Spector A, Grant FS (1970) Statistical models for interpretation aeromagnetic data. Geophys 35: 203-302.

13. Reid AB, Allsop JM, Granser H, Millett AJ, Somerton IW (1990) Magnetic interpretation in three dimensions using Euler Deconvolution. Geophys 55: 80-90.

14. Roest WR, Verhoef J, Pilkington M (1992) Magnetic interpretation using 3-D analytic Signal. Geophys 57: 116-125.

15. Egyptian General Petroleum Corporation (EGPC) (2009) Geologic information of the Qarun oilfield in the North Western Desert, Egypt.

16. Geometrics (2008) Proton Precision Magnetometers G-856 AX, USA.

17. GEM (2009) Over hauser Magnetometer, Canada.

22

Simulating Arsenic Mitigation Strategies in a Production Well

Dara A. Goldrath[1]*, John A. Izbicki[2] and Kathryn W. Thorbjarnarson[3]

[1]USGS, 6000 J St. Sacramento, CA 95819, USA
[2]USGS, 4165 Spruance Rd. Suite 200, San Diego, CA 92101, USA
[3]San Diego State University, 5500 Campanile Drive, San Diego, CA 92103, USA

Abstract

Water from well 5N/4W-31A1 in the regional aquifer in the Mojave River groundwater basin 97 kilometers north east of Los Angeles, California, occasionally exceeds the U.S. EPA Maximum Contaminant Limit for arsenic of 10 micrograms per liter (μg/L). Coupled well-bore flow and depth-dependent water-quality sampling for this well show arsenic concentrations less than 0.12 μg/L entering the well from the aquifer in the upper 163 meters (m) below land surface (bls). Arsenic concentration increase with depth to a maximum of 17.6 μg/L at 213 m bls. High arsenic in the deeper part of the well are associated with pH greater than 9 and dissolved oxygen concentrations less than 0.2 milligrams per liter. An axially-symmetric, radial groundwater flow simulation, developed using the computer program AnalyzeHOLE, was used to simulate flow to the well under pumping conditions. Simulations show that modifying the existing well design by eliminating the two deepest screened intervals below 189 m bls would reduce arsenic concentrations in the surface discharge of the well about 25 percent with a 30 percent reduction in yield. Such well modification may reduce or eliminate the need for costly arsenic treatment or blending of waters from different sources to reduce arsenic concentrations in water delivered to consumers.

Keywords: Arsenic; Groundwater; Well modification

Introduction

Arsenic (As) occurs naturally in rocks and water. Sources of arsenic in groundwater include the weathering of sulfide minerals, desorption from sediments under alkaline conditions, evaporation processes in closed and arid basins, volcanic rocks, and geothermal waters [1]. Consuming high concentrations of arsenic in drinking water can create health problems including bladder, lung, and skin cancers [2]. The U.S. Environmental Protection Agency Maximum Contaminant Level (MCL) for total arsenic in drinking water was reduced from 50 to 10 micrograms per liter (μg/L) in January 2001, and compliance with this standard was required beginning in January 2006.

Arsenic naturally sorbs onto oxide surfaces of mineral grains in sedimentary material. Two primary triggers are associated with arsenic mobilization into groundwater: (1) High pH values, particularly above pH 8.5, which can release arsenic sorbed to iron, manganese, and aluminum oxides on the surfaces of mineral grains under oxic conditions, and (2) Anoxic conditions, which can induce reductive dissolution of iron and manganese oxide coatings on mineral grains, releasing sorbed arsenic [3,4].

High concentrations of arsenic in groundwater have been identified in many areas of the USA, including Alaska, New England, some of the interior plains states, and the southwestern states of Nevada, California, and Arizona [3]. Arid regions of the southwestern United States often depend on groundwater resources to supply rapidly growing populations. Arsenic is in water from some wells in the Mojave River basin in the western Mojave Desert of southern California at concentrations in excess of the MCL for arsenic [5-8].

This study well is in the Mojave River basin and Mojave Desert near Victorville, California, about 97 km northeast of Los Angeles (Figure 1). The population of Victorville has increased from 64,029 in 2000 to 120,336 in 2008. The demand for groundwater has increased with population growth, and withdrawals exceed natural recharge. Pumping in excess of recharge since the mid-1940s has resulted in a decline in groundwater levels, degradation of groundwater quality, and land subsidence [9,10]. Arsenic concentrations from the Ground Water Ambient Assessment (GAMA) Program sampling, combined with California Department of Public Health (CDPH) sampling, in the

Mojave River basin 2000 to 2008) ranged from less than 2 μg/L to more than 50 μg/L and 21 wells had arsenic levels in excess of the USEPA MCL for arsenic of 10 μg/L. Well 5N/4W-31A1S near Victorville has a history of arsenic concentrations exceeding the MCL. In February 2008, the arsenic concentration in the surface discharge from this well was 17 μg/L.

Treatment strategies for water having high-arsenic concentrations include arsenic removal through coagulation/filtration and iron oxide adsorption [11,12]. Low-arsenic waters can also be blended with high-arsenic water from different sources to meet water quality standards. However facilities for arsenic mitigation using these methods are costly to construct and operate.

Another approach applicable in some groundwater settings is well modification [4,13]. The well-modification approach identifies zones of poor water-quality encountered by the well using coupled well-bore flow and depth-dependent water-quality data. This information is used to identify intervals having poor-quality water and modify the well construction to seal off those intervals from the well, thereby preventing poor quality water from entering the well and improving the quality of water yielded by the well. This method has been shown to inexpensively reduce high concentrations of contaminants such as chloride and arsenic in water from wells [14-16].

A study of high-arsenic concentrations in water from wells in the San Joaquin Valley near Stockton, CA by Izbicki et al. [4] demonstrated that well modifications in unconfined alluvial deposits could reduce arsenic concentrations in the surface discharge from wells to below the MCL. In comparison with the Mojave Desert, the alluvial deposits

*Corresponding author: Dara A. Goldrath, USGS, 6000 J St. Sacramento, CA 95819, USA, E-mail: dgold@usgs.gov

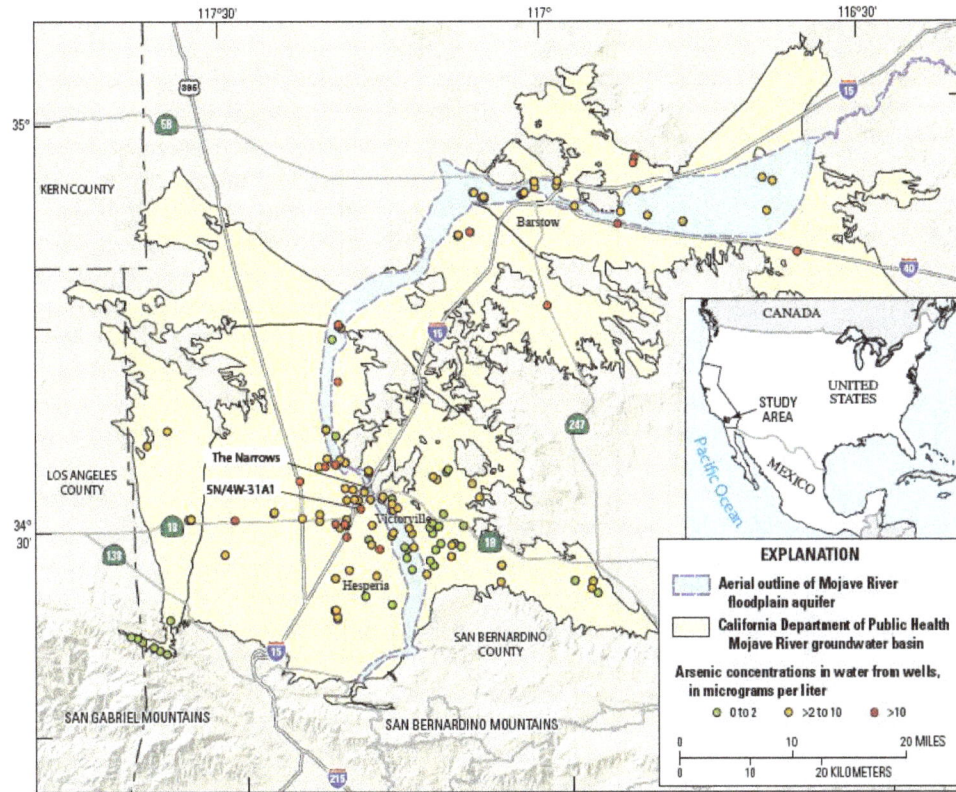

Figure 1: Study area, study well 5N/4W-31A1S, and arsenic concentrations in water from wells in the Mojave River groundwater basin, near Victorville, California 2000-2008; Arsenic data from the Department of Public Health Services and the Groundwater Ambient Monitoring and Assessment Priority Basin Project.

of the Stockton area are relatively more fine-textured and the wells were completed at shallower depths. Halford et al. [17] showed that well modification could reduce arsenic concentrations from more than 50 µg/L to 3 µg/L in water from a well in Antelope Valley, within the Mojave Desert about 80 km northwest of Victorville, CA. However, in the Antelope Valley, deeper high-arsenic water was separated from shallower low arsenic water by a thick, low-permeability lacustrine clay deposit that limited the upward movement of high-arsenic water to the modified well. This study examines a well drilled into unconfined alluvial deposits of the Mojave Desert that has high-arsenic water.

Hydrogeology

The Upper Mojave River Groundwater Basin has an arid climate, characterized by low humidity, low precipitation, and high summer temperatures [18,19]. Surface drainage is through the Mojave River, which originates in the San Bernardino Mountains, and flows north through Victorville [18]. The Mojave River flows only intermittently after winter storms; during the summer months, the river is dry [20]. Because of the lack of perennial streamflow, groundwater is the only dependable source of water supply in the area and is the focus of this study.

The Mojave River Groundwater Basin contains an unconsolidated alluvial aquifer along the Mojave River called the floodplain aquifer, Holocene to Pleistocene in age, consisting of sand and gravel weathered from granitic rocks in the San Gabriel and the San Bernardino Mountains [21]. The floodplain aquifer is typically less than 80 meters thick, surrounded and underlain by the more areally extensive regional aquifer, composed of basin fill and alluvial fan material deposits, Holocene to Miocene in age, eroded from the San Bernardino and San Gabriel Mountains [9,22]. Consolidation of the regional aquifer

deposits increases with depth [23]. Near Victorville, surrounding and underlying the floodplain aquifer, the regional aquifer includes deposits from the ancestral Mojave River. The ancestral Mojave River deposits are Pleistocene to Pliocene in age and range in depth from about 130 m to almost 200 m in thickness [9]. The ancestral Mojave River deposits are highly-permeable compared to the alluvial fan and basin fill material elsewhere in the regional aquifer. The regional aquifer contains a large amount of groundwater in storage; however, recharge is small in comparison with the floodplain aquifer [9]. As a consequence, some water in the regional aquifer was recharged more than 20,000 years ago [20,24,25]. This water has reacted extensively with minerals in aquifer deposits and is often highly alkaline with pH exceeding 8.0 in deeper deposits [24]. Many trace elements, such as arsenic, are soluble in groundwater under these conditions.

Well 5N/4W-31A1S was drilled in 2003 and screened within the regional aquifer, although most of the perforated interval is within ancestral Mojave River deposits. The well is 213 m deep and is screened from 151 to 154 m, 159 to 163 m, 166 to 171 m, 175 to 180 m, 189 to 201 m, and 209 to 213 m bls (Figure 2). The perforations from 151 to 201 m bls are within the ancestral Mojave River deposits. The deeper perforations are within the underlying alluvial fan and basin-fill deposits. During May 2008, the static water level was approximately 104 m bls. When pumped, the well yield was 190 L/s, and the pumping water level was approximately 130 m bls (26 m of drawdown).

Purpose and scope

The purpose of this study was to evaluate the well modification method as an alternative approach to arsenic mitigation in a public-supply well in relatively deep unconfined alluvial deposits of the

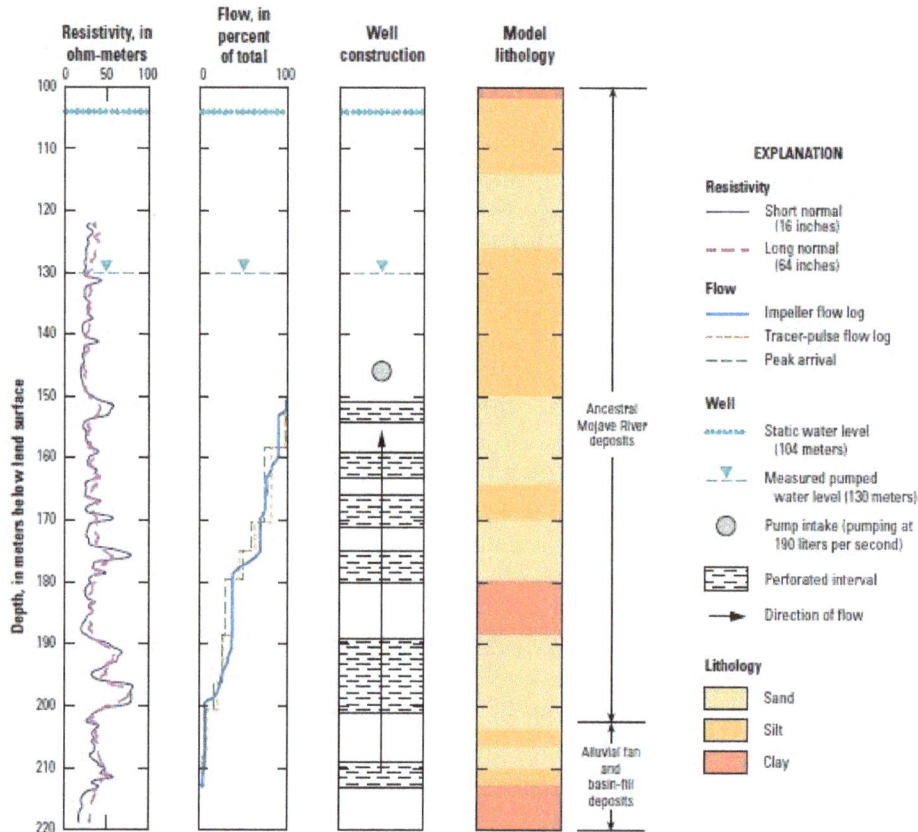

Figure 2: Resistivity log, flow from well impeller tool (May, 2008) and dye tracer tool (June, 2008), well construction, driller's log general lithology, and lithology for well 5N/4W-31A1S, near Victorville, California.

Mojave Desert. Well 5N/4W-31A1S, operated by the city of Victorville, was selected for this study because it contained arsenic concentrations in excess of the MCL of 10 μg/L. The scope of the study included collecting well-bore flow and depth-dependent water-quality. The data were related to aquifer property and hydraulic data, and interpreted using the computer program AnalyzeHOLE [13]. This study was completed as part of the USGS-GAMA Ambient Priority Basin Project in California.

Methods

Coupled well-bore flow and depth-dependent water samples were collected from well 5N/4W-31A1S in May and June of 2008. Well-bore flow was measured under pumping conditions using a commercially available impeller flowmeter and using the tracer-pulse method [26,27]. Water levels were measured in the well and the pumping rates were recorded throughout the logging of the well. Pumping and drawdown data collected over a 4-hour period were used to calculate aquifer transmissivity using the Cooper-Jacob method, a simplification of the Theis solution for unconfined aquifers [28]. Water at specific depth intervals within the well, selected on the basis of the well construction and velocity log data, was sampled under pumping conditions using a gas-displacement pump with a diameter of less than 2.5 cm [27].

The impeller-flow logs were collected by trolling the tool downward though the well at three rates: approximately 10, 20, and 30 m per minute (actual trolling rates were 9, 18, 27 m per minute). Access to the well was through a specially designed tube (commonly known as a camera tube) that entered the well below the pump intake.

Well-bore flow logs were examined for consistency. Data from the three trolling rates were used to calibrate the impeller tool and to evaluate the precision and sensitivity of the tool (Appendix A).

The dye tracer-pulse method measures flow and uses a high-pressure hose equipped with valves to inject rhodamine dye at known depths in a well [20,27]. The arrival of the dye at the surface discharge of the well is measured. For an interval within the well, a flow velocity is calculated from the difference in arrival times between the two injection depths that bracket the interval. A velocity profile was constructed from a series of such injections at different depths in the well (Figure 2). The dye tracer-pulse flow log measurements were collected at 11 depths above screened intervals and within screened intervals to determine the velocity profile within the well. For this study, dye tracer-pulse logs were analyzed using the first arrival and the peak arrival times (Figure 2). The first arrival time is when the injected rhodamine dye was first measured in the well water at surface distribution. The peak arrival is when the maximum amount of rhodamine dye was measured in in the well water at the surface distribution.

The impeller-flow logs were compared with the dye tracer-pulse flow logs to confirm estimates of flow into the well. It was useful to use both methods to identify intervals of low flow and high-yield flow into the well. Although results from both logs are similar, the impeller logs can define flow into the well with greater resolution than the tracer-pulse logs for aquifers where thin intervals contribute large amounts of flow into the well therefore the final interpreted log used for model simulations was derived from the impeller-flow log. The key depth-dependent water-quality parameters pH and dissolved oxygen were measured. These parameters were chosen as it is high pH values and very low dissolved oxygen that trigger desorption of arsenic from aquifer material. Groundwater samples collected to be analyzed for major and minor ions and trace elements, including arsenic, were filtered and preserved in the field and shipped within 24 hours to the USGS National Water Quality Laboratory (NWQL) in Denver, CO (Appendix A). The water-quality profile for arsenic was constructed from sample results from 7 depths including at land surface (Figure 3). The concentrations of arsenic (Ca) at the first sample depth (C_1) and the next sample depth (C_2) were used with velocity-log flow data at the first sample depth (Q_1) and the next sample depth (Q_2) to calculate the arsenic concentration in the water entering the well from the adjacent aquifer zone [27]:

Equation 1: $(Ca) = [(C_1Q_1 - C_2Q_2)/Qa]$; where $Qa = (Q_1 - Q_2)$

Results

Well-bore flow data

The impeller-flow log and dye tracer-pulse result profiles show two high-yield water-bearing zones within the well (Figure 2). Both first arrival and peak arrival dye tracer-pulse logs are plotted (Figure 2), the methods produced similar results. The first high-yield zone corresponds with the screen from 175-180 m bls and produces approximately 30 percent of flow; the second zone corresponds with the screen from 189-201 m and also produces approximately 30 percent of flow into the well. Both these zones are within the ancestral Mojave River. The upper three screened intervals within the well contribute only small amounts of flow (Figure 2). Two of the intervals correspond to relatively low resistivity values on the resistivity log collected at the time the well was

drilled. However, the uppermost interval is within a more resistive unit that may have been expected to yield more water to the well. Geologic and geophysical data collected from a well at the time of drilling are indirect measures of potential well performance, well-bore flow data are a more direct measure of well performance that integrates the hydraulic properties of the deposits encountered by the well and how these deposits are connected to deposits farther away from the well when the well is pumped. Within the basin-fill deposits, flow into the screen from 209-213 m bls was only 8 percent of the total yield (Figure 2).

Water-chemistry data

Depth-dependent samples collected within the well under pumping conditions show that pH and arsenic concentrations increase with depth while dissolved oxygen concentrations decrease with depth (Figure 3). Arsenic concentrations in samples from the well ranged from 6.4 to 17.6 µg/L. High concentrations of arsenic, greater than the MCL of 10 µg/L, enter the well from the aquifer below 166 m bls. Most of the arsenic enters the well from the screened intervals between 189 and 213 m bls where approximately 38% of the total inflow enters the well.

Arsenic concentrations begin to increase in the well at a depth of 166 m within the high-water yielding ancestral Mojave River deposits and above the low-yielding underlying alluvial fan and basin fill deposits. Dissolved oxygen concentrations begin to decrease at this depth (Figure 3) and arsenic concentrations in this area may be controlled by geochemical factors, such as pH and redox at deeper depths, rather than geologic factors, such as the source of the alluvial deposits. Increasing arsenic concentrations with depth agrees with water-quality data for arsenic concentrations reported in Mojave River basin monitoring wells [6].

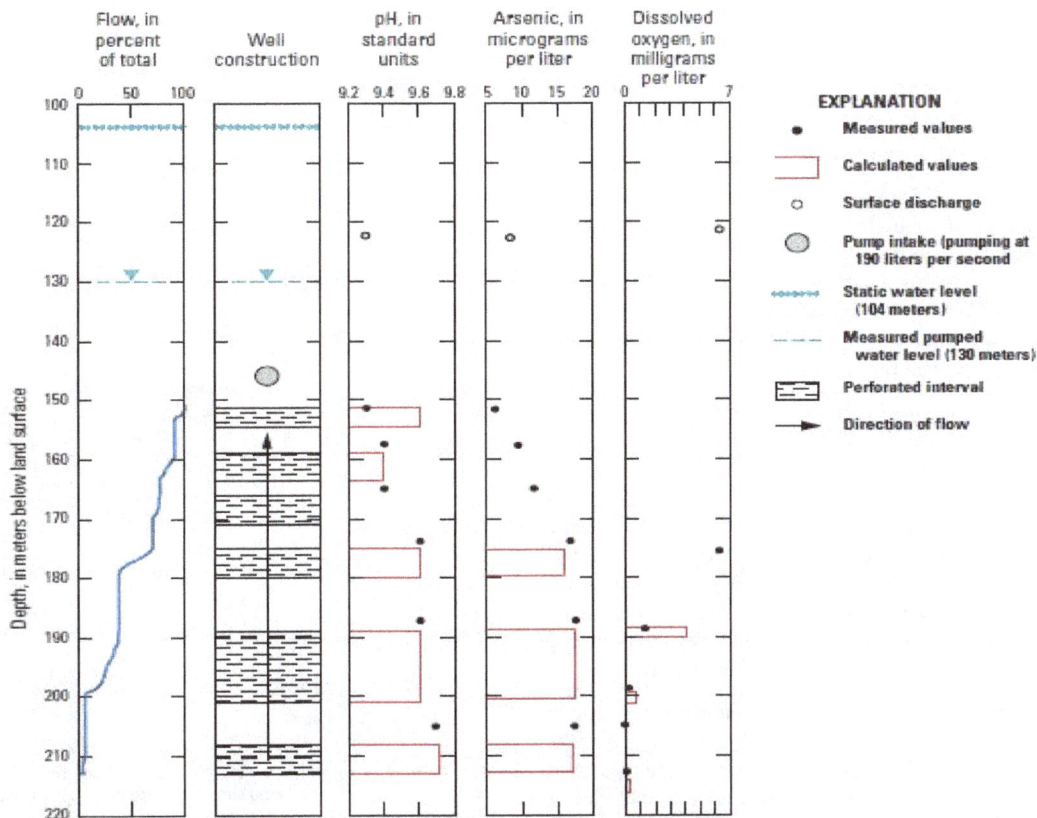

Figure 3: Selected well-bore flow (May, 2008) and depth dependent water-quality data (June, 2008) under pumping conditions from well 5N/4W-31A1S.

Simulation of well-bore flow

The computer program AnalyzeHOLE, a well-bore analysis tool, was used to simulate groundwater well-bore flow and arsenic concentrations in well 5N/4W-31A1S and the adjacent aquifer [13]. The program uses MODFLOW to simulate axially-symmetric, two-dimensional radial flow to the well in response to pumping. MODPATH a particle-tracking program embedded within AnalyzeHOLE, was used to calculate arsenic contributions before and after simulated well modifications [29,30].

The simulation consists of a cylinder of aquifer material with a radius of 6.1×10^4 m and a thickness of 143 m; the simulation grid has 76 variably-sized columns in the lateral direction and 100 rows of uniform thickness in the vertical dimension (Figure 4). This large simulated volume was used to ensure that lateral no-flow boundaries are beyond the pumping effects of the well. Hydrologic conductivities were initially assigned to the simulation on the basis of aquifer lithology described in the driller's log [31]. Aquifer transmissivity was estimated to be 1,200 m²/day using measured drawdown and well during sampling. The observed pumping rate of 190 L/s during sampling was the pumping rate used in the simulations.

The computer program MODPATH was used to simulate the movement of water particles within the simulation. For numerical purposes, particle movement was simulated as injection rather than withdrawal and the simulation assumes withdrawal is the mirror image of the injection [30]. To simulate pumping using this approach, the pumping water level within the simulation was approximated using the Theis equation and did not change during the simulation. The error induced by this approximation is believed to be small [17]. Movement of water to the well is shown as the particle pathlines that track particle movement in response to simulated pressure changes in the well from simulated pumping. Each cell of the simulation grid contained one particle that represents a discrete fraction of water contribution to the well with a unique water quality. The water quality produced by well 5N/4W-31A1S was calculated as the flow-weighted average of the particle concentrations. Arsenic concentrations for particles within the model ranged from 2 to18 µg/L.

The simulation was calibrated by adjusting the hydraulic conductivity of simulated aquifer material within reasonable ranges, from less than 1 to 15 m/day, to match the measured impeller-flow log data and observed drawdown while maintaining a constant transmissivity (Figure 5). The hydraulic conductivity of clay, silt, and sand are paired with lithology in Figure 5. The measured water-level decline (26 m) during pumping was simulated within 1 m using the estimated transmissivity value 1,200 m²/day and assigned hydraulic conductivities.

Simulations were generated to evaluate how well yield and arsenic concentrations in the surface discharge of the well would vary in response to changes in well construction. The first simulation used the existing well construction (213 m depth) to generate drawdown and particle movement; in each additional simulation, screened intervals were eliminated one at a time and the simulation was run to evaluate the arsenic concentrations to changes in the well construction. The time period for each simulation was 1,000 days of pumping.

The simulated drawdowns and particle movement generated for well 5N/4W-31A1S during a 1000 day period are shown in Figure 6. The simulated drawdown of 27 m compared closely to the measured drawdown of 26 m during well sampling. The simulated particles of water moved faster through more permeable deposits, and particles near the water table moved steeply downward until encountering

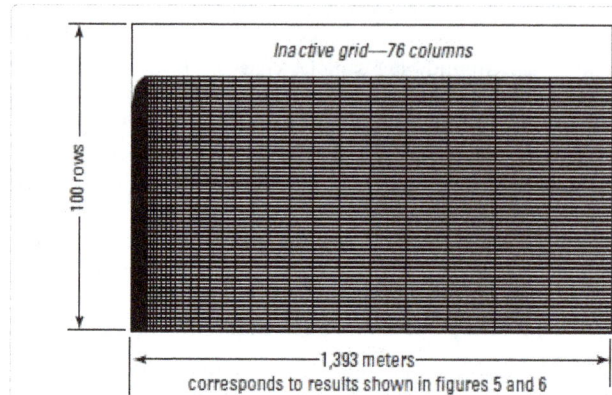

Figure 4: Model grid used to simulate flow to well 5N/4W-31A1S, near Victorville, California.

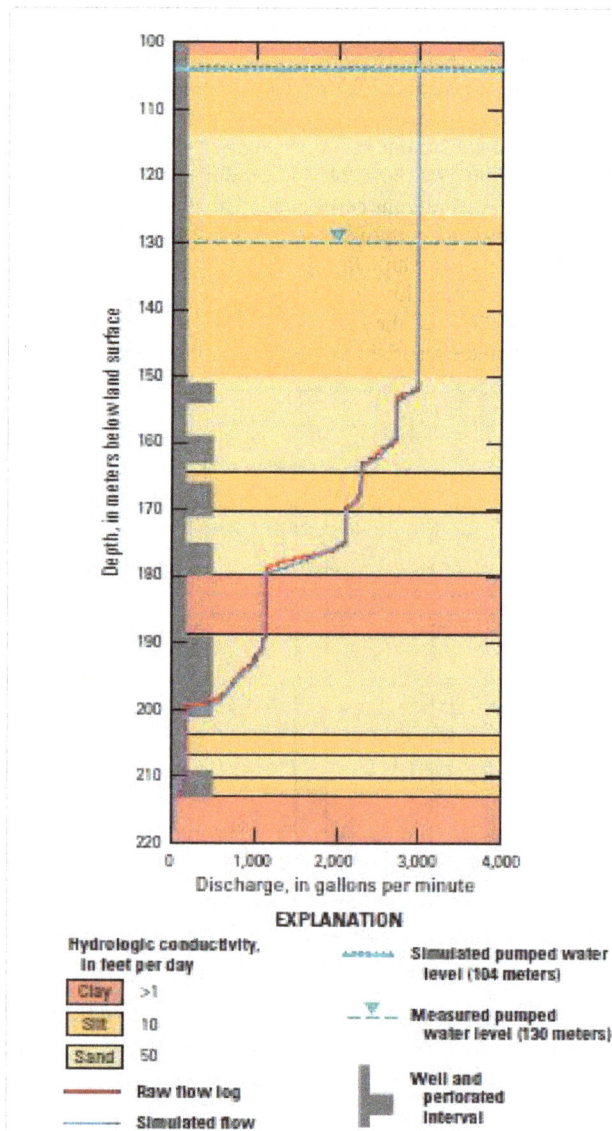

Figure 5: AnalyzeHOLE simulation results, from well 5N/4W-31A1S, near Victorville, California, showing simulated discharge compared to flow measured with impeller-flowmeter and lithological distributions for hydraulic conductivities based on calibrated simulation lithology, adjusted from initial estimates from resistivity log.

coarse-grained aquifer material where upon they moved toward the well Figure 6. The simulated arsenic concentration in well surface discharge water was 8.8 μg/L which compares closely to the sampled arsenic concentration of 8.4 μg/L.

Simulated drawdown and particle movement during pumping, after the two deepest screened intervals were removed from the simulated well are shown in Figure 7. High-arsenic water that had entered through the two deeper screened intervals no longer entered the well directly, but some of this water did move up through aquifer material and enter through the deepest remaining screen interval. Arsenic concentrations in surface discharge in the simulation of the modified well decreased to 6.7 μg/L and remained constant for the remainder of the 1,000 day simulation. This value represents about a 25% decrease in arsenic concentration. Elimination of the two deepest screened intervals resulted in a simulated decrease in well yield of 30 percent.

Additional simulations were run to calculate the change in arsenic concentrations in surface discharge after one, then two, then three of the lower three screened intervals were eliminated (Figure 8: scenarios 2-4). There was little change in arsenic concentration after removing the deepest screen interval, which is consistent with the low yield measured at that depth. Decreases in simulated arsenic concentrations in the surface discharge of the well were approximately 2 μg/L in scenario 3 when the second deepest screened interval was removed. The lowest arsenic concentrations were in scenario 4, in which the third deepest screened interval was sealed. However this simulation also resulted in the greatest yield reduction of 70%. The simulated arsenic concentrations in all tested scenarios were initially low then increased during the first 100 days of simulated pumping to a steady-state concentration for the remainder of the 1,000 day simulation as deeper water moved upward through the aquifer in response to pumping.

Simulation sensitivity analysis and limitations

The simulation sensitivity to changes in porosity, vertical anisotropy, specific storage, and specific yield was tested. Porosity, vertical anisotropy, specific storage, and specific yield were increased and/or decreased then the simulation was run to observe results. Changes in these input parameters had little to no effect on simulation results. The simulation sensitivity to bore-hole flow and drawdown was tested. The simulated bore-hole flow and drawdown were most sensitive to the changes in hydraulic conductivity. When lithological units with the conductivity values of sand were increased or decreased it produced deviations from the measured flow log and transmissivity.

The simulation developed to interpret well-bore flow and depth-dependent water-quality data from well 5N/4W-31A1S is a simplified two-dimensional radial representation of the surrounding regional aquifer flow system. The simulation assumes aquifer materials are flat-lying and areally extensive and does not account for no-flow boundaries, regional changes in subsurface geology, hydraulic variations, or interactions between surrounding pumping wells. The flow simulation is intended to be a simple tool useful for evaluating the effects of well design modifications on surface discharge water-quality and not an accurate representation of the regional groundwater flow field near the well.

Discussion

Coupled well-bore flow and depth-dependent water-chemistry data show that arsenic concentrations and pH values increase with depth in well 5N/4W-31A1S while dissolved oxygen concentrations

decrease with depth. Increases in arsenic concentrations in reducing conditions (dissolved oxygen less than 0.5 μg/L) below 175 m are consistent with reductive dissolution of iron hydroxide coatings on mineral grains and subsequent mobilization of arsenic. Arsenic concentrations change abruptly at the 175 m depth, increasing from less than the detection limit of 10 μg/L to almost 17 μg/L. This increase is controlled by changing redox conditions and does not occur at the geologic contact between the ancestral Mojave River deposits and the underlying basin-fill and alluvial fan deposits. Ancestral Mojave River deposits contribute most of the water to this well.

Data interpreted using AnalyzeHOLE to evaluate the effects of changes in the simulated well design on arsenic concentrations confirms that sealing off the bottom two screened intervals reduced

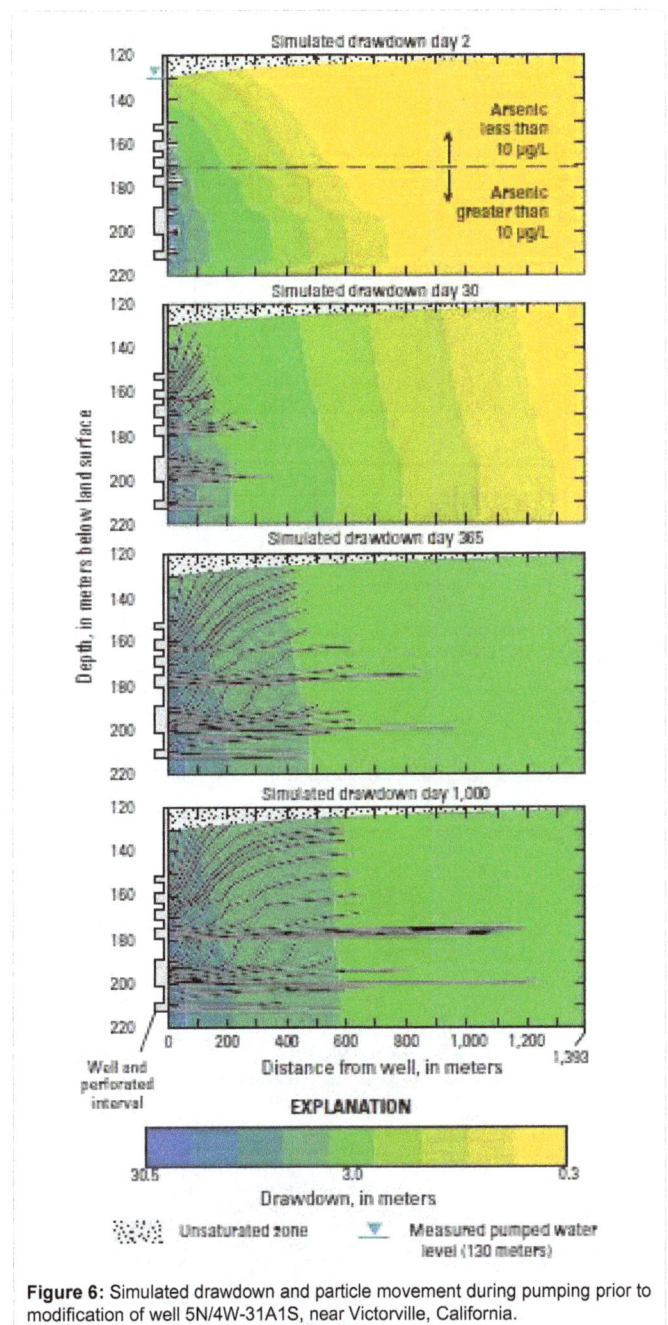

Figure 6: Simulated drawdown and particle movement during pumping prior to modification of well 5N/4W-31A1S, near Victorville, California.

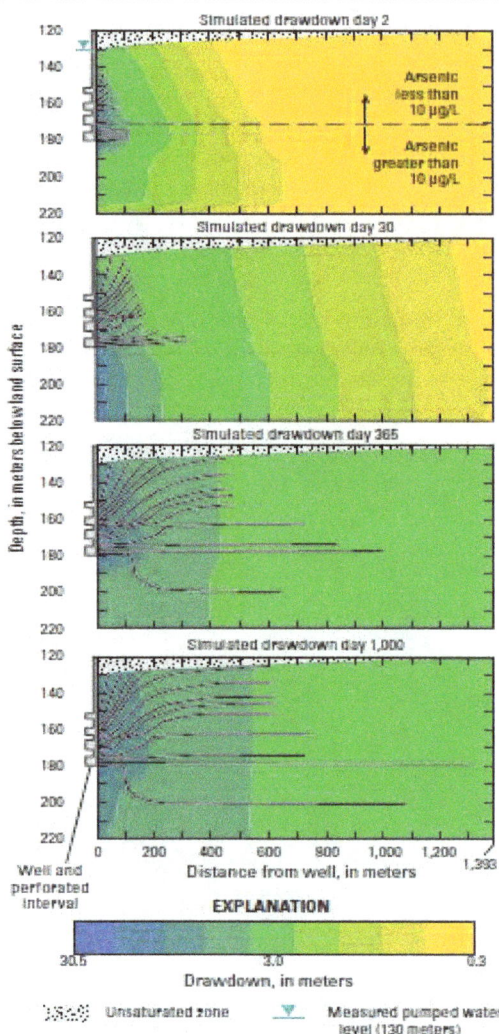

Figure 7: Simulated drawdown and particle movement during pumping after modification to well 5N/4W-31A1S, near Victorville, California.

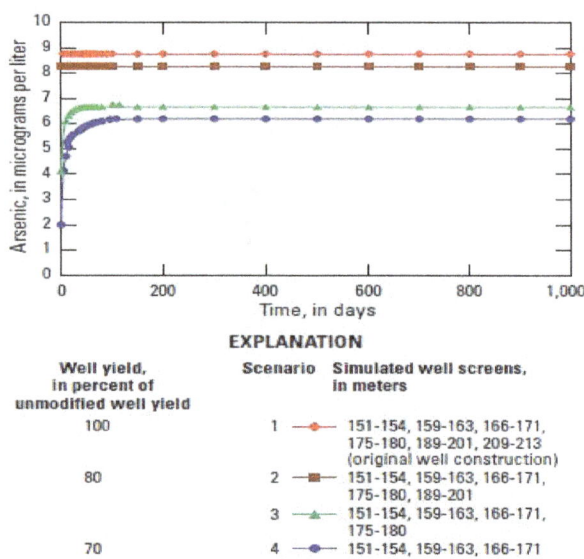

Figure 8: Simulated arsenic concentrations as a function of time after modification of well 5N/4W-31A1S, near Victorville, California.

simulated arsenic concentrations entering the well by about 25 percent to 6.7 μg/L and reduced the in well yield by 30 percent.

Water purveyors in the Mojave River groundwater basin near Victorville could benefit from modifying existing wells to reduce arsenic concentrations and from carefully designing future wells to ensure they do not penetrate depths containing high-arsenic groundwater. Results of this study show that this well yielding water exceeding arsenic concentrations above the MCL could be simply and cheaply modified to reduce arsenic concentrations to meet drinking water standards. Also high arsenic water could be avoided by drilling future wells to depths equal or lesser than 180 m bls. Such sampling methods, simulations, and well modification could also be used to target and address high concentrations of other trace elements such as chromium, in public supply wells.

Acknowledgements

The authors thank the GAMA Priority Basin Project for funding, and the Victor Valley Water District for well access and water-quality data. The authors also thank Christina Stamos, Nick Teague, Keith Halford, Joseph Montrella, and Gregory Smith of the USGS for assistance with collecting and data interpretation.

Appendix A: Comparison of well impeller trolling rates

Well-bore flow was measured under pumping condition using an impeller flowmeter in well 5N/4W-31A1S to provide well yield. To develop a calibration for well-bore flow, intercepts of the lines, 9 meters/minute and 18 meters/minute, were plotted against the difference in the trolling rates, 18 and 27 m/minute. This comparison of linear regression lines had a slope of approximately 1. This result indicates that the tool output was linear over the range of measured flow and suitable to develop a field calibration for the meter. The well-bore flow data was plotted with depth and adjusted by assuming zero flow into the well in blank (unscreened) casing intervals (Figure 2).

Water-quality samples were collected with a small diameter (less than 2.5 cm) gas-displacement pump. Water-quality samples collected at each well depth are a mixture of water from all the screened aquifer zones below the depth at which the sample was collected [14,27]. Samples were collected in accordance with the protocols established by the USGS National Water Quality Assessment (NAWQA) program and the USGS National Field Manual [32]. Arsenic was analyzed at the USGS NWQL by inductively coupled plasma mass spectrometry (ICP-MS) [33].

References

1. Welch AH, Stollenwerk KD (2003) Arsenic in ground water: Geochemistry and Occurrence. Kluwer Academic Publishers, Boston.

2. Lubin JH, Beane Freeman LE, Cantor KP (2007) Inorganic arsenic in drinking water: an evolving public health concern. National Cancer Institute 99: 906-907.

3. Smedly PL, Kinniburgh DG (2002) A review of the source, behavior, and distribution of arsenic in natural waters. Appl Geochem 14: 517-568.

4. Izbicki JA, Stamos CL, Metzger LF, Kulp TR (2008) Source, distribution, and management of arsenic in water from wells, eastern San Joaquin Ground-Water Subbasin, California. US Geol Surv Open File Rep 2008-1272.

5. Christensen AH, Fields-Garland LS (2001) Concentrations for total dissolved solids, arsenic, boron, fluoride, and nitrite–nitrate for wells sampled in the Mojave Water Agency Management Area, California, 1991–97. US Geol Surv Open-File Rep 2001-84.

6. Huff JA, Clark DA, Martin P (2002) Lithologic and ground-water data for monitoring sites in the Mojave River and Morongo ground-water basins, San Bernardino County, California, 1992-98. US Geol Surv Open-File Rep 02-354.

7. Stamos CL, Cox BF, Mendez GO (2003) Geologic setting, geohydrology and ground-water quality near the Helendale Fault in the Mojave River Basin, San Bernardino County, California. US Geol Surv Water-Resources Inves Rep 03-4069.

8. Mathany TM, Belitz K (2009) Groundwater quality data in the Mojave study unit, 2008: Results from the California GAMA Program. US Geological Survey Data Series 440: 80.

9. Stamos CL, Martin P, Nishikawa T, Cox BF (2001) Simulation of ground-water flow in the Mojave River basin, California. US Geol Surv Sci Inves Rep 01-4002: 129.

10. Sneed M, Ikehara EM, Stork SV, Amelung F, Galloway DL (2003) Detection and measurement of land subsidence using interferometric synthetic aperture radar and global positioning system, San Bernardino County, Mojave Desert, California. US Geol Surv Sci Inves Rep 03-4015.

11. Fields AK, Chen A, Wang L (2000) Arsenic removal from drinking water by coagulation/filtration and lime softening plants. National Risk Management Research Laboratory US, EPA.

12. Joshi A, Chaudhuri M (1996) Removal of arsenic from ground water by iron oxide-coated sand. Jou of Environ Eng 122: 769-771.

13. Halford KJ (2009) AnalyzeHOLE An integrated well-bore flow analysis tool. US Geological Survey Techniques and Methods 4-F2: 46.

14. Gossell MA, Nishikawa T, Hanson T, Izbicki JA, Tabidian MA, et al. (1999) Application of flowmeter and depth-dependent water quality data for improved production well construction. Ground Water 37: 729-735.

15. Ball JW, Izbicki JA (2004) Occurrence of hexavalent chromium in ground water in the Western Mojave Desert, California. Appl Geochem 19: 1123-1135.

16. Izbicki JA, Christensen AH, Newhouse MW, Smith GA, Hanson RT (2005) Temporal changes in the vertical distribution of flow and chloride in deep wells. Ground Water 43: 531-544.

17. Halford KJ, Stamos CL, Nishikawa T, Martin P (2010) Arsenic management through well modification and simulation. Ground Water 48: 526-537.

18. California Department of Water Resources (2004) Upper Mojave River Valley Groundwater Basin. Bulletin 118.

19. National Oceanic and Atmospheric Administration (2004) Monthly Station Climate Summaries, 1971-2000.

20. Izbicki JA (2004) Source and movement of the ground water in the western part of the Mojave Desert, southern California. US Geol Surv Sci Inves Rep 03-4313.

21. Lines GC (1996) Ground-water and surface-water relations along the Mojave River, southern California. US Geol Surv Sci Inves Rep 95-4189.

22. Cox BF, Hillhouse JW (2000) Pliocene and Pleistocene evolution of the Mojave River, and associated tectonic development of the Transverse Ranges and Mojave Desert, based on bore-hole stratigraphy studies near Victorville, California. US Geol Surv Open-File Rep 93-568.

23. California Department of Water Resources (1967) Mojave River Groundwater Basins investigation. Bulletin 84: 151.

24. Izbicki JA, Martin P, Michel RL (1995) Source, movement and age of groundwater in the upper part of the Mojave River basin, California. In: Adar, E.M., and Leibundgut, Christian, Application of tracers in arid zone hydrology: International Assoc of Hydrol Sci 232: 43-56.

25. Kulongoski JT, Hilton DR, Izbicki, JA (2003) Helium isotope studies in the Mojave Desert, California: Implications for groundwater chronology and regional seismicity. Chem Geol 202: 95-113.

26. Hill AD (1990) Production logging Theoretical and interpretive elements. Society of Petroleum Engineers 14: 154.

27. Izbicki JA, Christensen AH, Hanson RT (1999) U.S. Geological Survey combined well-bore flow and depth-dependent water sampler. US Geol Surv Fact Sheet 196-99.

28. Cooper HH, CE Jacob (1946) A generalized graphical method for evaluating formation constants and summarizing well field history. American Geophys Union Trans 27: 526-534.

29. Harbaugh AH, McDonald MG (1996) Programmer's documentation for MODFLOW-96, an update for the U.S. Geological Survey modular finite-difference ground-water flow model. US Geol Surv Open-File Rep 96-486.

30. Pollock DW (1994) User's guide for MODPATH/MODPATH-PLOT, version 3: a particle tracking post- processing package for MODFLOW, the U.S. Geological Survey's finite-difference ground-water flow model. US Geol Surv Open-File Rep 94-464.

31. Freeze AR, Cherry JA (1979) Ground Water. Prentice-Hall, Englewood Cliffs, N.J.

32. Koterba MT, Wilde FD, Lapham WW (1995) Groundwater data-collection protocols and procedures for the National Water-Quality Assessment Program Collection and documentation of water-quality samples and related data. US Geol Surv Open-File Rep 95–399.

33. Garbarino JR (1999) Methods of analysis by the U.S. Geological Survey National Water Quality Laboratory Determination of dissolved arsenic, boron, lithium, selenium, strontium, thallium, and vanadium using inductively coupled plasma-mass spectrometry. US Geol Surv Open-File Rep 99-093.

23

Lithofacies Superimposition in a Shallow Basin - Interplay of Tectonics and Sedimentation: Evidences from Kolhan Basin, Eastern India

Smruti Rekha Sahoo* and Subhasish Das

Department of Geology and Geophysics, IIT Kharagpur, 721302, India

Abstract

The 2.2-2.1 Ga pear shaped Kolhan basin show the development of a time transgressive group in a half-graben setting developed during the fragmentation of the Rodinia supercontinent. The overall style of sedimentation reflect a switchover from low-sinuous avulsed channels developed within a braided-fluvial-ephemeral streams to a lacustrine fan-delta complex during the later part of the sedimentation history.

The fan-delta facies indicate sediment dispersal by hyperconcentrated flows in the form of sheetfloods and channelized flows. These different-scale cycles are interpreted as the sedimentary response to pulses of deformation of the basin margin at variable frequencies, related to the contemporary thrusts (ca. 20 km away from the basin). The episodes of tectonism downwarped the basin margin sediments and made the basin shift periodically toward the margin, and created progressive lithological changes in the sedimentary succession. The immediate effects of a tectonic pulse included lake transgression and accentuation of the structural hinge of the basin margin, causing a decline of sediment supply from the source rock. As the basin margin was subsequently reduced by denudation, the fans prograded and fan deltas were formed in normal conditions of graben subsidence. The sediment geometries and the climate exerted a major control on the processes of sediment transfer.

Our results show that the fluvio-lacustrine strata show on lap-pinch out relationship at the centre of the basin but only on lap relationship along the lateral edges. This transition can be best explained by the fault growth models.

Keywords: Lithofacies; Tectonics; Kolhan basin

Introduction

Lacustrine fan deltas are lake margin depositional systems that occur in a wide range of tectonic settings, but their facies assemblages and stratigraphic architecture vary, depending primarily on the basin margin tectonic regime and sediment dispersal processes, which account for the climatic and catchment conditions [1]. Whereas, interpretation of provenance from sediment compositional data requires consideration of controlling factors such as transport distance, time, energy, and climate of the basin [2]. Composition of detrital sediments is controlled by various factors, including source rocks, modes of transportation, depositional environments, climate, and diagenesis [3]. This paper attempts to briefly contrast the salient characteristics of Proterozoic clastic sedimentation with an attempt on some aspects of the lithofacies study and paleotectonics of Kolhans in the Chaibasa- Noamundi basin.

The Kolhans were deposited in the intracratonic basins that developed within the Singhbhum-Orissa Iron Ore carton and are preserved as isolated outliers that spread over four detached basins – Chaibasa- Noamundi basin (type area), Chamkapur-Keonjhar basin, Mankarchua basin and Sarapalli-Kamakhyanagar basin. After the close of the Iron Ore Orogeny, there was a phase of extension. During this phase, the eastern side of the Iron Ore synclinorium was faulted giving rise to a halfgraben structure. Within this half-graben, fluviolacustrine sedimentation took place. Because of the continued tectonic instability the Kolhan basin was transversely segmented into four smaller sub basins by en-echelon fault systems. After the separation of the basins, sedimentation in each individual basin took place in their own way. The overall style of sedimentation reflects a change from braided fluvial ephemeral pattern to a lacustrine fan delta type. The sediment geometries and the climate exerted a major control on the processes of sediment transfer. Repeated fault controlled uplift of the source followed by subsidence and regression, generated multiple sediment cyclicity that led to the fluvio lacustrine- fan delta sedimentation pattern. The Kolhans represent more than a single phase of deposition and the internal erosion surfaces are indicative of channel avulsions. The Kolhan sandstones and shales of the Orissa-Jharkhand state can be described by a time-transgressive lithofacies model consisting of an earlier braided stream-channel levee complex subsequently superimposed by a fan delta complex. The field and sedimentological evidences show the paleobathymetry, the depositional history and the time-transgressive nature of the lithofacies assemblage.

The concept of geochemical proxies of mineral alteration (i.e., weathering indices) relies on the selective removal of soluble and mobile elements from a weathering profile compared to the relative enrichment of rather immobile and non-soluble elements. The advantage of using Chemical Index of Alteration (CIA) to estimate paleoweathering in paleoenvironmental studies has proved valuable. The CIA has been extensively used for understanding the continental weathering and denudation studying the variability of the CIA determined for the mean suspended solids load of large world rivers [4]. Many authors made an interesting comparison between weathering intensity, concluding that CIA values reflect more aggressive chemical weathering during Proterozoic basins since less sediment residence

*Corresponding author: Sahoo SR, Department of Geology & Geophysics, IIT Kharagpur, 721302, West Bengal, Inida, E-mail: smruti@iitkgp.ac.in

times due to the absence of vegetation cover and therefore faster transport time. The CIA actually reflects changes in the proportion of feldspar and various clay minerals in the weathering product [5]. As a consequence, CIA values of about 45-55 indicate virtually no weathering, whereas the value of 100 indicates intense weathering with complete removal of alkali and alkaline earth elements [4].

The present study demonstrates the field observations for lithofacies analyses, sandstone petrography and shale geochemistry to explain the tectonics and sedimentation interplay in Kolhan basin.

Geological setting

The Proterozoic Kolhan Group is the youngest unit in the Pre-Cambrian Singhbhum–Iron Ore stratigraphy. The Kolhans were deposited in the intracratonic basins that developed within the Singbhum-Orissa Iron Ore craton. The Kolhans unconformably overlie the Singhbum granite at the eastern margin and shows a faulted contact with Iron Ore Group of rocks at the western margin [6] (Figure 1). The Kolhans are shale dominated succession, and consists of northeast trending and gently westerly dipping, unmetamorphosed and undeformed strata of conglomerate and sandstones at the base overlain by extensive occurrences of shale with lenticular patches of limestone. The strata encompass dome and basin structure in westward part and show low dip near Singhbhum granite. Tectonically the Kolhan basin represents an epicontinental type with a NNE-SSW alignment, controlled by the similar trend of the Iron Ore Group. It is remarkable however that major part of the Kolhans does not show any appreciable effect of the younger Singhbhum Orogeny (905-934 Ma) [7]. This is probably partly due to the distance of these rocks from the Singhbhum shear zone and partly due to the blanketing effect of the basement granite which acted as a shield to absorb the southwards directed tectonic movements.

Methods of Study

Field work has been carried out to study different lithounits exposed along the road cutting, railway cutting and river cutting sections. On the basis of dominant structures, texture and lithology various lithofacies have been identified. Lithologs were prepared on the basis of sedimentary structures, textures and grain size taken from 18 different locations viz. Gangabasa, ITI College (two sections), Rajanka, Gumuagara, Kamarhatu, Singpokharia, Arjunbasa, Tunglei, Gutuhatu, Bringtopang, Bistampur, Dyliamarcha, Matgamburu, Rajanka, ITI hill top, Surjabasa (Figure 2). There are five diagnostic characters of sedimentary facies namely, geometry, lithology (grain size), sedimentary structures, paleocurrent patterns and fossils [8,9]. Based on these parameters, six lithofacies have been established for the sandstone of Kolhan Group. Detailed petrographic studies of 105 sandstones samples were done for modal composition and a variety of other petrographical features. For each thin section 300 points were counted using the Gazzi-Dickinson method [8]. Shale samples were analyzed for major oxides by using XRF (X-ray Fluorescence Spectrometer).

Facies description

Lithofacies study has been done following standard litholog technique [9]. Eighteen different vertical sections prepared from different locations in and around Chaibasa-Noamundi basin show the coarsening to fining upward sequence from conglomerate, sandstone and shale resting unconformably on the Singhbhum Granite (Figure 2). The identified six lithofacies for the sandstone of Kolhan Group are (a) granular lag facies (GLA), (b) granular sandstone facies (GSD),

(c) sheet sandstone facies (SSD), (d) plane laminated sandstone facies (PLSD), (e) rippled sandstone facies (RSD) and (f) thinly laminated siltstone-sandstone facies (TLSD). Primary sedimentary structures of varied scale and geometries recognized in the Kolhan sediments are trough cross bedding, symmetrical ripple, planar cross bedding, hummocky cross bedding, and graded bedding. Among the reported structures, the structures generated by flat beds are comparatively more noticeable than the structures related to the bedform migration. The layer thickness variation within the eighteen lithologs is non-uniform and shows asymmetricity (systematic thinning or thickening-upwards) with general thickening upward pattern which represents sand bar/delta mouth bar deposits.

The GLA facies is characterized by the occurrence of laterally impersistent, massive, ungraded, fine matrix supported conglomerate which is oligomict in character towards south and polymictic towards north. These conglomerates are mostly immature to sub-mature, and quite similar to the overlying sandstone. (Figure 3a-b). GSD facies is characterized by moderately to well sorting, moderate clast : matrix ratio, textural bimodality and development of normal grading with fining upward sequence. Planar cross-stratification is more commonly found as compared to trough cross- stratification (Figure 3c-d). The SSD facies is defined by sheets of subarkose-quartz arenite, sometimes intercalated with thin laminated siltstone with profuse development of planar cross bedding, and locally developed herringbone cross-bedding (Figure 3e-f). The PLSD facies is defined by thick amalgamated well sorted subarkose-quartz arenite, with a moderate-high grain: matrix ratio. The sandstone is medium to fine grained. The prominent structures are planar cross bedding, asymmetrical ripple (ripple laminations) (Figure 3g-h). The RSD facies is defined by predominance of packages of rippled sandstone with prolific development of both

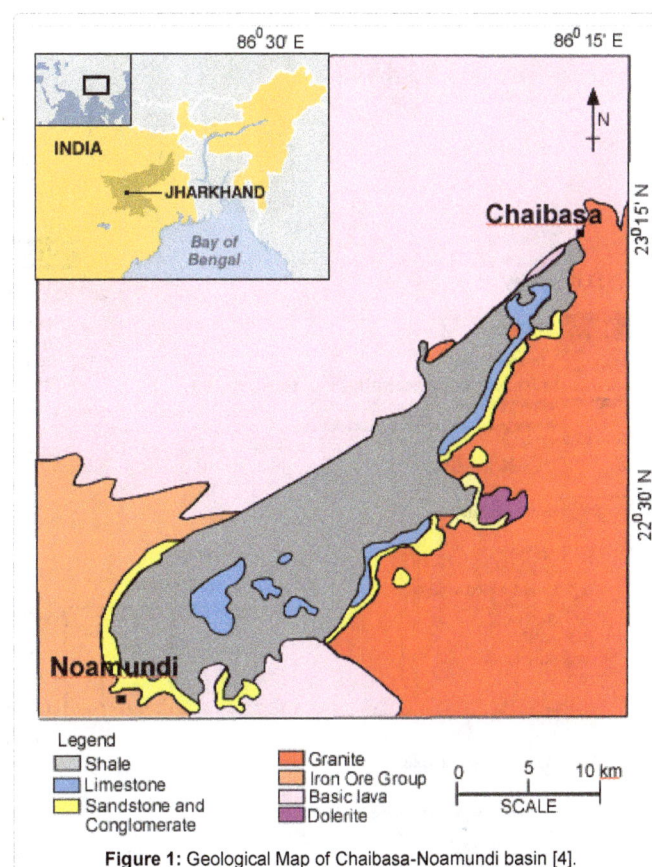

Figure 1: Geological Map of Chaibasa-Noamundi basin [4].

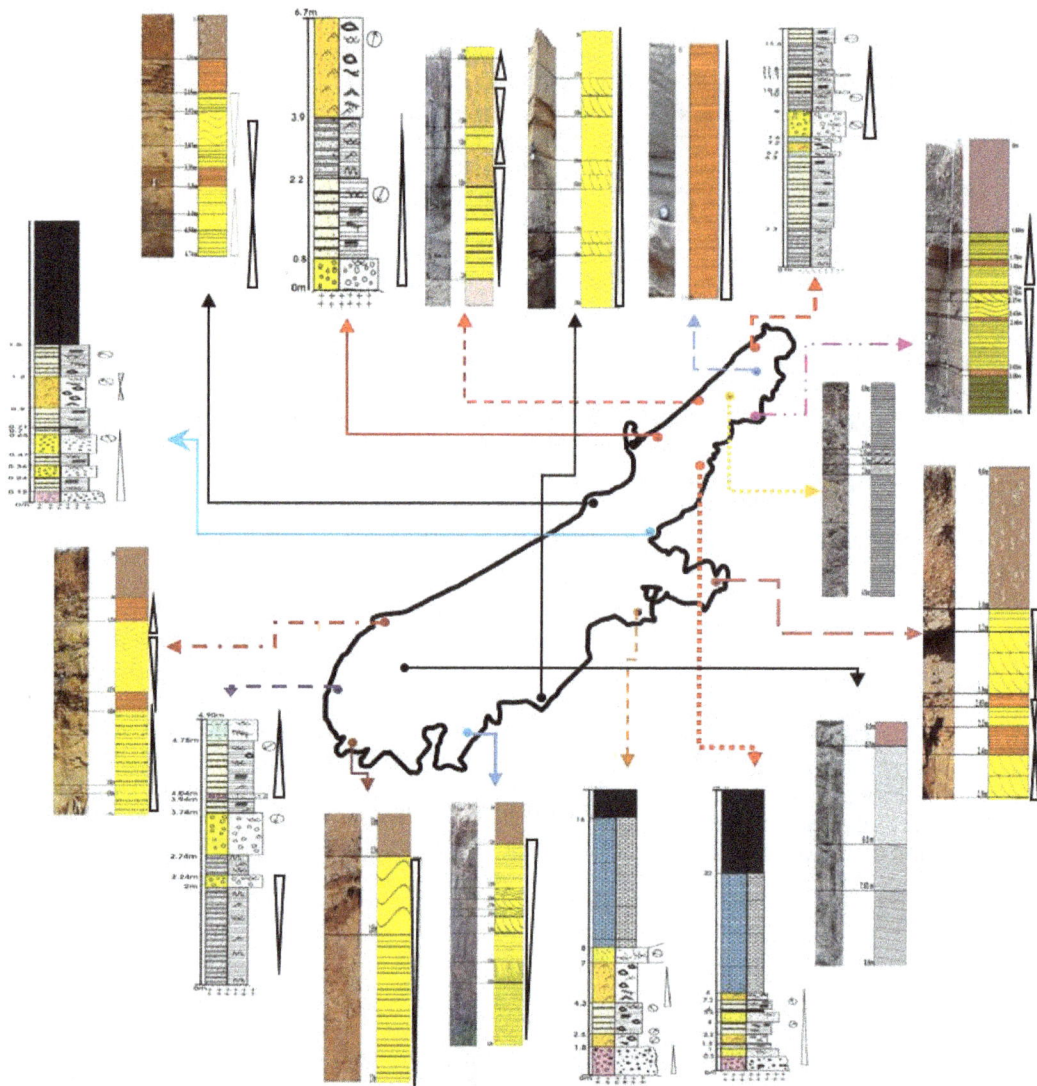

INDEX

(1) Lithology

▓	Shale
▦	Limestone
▤	Thin laminated siltstone sandstone /Rhythmic
▨	Wavy to ripple sandstone
▨	Pebbly sandstone
▤	Sheet sandstone
▦	Granite
◹	Crossbedded sandstone
▨	Soil
▥	Rhythmite
▤	Plane laminated sandstone
▤	Wavy laminated sandstone
▤	Granular sandstone

(2) Directions

⊘	Paleo Current Direction from directional structures
⊘	Paleo Current Direction from Grain orientation
▽	Fining downward sequence
△	Fining upward sequence

(3) Structures

▨	Planar cross-bedding
∿	Trough cross-bedding
◭	Symmetrical ripple
◹	Asymmetrical ripple
◺	Tabular cross-bedding
◿	Ripple drift cross lamination or climbing ripple lamination
▦	Even lamination/bedding
⬮	Hummocky cross-stratification
ℳ	Convolute lamination

Size scale

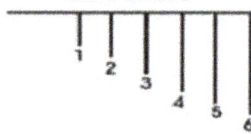

1- TLSD Very fine sand
2- RSD Fine sand
3- PLSD Medium sand
4- SSD Coarse sand
5- GSD Very coarse sand
6- GLA Conglomerate

Figure 2: Sampling locales for seventeen vertical sections.

Figure 3: (a) GLA facies: Conglomerate with angular pebbles of quartzite and jasper around Matgamburu (Hammer 15 cm for scale). (b) Extraformational conglomerate, Bistampur (Scale-15 cm). (c) GSD facies: Pebbly sandstone at Matgamburu (Scale –measuring tape, 40 cm) visible. (d) Granular sandstone at Rajanbasa (Scale –Hammer, 15 cm for scale). (e) SSD facies: Sheet likes structures in sandstone at Matgamburu (15 cm scale). (f) Planar cross-bedding in sheet sandstone at Matgamburu (Scale –Hammer, 15 cm for scale). (g) PLSD facies at Deoposi river. Pen 12 cm for scale. (h) Planar cross-bedding in PLSD facies at Deoposi river. Pen 12 cm for scale. (i) RSD facies: Assymetric ripple marks Bistampur. Scale 15 cm. (j) Cross bedded unit with multiple toe scour like structure in RSD. Pen 12 cm for scale. (k) TLSD facies: Rhythmic sandstone. Bistampur. Ruler 30 cm for scale. (l) Convolute lamination in rhythmic sandstone, Bistampur. Coin diameter 2.50 cm.

symmetrical and asymmetrical ripples. It is very commonly associated with thinly laminated sandstone facies and plane laminated sandstone facies (Figure 3i-j). TLSD facies is defined by the rhythmic alternation of sandstone and shale units, in which sandy layers are thicker than shale layers. Prominent structures are convolute lamination, trough cross-bedding and asymmetrical ripples (Figure 3k-l).

Petrological characters

The Kolhan sandstones are composed mainly of an aggregate of sub-angular to sub-rounded quartz embedded in siliceous-ferruginous matrix, with subordinate amounts of feldspar, jasper, muscovite, rock fragments of BHJ, chert, phyllite, recycled pebbles of quartzite and conglomerate (Figure 4). Composition of granite pebbles in the conglomerates shows a close similarity with the basement suggest that such pebbles are recycled erosional products of the Singhbhum granite basement over which the conglomerates were deposited [10].

The sandstones are coarse-grained and show considerable compositional variability, ranging from quartz arenite to subarkose [11,12]. The quartz grains are mainly monocrystalline with weak or absent undulatory extinction. Polycrystalline quartz occurs in two varieties: (a) grains with a polygonal fabric of interlocking grains and (b) grains with elongate, lenticular, interlocking, sutured crystals (Figure 5a). The feldspars are K-feldspars, microcline, and albite-rich plagioclase (in descending order of abundance). K-feldspars are commonly clouded with alteration products and also show microperthitic intergrowth with Na-plagioclase. Matrix quartz shows feeble recrystallization and fused contacts with the framework grains (Figure 5b). Quartz grains show bimodal distribution in quartz arenite (Figure 5c). The feldspar grains show rounded inclusions of quartz (Figure 5d-f). Sedimentary rock fragments are intrabasinal, intraformational and extraformational. During point-counting all efforts were made to identify replaced feldspar grains and to record them as feldspars. Because of the considerable influence of feldspar alteration, orthoclase and plagioclase are not reported separately. Partial replacement of feldspars by calcite has been observed, but is less common. The feldspar content ranges from 1.46 to 13.54 %, and rock fragments (0.49 to 7.82 %) embedded in ferruginous-siliceous cement (0.00-7.97%) (Fig. 6F) and cherty-sericitic matrix (1.41-10.21%) are the main constituents of those rocks (Figure 6a-b). Figure 6c-d shows the sparking colour muscovite laths and Figure 6e shows the sericite matrix or reconstituted complex aggregates of chert and sericite matrix. As

because the interest is more in the source area and the tectonic setting, only extrabasinal components have been considered.

Tectonic setting

Bivariate plot between Al_2O_3 and TiO_2 indicates the source rock to be granitic composition (Figure 7a). It has been shown for ancient as well as for modern mudstones that their SiO2 content and K_2O/Na_2O ratio can be used to discriminate between shales deposited in passive margin/cratonic, active margin, and island arc tectonic settings [13]. A prior knowledge on the tectonic setting of the Kolhans, a SiO_2 v. K_2O/Na_2O Figure 7b clearly show the plots of the Kolhan shales into the passive margin/cratonic field. In the K_2O vs.Na_2O plot, the Kolhan shale fall in quartz rich field suggestive of lithounits were deposited in plate interior either at stable continental margin or in the intracratonic basin [14] (Figure 7c). High SiO_2/Al_2O_3 and K_2O/Na_2O ratio of these shales imply their derivation from a granite dominated upper continental crust [4]. In the CaO-K_2O-Na_2O ternary plot the studied shale samples plot in passive margin field (Figure 7d) [15]. The rate of chemical weathering of source rock and the erosion rate of weathering profile are controlled by the prevailing climate, source rock composition and tectonics as well.

Paleoweathering-paleoclimate

A functional technique to assess the paleoweathering and tectonic history of the rock is the Chemical Index of Alteration (CIA)={Al_2O_3/(Al_2O_3+CaO+Na_2O)} × 100, to monitor the progressive alteration of plagioclase and K-feldspar to clay minerals [5]. CIA value increases with increasing weathering intensity, reaching 100 when all Ca, Na and K have been leached out from weathering residue. The CIA value for the Kolhan shale vary from 70.7 to 80.3, (average 75.2) indicating that the source rock underwent moderate to high degree of chemical weathering in humid tropical condition. The weathering intensity of sedimentary rock can be inferred from the concentration of Al_2O_3 and Na_2O [15]. In the discriminative diagram of Al_2O_3 and Na_2O, the plots are in the field of Amazon mud which indicates the high intensity of chemical weathering (feldspars have been altered to clay minerals) (Figure 7e). The weathering history of igneous rocks and the source for various clastic sedimentary sequences have been evaluated by using the A-CN-K (A=Al_2O_3; CN=CaO+Na_2O; K=K_2O) triangular diagram [5]. In A-CN-K plot Figure 7f; the compositional trends of various rocks during initial stage of weathering would be almost parallel to A-CN line from their respective fresh unweather points. These pathways

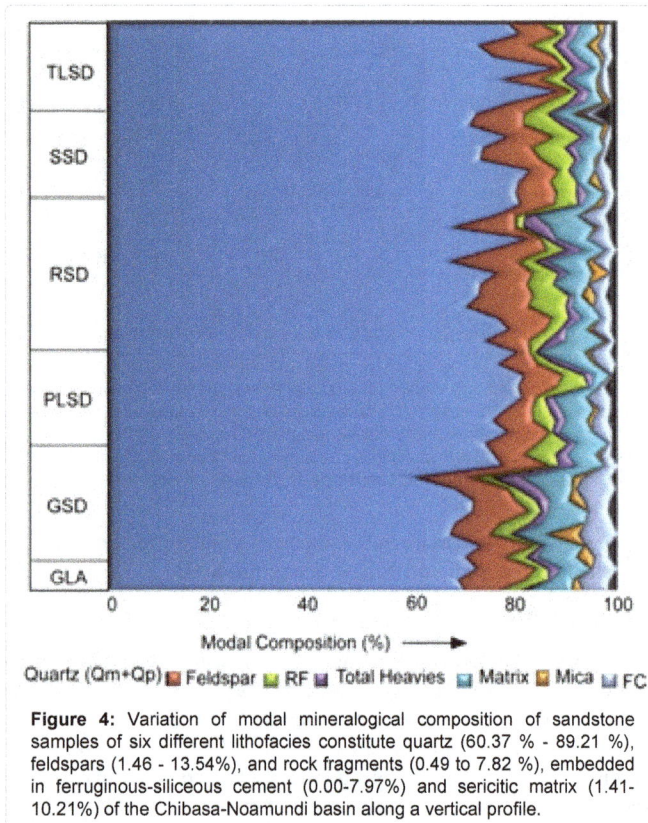

Figure 4: Variation of modal mineralogical composition of sandstone samples of six different lithofacies constitute quartz (60.37 % - 89.21 %), feldspars (1.46 - 13.54%), and rock fragments (0.49 to 7.82 %), embedded in ferruginous-siliceous cement (0.00-7.97%) and sericitic matrix (1.41-10.21%) of the Chibasa-Noamundi basin along a vertical profile.

of weathering for mafic and felsic igneous rocks are confirmed by weathering profile and thermodynamics/kinetic calculation [16]. The pathways are parallel to A-CN line because in the initial stage of weathering Na and Ca are removed from plagioclase and as the degree of weathering increases, K-feldspar are destroyed releasing K in preference to Al. During this process the residual bulk composition is enriched with Al_2O_3. All samples of Kolhan shale plot parallel and close to the Al_2O_3-K_2O boundary implying that their source area had undergone extensive weathering and produced shaly sedimentation. The A-CN-K triangular plots indicate potassium enrichments in the samples.

The CIA has been used to evaluate the intensity of weathering in the source area [16-18]. However, uncertainties exist because of possible post-depositional mobility of alkali and alkali earth elements. In near carbonate-free Kolhan shales, the CaO and Na_2O content should be always quite low. As because potassium probably experienced local redistribution (from detrital K-spar and K-mica to illite) during diagnosis, the CIA value of the Kolhan shale should only show minor post-depositional modification [19]. The average CIA value for the Kolhan shales is 75.2, suggesting moderate-to-intermediate chemical weathering [16].

Basin Model

The origin, characteristics, distribution and spatial arrangement of the various lithofacies, the predominance of stream flow over debris-flow deposits, the semi-radial, fan-like dispersion pattern of the paleocurrents, the associated subaerial and subaqueous depositional settings are indicative of a fan-delta system [20]. An interplay between several intrabasinal and extrabasinal controls probably determined the fan evolution. This is suggested by the occurrence of traction deposits, tectono-depositional intervals and their textural characteristics, and

by the evidence of synsedimentary extensional tectonism. The shallow water settings are highly sensitive to subsidence, and the presence of fine-grained sedimentation above coarse-grained deposits in a tectonically controlled sedimentary succession is the best indicator of renewed tectonic activity. As the rifting processes compartmentalized the basin into fault blocks, and grabens were consequently separated, subsidence lowered the floor of these grabens below base level, and lakes were formed as an immediate response to tectonics (Figure 8). A large volume of sediment was then available for erosion due to the differential relief between the uplifted source area and the subsided basin, and the fan-delta systems began to prograde into the lakes. In some compartments of the basin, where the lakes were probably relatively deeper due to a greater subsidence rate, the initial clastic progradation took place through sandy flows, formed when heavily sediment-laden, sandy gravity flows or stream flows were introduced into the lakes.

Conclusion

Singhbhum granitoid terrain and Iron Ore Group may be the source for Kolhan sedimentation. Evidences of such nearby provenance are indicated by clast to matrix ratio, textural and mineralogical maturity of SSD, PLSD and RSD facies, association of fresh feldspar grains, angularity and poor sorting in case of GLA and GSD facies and sediment structure. A three dimensional conceptual model has been created to discuss the depositional environment in high energy coastal complex near the margin of the Late Precambrian Epeiric Sea. The fluvial action persists throughout the deposition of the siliciclastics as GLA and GSD facies. Tidal conditions are abruptly superimposed on fluvial deposits with greater probability of storm event in near shore environment due to marine transgression [21,22]. This is proved by the scouring or erosional surfaces with granule layers in SSD, PLSD and diffused nature of contact between RSD and TLSD, megaripples, reactivation surfaces and weak linkage between SSD, PLSD and RSD facies suggesting partial obliteration and redistribution of coarse to fine sands during high energy, upper regime plane bed flow condition [23]. Dominance of mud in silts of TLSD facies and presence of asymmetric ripple marks on the top reflects waning phase during the stormy period in a shallow-coastal marine zone. Moreover, marine energy regimes are variable all along the coast, changing from wave and storm dominated in northern and north-central portion to tide dominated in the south and south-west. The superimposition of retrograding shore-line features on earlier prograding humid alluvial fans sand flat complex may be developed.

The sandstone petrography and the shale geochemistry clearly indicate that the Kolhans had both IOG and Singhbhum granite as the source rocks. In sandstones derived from low-to moderate relief source areas under humid climatic conditions there may be a depletion of feldspars and rock fragments and enrichment in quartz. Whereas sandstone petrography suggests a peneplained craton, dry climate, and very limited chemical weathering, the shale geochemistry (CIA) indicates a moderate-to-intermediate degree of chemical weathering. The abundance of shale itself poses another problem as because it is commonly assumed that the mud formation is very limited under conditions that favour arkoses. Apart from the fact that the CIA as used to determine weathering intensity in the source areas of shales is in need of reassessment and refinement, the conflict climate signals between sandstones and shales of the Kolhans may also be inherent in the way sandstones and shales are produced. As pointed out above, unaltered feldspars and well-rounded quartz and feldspar grains in conjunction with low latitude suggest that the Kolhan Formation was deposited in a

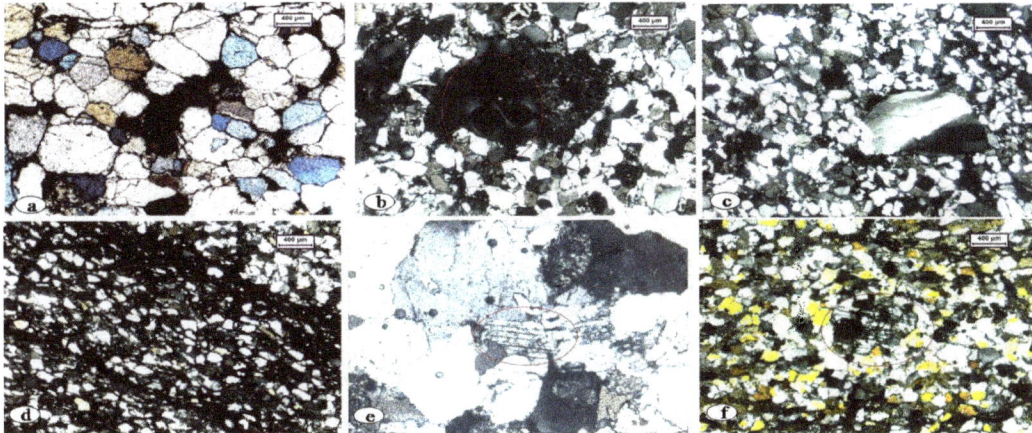

Figure 5: (a) Moderately well sorted, medium-coarse grained sandstone, well rounded quartz grains with long and concavo-convex contact. (b) Rounded to sub-rounded monocrystalline framework quartz grains in quartz arenite, Matrix quartz shows feeble recrystallization and fused contacts with framework grains. (c) Quartz grains showing bimodal distribution in quartz arenite. (d) Rhythmic sandstone thin section. (e & f) feldspar showing rounded grains.

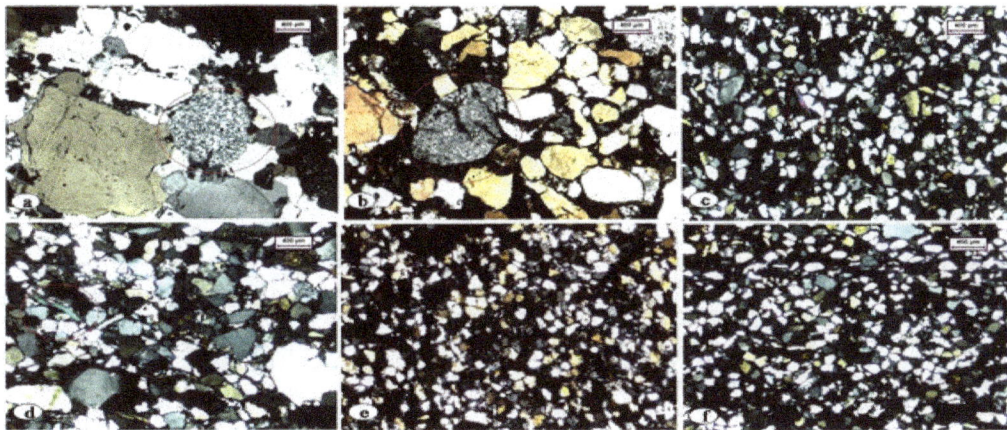

Figure 6: (a & b) Lithic fragment in sandstone shown inside red circle. (c & d) Sparking colour muscovite laths shown inside red circle. (e) Sericite matrix or reconstituted complex aggregates of chert and sericite matrix. (f) Silica and ferruginous cement.

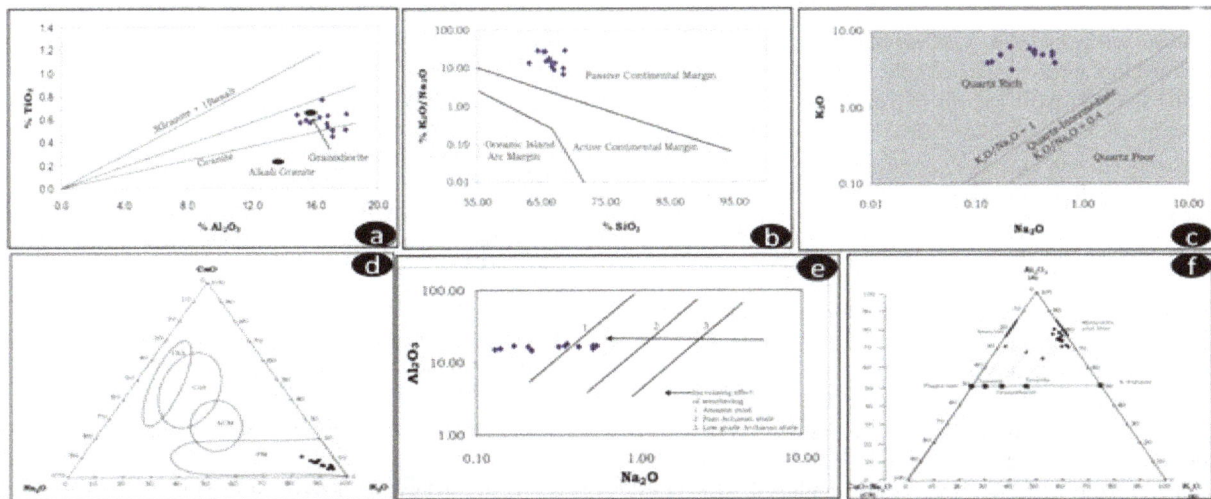

Figure 7: (a) Bivariate plot between Al_2O_3 and TiO_2 indicating the source rock to be granitic composition. (b) Tectonic discrimination diagram for the Kolhan shale indicating the passive margin setting. (c) K_2O-Na_2O diagram classifying Kolhan shale as quartz rich type [11]. (d) CaO-K_2O-Na_2O triangular diagram showing Kolhan shale in the passive margin (PM) field [13]; and also shown fields of different tectonic settings. CAN Active Continental Margin, OIA-Oceanic Island Arc, CIA-Continental Island Arc. (e) Discriminative diagram showing intensity of weathering [14]. (f) A-CN-K compositional space showing weathering trends for the Kolhan shale. The best fit line parallel to the A-CN join indicates the source to be granite to granitiod. CIA values have been represented in the vertical line the left of the A-CN-K compositional space

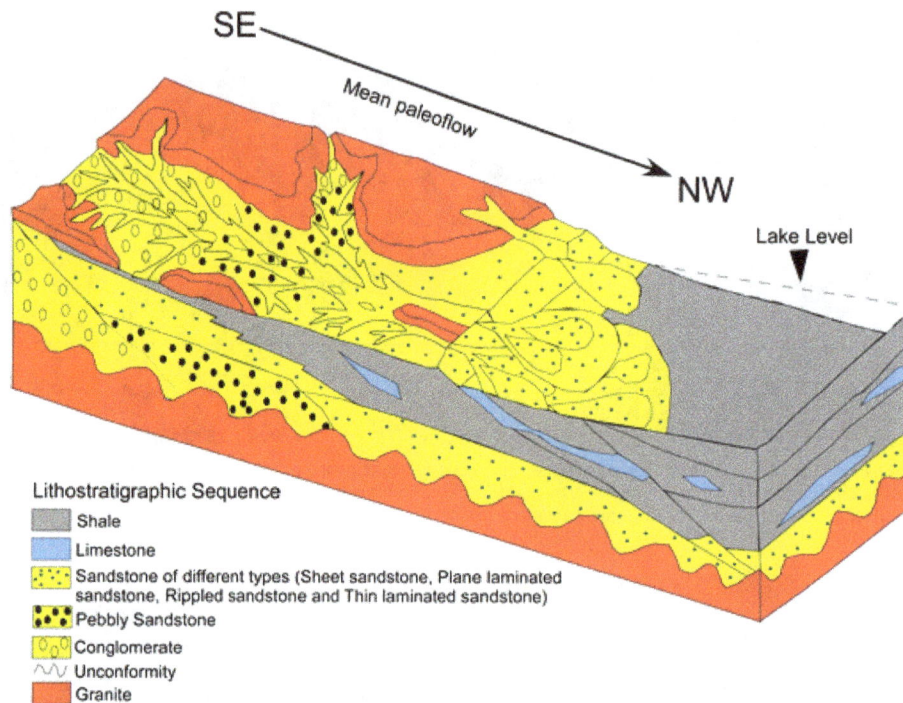

Figure 8: Block diagram to illustrate how erosion levels in the catchment area of the Kolhan basin have reached the proximal deposits supplying detritus for the Kolhan fan-delta system.

arid to semi-arid climate. In such a climatic setting, unaltered feldspars would become concentrated in sandy deposits of braided streams and may also undergo inland reworking, whereas fine detritus and clay (from feldspar weathering) would be carried to the basin as suspended load, thus leading to a separation between intensely (clay fraction) and incompletely weathered (sand fraction) material. Therefore intensities of chemical weathering indicated by shales will tend to be higher than those indicated by sandstones. It appears therefore that for realistic estimates of source area weathering conditions the data from shales and sandstones should be considered in conjunction.

The Kolhan basin activated as an asymmetric extensional basin, probably related to the reactivation of older thrusts in the chain. The sedimentation along the basin's northern margin was characterized by periodic transgressions of the lake, alternating with periods of fluvial fan progradation and fan-delta development. It is here suggested that syndepositional tectonism greatly overwhelmed a possible and concurrent climate forcing on the sedimentary dynamics recorded by the lithological successions. The episodes of thrusting forced the basin margin shifts onto the downwarped alluvial substrate and hindering temporarily the supply of sediment from fan catchments and causing sediment receive from the Iron Ore Group. The alluvial fans were coarse-grained, flood-dominated depositional systems, active during the post extension periods of intense denudation and slow normal subsidence. The alluvial sediment dispersal was predominantly by hyperconcentrated flows, generated by the sediment bulking in flash water flows, combined with their gravity transformation. The repetitive, high-frequency flood events are thought to have been generated by aerographic situations. The present case study demonstrates that an understanding of depositional processes is crucial to the reconstructions of sediment dispersal dynamics and basin-fill history.

Acknowledgement

The authors are grateful to Prof. D. Sen Gupta, Head of the Department, Department of Geology and Geophysics, IIT Kharagpur for providing necessary facilities to carry out the present investigation.

References

1. Nemec W, Steel RJ (1988) What is fan delta and how do we recognize it? In: Nemec W, Steel RJ Fan Deltas. Blackie, Glasgow.

2. Johnsson MJ (1993) The system controlling the composition of clastic sediments. In: Johnsson MJ, Basu A Processes Controlling the Composition of Clastic Sediments. Geological Society ofAmerica 284: 1–19.

3. Suttner LJ (1974) Sedimentary Petrographic Provinces: An evaluation. Soc Econ Paleontologists Mineralogists 21: 75-84.

4. McLennan SM, Hemming S, Mcdaniel DK, Hanson GN (1993) Geochemical approaches to sedimentation, provenance and tectonics. In: Johnsson MJ, Basu A Processes controlling the composition of clastics sediments. Geol soc Amer 284: 21-40.

5. Nesbitt HW, Young GM (1984) Prediction of some weathering trends of plutonic and volcanic rocks based on thermodynamics and kinetic consideration. Geochim Cosmo Acta 48: 1523-1534.

6. Geological Survey of India (GSI) (2006) Govt of India.

7. Sarkar SN, Saha AK (1977) The present status of the Precambrian stratigraphy, tectonics and geochronology of Singhbhum-Mayurbhanj-Keonjhar region, Eastern India. Ind Jour Earth Sci 37-65.

8. Ingersoll RV (1988) Tectonics of sedimentarybasins. Bull Geol Soc Amer 100: 1704-1719.

9. Miall AD (1984) Principles of Sedimentary Basin Analysis. (2nd eds), Springer-Verlag, NY.

10. Selley RC (1970) Ancient Sedimentary Environments and Their Sub-surface Diagnosis. (2nd eds), Routledge, Taylor and Francis, London.

11. Das S, Sahoo SR (2015) Coalesced time-transgressive Kolhans from the Chaibasa-Noamundi basin: implications on accommodation space-process

response sedimentation and graben tectonics. Recent advances in mineral development and environmental issues 274-214.

12. Sahoo SR, Das S (2015) The Paleoproterozoic supracrustal Kolhan basin: Provenance, Tectonic and Palaeoweathering histories. International Journal of Geology and Earth Sciences 1: 14- 25.

13. Roser BP, Korsch RJ (1986) Determination of tectonic setting of sandstone-mudstone suites using SiO_2 content and K_2O/Na_2O ratio. Jour Geol 94: 635–650.

14. Crook KAW (1974) Lithogenesis and geotectonics: the significance of compositional variations flysch arenites (greywakes). In: Dott RH, Shaver RH Modern and Ancient Geosynclinal Sedimentation. Soc Econ Paleo Min Spl Publ 19: 304-310.

15. Taylor SR, Rudnick RL, Mclennan SM, Erickson KA (1986) Rare earth element patterns in Archean high grade metasediments and their tectonic significance. Geochim Cosmo Acta 50: 2267-2279.

16. Nesbitt HW, Young GM (1982) Early Proterozoic climates and plate motions inferred from major elements of lutites. Nature 299: 715-717.

17. Wronkiewicz DJ, Kent CC (1987) Geochemistry of Archean shales from the Witwatersrand Supergroup, South Africa: source-area weathering and provenance. Geochimica et Cosmochimica Acta 51: 2401-2416.

18. Wronkiewicz DJ (1989) Geochemistry and provenance of sediments from the Pongola Supergroup, South Africa: evidence for a 3.0-Ga-old continental craton. Geochimica et Cosmochimica Acta 53: 1537-1549.

19. Aronson JL, Hower J (1976) Mechanism of burial metamorphism of argillaceous sediment: Radiogenic argon evidence. Geol Soc Amer Bull 87: 738-743.

20. Sahoo SR (2015) Paleocurrents and paleohydraulics studies of the Proterozoic Kolhan siliciclastic unit (India): a case study from the Chaibasa-Noamundi Basin, Jharkhand, India. J Geol Geophys.

21. Allen JRL (1982) Sedimentary Structures, Their Character and Physical Basis. Elsevier Amsterdam I, II: 515-645.

22. Reineck HE, Singh IB (1980) Depositional Sedimentary Environments. Springer-Verlag 549.

23. Harms JC, Fahnestock RK (1965) Stratification, bed forms and flow phenomena with an example from the Rio Grande. In: Primary Sedimentary Structures and their Hydrodynamic interpretation. Soc Econ Paleontologists Mineralogists 12: 84-115.

24

Oil and Gas Industrial and Ecosystem Mechanical Impacts of Environment

Elosta F*

Waha Oil Company, Tripoli, Libya

Abstract

Oil and gas from deuterons pollutant due to its hydrocarbon materials and toxic substances such as hydrogen sulfide and consist of organic compounds containing hydrogen and carbon and some parts non-carbon such as nitrogen sulfur oxygen and some small quantities of metals such as vanadium which organic compounds containing hydrogen and carbon where emit these gases when evaporation or degrade oil spill and these materials and gases threaten the ecological system the problem of pollution oil industries include pollution air by escalating gas such as hydrogen sulfide toxic as well as the oil spill on soil which contributes wind and flood water in a quantum population and agricultural projects special during irrigation and spill oil mean change in its chemical or quality of the components environment so distorting equilibrium in systems environment different water production processes associated with the oil that is produced in large quantities is also known that the production of barrels of oil offset producing four barrels of water.

The estimated production quantities of this water in most fields of Libya for example around 4,000,000 barrels per day almost a big problem as these waters are not taken to exploit only a few of them in the injections are exposed to the air which leads to the evaporation of deadly gases ones such as hydrogen sulfide and carbon dioxide and other gases and exploiting some of this water injection in large quantities may cause harm to the environment by groundwater contamination due to concomitant injections for engineering studies include hydrological and geological and that injections accompanied by compressions strength large create new depression may contribute to the ancient after cracks found in the region.

This process also earthquake movements may occur for plate movement on the edge of the continents because of the increased pressure on the layers of the earth which causes its human wrought in Environment, and carry out a study:

1. Study of the geological environment to see the possibility of cracks and fractures and the possibility of earthquake activity.

2. Hydrological studies surface water and agricultural projects and groundwater.

3. Carry out drainage water associated with oil scientific methods.

I hope to my God that this study will contribute to whatever is simple to learn influence the oil and gas industrial on environment regulations that give life.

Keywords: Pollution; Desertification; Water pollution; Oil reservoir

Introduction

Environment is a system dynamically complex includes and elements are intertwined and multiple where development of knowledge and information related to this system significantly in recent decade especially with its association largely being important dimension of sustainable of environment science he finishes mean basic to refer to the study of nature organic physical and chemical surrounding the living organism the ecosystem in nature and accordingly represents the result of a balance between elements with the factors and forces that interact with each other to balance occurs and imbalance in the ecosystem [1].

The oil and its derivatives-risk vehicles and high toxicity due to the decomposition of oil molecules to many toxic to all living organisms and oil begins harm since it relates to air [2].

Objective study

This study aims to find out the reasons for pollution real caused oil and gas industry in the world such as gases climatic contamination of soil groundwater surface injection process oil well without doing a study technological to develop geological area to be injection as well as the problem of water associated with petroleum diffusion H_2S.

Since this study aims real possible use techniques reduce water the spread of visible pollution due to oil and gas industry and oil companies' event to contribute to the process of treatment the causes of environment pollution.

Oil and gas pollution environmental

Is launching elements or compounds or mixtures of gas liquid or solid source of oil to the elements of the environment, which includes surface water and underground air, soil, causing a change in the presence of these elements and distributed damage oil pollution on all forms of life and human organisms sea and land plants and leads ultimately to death and extinction of millions of living organisms and marine of all genera and species and to disable most shipping services and the destruction of tourism through plants fish and destroys forest in addition to the destruction of the human diet [3,4].

The production oil combustion carbon monoxide dioxide sulfur, nitrogen oxides, hydrogen sulfide hydrocarbons and combustible sodium chloride salts and calcium and potassium salts that contain a special crude oil which contains the gases emitted by evaporation and

*Corresponding author: Elosta F, Waha Oil Company, Tripoli Libya
E-mail: hmbstf@yahoo.co.uk

fast-spreading oil pollutants from oil tankers and refineries and oil spills on soil and drilling fluids and associated water, this means that the refineries consume large amounts of water which is known into the sea rivers and oil where a major polluter of the environment and the investment exposure sea oil wells contaminated by leakage in both captain and production and if oil prices top concern of the world due to its location in the daily economy on the one hand and the decisions about consumption of hazardous emissions threaten the natural and human environment on the other [5-8].

That frequent consumption of fuel increases the temperature and climate caused by melting snow spoke harmful changes varied agricultural systems and threatens to soil erosion and forest extinction and damage types of crape and thus threaten human life [9].

The problem of pollution is the most important problems facing the oil industry has become in the world as a result of the oil companies polluting the environmental by increasing the proportion of special emissions of carbon dioxide in the atmosphere and there are many other ways including the occurrence of thoughts, desertification of large tracts of land the spread of infections, diseases in the world and also extinction of many organisms and disasters and the loss of some agricultural crops and occurrence of marine pollution (Figure 1).

Priorities for combating oil pollution

1. Maintain

2. Environmental protection

3. Resources protection and economic vitality

The most important problem that pollution in the oil industry

The pollution of sea water and rivers and soil in oil and petroleum products from the most dangerous pollutants in our time of its adverse effect on human and economic environment.

Sources of oil pollution are divided into four groups

1. **Unintentional pollution:** Includes accidents tankers and oil pipelines explosion that occasionally occur during drilling operations for the extraction of oil or stages of production transportation refining storage marketing and even get rid of emissions and waste.

2. **Intentional contamination:** Includes oil accidents as a result of wars in addition to empty the water balance of the ships.

3. **Pollution caused by negligence**: This group is close to 80% of the world's oil pollution occurs as a result of errors during process to extract the oil and injection wells.

4. **Natural pollution:** That natural sources of energy is a group of materials and energy in the environment include oil coal, shale a non-renewable sources and exposed to drain company resorted to producing technological world and the production process raising on the environment and human and food reports indicate that oil operations are responsible for most global carbon dioxide ,making it one of the largest sources of environmental pollution as transportation which estimated the worlds one billion car that roam the planet and burn fuel high octane pumped 100,000.000 of tons co_2 carbon dioxide in the most sensitive part our atmosphere and the total is synthesized annually for 6.000.000 cars and aviation cause 13% of air pollution while its share of the pollution of the universe in general amounts to 3% (Figures 2 and 3).

Impact of oil operation and environmental

˙Impact climate

˙Impact subsurface

˙Impact surface

Impact petroleum climate: The air pollution is subjected atmosphere of chemicals or particles physical or biological compounds cause damage to humans and organisms natural environment and the atmosphere is a system of natural gas reactants and complex, which is necessary to support life on earth and drained the ozone Layer of the most serious air and the most dangerous things in that threaten life earth environmental regulation [10].

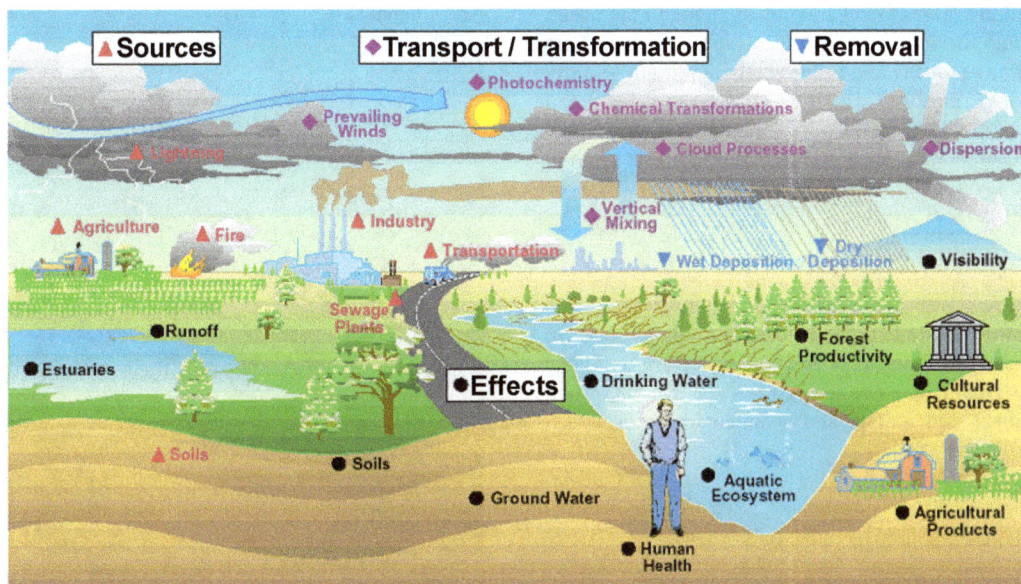

Figure 1: Operation environmental systems.

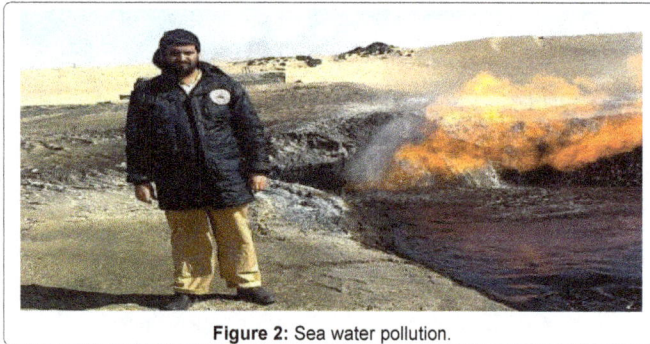

Figure 2: Sea water pollution.

Figure 3: Associate water oil in pet.

That the material associated with the oil which includes hydrocarbon materials and water threatening agricultural areas and human dangers of contamination of food and water and therefore diseases, the most important oil and petrochemical pollutants to climate sulfur dioxide-hydrogen sulfide gas –gases nitrogen oxides –carbon dioxide particles and volatile solid particles suspended [11].

Subsurface petroleum pollution: Is considered one of the most important source used by man even in countries that have large rivers the water table are those waters that saturation layer beneath the soil pores or rock crevices which underneath to form a loop water, and to fall unit the top layer is saturated with water and then up to the water underground the rock pores and cracks related to each other and store water in a complex geological layers which may be confined to a layer of clay or is confined to a layer of sand and moving water in the region strongly hydraulic into the area of low-lying and speed of up to 3 cm per year and is working porosity and permeability important to move the water ,but factor permeability which depends on the cracks and fractures the huge up to 100 km has been water pollution through these spaces and exposed r=this water pollution in many ways as a result of human activity such as oil spills on the surface of the soil fragile special leaks cracks (Figure 4).

Groundwater contamination

1. Pollution caused by natural dissolution of the composition of rocks, minerals underground rocks as well as analyzes the components of the aquifer and radon 222 and is highly soluble and radionuclide's resulting from the decomposition of environment and granite and sedimentary rocks which produces radioactive material.

2. Oil pipelines

3. Industrial pollution caused by industrials activity.

Injection wells of oil causes environmental pollution

Pressure is the class main force driving the oil towards the oil wells

and consists this pressure geostatic which exposed him granules and affects the spaces rock and the values of this pressure on porosity rock and thick and the quality of geological layers top of layer and increases usually increased depth and during the period of investment goes down the pressure with time leading to lower oil reservoirs productivity this decline varies from one according to the strength of the reservoir [12].

The pushing system with water more regulations effective and the process is done when it gets a drop in pressure class which leads to the movement of water in the saturated depends worker oil yield in the pushing system water in the regions oil reservoir regularly in reservoirs heterogeneous in terms of characteristics of storage but in the fact that reservoirs is in permeability high where up quickly bottoms producing wells and increasing proportions of water in these wells (Figure 5).

Characteristics of the water injection process

1. The class raise the pressure of the reservoir

2. Higher levels of fluid static, dynamic

3. Raise worker displacement

4. Improve the conditions of the reservoir.

5. Disposal of produced water with oil, but may cause damage in the case of the presence of cracks

Impact surface

Although human need for water and survival correlation water but did not improve the handling of water as a result of oil activities near

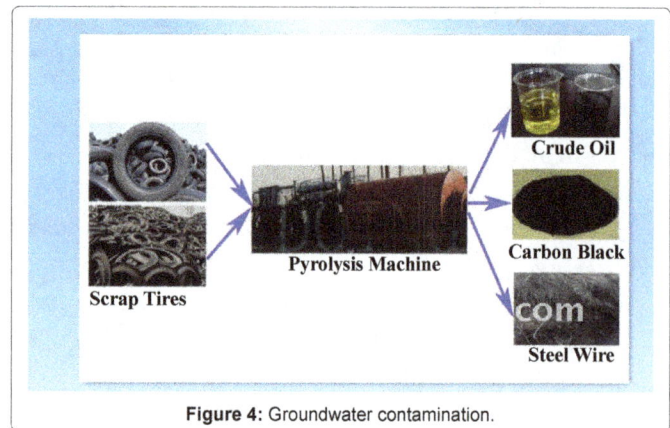

Figure 4: Groundwater contamination.

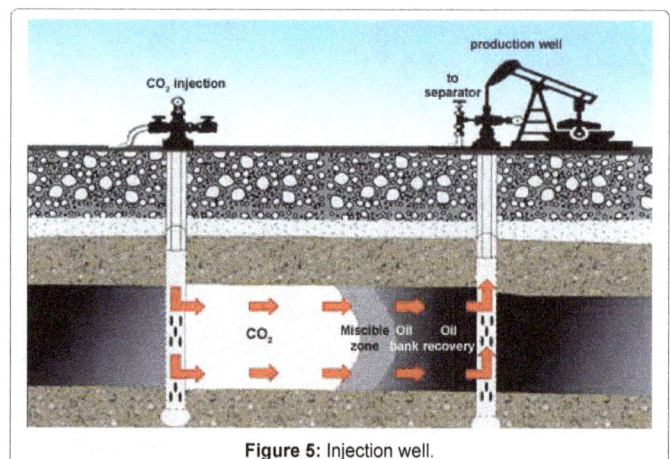

Figure 5: Injection well.

water sources, raising the concentration pollutants in the water, which reduced the physical, chemical properties.

The Main sources of pollution which affects the aquatic environment

1. Pollution physicist they change the temperature salinity and increased suspended solids and all that increases the rate of evaporation and transpiration.

2. Chemical pollution ,which change the shape of the water and increase the proportion of toxic metals such as barium cadmium lead mercury but non-toxic metals such as calcium, magnesium sodium, lead to disease.

3. Biological contamination, it increases micro-organisms causing diseases such bacteria, viruses parasites

4. Radioactive contamination is unclear reactors, all contaminated affect routes on surface water and irrigation operations because of transportation and shelf.

Marine pollution ,is a sea of risks that threaten marine life that what caused the oil spill is the result of the development and maintenance of facilities for loading and unloading on the beach than be caused by the same means of transport whether that serves the oil refinery or pipelines that the density oil its less than the density of water is floating on the surface component insulating layer between the water and the air and this layer spread over a large area of the surface of the water where prevent this layer gas exchange between the air and water which affects spigot soluble oxygen in sea water which affects the balance gas and stops the spread of oil on the water surface on the nature of the oil and the prevailing winds waves and ocean currents and strength the complicated sometimes the weather will turn crude oil leaking from the process of cleaning oil tanks crude oil leaking from the process of cleaning oil tanks and tankers to the emulsion becomes water more viscous and pollution four times the volume of crude oil, depending on the characteristics of the water ,also extends oil output to oil spills to the bottom of the sea after flying material plane remains heavy suspended matter insoluble substances and passage of time [13]. The wind is working to transfer oil spill which threatens beaches and marine life species which threatens. The human diet particular those that depend on the sea for food.

Petroleum pollution for soil

Is soil strong cohesive and dense consists of materials grainy or closed with small amounts of clay Silt and salt which is working on the coherence of these granules each other when the arrival of water to these materials in the soil become crumble and weaken contaminated soil usually the result of different processes during the exploration drilling and production example during currency drilling is used drilling fluids due to drilling gore technical side wells to collect drilling fluids and materials hydrocarbon excess and this drilling may take technical conditions urgent not seep into the groundwater and this fluid is usually mixed with soil and are transported by wind torrential (Figure 6).

Soil contamination

Soil contamination is divided into,

1. Soil contamination Balnviac and flammable materials.

2. Soil pollution factories manufacturing residues such as oil refineries refining which contain mud accumulated in the tanks and pipes.

3. Contamination radioactive Balnviac.

Pollution agriculture project

Acid rain can accept trees and destroy leaves of plants and can sneak out of the soil. Making it unsuitable for the purposes of feeding housing and hole ozone allows passage of ultraviolet radiation from the sun to enter the ground causing damage to trees and ozone prevents plants from breathing and may delay the process of photosynthesis (Figure 7).

Properties how the interaction of spilled oil and Gas

1. Spreading

2. Erosion

3. Evaporation

4. Disassembly-natural decomposition

5. Melting

6. Oxidation

7. Deposition

8. Bacterial decomposition

Sources of oil pollution

1. Natural sources (leaks from the ground).

2. Industrial sources (accidents, oil tankers, unloading loading, packing, oil injection wells)

Factors affecting pollution control

1. The quality and quantity of oil spilled.

2. Weather conditions Where spill oil.

3. Requirements plant.

Figure 6: Soil pollution.

Figure 7: Marine oil spill.

Environment protection and control of oil and gas pollution

Mechanical treatment

1. Rubber barriers to protect water.

2. Prevent the oil from spreading.

3. Change the direction of the oil spill.

4. Scraped oil.

Chemical treatment

1. Chemical pollution can be controlled by a biological solution using bacteria, so that it can be converted oil spills into very fine droplets easily.

2. Spray types of solvents and detergents to oil spills and oil emulsion converter in water and dissolves it.

3. Bio mediator used to accelerate the decomposition process bacteria adding and increasing the proportion of nutrients such as Nitrogen ,phosphorus and increase Nitrogen is necessary to increase the number of bacteria to do the decomposition process.

Burning the oil slick site soy

Is the removal of the oil from the surface of the water that collects oil and gas his fire-resistant. Barriers were burned spots.

Basics of environmental protection oil and gas

1. The nature and characteristics of the oil pollution.

2. Knowledge of contamination at all stages of exploration.

3. Knowledge of control and pollution control.

4. Prevention of occupation risks of the oil industry and health.

5. Prevention of pollution environmental pollution and oil pollution in water air and soil.

Ways combat oil and gas pollution

1. Bioremediation methods.

2. Way flooding barriers.

3. Way chemical spray types of chemical.

4. Physical treatment methods.

5. Specialized solvent washing methods.

6. Treatment oil environmental specifications.

7. Leaked oil absorption material in the areas of oil fields whish are separated from the water and contaminated materials, cleaned and processed and filtered material was sent to oil tanks.

8. Use of the separation process by centrifugation to separate and recycle water contaminated thanks and global environmental specifications.

Conclusion

The problem of environmental exposure to notify the oil and gas industry has become one of the most important problems dogging states as especially oil producing ecosystem processes carried out by some companies imbalance in this system causing problems not solved mostly and should work to establish an environmental management

through specialization to solve these problems which have become superficial and underground problems. The production processes and research and exploration of oil and gas has become threatening the world the dangers material carried by the hydrocarbon materials as well as the production process which are accompanied by such drilling fluid materials and water associated with oil during the production process that have become threatening the ecosystem.

References

1. Myers-Smith IH, Forbes BC, Wilmking M, Hallinger M, Lantz T, et al. (2011) Shrub expansion in tundra ecosystems: dynamics, impacts and research priorities. Environ Res Lett 6: 045509.

2. Callaghan TV, Tweedie CE, Akerman J, Andrews C, Bergstedt J, et al. (2011) Multi-decadal changes in tundra environments and ecosystems: synthesis of the International Polar Year-Back to the Future Project (IPY-BTF). Ambio 40: 705-716.

3. Stock P, Burton RJ (2011) Defining terms for integrated (multi-inter-trans-disciplinary) sustainability research. Sustain 3: 1090-1113.

4. Walker DA, Epstein HE, Raynolds MK, Kuss P, Kopecky MA, et al. (2012) Environment, vegetation and greenness (NDVI) along the North America and Eurasia Arctic transects. Environ Res Lett 7: 015504.

5. Pajunen AM, Oksanen J, Virtanen R (2011) Impact of shrub canopies on understorey vegetation in western Eurasian tundra. J Veg Sci 22: 837-846.

6. Kumpula T, Forbes BC, Stammler F, Meschtyb N (2012) Dynamics of a coupled system: multi-resolution remote sensing in assessing social-ecological responses during 25 years of gas field development in Arctic Russia. Rem Sens 4: 1046-1068.

7. Jones IL, Bull JW, Milner-Gulland EJ, Esipov AV, Suttle KB (2014) Quantifying habitat impacts of natural gas infrastructure to facilitate biodiversity offsetting. Eco Evol 4: 79-90.

8. Myers-Smith IH, Hallinger M, Blok D, Sass-Klaassen U, Rayback SA, et al. (2015) Methods for measuring arctic and alpine shrub growth: a review. Ear Sci Rev 140: 1-3.

9. Degteva A, Nellemann C (2013) Nenets migration in the landscape: impacts of industrial development in Yamal peninsula, Russia. Pastor Res Pol Prac 3: 15.

10. Baynard CW, Ellis JM, Davis H (2013) Roads, petroleum and accessibility: the case of eastern Ecuador. Geo J 78: 675-695.

11. Zeng H, Jia G, Forbes BC (2013) Shifts in Arctic phenology in response to climate and anthropogenic factors as detected from multiple satellite time series. Environ Res Lett 8: 035036.

12. Forbes BC (2013) Cultural resilience of social–ecological systems in the Nenets and Yamal-Nenets Autonomous Okrugs, Russia: a focus on reindeer nomads of the tundra. Eco Soc 18: 36.

13. Kivinen S, Kumpula T (2014) Detecting land cover disturbances in the Lappi reindeer herding district using multi-source remote sensing and GIS data. Int J Appl Ear Obs Geoinform 27: 13-19.

25

Lithology Investigation of Shaly Sand Reservoir by using Wire Line Data, "Nubian Sandstone" SE Sirt Basin

Ben Ghawar BM*

Geological Engineering Department, University of Tripoli, Tripoli, Libya

Abstract

Identify the rock lithology has important meaning for estimating the reserve of petroleum as reservoir capacity and storage ability. The lithology identification from well log based on not conventional cross plot proposed and studied, which is more easier instead of rock core data observation results. However, this work carried out comparison between chart and analytical solution of matrix parameter (ρ_{ma} and ΔT_{ma}) estimation values of producer shaly sand reservoir, and present the main depositional environment affects. In addition to influences of pyrite, ferruginous encrustations, organic material throughout this studied reservoir. Consequently, variety of matrix parameters values is contributed by clay minerals present in this reservoir type. This study based on wire line data measured over than 750 feet produced Upper Nubian Sandstone belong to two oil fields, SE Sirt Basin. This shaly reservoir divided into three main units (R, E and F), and each unit has been subdivided into three subunits (F3, F2, F1, E3, E2, E1, R3, R2 and R1) from bottom to top according to depositional and petrophysical properties.

Keywords: Wire line; Rock type; Minerals; Nubian sandstone; Sirte basin

Scope of Work

Most sandstone reservoirs, worldwide, contain varying amounts of clay and/or shale. Therefore, shaly sands are heterogeneous producer reservoir. Petrophysical evaluation of these reservoirs type is needed to deal carefully with lithology investigation. However, the wire line logs using different formulas and cross-plots techniques. Most of these techniques have to be directed toward clay mineral identifications. However, this work on four wells was selected from different oil fields at eastern Sirt basin to present main inspect lithology, which is effect on reservoir qualification (porosity, permeability, etc.). The available data were wire line logs recording over than 750 feet reservoir thickness and each half feet analyzed. These logs are include; Neutron, Density, Sonic, Photoelectric factor (PEF), Natural gamma ray (NGS), and Induction resistivity.

Previous work

Several techniques have been proposed for determining lithology from wire line log data, such as principal component and cluster analyses and discriminate analysis. Subsurface lithology is traditionally determined from core or cutting analysis. Cores are generally not identification from well log based on not conventional cross plot proposed and studied, which is more easy continuous and consequently do not provide a complete description of formations crossed by a well. Well logs give a practically continuous survey of the formations crossed by a well. They allow measurement of apparent thickness and of real thickness if dipmeter data are taken into account. Burke et al. and Clavier and Rust [1,2] have shown that well log responses can give a good idea of the lithology. With the increase of physical parameters recorded by modern logging tools e.g., parameters recorded by modern logging tools e.g., photoelectric cross section, natural or induced gamma ray photoelectric cross section, natural or induced gamma ray spectrometry (GRS), and dielectric constant it becomes more obvious that their combination can give a good lithologic description of the formations. This evidence was the basis for the concept of "electrofacies" (Serra and Abbott) defined as "the set of log responses which characterizes a bed and permits it to be distinguished from other beds" (Serra). Applied to open hole logs, this electrofacies is an equivalent of the lithofacies that, according to Moore, is the "total sum of the lithological characteristics (including both physical and biological characters) of a rock". It is not always obvious, however, to translate this electrofacies in terms of geologically meaningful rocks. A procedure combining modern wireline measurements with a lithofacies data base created from logs (strictly speaking, an electrolithofacies data base) has proved to be effective in this translation [3]. Akinyokun et al. [4] detection of Lithology and Fluid Contents from geophysical data has always been an important practical issue in the interpretation of geophysical oil prospecting data by Unsupervised Self Organizing Map (SOM) of neural networks of well log data obtained from the Niger-Delta region of Nigeria.

Geological review of the studied reservoir

Stratigraphic sequence of the Sirt Basin had been divided into into Pre-rift, Syn-rift and Post-rift. The Pre-rift and Post-rift sediments were dominated by clastic, whereas the Syn-rift sediments were dominated by carbonate, as shown on Figure 1 [5]. NW-SE faulting system of horst and graben patterns started in Early Cretaceous and culminated during the Tertiary. This dominant trend truncated by E-W trending in the southeastern part of the Basin (Figure 2). Also, this localized complexity at the junction of the Ajdabiya and Hamiemat Troughs, interpreted and caused of a local stress anomaly is unknown, as no earthquakes have been recorded for this region and active faulting has not been reported [6]. The eastern part of the Sirt Basin can be distinguished into five main structures which have predominantly east-west orientation. These structures are; Hamiemat Trough (Mar Trough, Metem depression),

*Corresponding author: Ben Ghawar BM, Geological Engineering Department, University of Tripoli, Tripoli, Libya, E-mail: gloriamuftah@yahoo.com

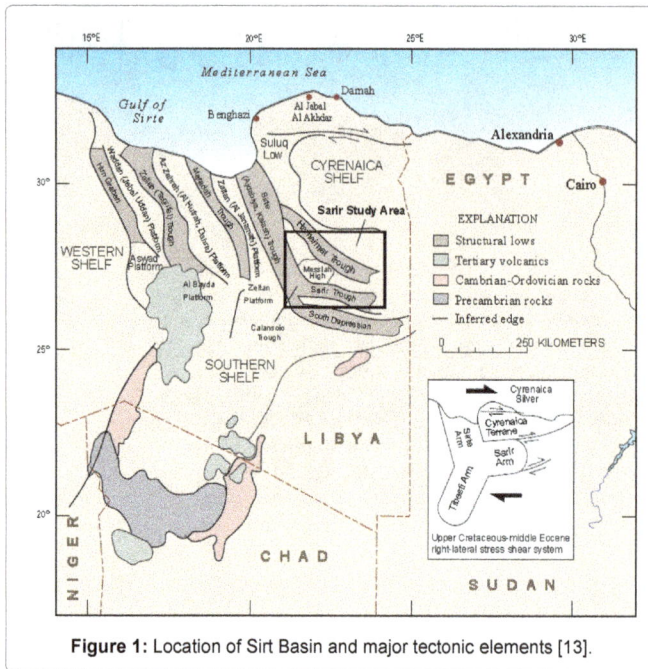

Figure 1: Location of Sirt Basin and major tectonic elements [13].

Messlah- Kalanshiyu (Wasat) High, North Sarir Trough, South Sarir Ridge, and South Sarir Trough.

The Nubian reservoir is deposited in deeper part of the Hameimat trough, deposited during late Jurassic to early Cretaceous age as basel part of the syn-rift sequence. It was preserved in grabens and is generally missing or very thin on the platforms and highs. The reservoir has a lateral and vertical type and character change of facies sequences. Ibrahim and El-Hawat [7,8] have recognized and divided the sequence into three units as Lower Sandstone member, Middle Shale member and Upper Sandstone member. Abdulghader [9] identified four lithofacies within this sequence. These lithofacies are: a meandering river facies with point bars, levees and over bank deposits, a relatively high-energy alluvial- plain association with low-sinuosity braided-streams, swamp facies and relatively deep-water lacustrine facies. This reservoir has been studied petrograhically by El-Bakush et al. [10] and in recently published paper [11] it was found that the Nubian Sandstone ranging from very to coarse grained, quartzitic and poorly sorted often with a clay matrix of fluvial and lacustrine sandstone. Therefore, variation of composition and shale content make the Nubian as a shaly sand reservoir.

Rock Types Determination

Lithology can be defined by rock samples taken from outcrops or core samples from drilled wells. Geophysical logging devices can be used instead when the rock samples are leaked or not found. Different techniques depend on these logs such as; M-N, MID, and GR- PEF plots. Therefore, the Shaly sand has a variation of clay content, so this rock type as not homogenous lithology. However, this studied reservoir rock divided into three main units (F, E and R) from bottom to top, and each unit include three subunit (F3, F2, F1, E3, E2, E1, R3, R2 and R1). Through the following sections explain each technique which applied to define lithology type for studied reservoir.

Øn – ρb and Øn – Δt cross – plots

The neutron porosity (Øn), bulk density (ρb) and travel time (Δt) reading values from each related log were plotted manually depend on

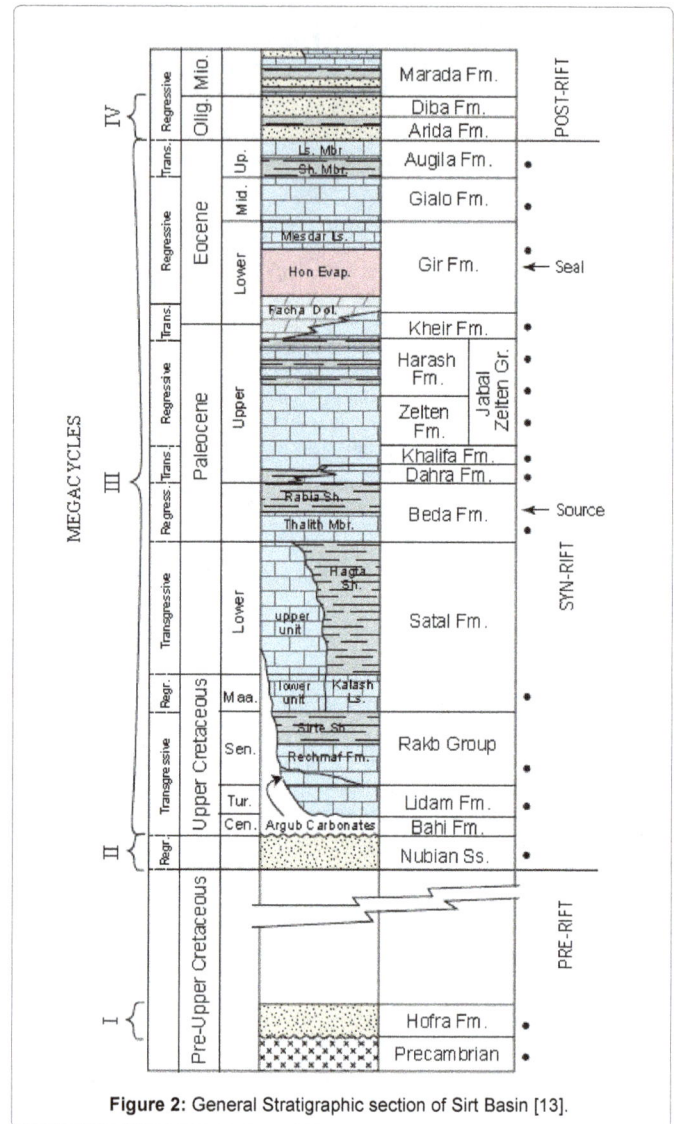

Figure 2: General Stratigraphic section of Sirt Basin [13].

base mud type (fresh)of Schlumberger CP14-a and CP-12 charts to determine matrix density (ρ_{ma}) and matrix travel time (Δtma). These manual plots were done before and after correction of the neutron porosity (Øn). Figures 3a and 3b present these plots of the well B. The average matrix density (ρ_{ma}) and average matrix travel time (Δt_{ma}) from these plots are summarized in Table 1 of the studied well B for R reservoir subunits. The plotted points fall below the clean sandstone line toward the limestone and dolomite lines. This situation could be related to the presence of shale content which take the same pattern or may be related to the existence of fraction of the cemented material. Therefore, the values of ρ_{ma} and Δt_{ma} are greater or less than those parameters of clean sandstone. These cross-plots demonstrate that the representative points generally fall below the sandstone line and virtually on the limestone line. These points may correspond to granite or granodiorite rocks of Serra [12]. Analytical solution of matrix parameters values of (ρma) and (Δtma) can be also calculated by knowing the volume of shale [13]. Formulas 1 and 2 are used for this procedure. Therefore, these reservoirs may be classified as weathered plutonic reservoirs (granite or base wash). Through Table 1 only presents results of three upper reservoir subunits of well B, because the results of whole reservoir units of different studied wells need more available space at this paper.

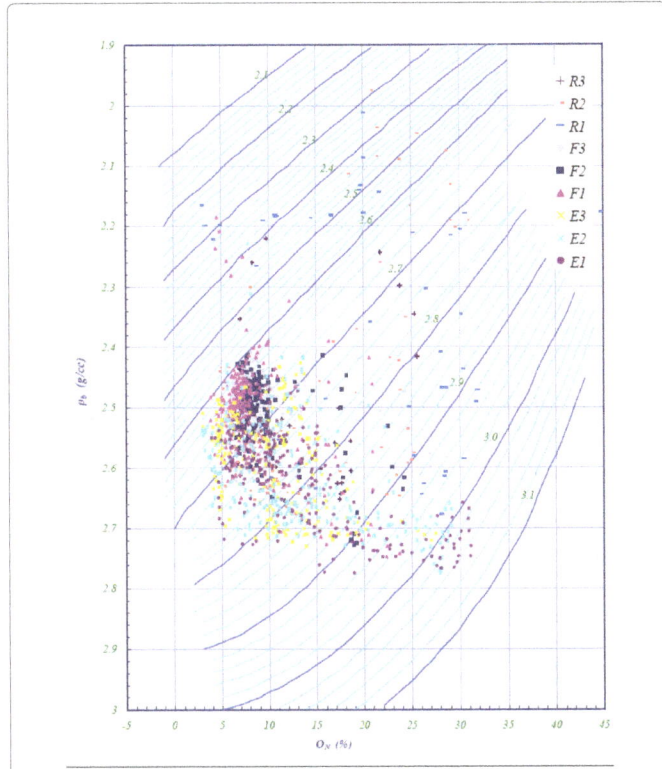

Figure 3a: Manual determination of (ρ_{ma}) a from FDC and CNL data, well B.

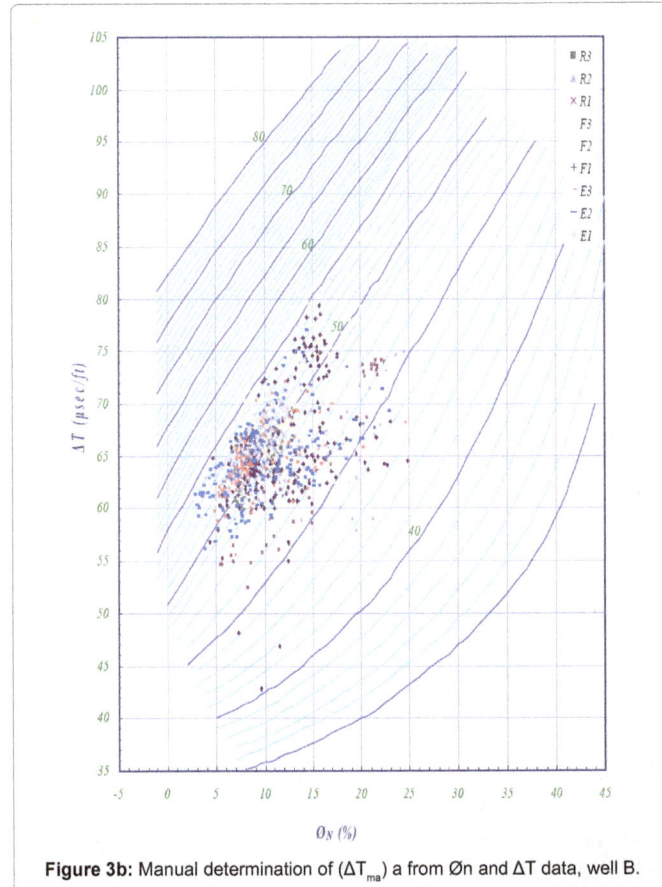

Figure 3b: Manual determination of (ΔT_{ma}) a from Øn and ΔT data, well B.

	Parameters							
Reservoir Unit	Chart				Analytical			
	Before Ø$_{ncor}$		After Ø$_{ncor.}$		Before Ø$_{ncor}$		After Ø$_{ncor.}$	
	ρ_{ma}	Δt_{ma}	ρ_{ma}	Δt_{ma}	ρ_{ma}	Δt_{ma}	ρ_{ma}	Δt_{ma}
R3	2.68	49.47	2.67	50.439	2.55	57.43	2.55	57.527
R2	2.71	48.5	2.67	50.89	2.567	57.91	2.567	49.437
R1	2.829	46.71	2.758	53.44	2.635	64.879	2.635	36.204

Table 1: Shows average ρ_{ma} and Δt_{ma} values of reservoir subunits (R) in well (B).

$$\rho_{ma} = \frac{V_{sh}[\rho_{sh}\phi_{nma} - \phi_{nma} - \rho_{sh} + \phi_{sh}] - (\rho_b\phi_{nma}) + \rho_b - \phi_n + \phi_{nma}}{V_{sh}(\phi_{sh} - 1) - \phi_n + 1} \quad (1)$$

$$\Delta t_{ma} = \frac{V_{sh}\left[-\Delta t_f\varphi_{sh} - \Delta t_{sh}\varphi_{nma} + \Delta t_{sh} + \Delta t_f\varphi_{nma}\right] + \Delta t_f\varphi_n + \Delta t\varphi_{nma} - \Delta t - \Delta t_f\varphi_{nma}}{\varphi_n - 1 + V_{sh}(1 - \varphi_{nsh})} \quad (2)$$

MID cross – plot

This technique was introduced by Clavier et al. [2], which combine the measurements of the same three tools (Neutron, Density and Sonic). An apparent matrix density (ρma) a and an apparent matrix travel time (Δtma) a are estimated from previous cross-plots. These parameters were plotted against each other. Figure 4 demonstrate that the data points form a cluster shape toward the decrease of matrix travel time value (Δtma), which displays the existence of secondary porosity. Whereas less data points fall below the mentioned cluster toward the anhydrite point indicate a shaly influence. However, minor data points were scattered in the direction of northeast of the main cluster point-out gas effect. These investigated what was incompliance with Schlumberger [14] interpretation.

M-N plot

This technique is very similar to the previous one and introduced by Burke et al. [1] for the study of complex lithologies. The authors apply this technique to compute parameters M and N independently of porosity assuming that all three tools (Neutron, Density and Sonic) respond linearly to porosity. These parameters (M and N) calculated by equation as follow:

$$M = \frac{\Delta t_f - \Delta t}{\rho_b - \rho_{nf}} \times 0.1 \quad (3)$$

$$N = \frac{\varphi_{nf} - \varphi_n}{\rho_b - \rho_f} \quad (4)$$

Figure 5 shows linear trend from anhydrite (below) to quartz (above) points in the northeast direction. The points close to the anhydrite express the shaly effect while the adjacent points to quartz demonstrate more clean sand [15,16]. It is reasonable to mention that the scattered points on the same trend refer to gas effect.

PEF – GR cross – plot

Gamma ray log data can be related to another log data in incompatible and compatible cross –plotting. Gamma ray –resistivity, gamma ray – Neutron porosity and photoelectron factor-gamma ray cross – plots are usually done to quantify lithology. This PEF-GR relation cross-plot illustrates the influences of pyrite, ferruginous encrustations, organic material and shale material (Figure 6). It is worth to notice from previous plotted figures that the points are concentrated between range 2-4 PEF values and gamma ray reading of 30 API should be considered

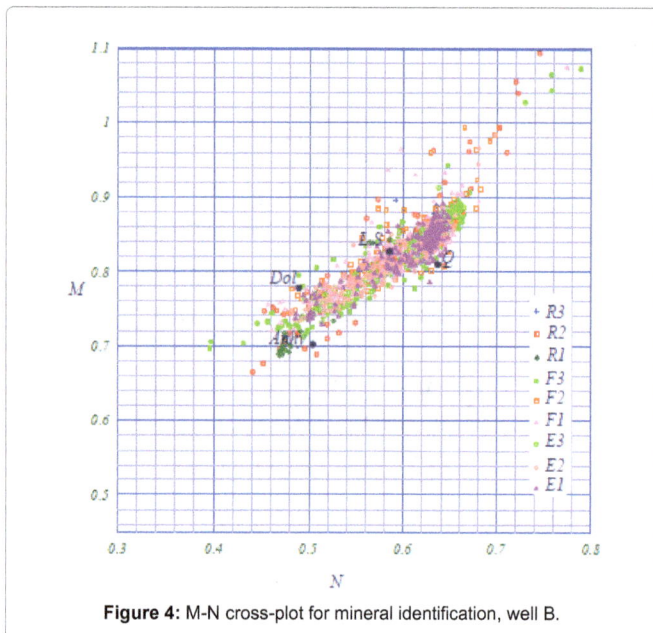

Figure 4: M-N cross-plot for mineral identification, well B.

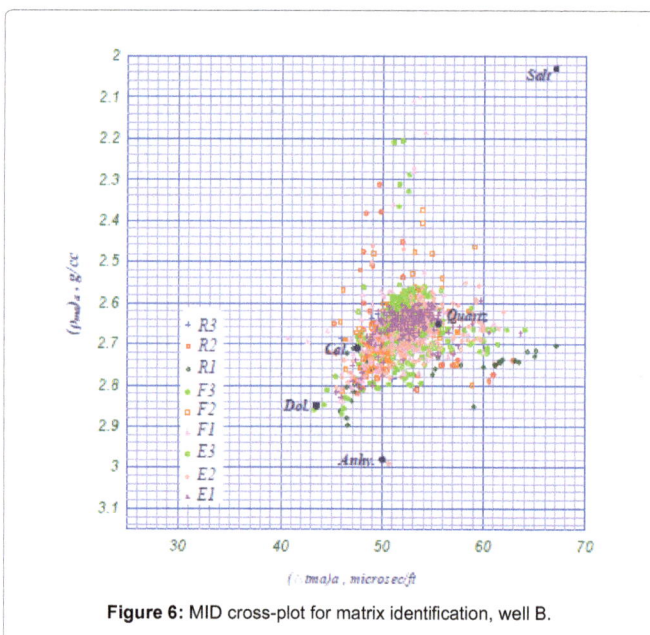

Figure 6: PEF-GR cross-plot for influence materials, well B.

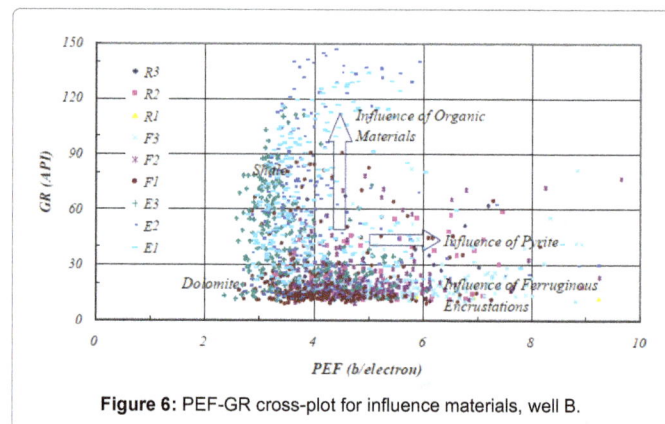

Figure 6: MID cross-plot for matrix identification, well B.

clean Sandstone. These values depend on bulk density, Travel time and Neutron measurements. It is clearly that this difference has affected the porosity calculation. The values of these parameters were estimated both by manual method plotted on Schlumberger charts and by analytical solution using derivative formulas for each reservoir subunit. The manually estimation gives better results when neutron porosity was corrected. Cross-plots demonstrate that the representative points generally fall below the sandstone line and virtually on the limestone line. These points may correspond to granite or granodiorite rocks of Serra [12]. Therefore, these reservoirs may be classified as weathered plutonic reservoirs (granite or base wash). While the MID and M-N techniques illustrate hydrocarbon effects and secondary porosity type on Nubian reservoir lithology. While GR-PEF technique shows the plot points are concentrated between range 2-4 B/e Pef values and gamma ray reading of 30 API should be considered and given more investigation due to its closeness to the dolomite point, additional to influences of ferruginous, pyrite and clay material.

References

1. Burke AJ, Campbell RL, Schmidt WA (1969) The Litho-Porosity Cross Plot A Method Of Determining Rock Characteristics For Computation Of Log Data. SPWLA, 10th Ann Log Symp Trans.

2. Clavier C, Rust DH (1976) MID-Plot: a new lithology Technique. The log Analyst 17.

3. Delfiner P, Peyret O, Serra O (1987) Automatic Determination of Lithology from Well Logs. Society of Petroleum Engineers. SPE Formation Evaluation 2: 303-310.

4. Akinyokun CO, Enikanselu AP, Adeyemo BA, Adesida A (2009) Well Log Interpretation Model for the Determination of Lithology and Fluid Contents. Pacific J Sci Tech 10: 1.

5. Saheel S, Samsudin AR, Hamzah UB (2010) Regional geological and tectonic structures of the sirt basin from potential field data. American J Sci Indus Res 1: 448-462.

6. Bosworth W (2008) North Africa–Mediterranean Present-day Stress Field Transition and Implications for Fractured Reservoir Production in the Eastern Libyan Basins. Geol East Libya 4: 123-138.

7. Ibrahim MW (1991) In: Salem MJ, Busrewil MT, Ben Ashour AM (eds.) Petroleum geology of the Sirt Group sandstone, eastern Sirt Basin. The Geology of Libya. Elsevier, Amsterdam 7: 2757- 2779.

8. El-Hawat AS (1992) The Nubian sandstone sequence in Sirte Basin, Libya: Sedimentary facies and events. In: Sadiq A (edn.) Geology of the Arab world Cairo University 317-327.

9. Abdulghader SG (1996) In Salem MJ, El- Hawat S, Sbeta AM (eds.) Depositional environment and diagenetic history of the Maragh formation NE Sirt Basin, Libya. Geology of the Sirt Basin: msterdam, Elsevier 2: 263-274.

10. El-Bakush SH, Al-Magdob TA, Al-Kafish KT (1997) Geology, Petrography and Petrophysical analysis of Rimal Oil Field "Nubian reservoir".

and given more investigation due to its closeness to the dolomite point [17,18]. This phenomena, in this study, may reveal that the shaly sand include one or more impurities of dolomite, potassium feldspar, etc. in matrix and/or cement.

Discussion and Conclusion

Oil fields are producing from inhomogeneous (shaly sand) reservoir needs more petrophysic parameters to present accurate reserve estimation. However, the studied reservoir has a multizonation from bottom to top (E3, E2, E1, F3, F2, F1, R3, R2 and R1) due to variety of shale content and clay minerals distribution. Whereas, this reservoir hold a significant hydrocarbon potential in Southeast Sirt Basin. Formation matrix parameters (ρ_{ma} and Δt_{ma}) values of this studied reservoir are different from the constant or standard values for

11. El- Bakush SH, Minas HA (2007) Petrophsical analysis of Nubian Reservoir, Rimal Oil Field- Sirte Basin.

12. Serra O (1986) Fundamentals of well log interpretation.

13. Ahlbrandt TS (2001) The Sirte Basin Province of Libya-Sirte-Zelten Total Petroleum System. U.S. Geological Survey Bulletin 2202–F, Version 1.0.

14. Schlumberger (1989) Log Interpretation Principles/Applications.

15. Hilchie WD (1982) Applied Open Log Interpretation for Geologists and Engineers.

16. Schlumberger (1982) Natural gamma ray spectrometry essentials of N.G.S interpretation, Applications of the Natural Gamma ray spectrometry.

17. Schlumberger (1986) Log Interpretation Charts.

18. Selley CR (1976) An Introduction to sedimentology. Academic Press, Inc. London 408.

26

Land Surface Temperature and Scaling Factors for Different Satellites Datasets

Mukesh Singh Boori[1,4]*, Heiko Balzter[3], Komal Choudhary[1] and Vit Vozenílek[4]

[1]*Samara State Aerospace University, Russia*
[2]*American Sentinel University, Colorado, USA*
[3]*University of Leicester England, UK*
[4]*Palacky University Olomouc, Czech Republic*

Abstract

Land surface parameters are highly integrated and have a direct effect on water and energy balance and weather predictions. Due to the difficulties in correcting the influences of the atmosphere absorbability and the earth surface emissivity diversification, the retrieval of land surface temperature (LST) from satellite data is a challenging task. To retrieve microwave land emissivity, infrared surface skin temperatures have been used as surface physical temperature. However, passive microwave emissions originate from deeper layers with respect to the skin temperature. So, this inconsistency in sensitivity depths between skin temperatures and microwave temperatures may introduce a discrepancy in the determined emissivity. In this research, six sample sites were chosen on the earth for 2013 and 2014 and then land surface temperature from AMSR-2, Landsat and ASTER brightness temperature values have been derived. The algorithm has been developed from a surface brightness temperature dataset, which has used as inputs surface parameters and atmospheric quantities. The retrieved LST has been compared within AMSR-2, Landsat and ASTER for the same period and area. Maximum time ASTER has shown higher temperature than other data and AMSR-2 has lower temperature on same area. Landsat and ASTER is closer to ground measured temperature than AMSR-2 data. It will be interesting to see how the satellite-derived surface temperature will behave in an assimilation scheme in a follow-up study.

Keywords: AMSR-2; ASTER; Landsat data; Brightness temperature

Introduction

Surface temperature is a key parameter in many energy balance applications, such as evaporation models, climate models and radiative transfer models. Land/use cover (in terms of temperature and soil moisture) affect the surrounding climate, such as the water vapours, the dust partials, the gas molecules and the clouds formation; so it directly affects rain and then land/use cover [1]. By this way, it affects the whole ecosystem and correlates with natural disasters. Land Surface Models (LSMs) are expected to lead to improved short-term to long-range forecasting models. Because the land/surface parameters are highly integrated, errors in land surface forcing, model physics and parameterization tend to accumulate in the land surface stores of these models, such as soil moisture and surface temperature [2-4]. This has a direct effect on the models' water and energy balance calculations, and may eventually result in inaccurate weather predictions. Timely monitoring of natural disasters is important for minimizing economic losses caused by floods, drought, etc. so it's important for emergency management during natural disasters. Consequently, remote sensing for land/surface temperature has become an important research subject to a global scale [5].

Ground observations are generally useful for local applications; however, they are highly intensive in man-power and equipment costs. Furthermore, ground observations of land surface temperature are point measurements and since variability can be high, especially in regions with discontinuous vegetation, scaling up to spatial averages is often difficult [6]. In other side satellite-generated brightness temperatures (BT) are more convenient on global level and temporal basis [7,8].

Instantaneous measures of microwave brightness temperature (BT) have been used in a variety of applications to estimate column water vapour abundance, rainfall rate, surface ocean wind speed, ocean salinity, soil moisture, freeze/thaw state, land surface temperature, inundation fraction, and vegetation structure [9-14]. Land surface properties can be inferred accurately if physical temperature and emissivity variations can be separated [15,16]. Diurnal synoptic and variations of land surface temperature, as well as the atmospheric temperature and water vapour profiles, affect the observed BT. More frequent are the observations of BT throughout the day, the better understanding of the variability of the retrieved parameters is obtained [17].

There are a very few studies dealing with the characterization of the BT diurnal variation over land. The diurnal variation of physical and brightness temperatures as a function of incident solar radiation has been modelled for the Tropical Rainfall Measuring Mission (TRMM) Microwave Imager (TMI) [18]. The characteristics of the skin temperature diurnal cycles as measured from IR over different land types were investigated. In densely vegetated areas with more moisture, skin temperature exhibits a smaller diurnal variation than in arid and desert areas [19].

The paper is organized in the sections below. Following the description of the study area in Section 2, brief discussion about

*Corresponding author: Mukesh Singh Boori, Department of Geo-informatics, Section of Earth Sciences Palacký University, Czech Republic
E-mail: mukesh.boori@upol.cz

datasets in section-3, the methodology of calculating LST, and a brief introduction to normalized mutual information measure are presented in Section 4. The analysis and conclusions are presented in Sections 5-6, respectively.

Study Area

The study area has six training sites on earth in different continent in order to compare the temperature in different satellite datasets in same location (Figure 1). Sample sites are located in the following places: (1) Karl Stefan Memorial Airport (USA); (2) Mondai, Santa Catarina (Brazil); (3) Belgrade, Serbia (Europe); (4) Khartoum, Sudan (Africa); (5) Chengdu, China; (6) Kalgoorlie, Australia.

Data Sets

This research work has used thermal bands of AMSR-2, Landsat and ASTER satellite data. Technical characteristics of used satellite data are listed as it follows:

AMSR2

The Advanced Microwave Scanning Radiometer-2 (AMSR-2) on board the GCOM-W1 satellite is a remote sensing instrument for measuring weak microwave emission from the surface and the atmosphere of the Earth. From about 700 km above the Earth, AMSR-2 provides highly accurate measurements of the intensity and scattering of microwave emissions and scattering. The antenna of AMSR-2 rotates once per 1.5 seconds and obtains data over a 1450 km swath. This conical scan mechanism enables AMSR-2 to acquire a set of daytime and night time data with more than 99% coverage of the earth every 2 days. It's had 7 frequencies with vertical and horizontal polarizations. Here 36.5 GHz vertical frequency was used for land surface temperature, get from Japan Aerospace Exploration Agency. The 36.5 GHz AMSR-2 footprint is an oval of 25 km square, where the derived surface temperature fields are resampled in a 0.25 degree grid. A subset covering the sample sites

is cut from the global dataset. The choice for these location is motivated by the presence of sets of observational data for the corresponding period of time (i.e. year 2013-14) and there availability. Observed data is collected from 6 stations located on earth, spatially representative of different types of land cover [20] (Figure 1).

Landsat

In this research work Landsat 8 satellite data was used. Landsat 8 carries two instruments: The Operational Land Imager (OLI) sensor and Thermal Infrared Sensor (TIRS) sensor. The TIRS sensor provides two thermal bands. These sensors provide improved signal-to-noise (SNR) radiometric performance quantized over a 12-bit dynamic range. This translates into 4096 potential grey levels in an image compared with only 256 grey levels in previous 8-bit instruments. Improved signal-to-noise performance enables a better characterization of land cover state and condition [21]. This product is delivered as 16-bit images (scaled to 55,000 grey levels). For this research work band number 11TIRS (11.5-12.51 μm) 100 m was used for surface temperature.

ASTER

ASTER data provides the user community with standard data products throughout the life of the mission. Algorithms to compute these products were created by the ASTER science team, and are implemented at the Land Processes Distributed Active Archive Centre (LP DAAC). Users can search and browse these products through GDS and NASA Reverb. For this research work band number 14 TIR (10.95-11.65 μm) was used for surface temperature. It's have 16 bit data with +/-8.55 telescope pointing capacity [22].

Methodology

In our previous study, we estimated microwave land surface emissivity over the globe from AMSR-E observations at all frequencies and polarizations [23]. The previous emissivity retrieval

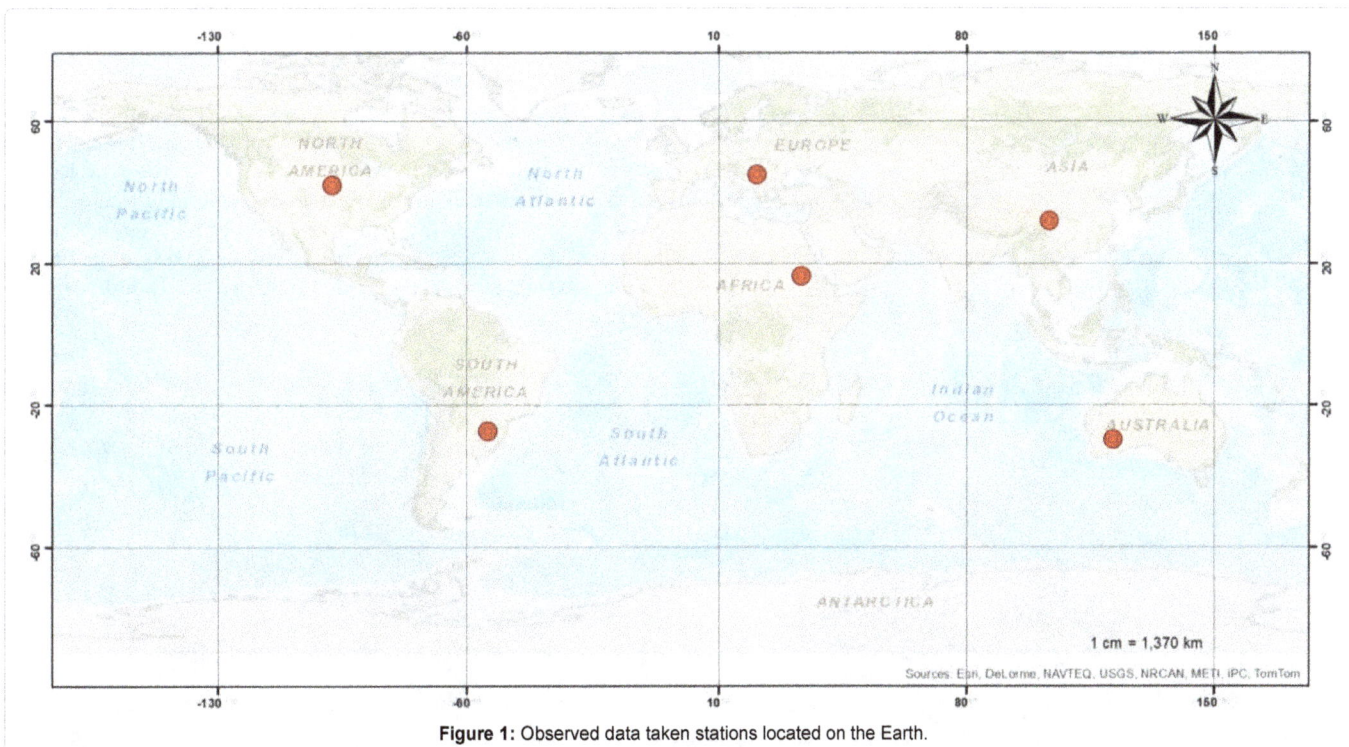

Figure 1: Observed data taken stations located on the Earth.

is based on the assumption that the infrared skin temperature is the effective physical temperature, which is equivalent to assume that the microwave brightness temperature originates from the skin. This assumption is not always true, but necessary due to the lacking of general globe information on penetration depths and temperature profiles.

A physical model was proposed to account for the effect of penetration depth on emissivity retrievals by revising the physical temperature [24]. In this research work, first take sample sites and then derive land surface temperature from satellite brightness temperature by the help of following scaling factors for different satellite data [25]:

AMSR-2

For AMSR2 data scale factor is set as 0.01 [K]. For instance, if brightness temperature data is stored as 28312, original value of the data is 283.12 [K].

ASTER

ASTER surface temperatures are computed from spectral radiance so begins by converting DNs to radiance and for that equation is following:

$L_\lambda = (DN-1) \times UCC$

Where L_λ is the spectral radiance, DN are the TIR band digital numbers, and UCC are the published Unit Conversion Coefficients (0.005225 w/m^{-2}/sr^{-1}/µm^{-1}).

Temperature (measured in degrees Kelvin) is then given by:

$T = K2 / (\ln (K1/ L_\lambda + 1)$

Where K1 (641.32) and K2 (1271.22) are constants derived from Planck's radiance function.

Landsat

OLI and TIRS band data can be converted to TOA spectral radiance using the radiance rescaling factors provided in the metadata file:

$L_\lambda = M_L Q_{cal} + A_L$

where:

L_λ = TOA spectral radiance (w/m^{-2}/sr^{-1}/µm^{-1})

M_L = Band-specific multiplicative rescaling factor from the metadata

(RADIANCE_MULT_BAND_x, where x is the band number)

A_L = Band-specific additive rescaling factor from the metadata

(RADIANCE_ADD_BAND_x, where x is the band number)

Q_{cal} = Quantized and calibrated standard product pixel values (DN)

TIRS band data can be converted from spectral radiance to brightness temperature using the thermal constants provided in the metadata file:

$T = K2/(\ln (K1/L_\lambda + 1)$

where:

T = At-satellite brightness temperature (K)

L_λ = TOA spectral radiance (w/m^{-2}/sr^{-1}/µm^{-1})

K_1 = Band-specific thermal conversion constant from the metadata

(K1_CONSTANT_BAND_x, where x is the band number, 10 or 11)

K_2 = Band-specific thermal conversion constant from the metadata

(K2_CONSTANT_BAND_x, where x is the band number, 10 or 11)

Analysis Results

In the following section, 1° × 1° spatial subsets of the products were co-registered so that the resulting subsets cover the same area as the Landsat, ASTER and AMSR-2 images in Figures 2-4. Where necessary, spatial re-sampling was performed using the 'nearest neighbour' method. The gravel plains are highly homogenous in space and time and outside globe two seasons (around September and June) the chances for precipitation are remote therefore, it is safe to assume that the land surface was completely dry for the results presented in this paper. The following section briefly discusses the actual land surface temperature on the same area and time from different satellite datasets (Figures 2-4).

AMSR-2

Figure 2 shows an example of the AMSR-2 surface temperature retrievals at day time in June 2014 and September 2013, when observational data have been sampled.

AMSR-2 June 2014 data show minimum 166.67 K and highest 305.34 K temperature and for September 2013 minimum is 170.78 K and highest is 305.34 K temperature. In June highest temperature is present in Sahara desert, western centre of North America and north-eastern part of Brazil, western sector of India and China. In September 2013 is the same like 2014 but high temperature is also present in Australian desert, in reduced areas of India and China but increased areas of South Africa and South America (Figure 2).

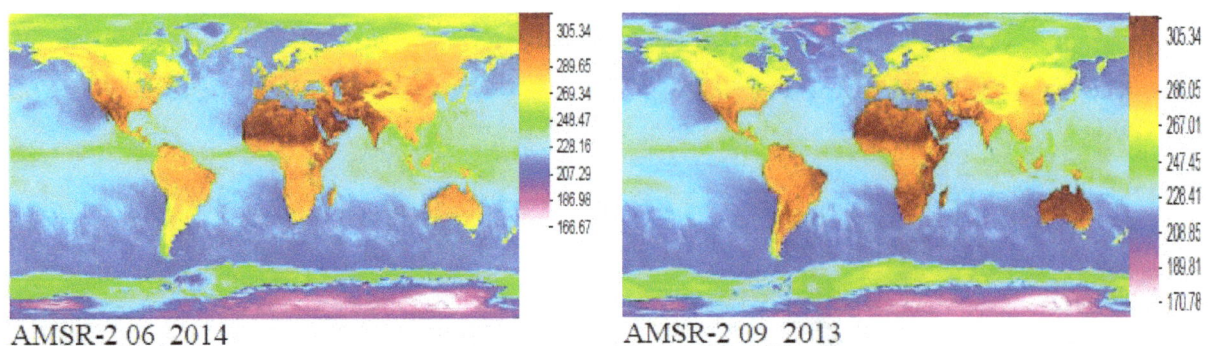

AMSR-2 06_2014   AMSR-2 09_2013

Figure 2: AMSR-2 data land surface temperature on globe.

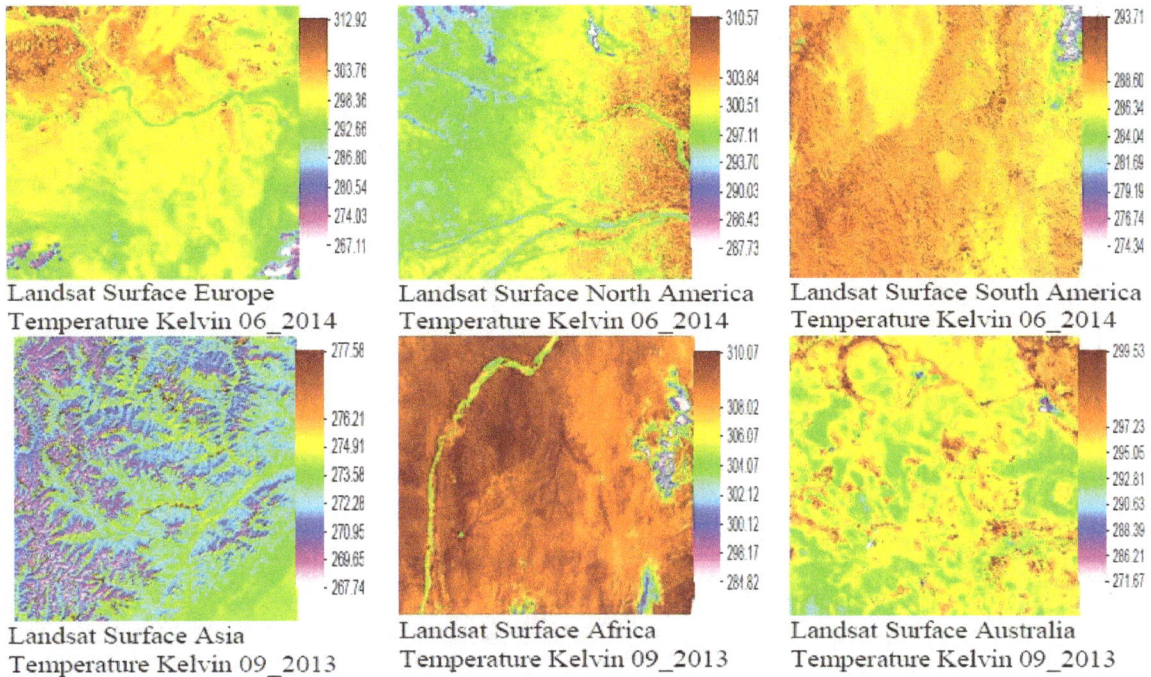

Figure 3: Landsat data land surface temperature on sample site on globe.

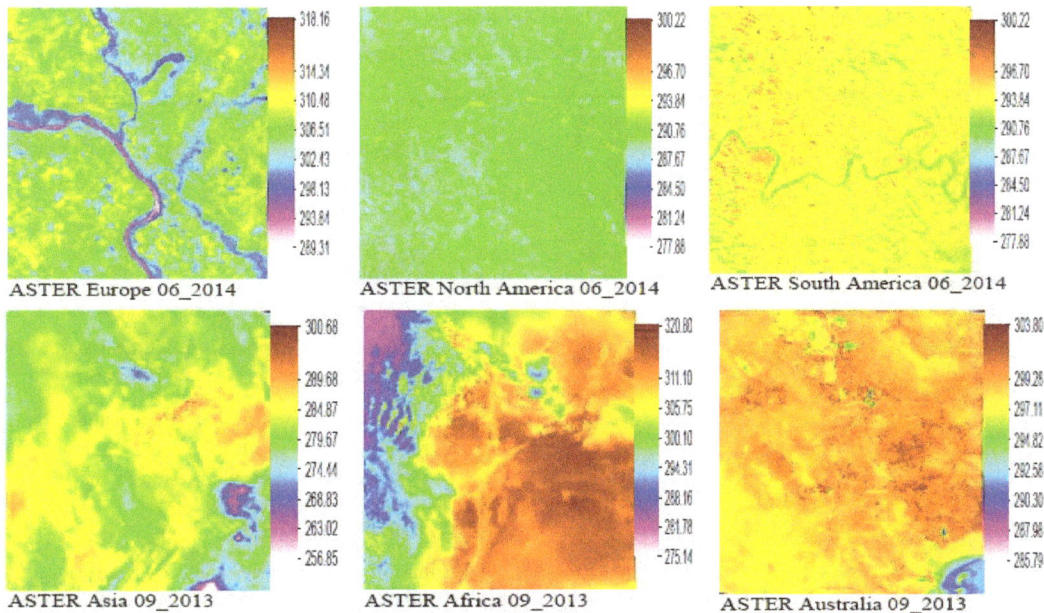

Figure 4: ASTER data land surface temperature on sample site on globe.

Landsat

In 2014 sample site in Europe, Landsat 8 band number 11 shows 267.11 K as lowest temperature and 312.92 K as the highest temperature. Maximum region show around 290 to 300 K temperatures. In North America its 287.73 to 310.57 K and in South America is 274.34 to 293.71 K but maximum region have high temperature, above than 285 K (Figure 3).

In 2013 in Australia the range of temperature is between 271.67 and 299.53 K. For Africa it is 284 to 310 K but a maximum area has above than 305 K temperature, which shows a high temperature. In Asia it's from 267.74 to 277.58 K and maximum area has average temperature (Figure 3).

ASTER

In ASTER data on same location in 2014 in Europe, temperature range is in between 289.31 to 318.16 K and for North America is from 277.88 to 300.22 K. In both regions, maximum area is cool. In South America and Asia temperature range is 277.88 to 300.22 K and 256.85 to 300.68 K respectively. Maximum area has average temperature. In Africa and Australia it's 275.14 to 320.80 K and 285.79 to 303.80 K respectively (Figure 4).

Comparing in land surface temperature in AMSR-2, Landsat and ASTER satellite data

In North America Landsat data showing highest and AMSR-

2 data is showing lowest temperature in all locations. ASTER data temperature values are in between in North America. In South America ASTER show highest temperature and AMSR-2 lowest and Landsat is in between (Figure 5).

For Europe is same like South America (Table 1). In Africa Landsat have highest and AMSR-2 lowest temperature and ASTER data temperature is in between. In Asia ASTER is highest, than Landsat and

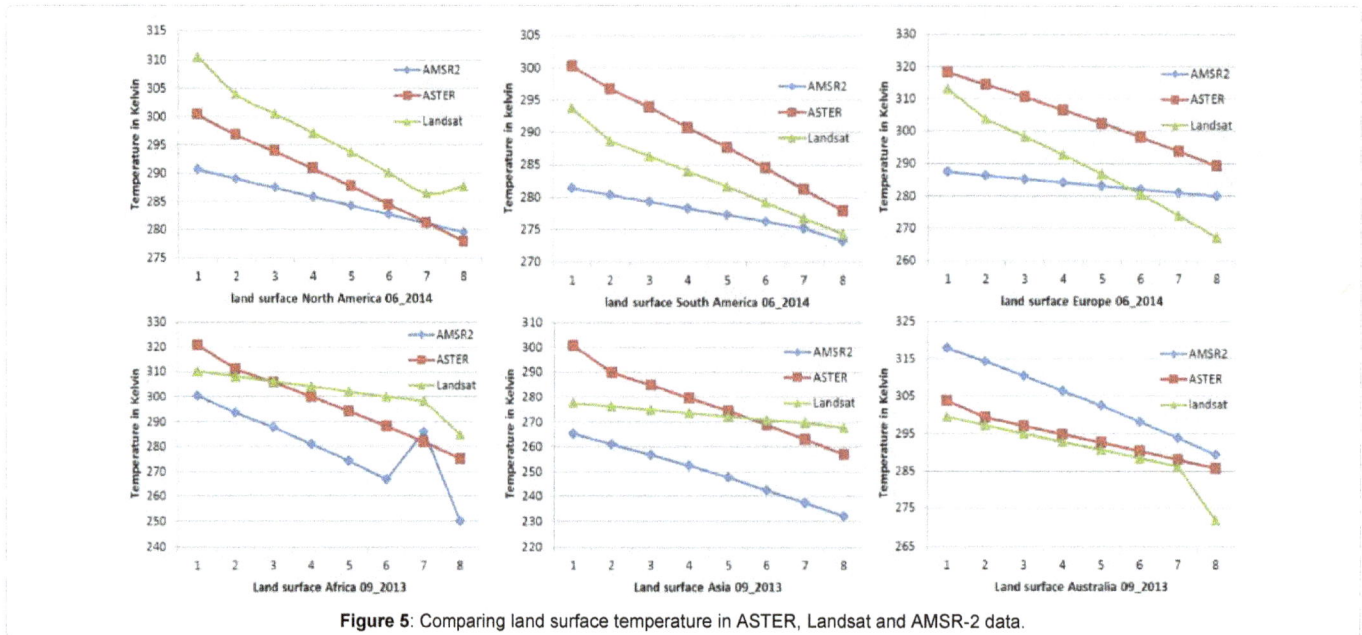

Figure 5: Comparing land surface temperature in ASTER, Landsat and AMSR-2 data.

AMSR2 Tem. K	ASTER Tem. K	Landsat Tem. K	AMSR2 Tem. K	ASTER Tem. K	Landsat Tem. K
Africa 09_2013			North America 06_2014		
300.22	320.8	310.07	290.63	300.22	310.57
293.6	311.1	308.02	288.99	296.7	303.84
287.6	305.75	306.07	287.43	293.84	300.51
281.07	300.1	304.07	285.83	290.76	297.11
274.3	294.31	302.12	284.27	287.67	293.7
266.75	288.16	300.12	282.67	284.5	290.03
285.87	281.78	298.17	281.11	281.24	286.43
250.07	275.14	284.42	279.55	277.88	287.73
Asia 09_2013			South America 06_2014		
265.13	300.68	277.58	281.39	300.22	293.71
260.93	289.68	276.21	280.33	296.7	288.6
256.75	284.47	274.91	279.32	293.84	286.34
252.36	279.67	273.58	278.28	290.76	284.04
247.71	274.44	272.28	277.27	287.67	281.69
242.63	268.83	270.95	276.23	284.5	279.19
237.47	263.02	269.65	275.45	281.24	276.74
232.05	256.85	267.74	273.16	277.28	274.34
Australia 09_2013			Europe 06_2014		
317.96	303.8	299.53	287.44	318.16	312.92
314.34	299.28	297.23	286.33	314.34	303.76
310.48	297.11	295.05	285.27	310.48	298.36
306.44	294.82	292.81	284.19	306.51	292.66
302.43	292.58	290.63	283.14	302.43	286.8
298.13	290.3	288.39	282.06	298.13	280.54
293.76	287.98	286.21	281.01	293.84	274.04
289.31	285.79	271.67	279.95	289.31	267.11

Table 1: Sensor measured temperature in kelvin.

North America	302.10-289.10
South America	293.1-288.10
Europe	303.10-289.10
Asia	300.10-268.10
Africa	311.10-296.10
Australia	302.10-285.10

Table 2: Ground measured temperature in kelvin.

AMSR-2 is the lowest temperature. For Australia, Highest is AMSR-2 than ASTER and Landsat is lowest temperature (Figure 5).

The analysis results show that Landsat and ASTER data temperature is close to ground measurements in compare of AMSR-2 data temperature due to their high resolution and scale (Tables 1 and 2).

Conclusions

Evaluation of different data sources of surface temperature indicates that satellite derived passive microwave surface temperature is not necessarily a superior estimate compared to simulated surface temperature, if evaluated against a data set of observed point measurements. In general, data assimilation systems take into account observational errors and are able, despite errors in the observations, to obtain improvement of land surface temperature results, as long as the temporal trends are well represented. This research work show that Landsat and ASTER is close to ground measurements in compare of AMSR-2 data due to their high resolution and scale.

It should be pointed out that the atmospheric forcing data sets used in the present study are mainly reanalysis data making use of *in-situ* observations. As a consequence, the simulations over the globe with a dense observation network are relatively accurate. However, no or little observation data are available for all sample sites and the quality of the forcing data and the simulations decreases. Hence, in these areas more scope is present for remote sensing data to constrain this type of research [26,27].

A further consideration is that the retrieval of passive microwave satellite derived surface temperature is hampered by weather conditions: frozen soil conditions in winter and (precipitating) clouds in summer. These phenomena appear to put some emphasis on 'near' in the assessment of the passive microwave retrieval of surface temperature as a 'near all-weather' technique. Consequently, this also applies to the passive microwave soil moisture algorithm retrieval, in case the passive microwave surface temperature estimate is used [28,29].

References

1. Maimaitiyiming M, Ghulm A, Tiyip T, Pla F, Carmona PL, et al. (2014) Effect of green space apatial pattern on the land surface temperature: Implications for sustainable urban planning and climate change adaptation. ISPRS J Photogramm Remote Sens 89: 59-66.

2. Norouzi H, Temimi M, Rossow WB, Pearl C, Azarderakhsh M, et al. (2011) The sensitivity of land emissivity estimates from AMSR-E at C and X bands to surface properties. Hydrol Earth Syst Sci 15: 3577-3589.

3. Boori MS, Netzband M, Vozenilek V, Choudhary K (2015) Urban growth in last three decades in Kuala Lumpur, Malaysia. *IEEE: Joint Urban Remote Sensing Event (JURSE)* 01-04.

4. McCollum JR, Ferraro RR (2005) Microwave rainfall estimation over coasts. J Atmos Ocean Tech 22: 497–512.

5. Min QL, Lin B, Li R (2010) Remote sensing vegetation hydrological states using passive microwave measurements. IEEE Journal of Selected Topics in Applied Earth Observations and Remote Sensing 3: 124–131.

6. Boukabara SA, Weng FZ, Liu QH (2007) Passive microwave remote sensing of extreme weather events using NOAA-18 AMSUA and MHS. IEEE Transactions on Geoscience and Remote Sensing 45: 2228-2246.

7. Entekhabi D, Njoku EG, O'Neill PE, Kellogg KH, Crow WT, et al. (2010) The Soil Moisture Active Passive (SMAP) Mission. Proceedings of the IEEE 98: 704–716.

8. Boori MS, Vozenilek V, Choudhary K (2015) Exposer intensity, vulnerability index and landscape change assessment in Olomuc, Czech Republic. *ISPRS: Int Arch Photogramm Remote Sens Spatial Inf Sci XL-7/W3*: 771 – 776.

9. Zhang LX, Zhao TJ, Jiang LM, Zhao SJ (2010) Estimate of phase transition water content in freeze-thaw process using microwave radiometer. IEEE Transactions on Geoscience and Remote Sensing 48: 4248-4255.

10. Prigent C, Jaumouille E, Chevallier F, Aires F (2008) A parameterization of the microwave land surface emissivity between 19 and 100 GHz, anchored to satellitederived estimates. IEEE Transactions on Geoscience and Remote Sensing 46: 344–352.

11. Norouzi H, Rossow W, Temimi M, Prigent C, Azarderakhsh M, et al. (2012) Using microwave brightness temperature diurnal cycle to improve imissivity retrievals over land. Remote Sensing of Environment 123: 470-442.

12. Boori MS, Vozenilek V, Choudhary K (2015) Land use/cover disturbances due to tourism in Jeseniky Mountain, Czech Republic: A remote sensing and GIS based approach. *The Egyptian Journal of Remote Sensing and Space Sciences* 18: 17 – 26.

13. Njoku EG, Jackson TJ, Lakshmi V, Chan TK, Nghiem SV (2003) Soil moisture retrieval from AMSR-E. IEEE Transactions on Geoscience and Remote Sensing 41: 215–229.

14. Papa F, Prigent C, Rossow WB, Legresy B, Remy F (2006) Inundated wetland dynamics over boreal regions from remote sensing: the use of Topex-Poseidon dual-frequency radar altimeter observations. Int J Remote Sens 27: 4847–4866.

15. Yang H, Yang Z (2006) A modified land surface temperature split window

16. Boori MS, Vozenilek V, Choudhary K (2015) Land Use / Cover Change and Vulnerability Evaluation in Olomuc, Czech Republic. ISPRS: Ann Photogramm Remote Sens Spatial Inf Sci II: 77-82.

17. Stephen H, Ahmad S, Piechota TC (2010) Land surface brightness temperature modeling using solar insolation. IEEE Transactions on Geoscience and Remote Sensing 48: 491–498.

18. Aires F, Prigent C, Rossow WB (2004) Temporal interpolation of global surface skin temperature diurnal cycle over land under clear and cloudy conditions. J Geophys Res 109.

19. Chen S, Chen X, Chen W, Su Y, Li D (2011) A simple retrieval method of land surface temperature from AMSR-E passive microwave data – A case study over Southern china during the strong snow disaster of 2008. Int J Appl Earth Obs 13: 140-151.

20. Boori MS, Vozenilek V (2014) Land-cover disturbances due to tourism in Jeseniky mountain region: A remote sensing and GIS based approach. *SPIE Remote Sensing* 9245: 01-11.

21. Fily M, Royer A, Goita K, Prigent C (2003) A simple retrieval method for land surface temperature and fraction of water surface determination from satellite microwave brightness temperatures in sub-arctic areas. Remote Sens Environ 85: 328–338.

22. Karbou F, Gerard E, Rabier F (2006) Microwave land emissivity and skin temperature for AMSU-A and -B assimilation over land. Q J Roy Meteor Soc 132: 2333–2355.

23. Boori MS, Vozenilek V, Burian J (2014) Land-cover disturbances due to tourism in Czech Republic. *Advances in Intelligent Systems and Computing. Springer International Publishing Switzerland* 303: 63-72.

24. Tedesco M, Kim EJ (2006) Retrieval of dry-snow parameters from microwave radiometric data using a dense-medium model and genetic algorithms. IEEE Transactions on Geoscience and Remote Sensing 44: 2143–2151.

25. Wilheit T, Kummerow CD, Ferraro R (2003) Rainfall algorithms for AMSR-E. IEEE Transactions on Geoscience and Remote Sensing 41: 204–214.

26. Boori MS, Ferraro RR (2013) Microwave polarization and gradient ratio (MPGR) for global land surface phenology. J Geol Geosci 2: 01 – 10.

27. Lambin EF, Ehrlich D (1997) Land cover change in sub-saharan Africa (1982-1991): Application of a change index based on remote sensing surface temperature and vegetation indices at a continental scale. Remote Sens Environ 61: 181-200.

28. Prigent C, Rossow WB, Matthews E, Marticorena B (1999) Microwave radiometric signatures of different surface types in deserts. J Geophysl Res Atmos 104: 12147–12158.

29. Boori MS, Amaro VE (2011) A remote sensing and GIS based approach for climate change and adaptation due to sea-level rise and hazards in Apodi-Mossoro estuary, Northeast Brazil. *International Journal of Plant, Animal and Environmental Sciences* 1: 14–25.

27

Utility of Large Scale Photogrammetric Techniques for 3-D Mapping and Precision Iron Ore Mining in Open Pit Areas

Murali Krishna G[1]* and Nooka Ratnam K[2]

[1]GIS Technology and Applications Development, Xinthe Technologies Pvt. Ltd. Visakhapatnam, India
[2]Department of Geology, School of Earth and Atmospheric Sciences, Adikavi Nannaya University, Rajah Rajah Narendra Nagar, Rajahmundry, India

Abstract

Precision mining and optimization of ore mining practise are gaining more importance as the global demand and competition for exploring the raw mineral material has grown manifold with increased industrialization world-wide. The overall process of mining activity involves ore identification, estimation, planning, excavation, transportation etc. The activity requires accurate mapping, monitoring and proper management of information pertaining to ore stockpile, mining pits and infrastructure of the areas on a regular basis. Of late, advancements in remote sensing techniques have paved a way for digital management of the mining activity. Especially, use of photogrammetric techniques for open pit mining are found to be highly accurate and effective in capturing, monitoring, mapping, managing the information pertaining to mining in a three dimensional (3D) space. Capturing of information pertaining to mines in 3-D perspective with respect to a specific location on the terrain is highly effective in accurate estimation of ore reserves, exploration and reclamation planning, ore continuity mapping and decision making. However, precision mining requires the use of large scale photogrammetric techniques with high resolution imagery of gigabytes size at mapping scales range from 1: 1000 to 1: 5000. In addition to that, the entire procedure involves the use of state-of-the-art software and hardware for fast processing of data and subsequent digital output generation. Above all, involvement of skilled photogrammetric experts with specialised knowledge on open pit mining is very much essential for accurate interpretation and delineation of the resources. A project was carried out for the 3D mapping of iron ore stockpiles, pits and infrastructure areas at various sites. The stockpile and pit mapping is used for very precise volume measurements and the infrastructure mapping is used for general mine planning activities. High accuracy is critical, as the data and volume reports are used to calculate the value of the ore being extracted, and any errors in the mapping data can result in incorrect payments of large amounts of money. Since, the timeline specified to complete the task is very short, it is crucial that the staff doing the final volume computations and downstream processing should receive the accurate data, correctly coded and mapped according to the standards outlined. The study has demonstrated a typical workflow for the effective use of close range photogrammetric techniques for 3D mapping and iron ore mining in open pit areas. The study also sees a brighter outlook and challenges of upcoming aerial and terrestrial photogrammetric technology for precision mining.

Keywords: Large scale photogrammetry; 3D mapping; Precision iron ore mining

Introduction

During the last decade, the photogrammetry and mapping industry has seen tremendous impetus in the implementation of innovative methods for accurate planning, 3D mapping and computation of mining ore and resources volumes. Notable studies by different authors particularly [1-3] have demonstrated the significance of accurate 3D shape reconstruction and volume estimation in many applications, for example, erosion studies, estimation of ore removed from a mine face and terrain assessment for construction etc. Particularly, digital photogrammetry has gained tremendous popularity due to the technology advancements in collecting detailed spatial information pertaining to the extent of mineral resource using high-resolution cameras or laser scanners and producing and rendering the digital 3D surface models [4-7]. Digital terrain models (DTMs) coupled with ancillary information, such as, dip direction and dip measurements, joint spacing etc., help in the assessment and characterization of mineral volumes [8,9]. Most importantly, digital photogrammetry is proven to be more reliable and cost effective when it is required to update the mapping of ore stockpiles, pits and infrastructure areas on a regular basis, generally at quarterly intervals for various sites, simultaneously [10]. On the other hand, it is also important to consider the geomorphic configuration and characteristics of the associated features as they may have certain degree of influence on open-pit mining [11,12]. A study has been undertaken to map and estimate the mineral ore stockpiles.

In some cases material may have been moved around on top of the stockpile itself which would also require re-mapping [13]. The stockpile and pit mapping is used for very precise volume measurements and the infrastructure mapping is used for general mine planning activities [14-16]. Data accuracy is very critical as it is used to asses stockpile information and for the generation of volume reports which will be used as the basis for calculating the payments to subcontractors [17]. Any errors in the mapping data can result in incorrect payments of large amounts of money. In majority of the cases, the projects need to be completed within the given timelines and it is very important that the work is done in accordance with the schedules defined by the concerned Project Manager. In general practice, unless and otherwise notified the workflow will always be the mapping and volumetric

*Corresponding author: Murali Krishna G, Head, GIS Technology and Applications Development, Xinthe Technologies Pvt. Ltd. Visakhapatnam-530003, India
E-mail: murali.krishna.gurram@gmail.com; ratna_k12@yahoo.com

computations of stockpiles first, then pits and finally infrastructure areas. As the timelines to complete the data updating task are very short, it is critical that the staff engaged in the computations of final ore volumes and downstream processing receive the data which is highly accurate, correctly coded and mapped according to the standards. One of the innovative techniques introduced in the study was to use only the alternate frames of the aerial photography received for mapping as the forward overlap was found to be 80% since use of such excessive overlap results in lower accuracy of DTM.

Objectives

Stockpile mapping requires the delivery of map data with high quality and information pertaining to ore volumes on time. There is a multitude of objectives to be met for each quarterly survey, but the primary goal is to provide the customer with an accurate measure of their production for the last three months, most specifically,

The amount of ore and product that is stockpiled; and

The amount of bulk earth excavated from the mine pits.

The secondary objectives are

To compare the current size of waste or overburden dumps to final design surface;

Measure the "in pit" dumping of waste or backfill; and

Provide updated site maps to be used for a variety of purposes.

Another important objective of the project is to reduce the density of data captured in certain areas of the mine, particularly within waste dumps and spoil. It is important that the data density in waste and spoil is kept to an absolute minimum so that the data loads quickly and efficiently into the database.

Data Inputs and Materials

High resolution aerial photographs of 7 cm to 10 cm GSD are used for the stockpile mapping study. Ground truth information collected by the supervisors is also used as the ancillary data to update the stockpiles. Feature symbology as specified by the client has been adopted for the mapping of stockpile, pits and infrastructure data. Similarly, Quality Assurance and Quality Control (QA/QC) plan is designed according to the recommendations of the client.

Methodology

The overall methodology involves a typical photogrammetry project workflow starting from aerial photo acquisition, data inventory, ground-truth verification in the mine area followed by Aerial Triangulation (AT). 3D Mapping of stockpile data starts with loading and updating the stockpile map data file that has been supplied by the client. Updating task involves identification of the areas of change (i.e. areas to be updated) and plotting them with an "update mapping polygon" around such areas. Once the areas which need to be updated have been defined and demarcated, the area inside this need to be mapped with sufficient breaklines on tops and toes of stockpiles along with spot heights on a notional 20 m grid to accurately define the stockpile surface. If new stockpiles are found in the image then the respective areas have to be identified and mapped accordingly. Subsequently typical QA/QC of the data has to be performed according to the standards specified for the data capturing task.

Stockpile mapping

The purpose of stockpile mapping exercise is to identify areas where the stockpile surface has changed since last aerial survey [18]. The change may be due to material being added or removed from the stockpile surface. Accuracy is critical to the stockpile mapping data and volumes supplied as they used to work out the amount of product material and therefore the value of the material pertaining to each mining site. The objective is to map three different categories of information pertaining to stockpiles which are subject to change continuously over a period of time. The three different categories which are to be mapped are stockpiles which have been changed, new stockpiles and stockpiles which no longer exist. The procedure for correctly mapping and updating stockpiles data is as mentioned below.

The stockpile areas which have been identified as the areas of change are to be mapped as "update mapping polygon". In case of stockpiles which are no longer existing, they have to be mapped using "update mapping polygon" around the original stockpile but no new data should be plotted. The update polygons should be placed very close to the area of change. Areas with no change can be identified easily with the presence of vegetation. However, make sure that the names of all stockpiles remain in the mapping data. Areas with minor changes, unchanged stockpile surface data and changes due to erosion processes around the stockpile toe need not be mapped. Stockpiles with very small or negligible areas which are in doubt for identification are captured as a stockpile with the addition of an explanatory note to the data. Once update areas have been defined, the areas inside these stockpiles are mapped with sufficient breaklines or tops and toes of stockpiles with necessary spot heights with a 20 m grid spacing to accurately define the stockpile surface. For the areas where there are rows of truck dumps there should be breaklines along the top and toe of each row.

This is done by properly defining and plotting the "edge of stockpile" boundary as the stockpile surface should be mapped or updated using the relevant stockpile mapping feature codes. In addition to this, a 10 m buffer of additional surface data is captured around the outside of the 'live' stockpile edge. This data is captured using the feature types relevant to the stockpile location within the mine site, e.g. if the stockpile is located in an area mapped with 'Pit' feature coding, then the 10 M buffer is captured with 'Pit' feature codes. This 10 M buffer of data is in the form of DTM points with spacing not greater than 5 m. This also affects the pit and infrastructure mapping process as the map data has to be voided (at 11 m from stockpile edge) to accommodate the additions in the stockpile map data.

Pit mapping

Pit mapping phase of the project involves update mapping of the remaining mine areas at each site. Generally the pit data consist of pit areas, spoil areas and surrounding natural surface feature data. The purpose of the pit mapping is to create a digital terrain model of the pit surface which we will use to calculate the volume of material removed from the pit since the previous aerial survey. This volume is then used to work out how much money the earth moving contractor gets paid, which is a function of how much crude ore they have removed from the pit.

Spoil areas are to be captured in such a way that the spacing between the points along the line strings are much less while the DTM spot spacing is optimized as per the natural surface. A fundamental requirement is that all top and toe strings must be on the terrain surface, except where toe strings are obstructed by broken material, in which case they have to be "pushed through" on a line where they would be had the broken material not obscured it. They are critical in defining both the pit floor level and bench level. In order to map the pit faces accurately, breaklines are to be captured to define the face between the top and the toe and should be very precisely positioned on the face. Breaklines are to be plotted down the face of the pits to define the corners accurately. Closed embankment features are defined by bench tops and toes that need to be properly snapped where appropriate. The top of bank (pit wall) should be at the same reduced level as the floor of the pit behind it, any material above and just behind the top of bank edge should be outlined by a material edge code and have material breaklines defining the top.

The spacing between consecutive vertices forming the line strings should not between 3 to 15 m. Figure 1 shows an example of far too many points captured along line strings. The original points, white dots, are in places less than a metre apart and are nearly all less than 3 m from their neighbour. The green and blue dots show the correct spacing. Green on top of bank and along breakline on face is between 5 to 8 m apart. Blue dots at back along material line are about 12 m to 14 m apart. In the case of both the blue and the red dots the surface is still defined accurately but with far less data. On the other hand, the blast site surface areas have been "coarsely" defined with breaklines.

Material heaps and safety bunds have been outlined with an edge of material boundary and material breaklines to define the surface of the heap. They form a closed polygon except where the material boundary snaps to a top of bank in which case the top of bank code takes precedence. When identifying areas of change in waste and spoil areas it is important to ignore changes that have been caused by erosion.

Infrastructure mapping

This phase of the project involves update mapping of the infrastructure areas at each mining site. The infrastructure mapping has many different users within the client's organisation but will mostly be used for planning purposes. The infrastructure data will usually consist of very detailed data in some areas and large areas of natural surface. It all needs to be checked, although there is not usually much change in the natural surface areas. For sites with multiple infrastructures it is important to make sure that the integration between the general infrastructure and the infrastructure to pit and stockpile area are checked and the relevant data is loaded as a reference to ensure a seamless integration. In areas where there are some preparatory

earthworks or drilling programs being done there will be occasions when an existing track is "cut off" from the rest of the network of roads and tracks. These "orphan" tracks must be recoded as breakline.

Quality analysis (QA)/quality control (QC) of the data

Once the data capture phase is complete the data must be cleaned in accordance with data cleaning standards defined for the stockpile mapping project. This includes, thorough review of entire data capturing process using various cleaning tools and processes custom designed for the task. Tools like 'review feature statistics' was used to ensure that valid codes have been used and to make sure that there are no unexpected codes in the data, e.g., to check the inclusion of any road feature codes in the stockpile data. There are other tools like "find crossing tool" that are used to ensure the edge matching of each dataset with reference to the adjoining datasets. There will be occasions when an update polygon clips a cell for a pole, post or other feature represented by a cell. When this happens the cell will "drop" to an ellipse or line feature. Running "review feature statistics" will highlight these as incorrect feature types.

QA check for control and check points are done as part of the AT procedures at each site. It is critical that the QA checks on control meets the required accuracy standards. If this does not occur then the Project Manager should be alerted immediately. Under no circumstances, should any data capture commence until the results of the control validation report are accepted.

GIS integration

All the data layers are to be integrated in GIS environment as entire stockpiles, pit and infrastructure map data should contain GIS ready centroids and connectors wherever applicable. GIS connectors are required for all road features, creeks and drains. GIS connectors should be used to close different road types from each other. Creeks should be connected where possible to create a continuous water flow network. The same is also required for drain features. The newly captured area features must have a centroid placed inside the feature boundary. Coding of each feature class should be done according to the feature code list which contains a corresponding centroid for every area feature.

Conclusions

The workflow adopted for the stockpile mapping study effectively demonstrates the utility of digital photogrammetry technology for the mapping of stockpiles, pits and infrastructure areas and for the estimation of stockpile volumes with high degree of accuracy. The study also highlights the high degree of reliability of digital photogrammetry technology in cases where the ore stockpiles have to be monitored, mapped and estimated very frequently on a timely basis without any interruption to mining activity.

References

1. Aguilar MA, Aguilar FJ, Agüera F, Carvajal F (2005) The evaluation of close-range photogrammetry for the modelling of mouldboard plough surfaces. Biosys Eng 90: 397-407.

2. Lee DT, Schachter BJ (1980) Two Algorithms for Constructing a Delaunay Triangulation. Int J Comp Inform Sci 9: 219-242.

3. Schulz T, Ingensand H (2004) Influencing variables, precision and accuracy of terrestrial laser scanners. Ingeo, Bratislava.

4. Atkinson KB (1996) Close Range Photogrammetry and Machine Vision. Whittles Publishing, Scotland.

5. Carbonell M (1989) In: Karara HM (ed.) Arquitectural photogrammetry. Non-topographic photogrammetry. Falls Church, Virginia: ASPRS.

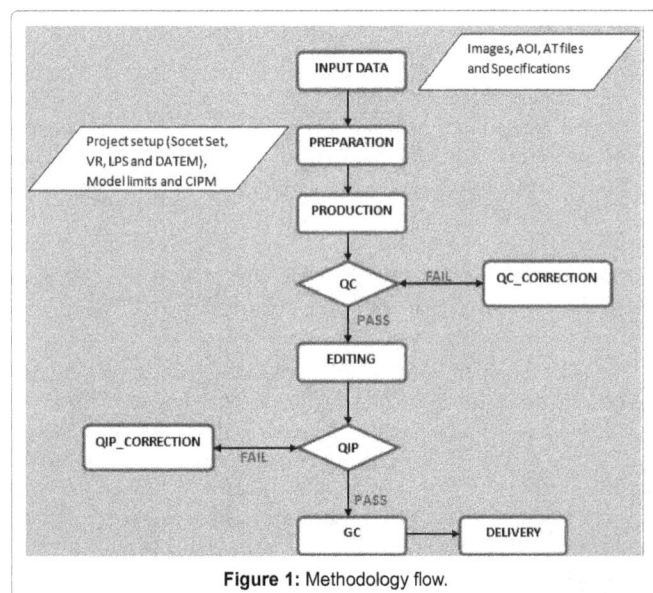

Figure 1: Methodology flow.

6. Patikova A (2004) Digital photogrammetry in the practice of open pit mining. Int Archiv Photogrammetry Remote Sensing Spatial Inform Sci 34: Part XXX.

7. Slama CC (1980) The Manual of Photogrammetry, 4th Edition. American Society of Photogrammetrists, Falls Church, VA.

8. Hao X, Pan Y (2011) Accuracy Analysis of Earthwork calculation based on Tringulated Irregular Network (TIN). Intell Automat Soft Comput 17: 793-802.

9. Wallace P (2008) New Topcon Imaging Station robotic Total Station Combines Imaging and Scanning.

10. Yakara M, Yilmazb HM (2008) Using In Volume Computing Of Digital Close Range Photogrammetry. Int Archiv Photogrammetry Remote Sensing Spatial Inform Sci XXXVII: Part B3b.

11. Jianping C, Ke Li, Kuo-Jen C, Giulia S, Paolo T (2015) Open-pit mining geomorphic feature characterisation. Int J Appl Ear Obser Geoinform 42: 76-86.

12. Labant S, Stsnkova H, Weiss R (2013) Geodetic Determining of Stockpile Volume of Mineral Excavated in Open Pit Mine. GeoSci Eng 59: 30-40.

13. Crotty JH, Ernst B, Machele A, Lategan J (2010) Measuring rock dumps from space? How new technology could change the way we measure rock dumps and tailings dams.

14. Chen C, Lin H (1991) Estimating pit-excavation volume using cubic spline volume formula. J Surv Eng 117: 51-66.

15. Easa SM (1988) Estimating pit excavation volume using nonlinear ground profile. J Surv Eng 114: 71-83.

16. Raeva PL, Filipova SL, Filipov DG (2016) Volume Computation of a Stockpile-A Study Case Comparing GPS and UAV Measurements in an Open Pit Quarry. Int Arch Photogram, Remot Sens Spatial Inform Sci XLI-B1, XXIII ISPRS Congress, 12–19 July 2016, Prague, Czech Republic.

17. Yanalak M (2005) Computing Pit Excavation Volume. J Surv Eng 131: 15-19.

18. Hamzah HB, Shaharuddin Md. S (2011) Measuring volume of stockpile using imaging station. Geoinform Sci J 11: 15-32.

28

Rare Earth Elements of the Arima Spring Waters, Southwest Japan: Implications for Fluid – Crust Interaction during Ascent of Deep Brine

Hitomi Nakamura[1,2]*, Kotona Chiba[2], Qing Chang[1], Shunichi Nakai[3], Kohei Kazahaya[4] and Hikaru Iwamori[1,2]

[1]Japan Agency for Marine–Earth Science and Technology, 2-15 Natsushima-cho, Yokosuka-shi, Kanagawa 237-0061, Japan
[2]Department of Earth and Planetary Sciences, Tokyo Institute of Technology, 2-12-1 Ookayama, Meguro-ku, Tokyo 152-8551, Japan
[3]Earthquake Research Institute, University of Tokyo, 1-1-1 Yayoi, Bunkyo-ku, Tokyo 113-0033, Japan
[4]Geological Survey of Japan, AIST, 1-1-1 Higashi, Tsukuba, Ibaraki 305-8567, Japan

Abstract

Rare earth elements (REEs) of the eight Arima spring waters in southwest Japan, including Arima-type brine that represents a specific type of deep-seated brine of up to 6 wt.% NaCl in the non-volcanic fore-arc region, have been investigated in order to discuss their upwelling processes and origins. We found four distinct patterns of REE composition of the spring waters within the Arima area of ~1 km^2, based on which two sources for REEs and two aquifers are inferred in the modification of the original deep-seated brine composition. On the basis of the REEs and isotopic compositions of the original deep brine, one of the two sources is thought to be slab-derived fluid dehydrated from the subducted Philippine Sea slab beneath the Arima area, represented by the 'Kinsen' hot spring water [1]. The convex-down REE pattern of most Arima spring waters, except for 'Kinsen' and 'Tansansen', suggests the presence of an oxidizing aquifer deeper than 160 m that causes co-precipitation of REEs with oxyhydroxides. CO_2 and He degassed from this aquifer flux the overlying shallow aquifer less than ~50 m in depth, producing highly carbonated water such as 'Tansansen' water that was originally derived from meteoric water. The carbonated water may dissolve a significant amount of REEs to the 'Tansansen' spring water from the host rocks, which are possibly silicic igneous rocks with Eu-negative anomalies. The four types of REE patterns with a wide concentration range, therefore, provide invaluable information concerning fluid–crust interaction during ascent of the deep brine.

Keywords: Brine; Spring water; Slab-derived; Fluid; Arima; Subduction

Introduction

Arima-type brine is non-volcanic hot spring water with high salinity found in a fore-arc region with no Quaternary volcanism [2]. This water has been geochemically characterized by a high Cl content at ~40000 ppm and specific O–H isotopic ratios that are similar to magmatic/metamorphic thermal waters. Despite its occurrence in a non-volcanic region, this water is distinct from meteoric water or buried sea water [2-4]. It also shows high $^3He/^4He$ ratios comparable or close to the mantle value and characteristic ratios of both stable and radiogenic isotopes (i.e., H, C, O, Sr, Nd, and Pb) that indicate slab-fluid derived from the subducted Philippine Sea Plate (PHS), suggesting a deep origin of the brine [1,5-7].

Arima springs in southwest Japan (Figures 1, 2A and 2B), the type locality of Arima-type brine, consist of hot springs ($\geqq$25°C) and associated cold springs (<25°C) including highly carbonated water. These springs generally exhibit a wide O–H isotopic range, which can be explained by mixing of meteoric waters and the deep brine component [3,4,7]. $\delta^{18}O$ is clearly correlated with δD, forming a straight mixing line between the meteoric water and the predicted slab-fluid generated by dehydration of PHS at ~50 km depth beneath the Arima area [7] (Figure 3A). The isotopic ratios also constitute linear variations with major solute elements such as Cl and Na (Figures 3B-3D), upon which the composition of the deep brine component has been estimated [4,7].

The deep-seated brine is relatively well characterized, as previously mentioned. It is thought to ascend along faults such as Atago-yama and Tenjin-yama faults associated with large tectonic lines such as the Arima–Takatsuki Tectonic Line (ATTL), as shown in Figure 2A, without undergoing significant geochemical modification [7,8]. However, the mechanism of encounter between the deep brine and the shallow meteoric waters to be mixed remains poorly understood.

In this study, we aim to constraining the geochemical processes at such shallow parts in which the deep component is mixed with the meteoric waters. For this objective, we analyze and utilize rare earth elements (REEs), particularly lanthanoids, of Arima spring waters. The geochemical behavior and partitioning of REEs between solid and fluid are sensitive to temperature, fO_2, fCO_2, and pH and may provide key information on the mixing processes at the shallow level where temperature, volatile fugacity, and pH are potentially variable [9,10]. First, we provide new data on REEs for Arima spring waters and compare the results with major solute elements and O–H isotopes, resulting in the identification of several types of spring water [9,10]. On the basis of these variations, we discuss the geochemical processes and mechanism related to the mixing of deep brine with meteoric waters.

Geological and Tectonic Setting of the Studied Area

In the Arima region, spring waters occur in a narrow area of approximately 1 km^2 (Figure 2B) and show compositional variation in major solute elements and gases [7]. These springs appear to upwell through the ATTL and subsidiary faults striking NW–SW or NW–SE with dextral slip associated with the Median Tectonic Line formed in the late mid-Miocene in the southwest Japan arc (Figure 2A) [11,12].

*Corresponding author: Hitomi Nakamura, Japan Agency for Marine–Earth Science and Technology, 2-15 Natsushima-cho, Yokosuka-shi, Kanagawa 237-0061, Japan, E-mail: hitomi-nakamura@jamstec.go.jp

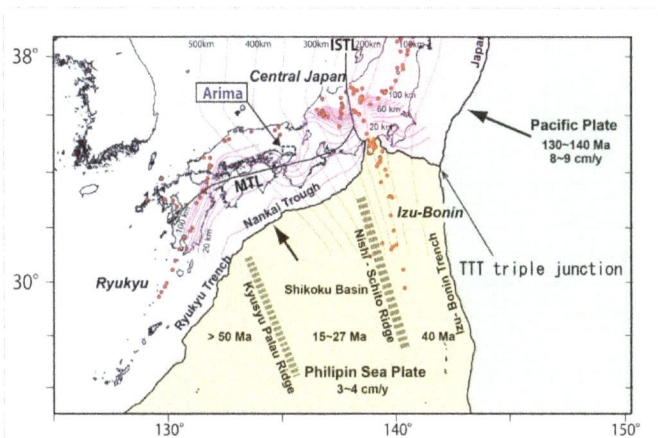

Figure 1: Tectonic setting in Japan and location of the study area. The map shows the distribution of Quaternary volcanoes (red circles) and the geometry of the subducting Pacific and Philippine Sea slabs relative to the Itoigawa–Shizuoka Tectonic Line (ISTL) and the Median Tectonic Line (MTL). The pinkish contour lines indicate the depth of the upper surface of the Pacific slab (50 to 300 km depth with 50 km intervals), whereas the purplish contour lines indicate that of the Philippine Sea slab (10 to 200 km depth with 10 km intervals). The aseismic parts are shown by the dotted line.

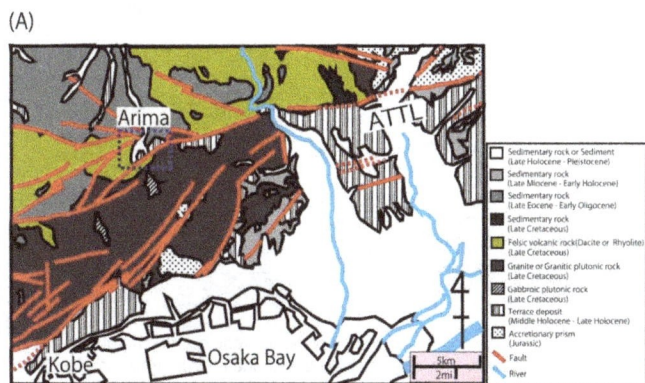

Figure 2: (A) Geologic map of the Arima area after Maruyama and Lin [12], shown as the square region in Figure 1.

The average slip rate for the eastern range-front segment of the tectonic line during the late Quaternary period is estimated to be 0.5–1.5 mm/year dextrally and 0.1–0.8 mm/year vertically [13]. The southwest Japan arc is associated with two oceanic plates, the Pacific plate and the PHS, which subduct beneath the area from the east at 9 cm/year and from the southeast at 4 cm/year, respectively. The slab surface depth beneath the Arima area is ~400 km for the Pacific slab but 50–80 km for the PHS, showing a large uncertainty [14,15]. Despite this active subduction, a Quaternary volcano has not formed in this region because the Pacific slab is too deep and the PHS is too shallow to fulfill the physiochemical conditions for arc magma generation (Figure 1) [16].

As shown in the geological map of this area (Figure 2A), the basement around the Arima area is composed of late Cretaceous felsic volcanic rocks such as rhyolite of the Arima Group, granitic rocks such as Rokko granite, and late Eocene to early Oligocene non-marine sedimentary rocks with rhyolitic tuff layers of the Kobe Group [17]. The Arima Group directly covers the Rokko granite south of the ATTL and the sedimentary rhyolitic rocks north of the ATTL in the Arima area [18].

Chemical Analysis

Sample description of Arima spring waters

The hot spring sources in the Arima area have been characterized by NaCl and hydrocarbon contents in addition to O–H–He isotopic composition [4,7], in most of which (Ginsen, Gokuraku, Gosho, Inari-Kinsen, Kinsen, Tansansen, Tenjin, and Uwanari waters) were investigated to understand their generation processes in this study (Figure 2B). During the dry season in December 2010, we collected 10 L samples of each of these waters directly from the well pipe before the water is pooled in tanks to be oxidized for commercial use. The Japanese names of these spring waters carry deep meanings. For example, "Kinsen" means gold-colored hot spring source, and "Tansansen" means high abundance of CO_2. Kusuda et al. [7] analyzed 12 solute elements/components and the isotopic ratios of H, He, C, and O in these spring waters and gases. Their results indicate a large range of elements, particularly in Cl (14–40,000 ppm) and hydrocarbon (12–1300 ppm) contents, adding to the wide range of O–H isotopic compositions. The He isotopic ratio has a smaller gap among these samples at 2–10 ($\times 10^{-6}$), which is similar to the mantle value, indicating deep origin [7]. The Li/Cl ratios for the spring waters are almost stable with a maximum of approximately 0.001, indicating a signature of Arima-type brine as suggested by Kazahaya et al. [19]. An exception is 'Tansansen' water, which has twice that value. These chemical signatures of spring waters can be explained by the mixing of meteoric water with slab-derived fluid of the PHS thought to be dehydrated at 50–80 km depths. The spring water with the lowest O–H isotopic composition close to a meteoric signature is identified with the lowest content of Cl and other cations such as Na, Ca, Br, Li, and Sr and the highest He^3/He^4 ratio, named as "Tansansen" which means saturated water with isolated carbonaceous gas. That with an opposite signature close to the end-member of 'Deep Brine', with the highest Cl and a higher He^3/He^4 ratio is identified as "Kinsen" (Figures 3A-3D). Other spring waters with moderate Cl and cation contents plotted between 'Tansansen' and 'Kinsen'.

Analytical method for REE composition in high-salinity brine

The high-salinity and solute elements in the brine may interfere with quantitative analysis of the REEs due to the matrix effect. In case of several water samples contained visible particles, we dissolved the visible particles in water by using acid. In order to consider the matrix effect with high-salinity conditions, we applied

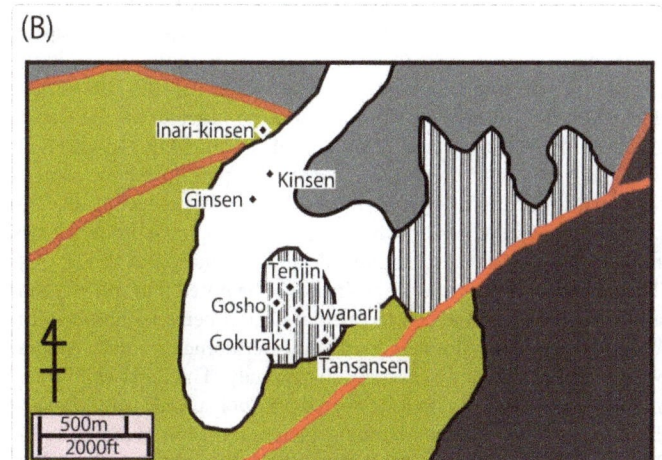

Figure 2: (B): Locations of the spring waters in the Arima area.

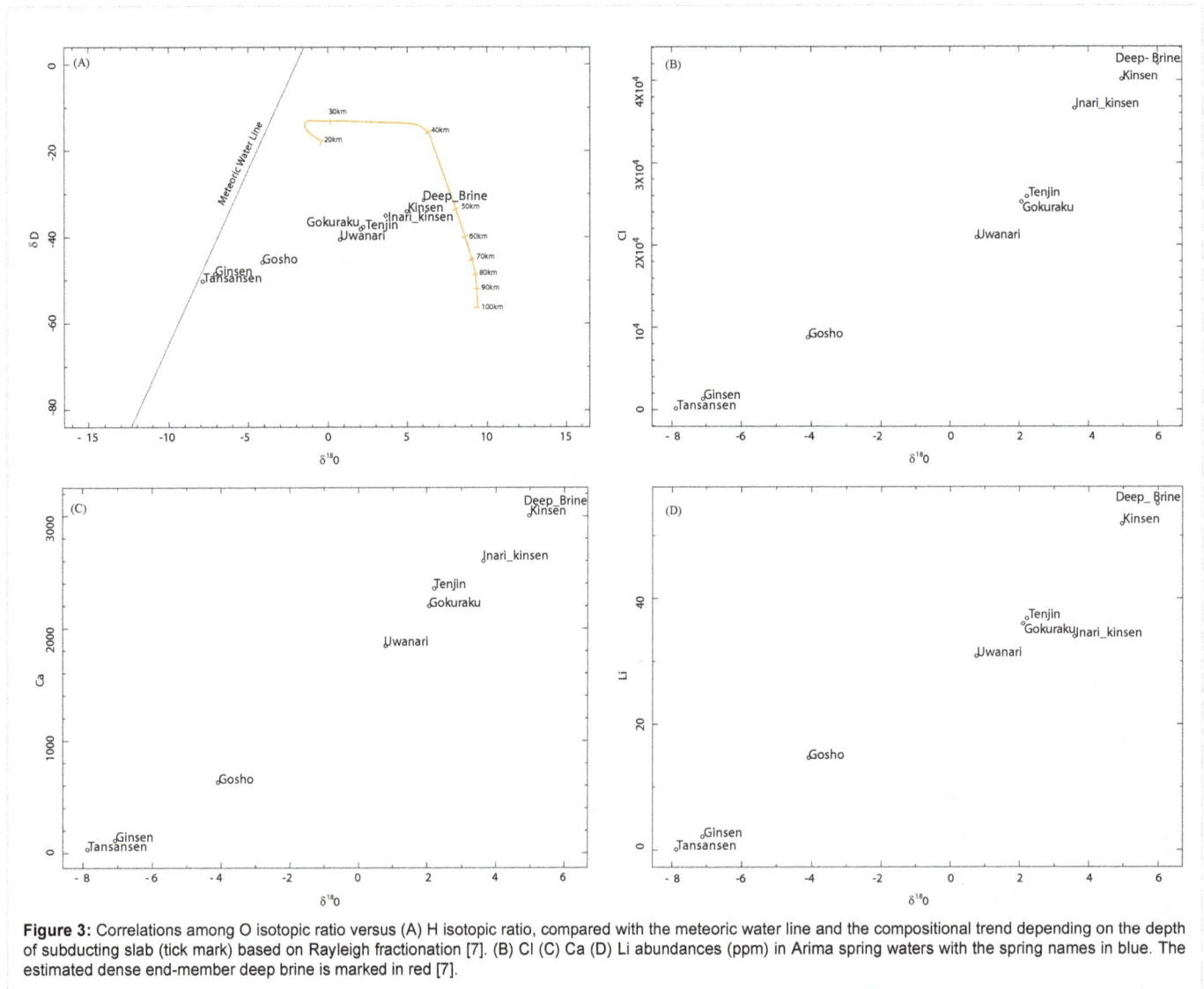

Figure 3: Correlations among O isotopic ratio versus (A) H isotopic ratio, compared with the meteoric water line and the compositional trend depending on the depth of subducting slab (tick mark) based on Rayleigh fractionation [7]. (B) Cl (C) Ca (D) Li abundances (ppm) in Arima spring waters with the spring names in blue. The estimated dense end-member deep brine is marked in red [7].

the standard addition method for solutions with strong matrix effects with 0.4 ml addition of XSTC-1 (SPEX CertiPrep Co. Ltd., USA) by 0.001 ppb, 0.01 ppb, and 0.1 ppb in each 2.0 ml of each sample [20]. The samples were carefully diluted with ultra-pure water (>18.3 MΩ milli-Q water, Millipore Corporation, Japan) by using a weight scale in a class 100 clean room to minimize possible environmental contamination. The ICP-MS (iCAP-Qc, ThermoFisher Scientific) analysis was conducted at the Japan Agency for Marine–Earth Science and Technology (JAMSTEC). The analytical conditions were constantly monitored by using a standard tuning solution and diluted 5% HNO_3 acid solution. The reproducibility standard deviation (RSD) was approximately 2–4% for five repeated data acquisitions, indicating analysis stability [21]. The detection limit was 0.1 ppt. It should be noted that Eu, Nd, Sm, and Gd were quantitatively re-calculated after being combined with ^{135}Ba for correction due to significant interference from Ba oxides, which enabled us to discuss the Eu-anomaly. The most significant interference on REEs signals come from Ba polyatomic ions, because Ba is one of the highest abundant elements in the brine samples. We carefully investigated the formation of these polyatomic ions in ICP-MS measurement by analyzing pure Ba solution, and estimated

for typical brine waters that Ba interfering ions contributed to La, Ce, Nd, Sm, Eu and Gd raw signals for 92%, 24%, 48%, 78%, 96% and 75%, respectively. Pr is free from any of Ba oxide and hydroxide interference which exist farthest to Gd. Intra REEs interfering, e.g., PrO, NdO and SmO overlapping on MREE and HREE signals, were found less than 2.7% (mostly <1%), thus ignorable for this study.

Results

REE abundances and patterns of Arima spring waters

The analytical results of REEs are listed in Table 1. The REE compositions are plotted in Figure 4 as spidergram patterns normalized by depleted mid-ocean ridge basalt (MORB) mantle (DMM). The REE abundances of Arima spring waters are approximately three to five times lower than those of DMM. We have detected four distinct patterns in these spring waters: (i) a flat pattern with a negative Eu-anomaly, represented in the figure by the light blue line; (ii) a slightly convex-down pattern of increasing concentration toward both the left and right, indicating light REEs (LREEs) and heavy REEs (HREEs), respectively, associated with a moderately positive Eu-anomaly (blue line); (iii) a strongly convex-down pattern with a strong positive Eu-anomaly (black line); and (iv) a slightly convex-down pattern with a moderately positive Eu-anomaly and a negative Ce-anomaly (dark blue line). Except for 'Tansansen' (light blue line), which is a highly carbonated cold spring, all spring waters have positive Eu-anomalies; 'Kinsen' (blue line) shows the highest REE content. Among 'Ariake', 'Gosho', 'Gokuraku', 'Tenjin', and 'Uwanari', the five samples classified as (iii), the degree of positive Eu-anomaly increased for samples with a stronger convex-down pattern. 'Kinsen', classified as (ii), maintained overall high concentrations for both LREEs and HREEs with a slightly convex-down pattern (Figure 4). 'Ginsen' (dark blue line), classified as (iv), is a cold spring exhibiting a relatively flat pattern and is in this sense similar 'Tansansen' (i). However, the former shows overall low concentrations (Figure 4).

Correlation of REEs and major solute elements

The solute major elements, Cl and Li abundances, and δ18O–δD–He isotopic ratios in the spring waters, the REEs of which are presented in Table 1, are summarized in Table 2 [7,22]. As previously mentioned, all spring waters are aligned along a single mixing line between the meteoric water and a dense end-member as the deep brine component (Figures 3A-3D), which approximately corresponds to a slab-derived fluid dehydrated at an approximately 50 km depth [7]. 'Kinsen' has high O and H isotopic ratios close to the dense end-member, indicating less involvement of meteoric water, whereas the two cold springs, 'Tansansen' and 'Ginsen', have nearly the same δ18O and δD as the meteoric water (Figure 3A). However, as shown in Table 2, 'Tansansen', the highly carbonated cold spring water, has the highest 3He/4He ratio, indicating deep origin; 'Kinsen' also has a high ratio indicative of its mantle origin. Other springs are plotted along a mixing line between 'Tansansen' and air-saturated water [7]. It is also noted that no clear relationship is observed between dissolved HCO3 and the major solute concentrations, particularly Cl, that represent the proportion of dense end-members involved in the samples (Figure 2 and Table 2).

Arima	La	Ce	Pr	Nd	Sm	Eu	Gd	Tb	Dy	Ho	Er	Tm	Yb	Lu	Dilution rate
Ginsen	0.2228	0.0093	0.0389	0.2089	0.0773	0.1467	0.0665	0.0161	0.1094	0.0411	0.1799	0.0429	0.3818	0.0769	10.3386
Gokuraku	2.0849	n.a.	0.0216	0.0299	0.1322	10.7723	0.3484	0.0576	0.0214	0.0051	0.0074	0.0474	0.7346	0.1454	1.0000
Gosho	0.4306	0.0013	0.0055	0.0113	0.0339	1.8843	0.0351	0.0061	0.0049	0.0014	0.0033	0.0043	0.0294	0.0099	1.0000
Inari-kinsen	18.7829	0.0062	0.0219	0.0115	0.0804	54.6893	0.3182	0.0301	0.0675	0.0184	0.0504	0.0422	0.2696	0.1118	9.6825
Kinsen	11.0772	5.2442	0.5676	2.5409	0.9609	27.8899	1.7651	0.3841	3.6653	0.8360	2.5692	0.3392	2.4161	0.2924	42.2693
Tansansen	4.1881	13.3532	1.9650	10.4090	2.7646	0.3982	4.2497	0.5709	4.1635	0.7909	2.6780	0.3121	2.2442	0.2057	49.9452
Tenjin	3.7923	n.a.	0.0300	0.0126	0.0244	30.1594	0.1479	0.0470	0.0628	0.0118	0.0107	0.0562	0.6806	0.3203	1.0000
Uwanari	1.0906	0.0705	0.0187	0.0599	0.0889	4.4356	0.0928	0.0162	0.0113	0.0038	0.0138	0.0162	0.0889	0.0421	1.0000

Table 1: Rare earth element (REE) abundances (ppb) of Arima spring waters. Abbreviation: n.a. = not analyzed.

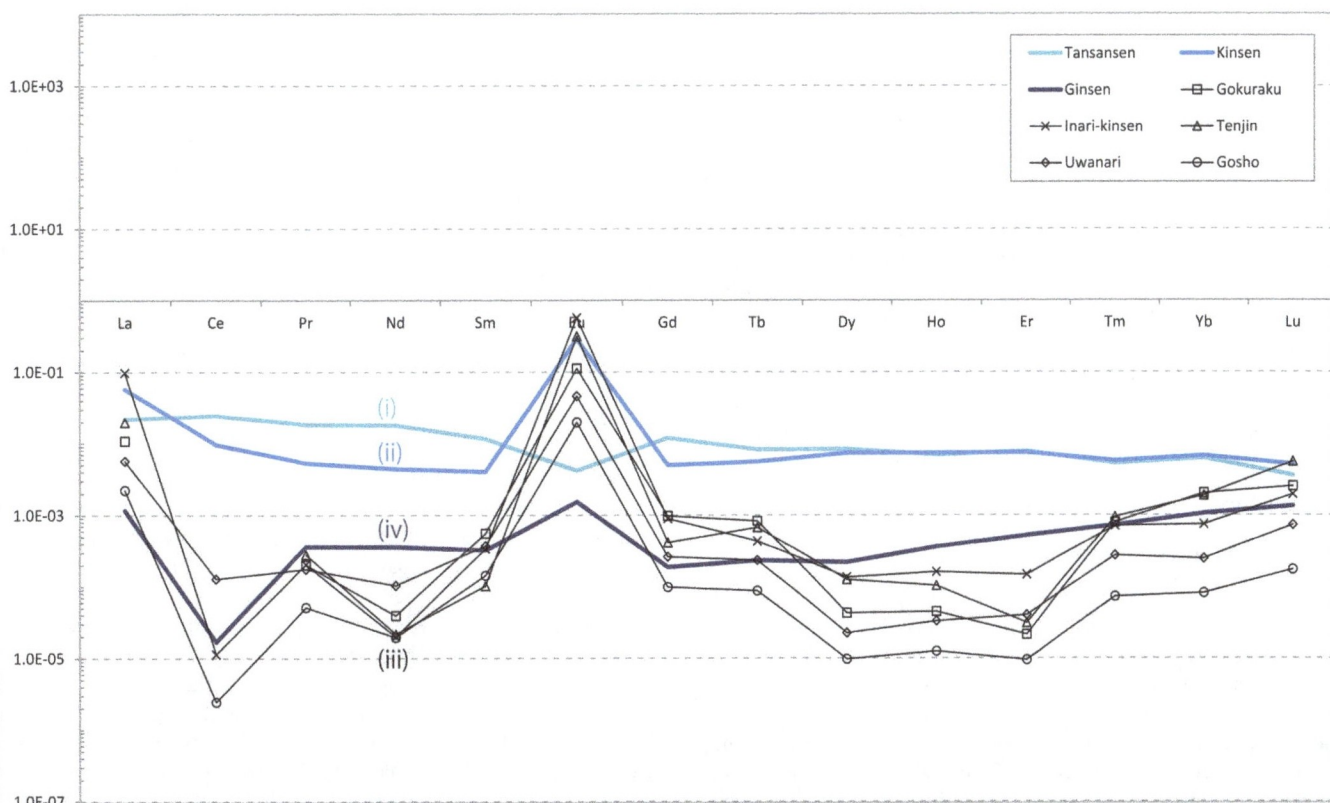

Figure 4: Depleted mid-ocean ridge basalt (MORB) mantle (DMM)-normalized rare earth element (REE) compositions of Arima spring waters analyzed in this study. The data are listed in Table 1.

Arima	Temperature (°C)	Depth (m)	Li (mg/L)	Cl (mg/L)	dD [SMOW]	d^{18}O [SMOW]	^{3}He/^{4}He [10^{-6}]	Li/Cl	HCO$_3$ (mg/L)	Ca (mg/L)	Sr (mg/L)
Ginsen	19.5	n.k.	2.08	1294	-48.56	-7.08	n.a.	0.0016	150	108	0.77
Gokuraku	95.3	240	35.78	25088	-38.15	2.07	3.26	0.0014	26	2192	43.49
Gosho	80.3	165	14.42	8742	-45.92	-4.09	3.84	0.0016	473	627	11.09
Inari-kinsen	32.2	n.k.	33.81	36602	-35.11	3.61	n.a.	0.0009	214	2598	93.94
Kinsen	29.2	n.k.	51.90	40033	-34.12	4.98	6.69	0.0013	1293	3006	80.35
Tansansen	19.1	40	0.03	14	-50.27	-7.88	10.29	0.0021	30	23	0.12
Tenjin	88.3	206	36.69	25828	-37.71	2.22	n.a.	0.0014	137	2356	41.01
Uwanari	88.3	187	30.76	20786	-40.50	0.77	3.34	0.0015	203	1845	37.08

Table 2: Major solute elements and isotopic composition discussed in this study. The data are from [7]. Abbreviation: n.k. = not known.

Concerning REE abundance in the spring waters, no clear correlation was noted with the Cl content and thus the proportion of deep brine component (Figures 5A-5C), implying that in addition to the mixing of the dense end-member with meteoric water, additional, independent processes controlling the REE content are required. The variability of the REE pattern shown in Figure 4 likely reflects such multiple processes (Figure 4). Because the dense, deep brine is thought to ascend through the fault system from depth without significantly reacting with the surrounding rocks [7], shallow crustal processes likely affect the REE behavior, where the dense brine encounters different environments with temperature and redox states distinct from those of the deep part. Such possibilities are discussed subsequently. As shown in Figure 5A-5C, 'Tansansen', being similar to the He systematics, again shows distinctly high REE concentrations despite a low Cl content. If 'Tansansen' is excluded, a broad and non-linear positive correlation exists between REE and Cl contents, with 'Kinsen' as an end-member of the highest concentration for most of the elements (Figure 5A-5C). This broad correlation is reflected in the REE pattern: 'Kinsen' plots at the highest concentrations, and the other five hot springs plot at lower concentrations with stronger convex-down patterns (Figure 4).

The convex-down pattern is related to the Eu-anomaly (Figure 4). In this section, we examine the relationship between the major solute elements and Eu abundance as shown in Figure 6A-6C. The behavior of Eu appears to be coupled with that of Cl, Ca, and Sr, although two broadly linear trends consisting of relatively cold and hot spring waters, respectively, are noted (Figure 6A-6C). This correlation implies that the Eu abundance is roughly related to the proportion of deep brine (Figure 6A) and is different from other REEs (Figure 5A-5C). As is graphically expressed in Figure 4, the five hot spring waters are depleted in REEs compared to 'Kinsen' and have decreased Cl contents, although the decrease in Eu abundance is less than that in other REEs. Figure 6 also shows that Eu behaves in correlation with Ca and Sr, suggesting the possible involvement of plagioclase, which may selectively buffer Eu among the REEs [23]. This possibility is examined subsequently.

Discussion

Origin and process of Eu-anomaly

In previous studies of the geochemical characterization of slab-derived fluid beneath volcanic arcs, no Eu-anomaly has been reported for slab-fluids as a possible source of the deep brine [24]. The estimated composition of slab-fluid from the Philippine Sea slab generally shows a smooth pattern with downward sloping toward HREEs [25]. The slab-derived fluid exhibits variable REE abundances depending on the temperature of dehydration. With slab dehydration at lower temperatures beneath the fore-arc region, including the Arima area, REEs in the slab-fluid exhibit significantly lower abundances due to the large temperature dependence of the partition coefficient between fluid and residual solids [26-28]. As a result of dehydration, the REE pattern

in a slab-derived fluid shows a smooth and slight convex-down pattern. Therefore, the positive Eu-anomaly observed for 'Kinsen' and other hot spring waters (Figure 4) is attributed to processes that occurred during their ascent from depth, likely at a shallow level when the deep brine encountered near-surface water under oxidizing conditions.

In this case with oxidizing conditions within a relatively shallow aquifer, if the temperature is decreased from the original high temperature state up to 500°C, REEs can coprecipitate as a form of Ln(OH)$_3$ with Fe oxyhydroxide [9,10]. The chemical reaction of Ln, involving slab-derived fluid and near-surface (e.g., meteoric) water as reactants with solid precipitates as reaction products, is expressed as $Ln^{3+}_{(aq)} + 3H_2O_{(aq)} \leftrightarrows Ln(OH)_{3(ss)} + 3H^+_{(aq)}$', where (aq) and (ss) denote aqueous solution and solid solution, respectively. It is also suggested that during this precipitation process, Eu in the spring waters has to be retained relative to the other REEs, resulting in a positive Eu-anomaly. If plagioclase abundant in Eu with a positive anomaly exists in the matrix of an aquifer, it may supply Eu for buffering, as is observed. Therefore, the low REE abundance with a strong positive Eu-anomaly in Figure 4, as is observed for 'Gokuraku', 'Gosho', 'Inari-Kinsen', 'Tenjin' and 'Uwanari' hot springs, is interpreted to be a result of overlapping effects associated with elution of Eu and precipitation of other REEs under relatively oxidized conditions. The difference in these five hot springs and 'Kinsen' could be attributed to the degree of the reaction progress due to retention time within the aquifer, although these springs may share a common deep brine.

Conversely, the negative Eu-anomaly observed in 'Tansansen' may have developed under relatively reductive conditions in presence of feldspar such as by reaction with surrounding rocks having negative Eu-anomalies such as Rokko granite or that by (organic) C in the sedimentary rocks exposed in this area [4,23]. The relationship between Eu behavior and other REEs concerning 'Tansansen' is discussed subsequently.

Behavior of REEs

Despite the negative Eu-anomaly and the lowest Cl content of the spring waters analyzed in this study, 'Tansansen' shows high REE concentrations and a flat pattern, except for Eu. The flat REE pattern is notably distinct from those of other Arima spring waters, basement igneous rocks, and river water [1,25] in particular, 'Tansansen' has higher HREE contents. This feature can be explained by the existence of CO$_2$ gas and carbonate species in aqueous fluid [26-28]. The gas-saturated fluid exhibits solubility of $Ln^{3+}(CO_3)_2^-$ approximately four times higher at 1 atm [29,30]. In addition, REEs are more preferentially partitioned into fluid than Fe oxyhydroxide with increasing NaHCO$_3$ concentration, and the dominant REE(III) species in aqueous solutions changes from $REE^{3+}_{(aq)}$ to $REECO^{3+}_{(aq)}$ and then to $REE(CO_3)^{2-}_{(aq)}$. This transition decreases Racah (E1 and E3) parameters for 4f electron repulsion of REEs and rapidly modifies the REE abundances in aqueous

flat pattern with a weaker positive Eu-anomaly compared than that in hot spring waters (type (iv) in Figure 4), suggesting a carbonate effect similar to 'Tansansen' but to a lesser extent. It is also noted that the cold springs are derived from a shallow aquifer seated at ~40 m in depth, yet their REE concentrations are significantly higher (up to 1000 times) than those in nearby river waters (Figure 5) [7]. This result indicates

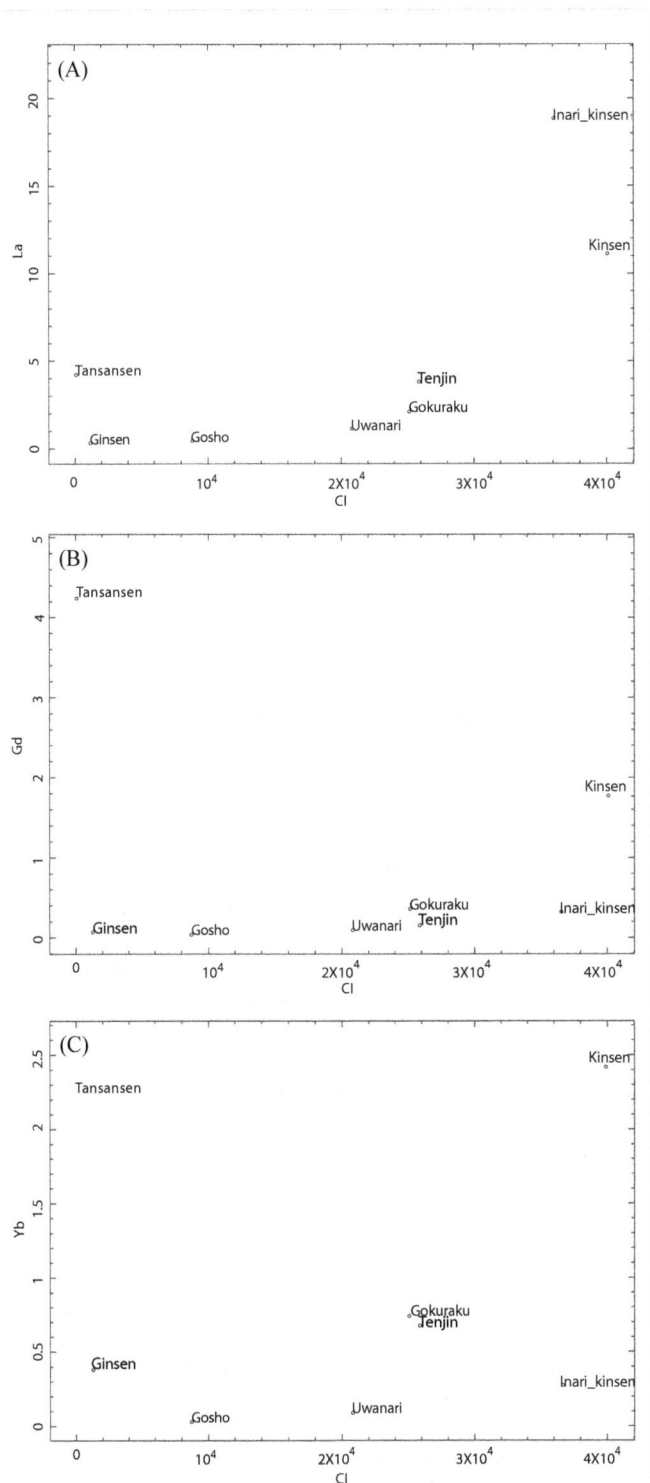

Figure 5: Correlation diagrams between Cl abundance (ppm) versus (A) La, (B) Gd, (C) Yb in ppb.

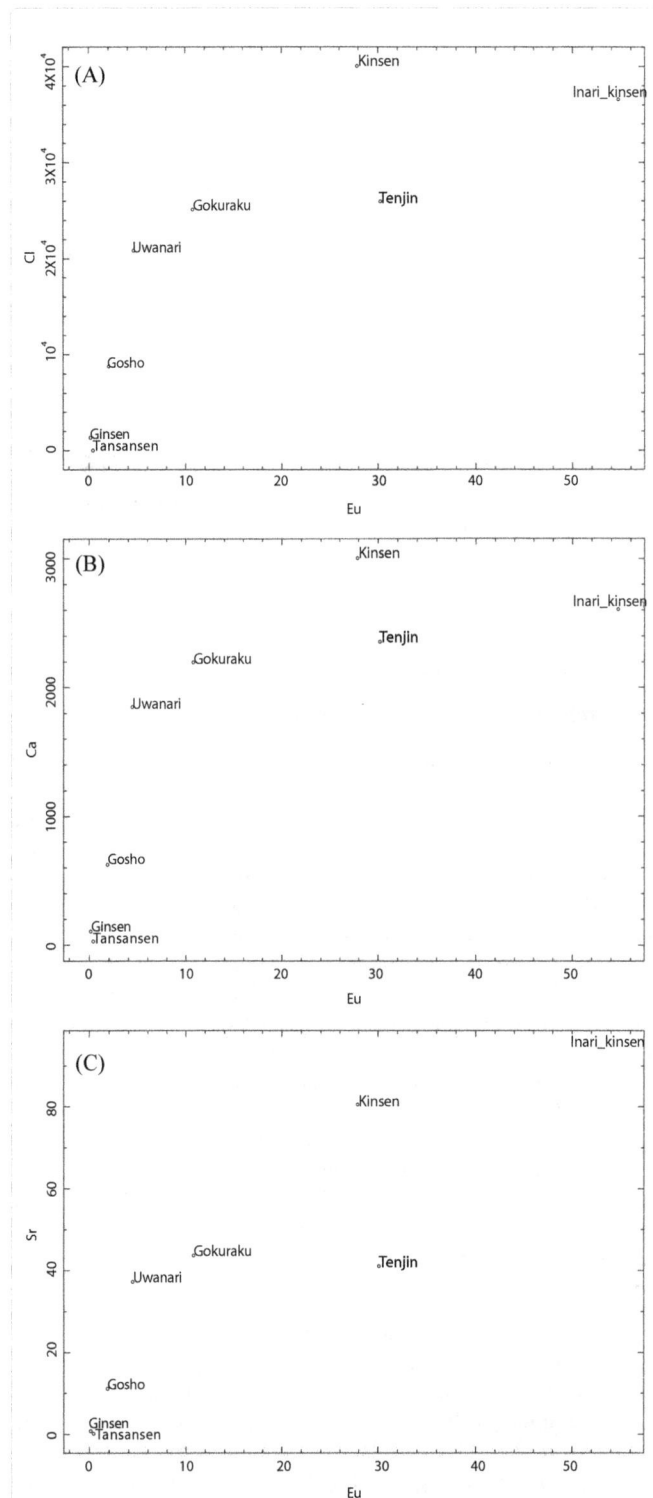

Figure 6: Correlation diagrams between Eu (ppb) abundance versus (A) Cl, (B) Ca, (C) Sr in ppm

solution from a relatively low level with a significant tetrad effect pattern to higher concentrations with a smooth pattern, depending on the carbonate abundance [9]. Therefore, highly carbonated water such as 'Tansansen' may have a high and flat REE pattern caused by reaction with solids. The 'Ginsen' cold spring, which has $\delta^{18}O$ and δD close to the meteoric water line (Figure 3A), also exhibits a relatively

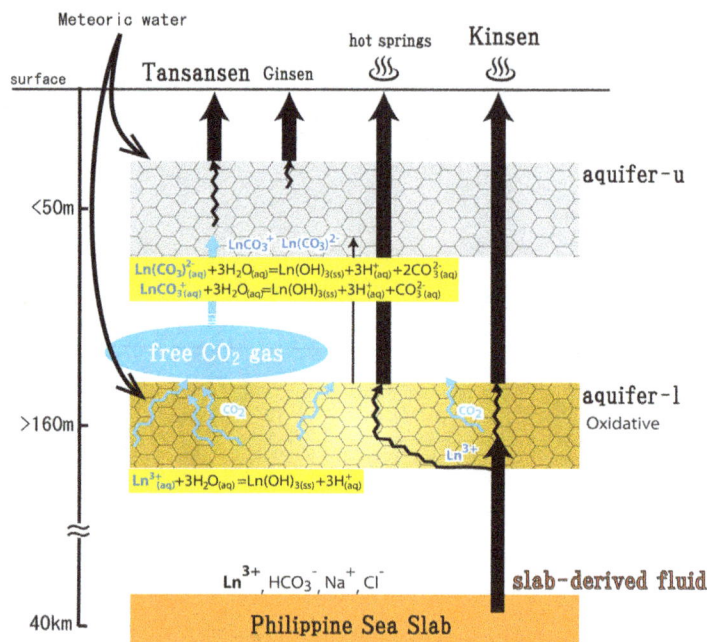

Figure 7: Geochemical evolution model for spring waters in the studied area. Meteoric aquifer-l is the same as lower meteoric aquifer (aq-l) in the main text. Meteoric aquifer-u is the same as upper meteoric aquifer (aq-u) in the text. The abbreviations used in the equations, (aq) and (ss), denote aqueous solution and solid solution, respectively.

that a significant uptake of REEs has occurred even during a shallow cycling of near-surface water.

Comprehensive model for Arima spring waters

Based on the observations and arguments discussed thus far, we propose a model for shallow-level chemical evolution of Arima spring waters, as shown in Figure 7. At least two types of meteoric aquifers occur beneath the study area. One correspond to lower aquifers (aq-l) seated deeper than 160 m, and the other represents upper aquifers (aq-u) shallower than ~40 m [7,8]. The slab-derived fluid dehydrated at a minimum temperature of 500°C from subducting PHS migrates upward and enters the upper aquifer, where it is mixed with meteoric waters to variable extents and is cooled to ~130°C [8]. Within the lower aquifer, the precipitation reaction, discussed above, occurs to induce degassing of CO_2 and He. The separated gas phase migrates upward and enters the upper aquifer to produce highly carbonated meteoric water with high $^3He/^4He$ ratios [2,7]. The precipitates remove REEs from the aqueous solution to variable extents, whereas the solution reacts with feldspar with a strong positive Eu-anomaly to produce the geochemical characteristics of 'Gokuraku', 'Gosho', 'Inari-Kinsen', 'Tenjin', and 'Uwanari' hot springs (Figures 4 and 7). The differences among these five hot springs may represent the degrees of such reactions reflecting retention time in aq-l.

Within aq-u, the host rock is thought to be composed of silicic igneous rocks with high REE contents with negative Eu-anomalies such as the Rokko granite or rhyolite exposed in this area [23]. When the CO_2-bearing gas phase that separates from aq-l dissolves into aq-u, dissolution of the REE elements from the rock matrix to the aqueous solution is enhanced, fingerprinting the Eu-negative anomaly of the host rocks. These processes within aq-u produced 'Tansansen' and 'Ginsen', the latter showing lesser dissolution. The 'Kinsen' hot spring water is nearly upwelled directly from the subducted slab through both

meteoric aquifers, even though a small degree of interaction is inferred from the weak positive Eu-anomaly, as shown in Figure 7.

In order to test this scenario, knowledge of the composition and reaction ratio of the meteoric waters and the host rocks is required. On the basis of these data, the model calculates qualitative and quantitative revisions. In any case, the abundances and patterns of REEs in the spring waters appear to provide unique and invaluable information on the chemical evolution of the deep brine at relatively shallow levels, which has been obtained by neither major solute elements nor isotopic compositions. The REE compositions will be useful for investigating such shallow chemical processes of other springs, especially cold–hot spring association, because the nature of the deep brine is clarified through removal of the shallow effects.

Conclusion

In addition to examining the major solute elements and isotopic compositions including $\delta^{18}O$–δD, He, and Sr–Nd–Pb systematics reported in previous studies, we have performed REE analyses for both hot and cold spring waters in the Arima area. We have shown that the simple binary mixing between the deep brine and meteoric water inferred from the previous studies cannot explain the REE variations. We have identified the following four distinct REE patterns according to isotope composition and major solute element and gas compositions: (i) high REE content with a flat pattern and negative Eu-anomaly, having the lowest $\delta^{18}O$ and δD isotopic ratios with abundant CO_2 gas and the highest He isotopic ratio; (ii) high REE content with a slightly convex-down pattern and positive Eu-anomaly, having the highest $\delta^{18}O$ and δD isotopic ratios with a high He isotopic ratio; (iii) lower REE content with a strong convex-down pattern and a positive Eu-anomaly, having lower $\delta^{18}O$–δD isotopic ratio with lower He isotopic ratios; and (iv) a slightly convex-down pattern with a moderately positive Eu-anomaly and a negative Ce-anomaly, having similar $\delta^{18}O$, δD, and He isotopic ratios as that in type (i).

We have presented the following model to explain the shallow-level chemical evolution of the deep brine resulting in distinct features of Arima spring waters. Two sources are present for REEs. The first is deep brine likely originating from the subducted Philippine Sea slab, and the other is basement crustal rocks serving as aquifers. The deep brine encounters meteoric water at the lower aquifer under relatively oxidizing conditions, in which the formed precipitates remove REEs from the solution and a CO_2-bearing gas phase is generated. In the upper meteoric aquifer, the gas from the lower aquifer dissolves in the meteoric water and enhances the solubility of REEs from the host rocks into the solution. This process produces cold spring waters with a flat REE pattern of relatively high concentrations. This information obtained by REE analysis will be useful for understanding the shallow geochemical processes for other spring waters that have not been identified by major solute or isotopic studies.

Acknowledgements

We would like to thank T. Yokoyama, K. Fujinaga, T. Ishikawa, M. Tanimizu, H. Sakuma, and H. Masuda for their help at various stages of this study, and M. Totani, the president of Arima Hot Springs Tourism Association, for permission and help with our field work and sampling. This work was supported by JSPS KAKENHI Grant Number 25400524 and the Cooperative Research Program of Earthquake Research Institute, The University of Tokyo.

References

1. Nakamura H, Fujita Y, Nakai S, Yokoyama T, Iwamori H (2014) Rare earth elements and Sr–Nd–Pb isotopic analyses of the Arima hot spring waters, Southwest Japan: implications for origin of the Arima-type brine. J Geol Geosci 3: 161.

2. Matsubaya O, Sakai H, Kusachi I, Satake H (1973) Hydrogen and oxygen isotopic ratios and major element chemistry of Japanese thermal water systems. Geochem J 7: 123-151.

3. Tanaka K, Koizumi M, Seki R, Ikeda N (1984) Geochemical study of Arima hot-spring waters, Hyogo, 408 Japan, by means of tritium and deuterium. Geochem J 18: 173-180.

4. Masuda H, Sakai H, Chiba H, Tasurumaki M (1985) Geochemical characteristics of Na-Ca-Cl-HCO₃ type waters in Arima and its vicinity in the western Kinki distinct, Japan. Geochem J 19: 149-162.

5. Nagao K, Takaoka N, Matsubaya O (1981) Rare gas isotopic compositions in natural gases in Japan. Earth Planet Sci Lett 53: 175-188.

6. Sano Y, Wakita H (1985) Geographical distribution of ³He/⁴He ratios in Japan – implications for arc tectonics and incipient magmatism. J Geophys Res 90: 8729-8741.

7. Kusuda C, Iwamori H, Nakamura H, Kazahaya K, Morikawa N (2014) Arima hot spring waters as a deep-seated brine from subducting slab. Earth Planets Space 66: 119.

8. Kozuki J (1962) Studies of Arima Spa. Nippon Shoin, Tokyo (in Japanese).

9. Ohta A, Kawabe I (2000) Rare earth element partitioning between Fe oxyhydroxide precipitates and aqueous NaCl solutions doped with NaHCO₃: Determinations of rare earth element complexation constants with carbonate ions. Geochem J 34: 439-454.

10. Ohta A, Kawabe I (2000) Theoretical study of tetrad effects observed in REE distribution coefficients between marine Fe-Mn deposit and deep seawater, and in REE (III)-carbonate complexation constants. Geochem J 34: 455-473.

11. Mitchell TM, Ben-Zion Y, Shimamoto T (2011) Pulverized fault rocks and damage asymmetry along the Arima-Takatsuki Tectonic Line, Japan. Earth Planet Sci Lett 308: 284-297.

12. Maruyama T, Lin A (2002) Active strike-slip faulting history inferred from offsets of topographic features and basement rocks: a case study of the Arima-Takatsuki Tectonic Line, southwest Japan. Tectonophysics 344: 81-101.

13. Sangawa A (1978) Fault topography and quaternary faulting along the middle and eastern parts of the Arima-Takatsuki tectonic line, Kinki district, central Japan. Geographical Review of Japan 51: 760-775 (Japanese with English abstract).

14. Nakajima J, Hasegawa A (2007) Subduction of the Philippine Sea plate beneath southwestern Japan: Slab geometry and its relationship to arc magmatism. J Geophys Res 112: B08306.

15. Hirose F, Nakajima J, Hasegawa A (2008) Three-dimensional seismic velocity structure and configuration of the Philippine Sea slab in southwestern Japan estimated by double-difference tomography. J Geophys Res 113: B09315.

16. Iwamori H (2007) Transportation of H₂O beneath the Japan arcs and its implications for global water circulation. Chem Geol 239: 182-198.

17. Fujita K, Kasama T (1983) Geological map. Geological Survey of Japan, National Institute of Advanced Industrial Science and Technology, Japan.

18. Arai T, Tainosho Y (2004) Lithologic variation and plutonic history of the Late Cretaceous granitoids in the Rokko Mountains, southwest Japan. J Geol Soc Jap 110: 452-462 (Japanese with English abstract).

19. Kazahaya K, Takahashi M, Yasuhara M, Nishio Y, Inamura A, et al. (2014). Spatial distribution and feature of slab-related deep-seated fluid in SW Japan. Journal of Japanese Association of Hydrological Science 44: 3-16 (in Japanese).

20. Uemoto M (2008) Principles and practices of ICP emission and ICP mass spectrometry, Ohmsha, Tokyo (in Japanese).

21. Chang Q, Shibata T, Shinotsuka K, Yoshikawa M, Tatsumi Y (2003) Precise determination of trace elements in geological standard rocks using inductively coupled plasma mass spectrometry (ICP-MS). Frontier Research on Earth Evolution 1: 357-362.

22. Teranishi K, Isomura K, Yano M, Chayama K, Fujiwara S, et al. (2003) Measurement of distribution of rare earth elements in Arima-type springs using preconcentration with chelating resin/ICP-MS. Bunseki Kagaku 52: 289-296.

23. Terakado Y, Fujitani T (1995) Significance of iron and cobalt partitioning between plagioclase and biotite for problems concerning the Eu²⁺/Eu³⁺ ratio, europium anomaly, and magnetite-/ilmenite-series designation for granitic rocks from the Inner zone of southwestern Japan. Geochim Cosmochim Acta 59: 2689-2699.

24. Nakamura H, Iwamori H (2009) Contribution of slab-fluid in arc magmas beneath the Japan arcs. Gondwana Res 16: 431-445.

25. Nakamura H, Iwamori H (2013) Generation of adakites in a cold subduction zone due to double subducting plates. Contrib Mineral Petrol 165: 1107-1134.

26. Pandler C, Hermann A, Arculus R, Mavrogenes J (2003) Redistribution of trace elements during prograde metamorphism from lawsonite blueschist to eclogite facies; implications for deep subduction-zone processes. Contrib Mineral Petrol 146: 205-222.

27. Kessel R, Schmidt MW, Ulmer P, Pettke T (2005) Trace element signature of subduction-zone fluids, melts and supercritical liquids at 120–180 km depth. Nature 437: 724-727.

28. Kimura JI, Hacker BR, van Keken PE, Kawabata H, Yoshida T, et al. (2009) Arc Basalt Simulator version 2, a simulation for slab dehydration and fluid-fluxed mantle melting for arc basalts: Modeling scheme and application. Geochem Geophys Geosyst 10: Q09004.

29. Michard A, Beaucaire G, Michard G (1987) Uranium and rare earth elements in CO₂-rich waters from Vals-les-Bains (France). Geochim Cosmochim Acta 51: 901-909.

30. Johannesson KH, Stentzenbach KJ, Hodge F (1997) Rare earth elements as geochemical tracers of regional groundwater mixing. Geochim Cosmochim Acta 61: 3605-3618.

Petrology and Geochemistry of Gabbros from the Andaman Ophiolite: Implications for their Petrogenesis and Tectonic Setting

Rasool QA[1]*, Ramanujam N[2] and Biswas SK[1]

[1]*Research Scholar, Department of Disaster Management Pondicherry University, Port Blair, Andaman islands-744112, India*
[2]*Professor and Head, Department of Disaster Management, Pondicherry University, Port Blair, Andaman islands-744112, India*

Abstract

The Andaman ophiolite has almost a complete preservation of ophiolite sequence with mantle section and comparatively less well-developed crustal section. The crustal part of the ophiolite comprises both ultramafic cumulates; dunite, wherlite and pyroxenite and mafic cumulates or gabbros. The gabbros are olivine gabbro, and pyroxene gabbro. They are distinguished into cumulates and non-cumulates. Geochemical features indicate that they are tholeiitic, fractionated and are formed in an arc-related tectonic setting. The large- ion lithophile (LIL) elements (Rb, Ba, Th, Sr) are relatively more enriched than Normal-mid-ocean ridge basalts (N-MORB). The enrichment of Large-ion lithophile (LIL) elements over the High-field strength (HFS) elements and the depletion of Nb relative to other High-field strength (HFS) - elements suggest involvement of subduction component in the depleted mantle source, and suggest that they are formed in a supra-subduction zone tectonic setting.

Keywords: Andaman ophiolite; Gabbros; Petrology; Geochemistry; Petrogenesis; Tectonic setting

Introduction

Gabbros are created by the injection of basaltic melt from the underlying rising mantle and are regarded as formed by slow crystallization in a magma chamber. They are an integral part of the crustal section in an ophiolite suite and may range at the base from layered gabbros to isotropic to foliated gabbros at the top. Some ophiolites have a well-developed gabbroic section e.g., Semail Ophiolite Oman, and Bay of Island ophiolite, Newfoundland, Canada [1,2]. Unlike these Ophiolites Andaman Ophiolite has less developed gabbroic sections and consists of succession of ultramafic-mafic cumulates at the base. In several parts of south Andaman, ophiolite occurrences have been described by many researchers [3-8]. However, no detailed study has been made of the geochemical characters of its gabbroic rocks. In this paper we are going to discuss the petrogenesis and tectonic setting of the gabbros from Andaman ophiolite using their field features and petrography and geochemistry.

Andaman Ophiolites

The Andaman Ophiolite Belt marks the southern extension of the Manipur and the Burmese Arakan Yoma Belt, which is the easternmost continuation of the Tethyan Belt (Figure 1). The Tethyan Belt extends from Baltic Cordillera and rift of Spain and Africa eastwards through Alps, the Denirides in Yugoslavia, through Greece, Turkey, Iran, Oman, Pakistan and the Himalayas, Burma, Andaman-Nicobar islands and Indonesia [3]. The Andaman Ophiolite Belt belongs to a region of distinct structural and topographical belt that trends north-south and then curves eastward from Sumatra towards Java [5]. Further, the Andaman islands, the central part of Burma-Java subduction complex is also believed to expose tectono stratigraphic units of accretionary prism in an outer-arc setting [9]. From east to west there are four such structural cum topographic zones which are:

- Peripheral eastern massif of Shan Plateau, the Malay Peninsula and its western shelf, the Malacca strait and Sumatra.

- A zone of topographic lows including Irrawaddy Valley of Burma, the Andaman Basin and Mentwai through between Sumatra and Mentwai Islands.

- A zone of high relief including the Arakan Yoma of Burma, the Andaman-Nicobar Islands.

- The Java Trench which probably does not extend to the latitude of the Andaman Basin.

The overall physiographic trend of Andaman-Nicobar islands is the continuation of Arakan Yoma of western Burma, which is a southward trending branch of the eastern Himalayas. The Mentwai islands (south and west of Sumatra) are considered to be a southenly continuation of the Andaman-Nicobar trend. The Andaman-Nicobar group of islands form an arcuate chain extending for about 850 km bounded by latitude 6°45'N to 13°45'N and by longitude 92°15'E to 94°00'E. In some of these literatures, this ophiolite occurrence has been reported to be a dismembered ophiolite [8]. Although Saha reported the complete preservation of ophiolite suite from Port Blair (11°39'N: 92°47'E) to Chadiyatapu (11°30'24": 92°43'35"E) (Figure 2(a)).

The Andaman ophiolite consists of a plutonic complex, a volcanic sequence and pelagic sedimentary rocks (Figure 2(b)) Continuous ophiolite sections are rare owing to a thick weathering profile, tropical forest cover, and pervasive east-west and north-south fault systems. Regional mapping over a large part of the Andaman Islands has determined local contact relationships and allowed development of a regional ophiolite stratigraphy. The lower part (80% of the total ophiolite outcrop) comprises foliated and highly serpentinized peridotite. The upper part comprises a layered sequence of ultramafic-mafic rocks, an intrusive section of homogeneous gabbro-plagiogranite-diorite-dolerite and an extrusive section of boninite and tholeiitic basalt lavas Figure 2(c).

An assemblage of plutonic and extrusives represents the crustal

---

*Corresponding author: Rasool QA, Research Scholar, Department of Disaster Management Pondicherry University, Port Blair, Andaman Islands-744112, India E-mail: akhter81@gmail.com

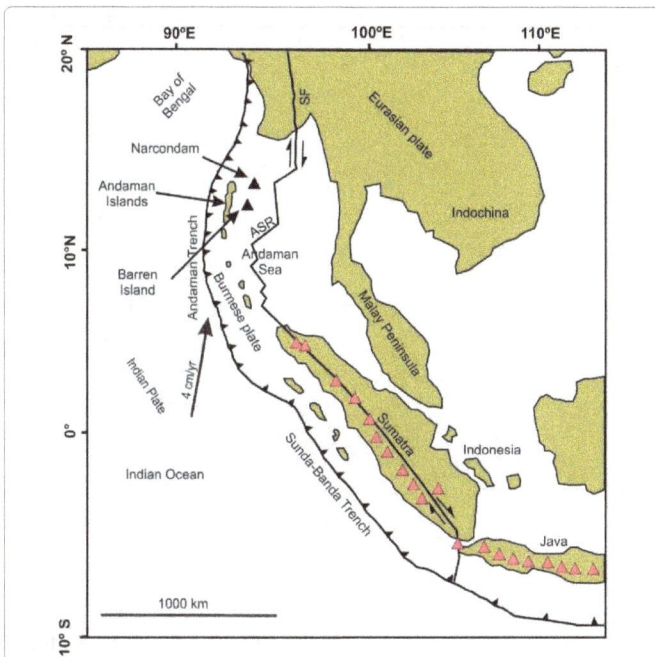

Figure 1: Map of Southeast Asia and Northeastern Indian Ocean showing major geological and tectonic features, ASR: Andaman Sea Ridge; SF: Saigang Fault.

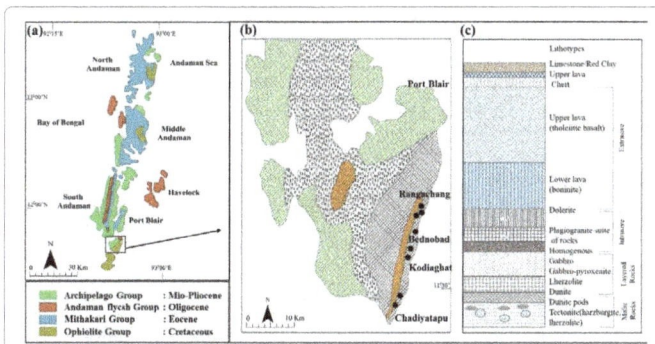

Figure 2: (a) Generalized geological map of Andaman Islands showing distribution of the ophiolites and sedimentary units with their stratigraphic relationship and tectonic setting. (b) Detailed geological map of part of South Andaman (modified after Pal [9]), with sample location. (c) Composite stratigraphic section of the Andaman ophiolite showing lithological units and mineral occurrences.

sequence of Andaman Ophiolites. The plutonic unit of crustal sections is represented by layered cumulates of peridotite-gabbro, overlain by non-cumulate gabbro and high level intrusives of a plagiogranite-diorite andesite suite of rocks with thin dykes of basalt and diabase. It comprises of layered sepentinized dunite, harzburgite, lherzolite wehrlite, pyroxenite and gabbro with pods and stingers of chromitite.

Methodology

Petrology

The mafic cumulates (Gabbro) are mainly located at and around the road sections of Kodiyaghat (11^{0}31'16.8"N:92^{0}43'02.8"E)- Bednabad (11^{0}34'42"N:92^{0}43'21.2"E) and Rangachang (11^{0}34'13.6"N: 92^{0}43'36"E) Figure 2(b). The mafic cumulates shows the repetitive and undeformed parallel layers of alternating olivine gabbro, norite, gabbro norite and pyroxenite. The gabbroic rocks can be divided into layerd and

homogenous gabbro. The gabbro has anhedral granular texture and consists of large clinopyroxen and plagioclase crystals.

Petrographically gabbroic rocks are holocrystalline, fine to coarse grained and subophitic texture. In gabbro's preferential arrangement of mineral grains is clearly evident which is indicative of primary cumulus texture. Plagioclase occurs as stout, latch shape crystals that show variable degree of sausitirizaton. Plagioclase occurs as both a cumulus and post cumulus phase with an An$_{95}$ composition. Complete sausitirization of plagioclase appears as dusty (Figure 3). The pyroxenes occur as small (20-30 µm) subhedral crystals with two sets of cleavage. Quartz occurs as high relief subhedral crystal within gabbros. The patches of anhedral to subhedral crystals of chlorite are seen throughout gabbros which are alteration products of clinopyroxene. Highly brecciated and mylonitised gabbro indicating post emplacement shearing related to tectonic evolution of the region.

Cumulate gabbro grades from olivine gabbro to olivine norite to gabbro, with 1-12 vol% olivine and 30-68 vol% of plagioclase. In olivine gabbro, clinopyroxene occurs as large adcumulus crystals and large anhedral grains that contain many olivine inclusions. Homogenous gabbros are coarse to medium garined, hypiodiomorphic granular texture. Clinopyroxene are mostly altered to amphibole and plagioclase to sausurite. Plagioclase has an An$_{93}$ composition, similar to cumulate gabbro magnetite and ilmenite are present as accessory phase.

Geochemistry

Analytical methods: Eight samples of gabbros were analysed for major and trace elements and five for rare earth elements (REE). The samples were crushed in a jaw crusher. The weathered surfaces were removed and the fresh parts were powdered using a tungsten carbide mill to the size of <200 mesh. Then the required number of grams of the powder of each sample was heated in a porcelain crucible to 9000C for 2 hr to determine the loss on ignition (LOI).

For major elements, the sample powder was thoroughly mixed with lithium tetra-borate (flux) with a 1:5 sample flux ratio and the glass beads were formed. Then the fused beads were analysed for major elements composition using a X-Ray Fluorescence (XRF) at the Central Instrumentation Facility, Pondicherry University, Pondicherry. For

Figure 3: (a) Field photographs of gabbro with tachylite vein. (b) Repitative cumulate layers of olivine gabbro or norite and pyroxenite. (c & d) Pictomicrograph of gabbro shows plagioclase as stout latch shape and anhedral to subhedral crystal of chlorite.

trace elements the powdered pellets of all the samples were prepared by putting 5-7 grams powdered sample (< 200 mesh) in an aluminium cup and compressed between two tungsten carbide plates (within circular briquettes) at about 20 ton per square inch pressure in a hydraulic press. Then these pellets were analysed for trace elements composition using a XRF at Central Instrumentation Facility, Pondicherry University, Pondicherry. For REE study the powder was dissolved in 30 ml of 10% HNO_3 and 20 ml of deionized water. Samples for REE analysis were prepared by a standard Teflon vial acid-digestion procedure using a mixture of HF ± $HClO_4$-HNO_3. All samples were spiked to 50 mg/ml with indium to serve as an internal standard. REE analyses for this study were calibrated against a set of multi-element working standard solutions. All the solutions were introduced via a peristaltic pump, and analyses were performed by Inductively Coupled Plasma Mass Spectrometer (ICP-MS) at the Department of Earth Science, Pondicherry University, Pondicherry.

Classification: The Andaman ophiolite gabbros are classified by plotting on total alkali versus SiO_2 diagram [10]. All the samples plot in the gabbro field confirming their gabbroic character (Figure 4). Most of the gabbro samples have very low values of total alkali and SiO_2 may be due to their cumulate nature or it could be due to alteration. The concentration of major, trace and REE of all the gabbros samples are reported in Table 1.

Major element characteristics: The Andaman gabbros have low concentration (wt%) of SiO_2 (45.67-49.58), TiO_2 (012-1.16), Na_2O+K_2O (.53-4.6), P_2O_5 (.01-.14) and wide range of Al_2O_3 (7.44-16.91), Fe_2O_3 (5.03-11.69), MgO (5.06-15.41) and CaO (11.81-18.16). The high concentration of CaO and low concentration of SiO_2 are due to the ateration and the presence of small vinelets of Calcite. The very low Total Alkali content can be explained by cumulate nature of the rocks .the cumulate nature becomes more evident when SiO_2 and MgO are plotted against the selected major elements as fractionation index (Figure 5). In most of the plots the samples cluster together with only minor degree of scattering. The clustering of samples indicates that the most of these gabbros are less affected by fractionation and where they

	GB/AN-1	GB/AN-15	GB/AN-11	GB/AN-10	GB/AN-3	GB/AN-7	GB/AN-6	GB/AN-14
SiO_2	49.38	49.58	47.12	46.21	46.12	45.91	48.38	45.67
TiO_2	1.11	0.64	0.54	0.28	0.12	0.33	1.16	0.24
CaO	15.47	15.89	18.16	18.05	14.6	15.56	16.14	11.81
Al_2O_3	11.65	11.71	12.17	15.67	16.91	7.44	11.55	13.76
Fe_2O_3	10.54	8.82	11.27	8.59	5.03	11.69	9.49	8
MnO	0.22	0.19	0.34	0.18	0.11	0.16	0.19	0.12
MgO	5.06	8.25	8.02	9.71	14.77	14.74	6.45	15.41
Na_2O	3.94	2.81	1.49	0.8	0.56	1.17	4.14	0.45
K_2O	0.48	0.67	0.03	0.04	0.56	1.33	0.46	0.08
P_2O_5	0.14	0.14	0.09	0.01	0.03	0.01	0.14	0.14
LOI	1.99	1.29	0.17	0.45	1.74	1.75	1.9	4.34
Total	99.98	99.99	99.4	99.99	100.55	100.09	100	100.02
Sc	32.5	17.6	39.6	43.4	40.2	52.4	35.6	51
V	28.9	42.6	159.3	171.4	89.4	172.3	245	110
Cr	28.9	21.4	14.3	105.2	888.8	61.3	59	1874
Co	50	29.3	66.3	70	59.9	67.7	35	70
Ni	25	22.7	14.2	39.7	194.1	49.4	50	303
Zn	85	5.6	100	32.7	14.7	21.1	40	46
Ga	17.5	25.6	28.1	25.3	19.4	23.8	15.7	8
Sr	251.3	394.7	297	237.9	194	196	289	165
Y	5.9	6.9	7.8	6.3	7	4.9	4.5	5
Zr	34.1	20	15.1	13.1	31.1	10.8	25.2	26
Nb	0.05	0.5	0.4	0.3	0.5	0.2	0.11	5
Ba	227.5	44.7	44.7	33.5	12.6	20.8	113.8	-
La	3.5	8.9	4.2	4	0.8	11.9	0.17	-
Ce	7.1	8.1	8.4	7.7	15.9	0.5	0.78	-
Nd	8.5	0.1	3	7.3	36.8	7.8	1.24	-
Yb	2.7	0.66	0.8	0.64	14.1	0.37	0.55	0.44
Sm	3.3	-	0.96	2.8	9.6	-	0.56	-
Th	2.4	1.9	2.1	3.6	0.9	0.94	0.01	2.5

Table 1: Major elements (wt %), trace and REE elements (ppm) composition of the gabbros from the Andaman ophiolite.

scattered, they show their cumulate nature.

Trace and REE characteristics: The Andaman gabbros have variable concentration of both large ion lithophile (LIL) elements (in ppm) such as Ba (12-227.5), Sr (165-394.7), Ce (0.5-15.9) and HFS elemnts like Zr (10.8-34.1), Y (4.5-7.8) and Nb (0.05-5). The alteration is obvious from the wide range of CaO (11.8-18.16) and LOI content (0.17-4.34). So it is expected that the LIL elements (Rb, K, Ba, Sr) have been remobilized to variable degree during the ateration process [11,12]. Zr s plotted against the high field strength (HFS) elements (Nb, Y, Ti, P) in (Figure 6). These plots shows a linear relationship with some degree of scatter.These trends shows that all the gabbroic rocks are probably formed through the process of fractional crystallization. These gabbros also show high (FeO^T/MgO) ratio (0.88-2.08) which is considered as indicator of advanced fractional crystallization [13,14]. It is further confims that the HFS elements are less affected by alteration so the data for these elements will be reliable to use for the petrogenesis and tectonic setting of these rocks.

Discussion

The Andaman ohiolite gabbros include olivine gabbro, gabbro-norite and gabbro. They show the charcteristics of theolitic Igneous rocks as evident by low ration of Zr/Ti (0.0036-.043) and Nb/Y (0, 0085-.080) and their plot as basaltic rocks on Zr/Ti-Nb/Y diagram (Figure 7a) of Pearce [15] and a theolitic with mild alkaline affinity on

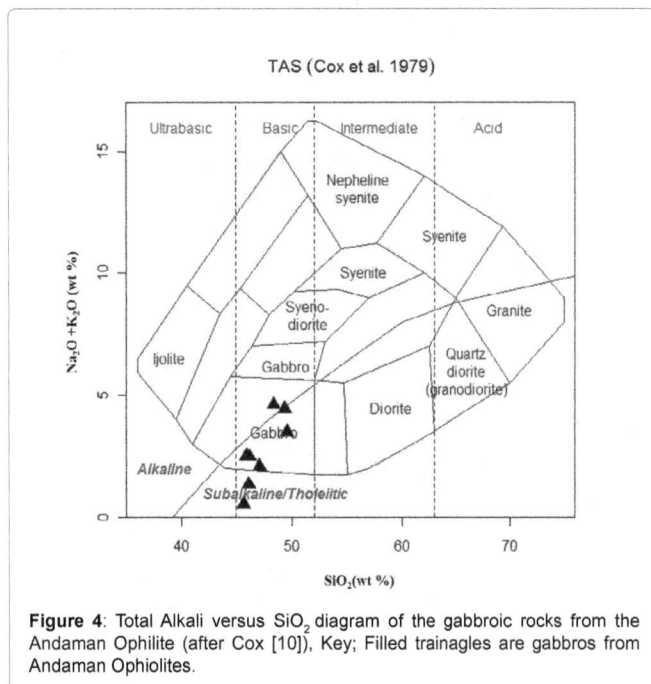

Figure 4: Total Alkali versus SiO_2 diagram of the gabbroic rocks from the Andaman Ophilite (after Cox [10]), Key; Filled trainagles are gabbros from Andaman Ophiolites.

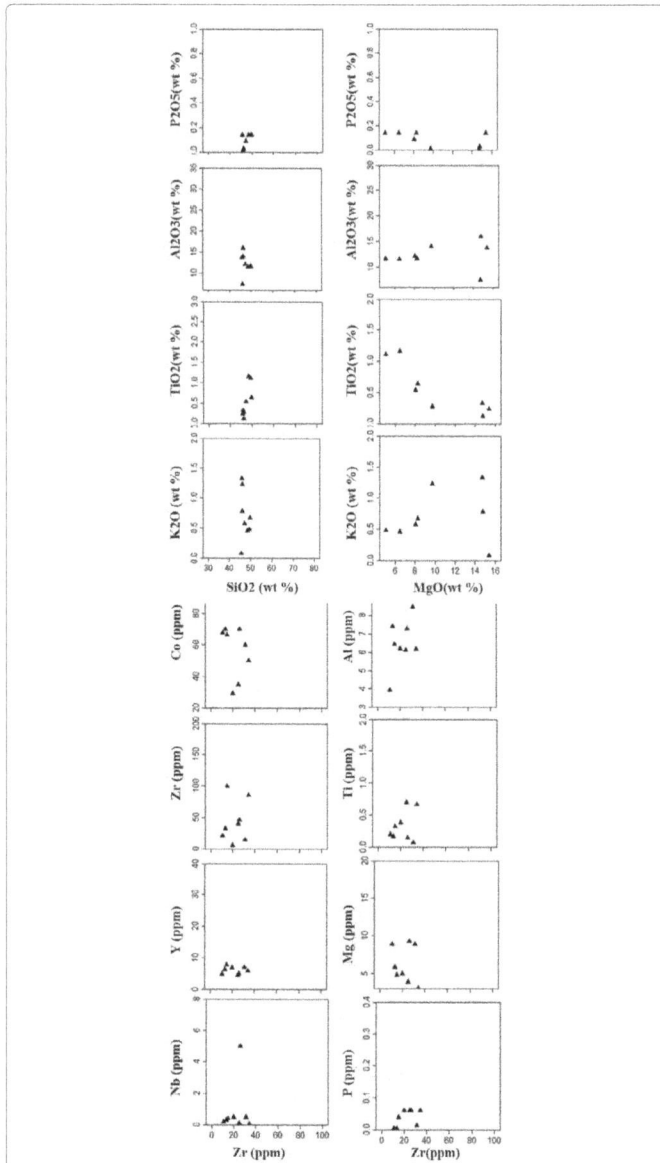

Figure 5: SiO$_2$ and MgO versus selected major elements plots of Andaman Ophiolites. Key: same as in Figure 4.

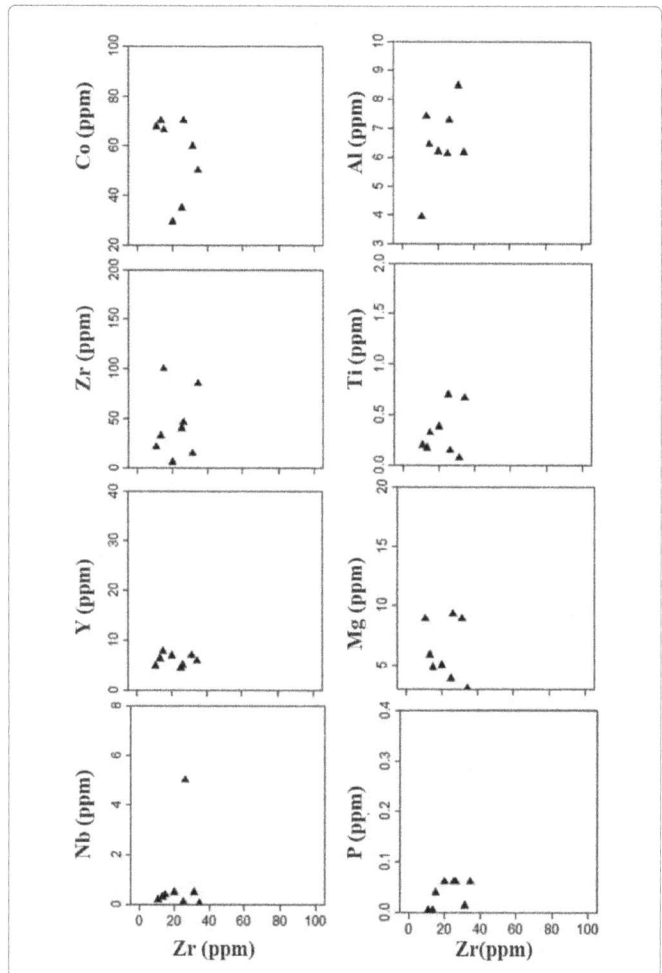

Figure 6: Zr versus major and trace elements plots of the Andaman Ophiolite. For symbols see Figure 4.

TiO$_2$ versus Zr/P$_2$O$_5$*10^4 (Figure 7b) by Winchester and Floyd [16].

Several attributes of gabbroic rocks of Andaman ophiolites point out the tectonic setting of magma generation responsible for their formation. The tectonic discrimination diagram e.g., Ti/100-Zr-Y*3 diagram (Figure 7c) of Pearce and Cann [17] classify the gabbros as Island arc theolittic (IAT) with mild affinity to depleted mantle oceanic fllor basalts (N-MORB). There is apossibly that these gabbros have a transitional charcter between the IAT and N-MORB and so it is likely that these gabbros have formed by fractionation directly over a subduction zone or in the supra-subduction zone. The Supra subduction setting of the gabbroic rocks from the Andaman ophiolites is further confirmed when major elements composition is plotted on the AFM diagram of Beard. Most of the samples plot in the arc related mafic cumulate field, where as some some samples plot in the arc related non cumulates field (Figure 8). This implies that the gabbroic rocks are formed in suprsubduction zone tectonic setting.

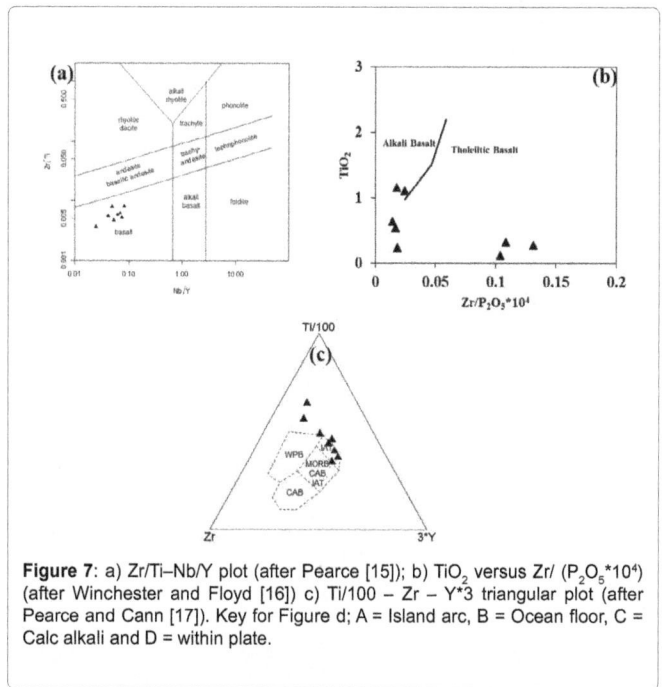

Figure 7: a) Zr/Ti–Nb/Y plot (after Pearce [15]); b) TiO$_2$ versus Zr/ (P$_2$O$_5$*10^4) (after Winchester and Floyd [16]) c) Ti/100 – Zr – Y*3 triangular plot (after Pearce and Cann [17]). Key for Figure d; A = Island arc, B = Ocean floor, C = Calc alkali and D = within plate.

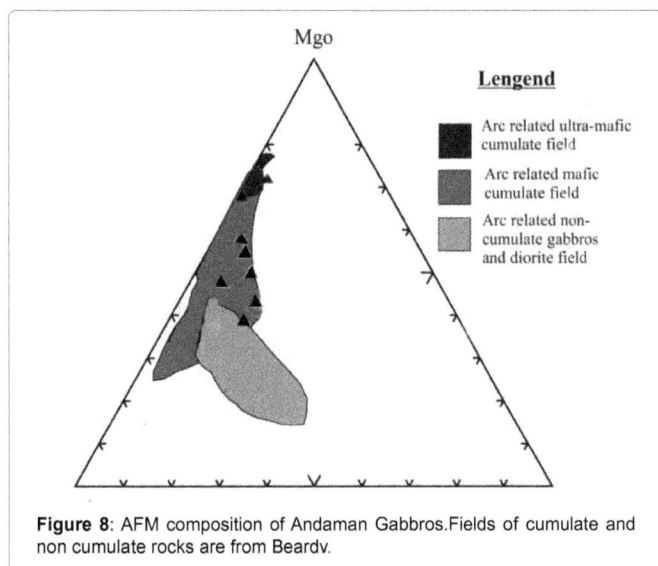

Figure 8: AFM composition of Andaman Gabbros. Fields of cumulate and non cumulate rocks are from Beardv.

Conclusion

Tectonically, the Andaman -Nicobar Island are know to be the part of the outer sedimentary arc of the Sunda-Burmese double chain arc system, where the Indian Oceanic plate is being subducted. Geochemical data suggest that the Andaman gabbros are theoleittic with mild alkaline affinity. The crustal part of the Andaman Ophiolite has less developed gabbroic section with average thickness of ultrmafic-mafic cumulate section. The agbbros are olivine gabbro, gabbro norite. They are identified both cumulates and non cumulates. The enrichments of LIL elemnets and depletion of Nb, transitional characterization between the N-MORB and IAT suggest Supra-subduction zone tectonic setting of Andaman gabbros.

References

1. Nicolas A (1989) Structures of Ophiolites and Dynamics of Oceanic Lithosphere. Kluwer, Dordrecht.

2. Bedard JH (1991) Cumulate recycling and crustal evolution in the Bay of Islands ophiolite. J Geol 99: 225-249.

3. Haldar D (1984) Some aspects of the Andaman Ophiolite Complex. *Re Geol Surv India* 119: 1-11.

4. Ray KK, Sengupta S, Van den Hul HJ (1988) Chemical characters of volcanic rocks from Andaman Ophiolite, India. *J Geol Soc London* 145: 393-400.

5. Vohra CP, Haldar D, Ghosh Roy AK (1989) The Andaman–Nicobar Ophiolite Complex and associated mineral resources - current appraisal; Phenerozoic Ophiolite of India. Ghose NC Sumna Publ, Patna.

6. Shastry A, Srivastava RK, Chandra R, Jenner GA (2001) Fe-Ti enriched mafic rocks from south Andaman ophiolite suite: Implication of late stage liquid immiscibility. *Curr Sci* 80: 453-454.

7. Shastry A, Srivastava RK, Chandra R, Jenner GA (2002) Geochemical characteristics and genesis of oceanic plagiogranites associated with south Andaman ophiolite suite, India: A late stage silicate liquid immiscibility product. *J Geol Soc India* 59: 233-241.

8. Srivastava RK, Chandra R, Shastry AK (2004) High-Ti type N-MORB parentage of basalts from the south Andaman ophiolite suite, India. *J Earth Syst Sci* 113: 605-618.

9. Pal T, Chakraborty PP, Dutta Gupta T, Singh CD (2003) Geodynamic evolution of the outer-arc-forearc belt in the Andaman Islands, the central part of the Burma-Java subduction complex. *Geol Mag* 140: 289-307.

10. Cox KG, Bell JD, Pankhurst RJ (1979) The Interpretation of Igneous Rocks. Allen and Unwin, London.

11. Humphris SE, Thompson G (1978) Trace element mobility during hydrothermal alteration of oceanic basalts. Geochimica et Cosmochimica Acta 42: 127-136.

12. Staudigel H (2003) Hydrothermal alteration processes in the oceanic crust. In: Holland HD, Turekian KK Treatise on geochemistry 3: 511-535.

13. Miyashiro A, Shido F (1975) Tholeiitic and calc alkalic series in relation to the behaviour of titanium, vanadium, chromium, and nickel. Am J Sci 275: 265-277.

14. Tatsumi Y, Ishizaka K (1982) Origin of high magnesian andesites in the Setouchi Volcanic belt, Southwest.

15. Pearce JA (1996) A user's guide to basalt discrimination diagrams. In: Bailes AH., Christiansen EH, Galley AG, Jenner GA, Keith Jeffrey D, et al. Trace element geochemistry of volcanic rocks; applications for massive sulphide exploration. Short Course Notes-Geological Association of Canada 12: 79-113.

16. Winchester JA, Floyd PA (1976) Geochemical magma type discrimination: Application to altered and metamorphosed igneous rocks. Earth and Planetary Science Letters 28: 459-469.

17. Pearce JA, Cann JR (1973) Tectonic setting of basic volcanic rocks determined using trace element analysis. Earth and Planetary Science Letters 19: 290-300.

Permissions

List of Contributors

Jiren Xu and Zhixin Zhao
Institute of Geology, Chinese Academy of Geological Sciences, Beijing 100037, China

Sajwan KS and Sushil K
Disaster Mitigation and Management Centre (DMMC), Department of Disaster Management, Government of Uttarakhand, Uttarakhand Secretariat, Rajpur Road, Dehradun 248001, Uttarakhand, India

Jeyavel Raja Kumar T, Dushiyanthan C, Thiruneelakandan B, Suresh R, Vasanth Raja S, Senthilkumar M and Karthikeyan K
Department of Earth Sciences, Annamalai University, Annamalai Nagar – 608002, India

Alam I
Atomic Energy Minerals Centre, Lahore

Majid K and Muhammad H
Institute of Geology and Geophysics, Chinese Academy of Sciences, 100029, Beijing, China
University of Chinese Academy of Sciences, 100049, Beijing, China

Shahid N
National Centre of Excellence in Geology, University of Peshawar, 25000, Khyber Pakhtunkhwa, Pakistan

Munawar S
University of Chinese Academy of Sciences, 100049, Beijing, China
Shanghai Astronomical Observatory, Chinese Academy of Sciences, 200030, Shanghai, P.R. China

Khashaba A and Takla EM
National Research Institute of Astronomy and Geophysics, Egypt

Soliman SA, Farouk S and Mostafa R
Egyptian Petroleum Research Institute, Egypt

Fkirin MA and Badawy S
Electronic Engineering, Department of Industrial and Control Engineering, Menoufia University, Egypt

El deery MF
Hanwha Chemical Research and Development Center, Tukh, Egypt

Shilenje ZW, Thiong'o K and Philip SO
Kenya Meteorological Department, Nairobi, Kenya

Ongoma V
College of Atmospheric Science, Nanjing University of Information Science and Technology, Nanjing, Jiangsu, P.R. China
Department of Meteorology, South Eastern Kenya University, Kitui – Kenya

Nguru P and Ondimu K
National Environment Management Authority, P.O. Box 67839-00200, Nairobi – Kenya

Williams B
Retired, Gillette, Wyoming, United States

Abu-Hashish MF
Geology Department, Faculty of Science, Menoufiya University, Egypt

Ahmed Said
Qarun Petroleum Company, Cairo

Ibrahim Khashaba A and Essam Ghamry
National Research Institute of Astronomy and Geophysics, Geomagnetism, Egypt

Sen S
Council of Scientific and Industrial Research, Hemantika, O-26, Patuli, Kolkata, India

Abir B and Chokri Y
Laboratory of Sedimentary Dynamic and Environment, National Engineering School of Sfax, Street of Soukra km 4, 3038 Sfax, Tunisia

Mohamed B
Geotechnical Company Thynasondage and Geotechnical Engineer, Tunisia

Samir M
High Institute of Technological Studies, Sfax, BP46, Sfax, 3041, Tunisia

Youdeowei PO
Institute of Geosciences and Space Technology, Rivers State University of Science and Technology, Port Harcourt, Nigeria

Nwankwoala HO
Department of Geology, University of Port Harcourt, Nigeria

Adelekan AO and Oladunjoye MA
Department of Geology, University of Ibadan, Nigeria

Igbasan AO
Department of Applied Geophysics, Federal University of Technology, Akure, Nigeria

Kotona Chiba and Qing Chang
Japan Agency for Marine–Earth Science and Technology, 2-15 Natsushima-cho, Yokosuka-shi, Kanagawa 237-0061, Japan

Hitomi Nakamura and Hikaru Iwamori
Japan Agency for Marine–Earth Science and Technology, 2-15 Natsushima-cho, Yokosuka-shi, Kanagawa 237-0061, Japan
Department of Earth and Planetary Sciences, Tokyo Institute of Technology, 2-12-1 Ookayama, Meguro-ku, Tokyo 152-8551, Japan

Noritoshi Morikawa and Kohei Kazahaya
Geological Survey of Japan, AIST, 1-1-1 Higashi, Tsukuba, Ibaraki 305-8567, Japan

Shaaban FF and Al-Rashed AR
College of Basic Education, PAAET, Kuwait

Nwankwoala HO
Department of Geology, University of Port Harcourt, Nigeria

Adiela UP
Geosciences Department, Nigerian Agip Oil Company, P.O Box 923, Port Harcourt, Nigeria

Nazeer A and Habib Shah S
Directorate of Asset Operations, Pakistan Petroleum Limited (PPL), Islamabad, Pakistan

Abbasi SA and Solangi SH
Center of Pure and Applied Geology, University of Sindh, Jamshoro, Sindh, Pakistan

Ahmad N
Ex GM Exploration, Pakistan Petroleum Limited (PPL), Islamabad, Pakistan

Reshma B, Sakthivel R and Mohanty JK
CSIR-Institute of Minerals and Materials Technology, Bhubaneswar, Odisha, India

Khashaba A, Mekkawi M, Ghamry E and Abdel Aal E
National Research Institute of Astronomy and Geophysics (NRIAG), Helwan, Cairo-Egypt

Dara A. Goldrath
USGS, 6000 J St. Sacramento, CA 95819, USA

John A. Izbicki
USGS, 4165 Spruance Rd. Suite 200, San Diego, CA 92101, USA

Kathryn W. Thorbjarnarson
San Diego State University, 5500 Campanile Drive, San Diego, CA 92103, USA

Smruti Rekha Sahoo and Subhasish Das
Department of Geology and Geophysics, IIT Kharagpur, 721302, India

Elosta F
Waha Oil Company, Tripoli, Libya

Ben Ghawar BM
Geological Engineering Department, University of Tripoli, Tripoli, Libya

Komal Choudhary
Samara State Aerospace University, Russia

Mukesh Singh Boori
Samara State Aerospace University, Russia
Palacky University Olomouc, Czech Republic

Heiko Balzter
University of Leicester England, UK

Vit Vozenílek
Palacky University Olomouc, Czech Republic

Murali Krishna G
GIS Technology and Applications Development, Xinthe Technologies Pvt. Ltd. Visakhapatnam, India

Nooka Ratnam K
Department of Geology, School of Earth and Atmospheric Sciences, Adikavi Nannaya University, Rajah Rajah Narendra Nagar, Rajahmundry, India

Qing Chang
Japan Agency for Marine–Earth Science and Technology, 2-15 Natsushima-cho, Yokosuka-shi, Kanagawa 237-0061, Japan

Hitomi Nakamura and Hikaru Iwamori
Japan Agency for Marine–Earth Science and Technology, 2-15 Natsushima-cho, Yokosuka-shi, Kanagawa 237-0061, Japan
Department of Earth and Planetary Sciences, Tokyo Institute of Technology, 2-12-1 Ookayama, Meguro-ku, Tokyo 152-8551, Japan

Kotona Chiba
Department of Earth and Planetary Sciences, Tokyo Institute of Technology, 2-12-1 Ookayama, Meguro-ku, Tokyo 152-8551, Japan

Shunichi Nakai
Earthquake Research Institute, University of Tokyo, 1-1-1 Yayoi, Bunkyo-ku, Tokyo 113-0033, Japan

Kohei Kazahaya
Geological Survey of Japan, AIST, 1-1-1 Higashi, Tsukuba, Ibaraki 305-8567, Japan

Rasool QA and Biswas SK
Research Scholar, Department of Disaster Management Pondicherry University, Port Blair, Andaman islands-744112, India

Ramanujam N
Professor and Head, Department of Disaster Management, Pondicherry University, Port Blair, Andaman islands-744112, India

Index

A

Abia State, 99-100
Abu Gharadig Basin, 61-63, 71-72
Air Blast, 36
Air Pollutants, 47-48
Ambient Air Quality, 47, 55
Amelioration, 47
Amplitude, 30, 41, 103-104, 108, 162-163
Aquifer, 16, 19-20, 99, 102, 113, 121, 124-125, 140-141, 166-172, 184, 202, 207-209
Arima, 113-115, 117-119, 121, 123-126, 202-205, 208-209
Arsenic, 166-173
Artificial Reef, 88, 94-95

B

Bayelsa State, 136-137
Beach Nourishment, 88
Bedrock, 16, 103, 108-109, 111
Borehole, 27, 64, 100, 128, 136-137
Borrow Site, 88, 98
Brine, 30, 113-115, 117, 121, 123-126, 202-205, 208-209

C

Causative Factors, 9-10
Celestial Body, 85
Coalbed Methane, 141, 143, 152-153
Coastal Flooding, 88

D

Data Processing, 37, 41, 112
Deformation Style, 21
Diagenetic Fractures, 21

E

Eastern Margin of The Tibetan Plateau, 1-4, 7
Ecosystem, 47, 182, 186, 192
Ee Index, 73
Eej-magdas, 73
Egypt, 36-37, 39-41, 46, 61, 72-73, 111, 130, 135, 160-161, 165
Engineering Geology, 136
Equatorial Electrojet, 73-74, 78, 82

F

Fluid, 21, 23-24, 26, 35, 58, 64, 70, 83-87, 113, 123-126, 129, 184-187, 190, 202-203, 205, 208-209
Foundation, 7, 40, 99, 102-103, 111, 136, 139-140

G

Geospheres, 83, 85-87
Geostatic Model, 61-62, 64
Geotextile, 88, 91, 94, 98
Geothermal Forcing, 57
Ground Vibration, 36
Groundwater, 16-20, 41, 46, 99-100, 102, 112, 126, 166-167, 169-173, 182, 184-185, 209
Groundwater Potential, 16-20

I

Impedance, 27-31, 33-35

K

Kii, 113-115, 117-119, 121, 123-126

L

Land Magnetic Survey, 160-161, 163
Landslide, 9-15
Layers, 16-18, 27, 29, 41, 43-46, 68, 92, 103, 107-108, 114, 177-178, 182, 184, 192, 200, 203, 211
Layers Samples Analysis, 41
Lithostratigraphic Breaks, 128-135
Low Grade Natural Emerald, 154, 156-158
Lower Indus Basin, 27-29, 31, 35, 141-143, 146, 151-153

M

Magnetic Poles, 57-58, 85

N

Normal Faulting Type Event Region, 1-4, 6-7

O

Obehie, 99
Open Pit Mine, 36, 201

P

Peak Particle Velocity, 36
Permeability, 18, 21-23, 25, 61, 64-66, 71-72, 91, 99-100, 124, 141, 167, 184, 187
Petrophysics, 61
Photoluminescence, 154, 157-158
Plate Tectonics, 3, 83-85
Principle Compressive P-axis, 1

R

Rain, 13, 88-89, 91, 97, 185, 192
Raman Spectra, 154, 159

Reservoir Potential, 21-25
Rock Blasting, 36

S

Sakesar Limestone, 21-26
Seismic Activity, 7, 57, 60, 160
Seismic Interpretation, 27, 29, 35, 61, 65, 67-68, 71
Seismic Refraction, 41, 45-46, 103
Seismic Waves, 35, 41, 43, 46, 84
Seismograph, 41, 43-44, 46
Shear Waves, 27, 33-35
Silicate Rocks, 35, 83-84
Sindh, 27, 141-144, 146-149, 151-153
Soil Characterization, 99
South China Block, 1-3, 6-7
Spring Water, 113, 115, 117-118, 121, 123, 125-126, 202-203, 205
Stratigraphy, 21, 26-29, 40, 61, 72, 128, 135-137, 139, 173, 175, 180, 210
Structural Interpretation, 27, 31
Sub-soil, 99-100, 102, 136, 139-140
Subducting Slab, 113, 126, 204, 209
Subsoil, 41, 43-44, 46, 136, 139
Subsurface, 16-17, 21, 24, 27, 30-31, 33-36, 46, 103-104, 107-109, 111-112, 128, 130, 133, 135, 140-141, 151, 153, 160-161, 163, 165, 171, 183-184, 187
Subsurface Structures, 46, 160-161, 165

Surghar Range, 21-26

T

Tambaraparani River, 16, 20
Tectonic, 1-3, 6-8, 10-11, 14, 16-17, 21-22, 24-29, 32, 35, 43, 57, 61, 67, 72, 87, 113-115, 124, 126, 129, 160-161, 173-175, 177-179, 181, 188, 190, 202-203, 209-214
Topsoil, 103-108
Triggering Factors, 9-11

U

Undifferentiated Rock Units, 128, 133, 135
Upper Alaknanda Valley, 9-12, 14
Uppodai, 16-17, 20
Urban Coastal Zone, 88
Usage, 41
Uttarakhand Himalaya, 9, 12

V

Vertical Electrical Sounding, 16, 20, 103-104

W

Wave Run Up, 88-90, 92, 97-98
Well Logs, 27, 61-62, 64-65, 70, 128, 135, 187, 190
Well Modification, 166-167, 172-173
West Qarun Oil Field, 160-161, 164
Western Desert, 36, 40, 61, 72, 160-161, 165

www.ingramcontent.com/pod-product-compliance
Lightning Source LLC
Chambersburg PA
CBHW080624200326
41458CB00013B/4500